AGJ
4608

VIBRATION TESTING

VIBRATION TESTING
Theory and Practice

Kenneth G. McConnell
Professor of Aerospace Engineering and Engineering Mechanics
Iowa State University of Science and Technology
Ames, IA

A WILEY-INTERSCIENCE PUBLICATION

JOHN WILEY & SONS, INC.

New York • Chichester • Brisbane • Toronto • Singapore

This text is printed on acid-free paper.

Copyright © 1995 by John Wiley & Sons, Inc.

All rights reserved. Published simultaneously in Canada.

Reproduction or translation of any part of this work beyond that permitted by Section 107 or 108 of the 1976 United States Copyright Act without the permission of the copyright owner is unlawful. Requests for permission or further information should be addressed to the Permissions Department, John Wiley & Sons, Inc., 605 Third Avenue, New York, NY 10158-0012.

This publication is designed to provide accurate and authoritative information in regard to the subject matter covered. It is sold with the understanding that the publisher is not engaged in rendering legal, accounting, or other professional services. If legal advice or other expert assistance is required, the services of a competent professional person should be sought.

Library of Congress Cataloging-in-Publication Data

McConnell, Kenneth G.
  Vibration testing: theory & practice / by Kenneth G. McConnell,
    p.   cm.
  Includes index.
  ISBN 0-471-30435-2 (cloth: alk. paper)
  1. Vibration tests. 2. Vibration—Measurement. I. Title.
TA418.32.M4 1995
620.3'028'7—dc20   95-16398

Printed in the United States of America

10 9 8 7 6 5 4 3 2 1

*The author is deeply indebted to Dee,
who has been my loving companion and best friend for over 35 years,
as well as to our four sons, Christopher, John, Thomas, and Mark,
who have been sources of real joy to their parents
by being responsible free spirits
who are making their own contributions to society
in their own unique ways*

CONTENTS

1 An Overview of Vibration Testing   1

    1.1   Introduction   2
    1.2   Preliminary Considerations   5
    1.3   Overview of Equipment Employed   8
    1.4   Summary   10

2 Dynamic Signal Analysis   11

    2.1   Introduction   12
          2.1.1   Signal Classification   12
          2.1.2   Temporal Mean Value   15
          2.1.3   Temporal Mean Square and Temporal Root Mean Square   15
          2.1.4   The Frequency Spectrum   16
          2.1.5   Analysis of a Single Sinusoid   16
    2.2   Phasor Representation of Periodic Functions   19
          2.2.1   The Phasor   19
          2.2.2   The Phasor and Real Valued Sinusoids   20
    2.3   Periodic Time Histories   23
          2.3.1   Periodic Fourier Series   24
          2.3.2   The Mean, Mean Square, and Parseval's Formula   26
          2.3.3   Analysis of a Square Wave   27
    2.4   Transient Signal Analysis   29
          2.4.1   Difference Between Periodic and Transient Frequency Analysis   29
          2.4.2   The Transient Fourier Transform   32
          2.4.3   Transient Mean, Mean Square, and Parseval's Formula   34
    2.5   Correlation Concepts—A Statistical Point of View   35
    2.6   Correlation Concepts—Periodic Time Histories   38
          2.6.1   Cross-Correlation   38
          2.6.2   Auto-Correlation   42
    2.7   Correlation Concepts—Transient Time Histories   44
          2.7.1   Cross-Correlation   44
          2.7.2   Auto-Correlation   46
    2.8   Correlation Concepts—Random Time Histories   48

viii CONTENTS

       2.8.1 Auto-Correlation and Auto-Spectral Density 48
       2.8.2 Cross-Correlation and Cross-Spectral Density 54
       2.8.3 Correlation and Spectral Densities of Multiple Random Processes 55
       2.8.4 Statistical Distributions 57
  2.9 Summary 60
       References 61

3 Vibration Concepts   63

  3.1 Introduction 64
  3.2 The Single DOF Model 64
       3.2.1 Equations of Motion 65
       3.2.2 Free Undamped Vibration 66
       3.2.3 Free Damped Vibration 67
       3.2.4 Structure Orientation and Natural Frequency 71
  3.3 Single DOF Forced Response 72
       3.3.1 The Viscous Damping Case 73
       3.3.2 Common FRFs 74
       3.3.3 Damping Models in Forced Response 75
       3.3.4 The Structural Damping Response 77
       3.3.5 The Bode Diagram 79
       3.3.6 Real and Imaginary Plots and Nyquist Diagrams 84
  3.4 General Input-Output Model for Linear Systems 91
       3.4.1 The Frequency-Domain (Fourier Transform) Approach 92
       3.4.2 The Time-Domain Impulse Response Approach 94
       3.4.3 Receptance FRF Versus Impulse Response Function 96
       3.4.4 Random Input-Output Relationships 98
       3.4.5 Shock Response Spectra 98
  3.5 Nonlinear Behavior 103
       3.5.1 The Phase Plane 104
       3.5.2 The Simple Pendulum 105
       3.5.3 The Duffing Equation of Forced Vibration 107
       3.5.4 The van der Pol Equation and Limit Cycles 110
       3.5.5 The Mathieu Equation 113
       3.5.6 Chaotic Vibrations 113
  3.6 The Two DOF Vibration Model 114
       3.6.1 Equations of Motion 115
       3.6.2 Undamped Natural Frequencies and Mode Shapes 117
       3.6.3 Steady-State Forced Vibration Response (Direct Method) 119

        3.6.4 Steady-State Forced Vibration Response (Modal Method)   120
        3.6.5 Comparison of Direct and Modal Response FRFs   122
3.7 The Second Order Continuous Vibration Model   127
        3.7.1 The Fundamental Equation of Motion   127
        3.7.2 Separation of Space and Time Variables   130
        3.7.3 Orthogonality Conditions   133
        3.7.4 The Modal Model and Forced Vibrations   135
        3.7.5 The Generalized Excitation Force for Distributed Loads   137
        3.7.6 Continuous Model FRFs   138
3.8 The Fourth Order Continuous Vibration System—The Beam   140
        3.8.1 The Fundamental Equation of Motion   140
        3.8.2 Natural Frequencies and Mode Shapes   141
        3.8.3 Natural Frequencies and Boundary Conditions   144
        3.8.4 The Modal Model   149
        3.8.5 The Beam Under Tension   150
3.9 Summary   153
    References   158

4 Transducer Measurement Considerations   161

4.1 Introduction   161
4.2 Mechanical Model of Seismic Transducers—The Accelerometer   163
        4.2.1 The Basic Mechanical Model   164
        4.2.2 Gravity Forces and Acceleration Measurements   166
4.3 Piezoelectric Sensor Characteristics   169
        4.3.1 Basic Circuits and Operational Amplifiers   170
        4.3.2 Charge Sensitivity Model   172
        4.3.3 The Charge Amplifier   173
        4.3.4 Built-in Voltage Followers   177
        4.3.5 The Overall Accelerometer FRF   181
4.4 Combined Linear and Angular Accelerometers   182
        4.4.1 Using Multiple Accelerometers to Measure Combined Motions   182
        4.4.2 The Combined Linear and Angular Accelerometer   186
4.5 Transducer Response to Transient Inputs   189
        4.5.1 Mechanical Response   189
        4.5.2 Piezoelectric Circuit Response to Transient Signals   195
        4.5.3 Field Experience with Shock Loading   198

- 4.6 Accelerometer Cross-Axis Sensitivity 201
  - 4.6.1 The Single Accelerometer Cross-Axis Sensitivity Model 202
  - 4.6.2 The Tri-Axial Accelerometer Cross-Axis Sensitivity Model 203
  - 4.6.3 Correcting Tri-Axial Acceleration Voltage Readings 204
  - 4.6.4 FRF Contamination and Its Removal 206
  - 4.6.5 Cross-Axis Resonance 210
- 4.7 The Force Transducer—General Model 211
  - 4.7.1 General Electromechanical Model 211
  - 4.7.2 Force Transducer Attached to a Fixed Foundation 215
  - 4.7.3 Force Transducer Attached to an Impulse Hammer 216
  - 4.7.4 Force Transducer Used with Vibration Exciter and Structure 217
  - 4.7.5 The Impedance Head 222
- 4.8 Correcting FRF Data for Force Transducer Mass Loading 223
  - 4.8.1 Consistent Force Transducer Model 224
  - 4.8.2 Correcting Driving Point Accelerance FRF in Frequency Domain 226
  - 4.8.3 Correcting Transfer Accelerance FRFs in Frequency Domain 228
  - 4.8.4 Electronic Compensation Using Seismic Acceleration 230
  - 4.8.5 Errors due to $HI_{pp}(\omega)$ Being Nonunity 232
- 4.9 Calibration 234
  - 4.9.1 Accelerometer Calibration—Sinusoidal Excitation 235
  - 4.9.2 Accelerometer Calibration—Transient Excitation 237
  - 4.9.3 Force Transducer—Sinusoidal Excitation 240
  - 4.9.4 Force Transducer—Transient Excitation 243
  - 4.9.5 Effects of Bending Moments on Measured Forces 247
- 4.10 Environmental Factors 251
  - 4.10.1 Base Strain 251
  - 4.10.2 Cable Noise 253
  - 4.10.3 Humidity and Dirt 254
  - 4.10.4 Mounting the Transducer 254
  - 4.10.5 Nuclear Radiation 255
  - 4.10.6 Temperature 255
  - 4.10.7 Transducer Mass 256
  - 4.10.8 Transverse Sensitivity 256
- 4.11 Summary 256
- References 259

5 The Digital Frequency Analyzer 261

- 5.1 Introduction 261
- 5.2 Basic Processes of a Digital Frequency Analyzer 263
  - 5.2.1 The Time Sampling Process 264
  - 5.2.2 Time Domain Multiplication and Frequency Domain Convolution 266
  - 5.2.3 Sample Function Multiplication Gives Aliasing 267
  - 5.2.4 The Window Function Creates the Digital Filter Characteristics 271
  - 5.2.5 Filter Leakage 273
- 5.3 Digital Analyzer Operating Principles 278
  - 5.3.1 Operating Block Diagram 279
  - 5.3.2 Internal Calculation Relationships 280
  - 5.3.3 Display Scaling 282
- 5.4 Factors in the Application of a Single Channel Analyzer 285
  - 5.4.1 Filter Performance Characteristics 286
  - 5.4.2 Four Commonly Employed Window Functions 288
  - 5.4.3 Window Comparison for Use with Sinusoidal Signals 292
  - 5.4.4 Spectral Line Uncertainty 297
  - 5.4.5 Recommended Window Usage 302
- 5.5 Overlapping Signal Analysis to Reduce Analysis Time 303
  - 5.5.1 Overlapping and Ripple 304
  - 5.5.2 Effective Bandwidth Time Product and Measurement Uncertainty 309
  - 5.5.3 Real Time Analysis 310
- 5.6 Zoom Analysis 313
  - 5.6.1 Zoom FFT Analysis Using the Heterodyning Method 315
  - 5.6.2 Long Time Record Zoom FFT Analysis 317
  - 5.6.3 Zoom Analysis with and Without Sample Tracking 321
- 5.7 Scan Analysis, Scan Averaging, and More on Spectral Smearing 323
  - 5.7.1 Scan Analysis 324
  - 5.7.2 Scan Averaging and Resulting Frequency Spectra 326
  - 5.7.3 Scan Average Analysis of a Transient Signal 327
  - 5.7.4 More on Spectral Smearing 329
- 5.8 The Dual Channel Analyzer 333
  - 5.8.1 Ideal Input-Output Relationships 335
  - 5.8.2 Actual Input-Output Estimates for a Digital Analyzer 338
  - 5.8.3 Auto-Spectra and Cross-Spectra Averaging 340

        5.8.4 Some Reasons Coherence Is Less Than Unity 341
        5.8.5 Operating Block Diagram 344
  5.9 The Effects of Signal Noise on FRF Measurements 345
        5.9.1 Noise in the Input Signal 345
        5.9.2 Noise in the Output Signal 348
        5.9.3 Noise in the Input and Output Signals 350
        5.9.4 More Than One External Input 352
  5.10 Summary 355
       References 359

6 Vibration Exciters   363

  6.1 Introduction 363
        6.1.1 Static Excitation Schemes 364
        6.1.2 Dynamic Impulse Loading Schemes 366
        6.1.3 Controlled Dynamic Loading Schemes 370
  6.2 Mechanical Vibration Exciters 371
        6.2.1 The Direct-Drive Exciter Model 372
        6.2.2 The Direct-Drive Vibration Exciter Table 377
        6.2.3 The Direct-Drive Reaction Type Vibration Exciter 379
        6.2.4 The Rotating Unbalance Exciter 379
        6.2.5 Driving Torque Considerations 381
  6.3 Electrohydraulic Exciters 382
        6.3.1 Electrohydraulic System Components 383
        6.3.2 Application to Earthquake Simulator 387
  6.4 The Modeling of an Electromagnetic Vibration Exciter System 391
        6.4.1 Exciter Support Dynamics 393
        6.4.2 Armature Dynamics 399
        6.4.3 Electromechanical Coupling Relationships 403
        6.4.4 Power Amplifier Characteristics 405
  6.5 An Exciter System's Bare Table Characteristics 407
        6.5.1 The Electrodynamic Model 407
        6.5.2 Current Mode Power Amplifier 408
        6.5.3 Voltage Mode Power Amplifier 410
        6.5.4 Comparison of Bare Table Armature Responses 411
  6.6 Interaction of an Exciter and a Grounded Single DOF Structure 414
        6.6.1 The Single DOF SUT and Electrodynamic Exciter Model 415
        6.6.2 The SUT's Accelerance Response 417
        6.6.3 The Force Transmitted to the SUT and Force Dropout 422

- 6.7 Interaction of an Exciter and an Ungrounded Structure under Test   425
  - 6.7.1 A General Dynamic Model Using Driving Point and Transfer Accelerance   427
  - 6.7.2 Ungrounded Test Structure and Exciter Accelerance Characteristics   429
  - 6.7.3 Test Responses for Current and Voltage Mode Power Amplifier Inputs   431
- 6.8 Summary   435
  References   439

7 The Application of Basic Concepts to Vibration Testing   441

- 7.1 Introduction   442
- 7.2 Sudden Release or Step Relaxation Method   444
  - 7.2.1 Theoretical Modal Model   445
  - 7.2.2 Midpoint Excitation and Response   446
  - 7.2.3 Resolving the Measurement Dilemma   449
  - 7.2.4 Results from an Actual Test   454
- 7.3 Forced Response of a Simply Supported Beam on an Exciter   461
  - 7.3.1 Modal Model of the Test Environment   462
  - 7.3.2 The Effect of Accelerometer Mass on Measurements   467
  - 7.3.3 The Limits for Modal Mass Correction to Natural Frequencies   468
  - 7.3.4 Added Mass at the Quarter Point   471
- 7.4 Impulse Testing   475
  - 7.4.1 Impulse Requirements   476
  - 7.4.2 The Input Noise Problem   477
  - 7.4.3 The Output Leakage Problem   480
  - 7.4.4 Application of Impulse Testing to a Free-Free Beam   482
- 7.5 Selecting Proper Windows for Impulse Testing   487
  - 7.5.1 Window Parameters   487
  - 7.5.2 Modeling the Data Process   489
  - 7.5.3 Truncation and Exponential Window Effects   493
  - 7.5.4 The Effects of the Input Transient Window   498
  - 7.5.5 Recommended Procedure for Setting Window Parameters   500
- 7.6 Vibration Exciter Driving a Free-Free Beam with Point Loads   501
  - 7.6.1 Selecting the Excitation Signal   501
  - 7.6.2 Test Setup   504

- 7.6.3 Theoretical Exciter Structure Interaction 505
- 7.6.4 Comparison of Experimental and Theoretical Results 507
- 7.7 Windowing Effects on Random Test Results 510
  - 7.7.1 A Model of Window Function Filter Leakage Characteristics 512
  - 7.7.2 Estimating FRF Errors Due to Leakage 514
  - 7.7.3 Theoretical Simulation of Leakage and Its Effects 517
  - 7.7.4 Recommendations to Check for This Filter Error 522
- 7.8 Low Frequency Damping Measurements Reveal Subtle Data Processing Problems 522
  - 7.8.1 The Test Setup 523
  - 7.8.2 The Hardware Error 525
  - 7.8.3 A Software Problem 526
  - 7.8.4 Another Common Measurement Error Source 529
- 7.9 Summary 530
  References 532

8 General Vibration Testing Model: From the Field to the Laboratory — 535

- 8.1 Introduction 535
  - 8.1.1 General Linear System Relationships 538
  - 8.1.2 Three Structures Involved in the Process 541
- 8.2 A Two Point Input-Output Model of Field and Laboratory Simulation Environments 542
  - 8.2.1 Notation Scheme 542
  - 8.2.2 The Field Environment 544
  - 8.2.3 Laboratory Environment 546
  - 8.2.4 Discussion of Elementary Results 548
- 8.3 Laboratory Simulation Schemes Based on the Elementary Model 548
  - 8.3.1 Test Scenario Number 1—Matched Interface Motions and No External Forces 549
  - 8.3.2 Test Scenario Number 2—Matched Interface Forces and No External Forces 551
  - 8.3.3 Test Scenario Number 3—Matched Test Item Motion and No External Forces 552
  - 8.3.4 Test Scenario Number 4—Matched Interface Motion with Field External Force but No Laboratory External Force 553
  - 8.3.5 Test Scenario Number 5—Matched Interface Forces with Field External Force but No Laboratory External Force 555

- 8.3.6 Test Scenario Number 6—Matched Test Item Motion with Field External Force but No Laboratory External Force 556
- 8.3.7 Summary of Six Test Scenarios 557
- 8.4 An Example Using a Two DOF Test Item and a Two DOF Vehicle 558
  - 8.4.1 Test Item and Vehicle Dynamic Characteristics 559
  - 8.4.2 Laboratory Test Setup Employed During Tests 562
  - 8.4.3 Field Simulation Results 563
  - 8.4.4 Laboratory Simulation 568
  - 8.4.5 Predicting Interface Forces and Accelerations from Bare Vehicle Interface Acceleration ASD Data 575
  - 8.4.6 Summary and Conclusions for This Simple Example 575
- 8.5 The General Field Environment Model 577
  - 8.5.1 Basic Motions Due to Interface and Noninterface Forces 578
  - 8.5.2 Interface Boundary Conditions and Resulting Forces and Motions 580
  - 8.5.3 Test Item Motions 582
- 8.6 The General Laboratory Environment 584
  - 8.6.1 Basic Motions Due to Interface and Noninterface Forces 585
  - 8.6.2 Interface Boundary Conditions and Resulting Forces and Motions 587
  - 8.6.3 Test Item Motions 588
- 8.7 Comparison of Field and Laboratory Environments 590
  - 8.7.1 Comparison of Theoretical Results for the General Case 591
  - 8.7.2 The Four Point Test Item Model 593
- 8.8 Summary 595
  - References 596

Index 599

PREFACE

This book is dedicated to obtaining the maximum possible benefit when applying vibration testing techniques to a wide range of practical vibration problems. It is the culmination of over 30 years of experimental testing experience of working with students, other researchers, and engineers in industry. Each individual has taught me new insights because of his or her unique talents or points of view. I owe a lot to these persons for the lessons given. I am sure there are insights and ideas that I present in this book that came from a forgotten comment, or reference, or person. One of the advantages of teaching advanced courses in vibrations and laboratory courses in experimental dynamics is the constant review of the course concepts with new persons who have different experiences and viewpoints. Another advantage of academic research is that we usually test our ideas on simple structures so that measurement errors and test procedural errors are usually evident. This situation is contrasted with the common industrial case, in which the structure is very complicated and any measurement errors are easily lost in our lack of understanding of the structure's characteristics.

I am reluctant to start to name the persons who have influenced me over the years for fear of leaving someone out. However, I want to acknowledge a few individuals, as well as several groups of persons who made major contributions to my understanding of vibrations. First, all of the graduate students who cast their lot with me during their graduate studies have made significant contributions by their dissertation work and resulting testing experiences. Second, Mr. Tom Priddy of Sandia Laboratories gave me a most interesting assignment during the summers of 1988 and 1989, when I reviewed Sandia's shock and vibration testing procedures. The competent Sandia Vibration Testing group was most cooperative and open about what appeared to work well when there were significant questions about technique, theory, and methods. Third, the Iowa State University Faculty Improvement Leave of Absence provided a key year in order to get this book started while I was working in England at Imperial College (London) with Professor David Ewins, who helped me with stimulating discussions and helpful suggestions for the book's outline, as well as introduced me to other persons in England who were of great assistance during that year. Forth, Mr. Paulo S. Varoto, who is a current graduate student from Brazil, has been a most enthusiastic participant in conducting special tests and reading the many sections for clarity of content. Finally, there are the important persons who edit and put the book together for the publisher. Of these, I want to acknowledge the help of Associate Managing Editor Donna Conte, who got us organized and kept the project on schedule, and Linda Grady-Troia, who did an outstanding job of catching me in my usual inconsistencies and expertly challenged my perfect—well nearly perfect—manuscript.

Finally, as all human efforts contain errors and significant omissions and oversights, I would appreciate comments and suggestions for improvement of this book so that others can do their jobs more efficiently, even if it means pointing out an error or two on my part.

KENNETH G. MCCONNELL

Ames, Iowa

1 An Overview of Vibration Testing

Rotating machinery diagnostics are being conducted using a portable vibration frequency analyzer specifically designed for such field in situ testing tasks. Such machinery monitoring is important to achieve optimum machine performance over long periods of time with minimum unscheduled maintenance. (Photo Courtesy of *Sound and Vibration*).

1.1 INTRODUCTION

This book is dedicated to obtaining the maximum possible benefit when applying vibration testing techniques to a wide range of practical vibration problems. I believe that maximum benefits are obtained when the instrumentation and analysis techniques used are appropriate for and consistent with the desired test objectives for a given machine and/or structure. The most sophisticated and brilliant analysis methods are easily defeated by poor, inaccurate, or inappropriate data. Thus one needs to understand the fundamental principles that are involved in vibration testing, so that an informed skepticism can be brought to bear on the test results. "Why is this glitch here?" is often an important starting point in understanding that there may be problems with a certain set of test data.

Vibration tests are run for a number of reasons. Among them are:

1. Engineering development testing
2. Qualification testing
3. Reliability qualification testing
4. Production screening testing
5. Machinery condition monitoring

The same types of instruments, frequency analyzers, and analysis methods are employed in nearly all type of tests. Various types of vibration exciters, however, are employed for various tests. In some cases the tests are run in the laboratory. In other cases the only way to run a needed test is to run it in the field, under actual operating conditions, thus determining why a particular part of the structure is deteriorating too quickly, or why it is failing to function correctly under service conditions. Because of this wide range of techniques and goals, the intent of this book is to cover a wide range of fundamental concepts that are useful in most vibration testing situations.

I don't recall just when I became aware of vibration testing as an engineering activity, but I know that it was not during the laboratory exercises during my first vibrations course. However, when I worked at the Naval Ship Research Center at Carterrock, MD, I began to sense that getting good field vibration data was not a simple matter. At any rate, over the years a number of experiences have led me to attempt to identify what vibration testing is and how vibration testing fits into the engineering function. One thing is evident to me: in attempting to identify what vibration testing is, a number of interesting facets come to light.

I clearly remember one interesting event that occurred in the late 1960's at a Shock and Vibration Symposium. The U.S. Navy wanted to mount an electronic device on top of a large gun turret such as the one shown

INTRODUCTION 3

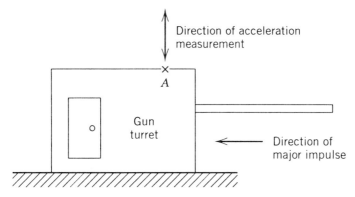

Fig. 1.1.1. Schematic of a gun turret showing location of acceleration measurements.

in Fig. 1.1.1. Apparently, navy personnel measured the vertical acceleration at point A, where the electronic device was to be mounted, while several shells were fired from the gun. These acceleration records were used to develop the device's *dynamic environment*. The electronic device was designed, tested, and passed this dynamic environment. Yet when the device was installed, it failed to function after the first firing. Obviously, there was a problem in translating the measured dynamic environment into an adequate test specification. What happened?

First, the vertical acceleration at point A is probably significantly less than the horizontal acceleration resulting from the reaction to firing the gun in a nearly horizontal direction. Second, the electronic device was mounted so that it presented a significant frontal area to the gun's air blast. It is clear that either of these two inputs could be the source of significant vibration levels. Obviously, the procedures employed were inadequate and caused unnecessary grief to all parties concerned.

This story illustrates that significant questions need to be answered in planning vibration tests, questions such as:

1. What field data should be taken?
2. Where should the transducers be placed on the test item to achieve the maximum useful information over a given frequency bandwidth?
3. How should the field data be stored for future reference and recall?
4. What is the effect of changing boundary conditions between field and test environments due to test fixtures?
5. Which testing procedures are best suited to simulating a given field environment?

These and certainly other questions need to be answered in order to properly plan the field test as well as a corresponding laboratory simulation.

In another instance, while preparing to conduct vibration tests on a company's product, I was given two sets of test specifications, which I found to be quite contradictory. If I followed one set of specifications, it appeared that nothing would fail because the inputs were so low, but if I followed the other set, premature failure could result because the inputs were so large. In struggling with these contradictory requirements, it became evident that the test specifications were quite arbitrary, since no one could provide me with actual field data. Thus no clear means was available to determine what the test specifications should be, and I was left to make my own judgments. Obviously, I chose to be conservative, which in turn, added unnecessary weight to the product.

I had another enlightening experience during two summers I spent reviewing the shock and vibration testing procedures employed at Sandia National Laboratories. This review involved interviews with over a dozen of Sandia's very experienced vibration test personnel. It became evident that some underlying rules for conducting these tests were not being properly addressed by the test standards they were required to use. Fundamental questions about how field data is obtained, stored, and converted to useful test specifications were often discussed during these interviews. Clearly, a better framework is required by which to judge what procedures should be used. While attempting to understand and explain what these rules should be, the basic thoughts contained in Chapter 8 of this book were developed, in order to establish some framework to guide the processes used in setting up a vibration testing program for a given application. This area requires considerably more thought and research in order to understand implications of the many choices we can make in writing a test specification from field vibration data.

Subsequent to the Sandia experience, I surveyed practitioners of vibration testing at over a dozen sites in the United States in order to determine a broader view of the current state of affairs in the vibration testing industry. A striking result from this survey is that there is no general theory for vibration testing, so that test specifications are often modified based on the tester's experience, much as I had done when testing products. The survey also revealed that if any model is referred to, it is most often that of a single degree of freedom (DOF) system, an inappropriate model for most testing cases that involve rather complicated structural systems.

A small sample of the type of comments I heard during my survey is summarized as follows:

1. Writers of test specifications do not recognize test equipment limitations.

2. Required test spectra do not reflect actual vehicle (field) environments.
3. The relation between test level and test item endurance seems to be quite arbitrary.
4. Stress level and duration need to be sensibly related to reliability and endurance testing.
5. Test fixtures are usually much stiffer than the host vehicle.
6. Unless a person oversees instrumentation installation and subsequent data reduction, there is usually insufficient data available to separate sources of excitation.
7. When data is collected by others, sensor location is often poorly defined.
8. "Many people operate in a particular way because that is what they learned to do, without necessarily knowing why or how to predict results."[1]
9. "It takes much more competence to do it correctly than it does to do it the 'standard' way!"[1]

It is hoped that this book will illuminate many of the required fundamental concepts so that test personnel will become aware of opportunities to conduct tests better; and hence, obtain maximum possible benefits.

1.2 PRELIMINARY CONSIDERATIONS

It is hard to decide where to start this adventure, but experience has shown that an overview and precise definitions significantly improve the communication process. Fig 1.2.1 shows a general model of a structure where it is assumed that a transducer (usually an accelerometer) measures the motion X_p at location p. X_p is used to represent displacement, velocity, or acceleration. A common notation used throughout this book is that lower case letters represent a time domain variable so that $x = x(t)$, while capital letters represent a frequency domain variable so that $X = X(\omega)$ is a frequency spectra. This notation is most convenient since we often have frequency spectra information available from our frequency analyzers.

Motion X_p is caused by numerous excitation sources. These sources can be *internal* forces (rotating unbalances), which are represented by S_q ($= S_q(\omega)$), as well as *external* forces. The external forces are broken into two subgroups, those that are due to external sources and are represented by F_q ($= F_q(\omega)$) and those that are due to the presence of a *boundary* and

---

[1]Personal communication from E. A. Szymkowiak, July 1990.

6 AN OVERVIEW OF VIBRATION TESTING

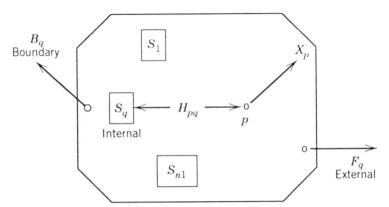

Fig. 1.2.1. General structure showing internal, external, and boundary types of excitation forces.

are represented by B_q $(= B_q(\omega))$. Then, if we use the linear input–output frequency domain representation H_{pq} $(= H_{pq}(\omega))$, we can express the frequency domain output motion as

$$X_p = \sum_{q=1}^{n_1} H_{pq} S_q + \sum_{n_1+1}^{n_2} H_{pq} F_q + \sum_{n_2+1}^{n_3} H_{pq} B_q \qquad (1.2.1)$$

$$\text{internal} \qquad \text{external} \qquad \text{boundary}$$

where H_{pq} is the frequency response function (FRF) that is a function of frequency ω.

n_1 is the number of internal sources.

$(n_2 - n_1 - 1)$ is the number of external forces.

$(n_3 - n_2 - 1)$ is the number of boundary forces.

FRF H_{pq} can represent receptance, mobility, or acceleration, depending on whether X_p represents displacement, velocity, or acceleration, respectively.

The situation shown in Fig. 1.2.1 is characteristic of the general vibration testing situation where the vibrations of a structure (or machine) are monitored in order to determine either the structure's dynamic characteristics or its mechanical health. The purpose of the test depends on the end use of the test data. In order to determine the structure's dynamic characteristics, we might do a modal analysis. On the other hand, we may measure the machine's dynamic characteristics in order to determine its operating condition, a process that is often called machine monitoring. The test structure could be either an aircraft engine or a steam turbine. Obviously, the boundary conditions, test conditions, and instrumentation requirements are strikingly different for each of these two machines.

Now suppose that a critical circuit board is involved in controlling the jet engine, and that this circuit board must not fail during operation. We need to ensure that the circuit board will survive the dynamic environment of the aircraft/jet engine. Now, the question is how do we go about achieving this goal?

We begin by breaking the problem down, as illustrated in Fig. 1.2.2, where we identify several major components that are involved in the process of gathering data, interpreting this data, and communicating the results to the major participants. The components are: the aircraft, which we call the *vehicle*; the circuit board, which we call the *test item*; the vibration laboratory, which we call the *laboratory*; the finite element computer program or other design tool, which we call simply *design*; and the *process*. In the process part we need:

1. To obtain experimental field data, which may involve installation of a prototype test item, a different test item, or no test item.
2. To analyze the field data, taking the field test conditions into account.
3. To communicate the field data in the proper form to the designer for use during the design stage.
4. To provide vibration test personnel with realistic data so that adequate laboratory testing is performed to achieve a realistic dynamic environment.

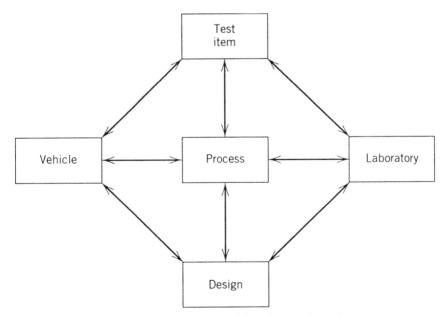

Fig. 1.2.2. Definition of major elements and data flow under various test configurations.

8 AN OVERVIEW OF VIBRATION TESTING

This process often requires communications between persons from different government and/or industrial organizations, who often have their own vested interests, which may not be completely compatible. Consequently, interesting personal relations situations often occur in large projects so that the importance of skills for working together becomes evident. However, if everyone can keep his or her eye on the target and cooperate for the common good of the project, a successful product that is reliable, is cost effective, and performs as expected will result. The full implications of Fig. 1.2.2 will be evident as the reader proceeds through the book.

1.3 OVERVIEW OF EQUIPMENT EMPLOYED

There are many different types of vibration tests. Some involve field measurements while the structure is in its normal operational state, while others involve situations where the structure is excited by some external means, either in place in the field or in a laboratory setting. These tests can be performed for a wide range of reasons such as vibration monitoring in order to determine a machine's suitability for operation, a general vibration survey to find out what is happening, a complete modal analysis to determine the structure's dynamic characteristics, and so on. However, in each instance, commonly used concepts and equipment are involved.

Fig. 1.3.1 shows a generic test item with its motion measured by one transducer system (usually an accelerometer) and the input force measured by a second transducer system (usually a force transducer). In either case, these transducers often employ either the piezoelectric or the strain gage type of sensing element. The electronic signals from these transducers are amplified electronically and analyzed. The frequency analyzer is commonly employed for analysis purposes. It then becomes the engineer's job to interpret the resulting frequency spectra and to store the data in a suitable form. A computer is often used for this purpose.

As one thinks about the processes shown in Fig. 1.3.1, it becomes evident that certain common concepts and physical laws are employed. The task is then to organize these concepts and laws in an orderly fashion so that we can explore their importance. This information is organized into seven additional chapters, as follows.

Chapter 2 *Dynamic Signal Analysis* This chapter is devoted to the basic concepts involved in dynamic signal analysis, since these concepts underlie much of what we do in later chapters. Periodic, transient, and random signals are considered on a theoretical basis.

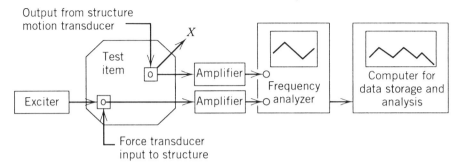

Fig. 1.3.1. Generic test items showing force and motion transducers, amplifiers, frequency analyzer, and computer data storage.

Chapter 3   *Vibration Concepts*   This chapter is a review of basic vibration concepts. While we could cite this material from other texts, it was felt that a certain set of fundamental ideas and viewpoints are required in order to understand what is happening in vibration testing and why procedures need to be done in a certain way.

Chapter 4   *Transducer Measurement Considerations*   This chapter is concerned with how transducers behave. Models for the mechanical and electrical characteristics of common transducers are developed. Experience indicates that transducers can do interesting and unpredictable things, depending on the environment into which we put them. Often the user can dramatically change the transducer's behavior so that an informed user is required to obtain reliable data.

Chapter 5   *The Digital Frequency Analyzer*   This chapter carefully examines the characteristics of the digital frequency analyzer. The rules by which a digital analyzer works are important to understanding what the measured frequency spectra mean. If the analyzer is improperly set, the calculated results may contain grossly distorted information.

Chapter 6   *Vibration Exciters*   This chapter is concerned with the dynamic characteristics of vibration exciters. These excitations include static release, mechanical, and electromagnetic type devices. The interaction of these exciters with their test environment is explored.

Chapter 7   *The Application of Basic Concepts to Vibration Testing*   This chapter looks at what happens when we put everything together while attempting to conduct an actual test. Several simple examples are used to illustrate significant factors that must be considered and how these factors can influence the experimental results.

Chapter 8 *General Vibration Testing Model: From the Field to the Laboratory* This chapter explores the framework for conducting field tests and converting the results into a meaningful laboratory test specification.

1.4 SUMMARY

This introduction suggests that a broad range of physical laws and mathematical concepts are involved in the practical execution of a successful vibration test. This book is dedicated to exploring these laws and concepts so that the practicing test engineer, as well as graduate students, will be able to understand some of the subtleties that can occur. No results are better than the instruments employed. As will be seen, recent research shows that instruments can be the Achilles' heel of the entire process.

As a closing to this first chapter, I would like to offer the following definition:

Vibration Testing The art and science of measuring and understanding a structure's response while exposed to a specific dynamic environment; and if necessary, simulating this environment in a satisfactory manner to ensure that the structure will either survive or function properly when exposed to this dynamic environment under field conditions.

As with all definitions, this one has its limitations. If any reader can find a more inclusive statement, I would appreciate your sharing it with me.

2 Dynamic Signal Analysis

A real-time frequency analyzer is being used to monitor the dynamic signals contained in noise from aircraft taking off from Sea-Tac International Airport, Seattle, WA. The analysis of vibration signals is important in conducting vibration tests and for understanding the test results. (Photo Courtesy of *Sound and Vibration*).

2.1 INTRODUCTION

Signal analysis is fundamental to vibration testing. Consequently, understanding it and its proper use should be a high priority to any practitioner. The objective of this chapter is to identify the different signal types that are analyzed in vibration testing and their characteristics. Periodic and transient Fourier transforms (FT), auto- and cross-correlation, mean square spectral density (MSSD) or auto spectral density (ASD) (also called power spectral density PSD), and cross-spectral density (CSD) are used to describe various signal characteristics. The periodic Fourier transform is used in Chapter 5 to develop digital frequency analyzer characteristics. The distinction between these formulations and their uses will become evident as the underlying ideas and concepts are developed in this chapter and Chapter 5. Understanding these concepts and their interrelationships makes frequency analysis more meaningful to the user of this technology.

2.1.1 Signal Classification

Dynamic signals are generally classified as *deterministic* and *random*, as shown in Fig. 2.1.1. The *chaotic* signal is a recently recognized phenomenon where a random appearing signal is controlled by a deterministic process. Chaotic signals are receiving more attention in an effort to understand the processes that create them as well as how to identify and analyze them. Just how this research will impact future signal classification is not clear at this time; thus, the question marks in Fig. 2.1.1. However, it must be recognized that some random appearing signals may be chaotic, but they will be misjudged as random for signal analysis purposes until ways are found to clearly distinguish chaotic signals from random signals.[1] Chaotic signals are not considered in this book beyond cursory reference to them.

Deterministic signals are further classified as either *periodic* or *transient*. A periodic signal is one that repeats itself in time and is a reasonable model for many real processes, especially those associated with constant speed machinery. A transient signal is one that has no significant variation occurring for long periods of time with short periods of intense activity. Ideally, infinite time occurs before and after the transient event. In practical terms, it is necessary only that all vibration ceases before another event occurs for a signal to be classified as a transient.

---

[1]F. C. Moon, *Chaotic Vibrations: An Introduction to Chaotic Dynamics for Applied Scientists and Engineers*, John Wiley & Sons, New York, NY, 1987.

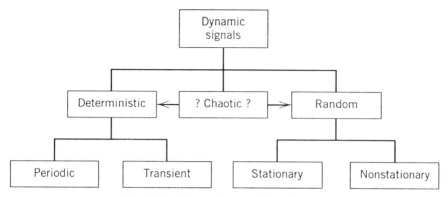

Fig. 2.1.1. Dynamic signal classification.

Signals can also be classified as either *stationary* or *non-stationary*. Stationary signals are ones that have constant parameters to describe their behavior, while nonstationary signals have time dependent parameters. This is easily seen when one thinks of an engine excited vibration where the engine's speed varies with time; in this case, the fundamental period changes with time as well as with the corresponding dynamic loads that cause vibration.

Random signals are characterized by having many frequency components present over a wide range of frequencies. The amplitude versus time appears to vary rapidly and unsteadily with time. The "shhhhh" sound is a good example that is rather easy to observe using a microphone and oscilloscope. If the sound intensity is constant with time, the random signal is stationary, while if the sound intensity varies with time the signal is nonstationary. One can easily see and hear this variation while making the "shhhhh" sound.

Random signals are characterized by analyzing the statistical characteristics across an ensemble of records. Then, if the process is *ergodic*, the time (temporal) statistical characteristics are the same as the ensemble statistical characteristics. It is assumed in this book that all random processes are stationary, ergodic, and have a Gaussian amplitude statistical distribution so that temporal definitions may be used. These are the common operating assumptions used in dealing with practical signal analysis problems.

Two common single valued figures of merit (or parameters) to describe a signal are the *temporal mean* and the *temporal mean square* or the *root mean square*. The word *temporal* means that a time average definition is used in place of an ensemble statistical definition.

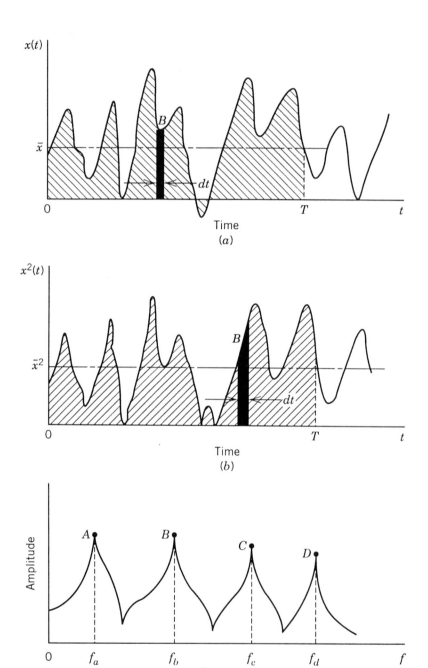

Fig. 2.1.2. Methods of signals analysis. (*a*) Signal $x(t)$ and definition of temporal mean. (*b*) Signal $x^2(t)$ and definition of temporal mean square. (*c*) Frequency spectrum of signal $x(t)$. (From J. W. Dally, W. F. Riley, and K. G. McConnell, *Instrumentation for Engineering Measurements*, 2nd ed., Copyright © 1993 by John Wiley & Sons, New York. Reprinted by permission.)

2.1.2 Temporal Mean Value

Consider the time history $x(t)$ shown in Fig. 2.1.2a. The temporal mean of this signal is a statistical quantity and is defined as the signal's time average value, which is described mathematically by

$$\bar{x} = \lim_{T \to \infty} \frac{1}{T} \int_0^T x(t)\, dt \qquad (2.1.1)$$

where $\bar{x}$ is the temporal mean value of signal $x(t)$.

T is the integration and averaging time.

Figure 2.1.2a illustrates this time averaging concept where $x(t)\, dt$ is the *differential area* under the time history curve at point B. The addition of these differential areas gives the total crosshatched area, which when divided by time T, gives a time average that is the mean height. Thus the rectangular area given by $\bar{x}T$ is the same as the area under the signal from 0 to T. The time-average fluctuations are seen to be smaller and smaller as time T increases. These fluctuations disappear as T increases indefinitely, so that long averaging times give good time average estimates. The time average value is often referred to as the *static* or *DC* frequency *component*.

2.1.3 Temporal Mean Square and Temporal Root Mean Square

The temporal *mean square* (MS) is defined as the time average of the time history squared and is described mathematically by

$$\text{MS} = \overline{x^2} = \lim_{T \to \infty} \frac{1}{T} \int_0^T x(t)^2\, dt \qquad (2.1.2)$$

Equation (2.2.1) is interpreted the same as Eq. (2.1.1); that is, the time average height of the square of $x(t)$ as shown in Fig. 2.1.2b where $x^2(t)\, dt$ is the differential area. Note the notation difference between $\bar{x}$ for the mean value and $\overline{x^2}$ with a bar over the entire term for the mean square.

The *root mean square* (RMS) is the square root of the mean square. Thus, from Eq. (2.1.2), the RMS value becomes

$$A_{\text{RMS}} = \sqrt{\lim_{T \to \infty} \frac{1}{T} \int_0^T x(t)^2\, dt} \qquad (2.1.3)$$

where A_{RMS} represents a meaningless amplitude without further information. Many different time histories can give the same $\bar{x}$ and A_{RMS}

16 DYNAMIC SIGNAL ANALYSIS

values, so these values are not unique to any given signals. Additional information is required before these values have meaning.

2.1.4 The Frequency Spectrum

A third possibility for characterizing $x(t)$ is to use a frequency spectrum as shown in Fig. 2.1.2c. A frequency spectrum is a plot of component amplitudes as a function of frequency f. In this plot, four peak amplitudes stand out at points A through D. It may be possible to relate the peak frequencies (f_a, f_b, etc.) to rotating unbalance of one or more rotating shafts, blade passage rates, gear mesh rates, bearing noise or instability, structural resonances, and so on. A frequency plot is more useful than either a mean value, or an RMS value, since these plots often indicate discrete frequencies that are related to specific machine components and operating characteristics.

The frequency spectrum also contains significant amplitude information that may be useful to the engineer in judging significant system behavior in order to rate operating condition, to make a redesign, to plan corrective action, and so on. In the following sections, we attempt to give meaning to *frequency spectra*.

2.1.5 Analysis of a Single Sinusoid

At this point, it is informative to see how the above ideas work with an offset sinusoidal signal described by

$$x(t) = D + B\cos(\omega t + \phi) \qquad (2.1.4)$$

where D is the offset.
    B is the sinusoidal amplitude.
The corresponding mean, mean square, and frequency spectrum of this signal are now examined.

Temporal Mean The temporal mean value is obtained when Eq. (2.1.4) is substituted into Eq. (2.1.1) to obtain the temporal mean value expression of

$$\bar{x} = D + B\left[\frac{\sin(\omega T + \phi) - \sin(\phi)}{\omega T}\right] \cong D \qquad (2.1.5)$$

Equation (2.1.5) shows that the sinusoidal part decreases with increasing time T while D is the mean value. The mean value is obtained directly when (ωT) is a multiple of 2π. It is seen that sinusoidal functions have a zero mean when averaged over one time period, a result that is used in selecting a voltmeter's averaging time so as to remove power line frequency components from an instrument's reading. However, long averaging times are recommended so that $B/\omega T$ is small compared to D.

Temporal Mean Square The temporal mean square is obtained by substitution of Eq. (2.1.4) into Eq. (2.1.2). The integration gives

$$A_{\text{RMS}}^2 = D^2 + 2BD\left[\frac{\sin(\omega T + \phi) - \sin \phi}{\omega T}\right]$$
$$+ \frac{B^2}{2}\left[1 + \frac{\sin 2(\omega T + \phi) - \sin 2\phi}{2\omega T}\right] \quad (2.1.6)$$

which, in the limit as $\omega T \to \infty$, reduces to

$$A_{\text{RMS}}^2 = D^2 + \frac{B^2}{2} = D^2 + B_{\text{RMS}}^2 \quad (2.1.7)$$

Equation (2.1.6) shows two terms that contaminate the RMS value if ωT is too small. The first contaminant comes from the cross product involving BD. This term cancels when ωT is a multiple of 2π or is insignificant when ωT is sufficiently large to make the BD product insignificant compared to either D^2 or $B^2/2$. The second contaminant term cancels when ωT is a multiple of π. This term becomes insignificant compared to unity when $2\omega T$ is on the order of 100 or larger. Of these two terms, the more subtle one is the cross product term, which is often overlooked.

Equation (2.1.7) indicates three additional important points. First, when B is zero, it is seen that overall RMS amplitude, A_{RMS}, is equal to the mean value D; that is, $A_{\text{RMS}} = D$. Second, when D is zero, Eq. (2.1.7) becomes

$$A_{\text{RMS}} = B_{\text{RMS}} = \frac{B}{2} = 0.707B \quad (2.1.8)$$

where B_{RMS} is the *sinusoidal RMS amplitude*. Third, the signal's overall RMS amplitude, A_{RMS}, is the root mean square of the total signal that includes the square of the mean plus the square of the sinusoidal RMS amplitude. Thus in general, A_{RMS} and B_{RMS} represent two completely different terms.

The peak amplitude, B, and the RMS amplitude, B_{RMS}, are related by Eq. (2.1.8) for sinusoidal functions only. *Under no circumstances can the A_{RMS} amplitude for an arbitrary signal be converted to an equivalent peak*

18 DYNAMIC SIGNAL ANALYSIS

or RMS sinusoidal amplitude through use of Eq. (2.1.8). An examination of Eqs. (2.1.7) and (2.1.8) should make this point sufficiently clear, since the mean square also includes the square of the mean. Unfortunately, this distinction is often overlooked.

These results have implications as to what a DC and AC coupled voltmeter can read. Equation (2.1.7) indicates that an AC coupled RMS voltmeter measures only a B_{RMS} amplitude since AC coupling eliminates the mean value D. Similarly, a DC coupled voltmeter must have the signal filtered to remove any AC (sinusoidal) components. Otherwise, the DC coupled instrument will give the overall RMS value of A_{RMS}, not D^2. Consequently, two separate voltmeters are needed, one AC coupled to measure B_{RMS} and the other a highly filtered DC voltmeter to measure the mean value. This voltmeter combination provides more information, since both B_{RMS} and D are available to describe the signal. Then the overall RMS amplitude can be computed from Eq. (2.1.7).

Frequency Spectrum The frequency spectrum of Eq. (2.1.4) can be presented in a number of ways, as shown in Fig. 2.1.3, since there is a choice of using amplitudes, mean square amplitudes, or RMS amplitudes. Figure 2.1.3a shows an amplitude spectrum, Fig. 2.1.3b shows a mean square amplitude spectrum, and Fig. 2.1.3c shows a RMS amplitude spectrum. The mean value of D always occurs at zero frequency. The sinusoidal amplitudes always plot at frequency ω with magnitudes that are dependent on the type presentation. While it is necessary to know the amplitude and frequency of a sinusoid in order to describe it, Fig. 2.1.3 illustrates that this information can be displayed in several different ways. These equations and plots suggest that the mean value and the sinusoidal amplitudes

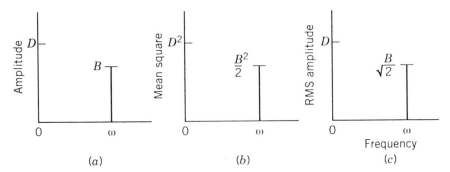

Fig. 2.1.3. Three different frequency spectra to represent the same sinusoidal time history with DC offset. (*a*) Amplitude. (*b*) Mean square amplitude. (*c*) RMS amplitude. (From J. W. Dally, W. F. Riley, and K. G. McConnell, *Instrumentation for Engineering Measurements*, 2nd ed., Copyright © 1993 by John Wiley & Sons, New York. Reprinted by permission.)

behave in quite different ways. This difference in behavior also occurs in more complicated time histories, as will be seen in the next section.

2.2 PHASOR REPRESENTATION OF PERIODIC FUNCTIONS

This section is concerned with using phasors to represent periodic time histories. Two different forms are used. One form uses a single rotating vector in the complex plane that has real and imaginary parts. The other form uses counterrotating vectors so that the time history is real valued with no imaginary parts.

2.2.1 The Phasor

Phasor $\bar{A}$ is a vector that rotates in a counterclockwise direction with angular velocity ω in the complex plane, as shown in Fig. 2.2.1. It can be written in terms of its real (R) and imaginary (Im) components as

$$\bar{A} = A\{\underbrace{\cos(\omega t)}_{\text{real}} + j\underbrace{\sin(\omega t)}_{\text{imaginary}}\} = A\,e^{j\omega t} \qquad (2.2.1)$$

where $j = \sqrt{-1}$
 A is the vector's magnitude.

Equation (2.2.1) is the well known Euler equation that relates the exponential function to its sine and cosine components multiplied by a constant magnitude. In general, this formula can be written as

$$e^{\pm j\theta} = \cos(\theta) \pm j\sin(\theta) \qquad (2.2.2)$$

Similarly, a second vector $\bar{A}_1$ can be written as

$$\bar{A}_1 = A_1\,e^{j(\omega t + \phi)} = \{A_1\,e^{j\phi}\}\,e^{j\omega t} \qquad (2.2.3)$$

where ϕ is the phase angle between $\bar{A}_1$ and $\bar{A}$ as shown. In this case, the reference phasor is $\bar{A}$ since it has a zero phase angle. A positive phase angle means that phasor $\bar{A}_1$ leads phasor $\bar{A}$ by angle ϕ. A negative phase angle means that phasor $\bar{A}_1$ lags behind phasor $\bar{A}$ by angle ϕ.

The bracketed quantity in Eq. (2.2.3) is a vector (magnitude and phase) relative to the orthogonal directions of reference vector $\bar{A}$. This relative position characteristic is reflected by a phase shift with respect to $\bar{A}$ as shown by the projection of $\bar{A}_1$ onto vector $\bar{A}$ and direction OC that is perpendicular to $\bar{A}$ as shown in Fig. 2.2.1. This projection is seen more

20 DYNAMIC SIGNAL ANALYSIS

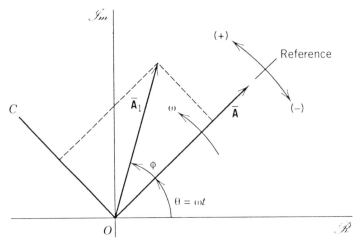

Fig. 2.2.1. Rotating phasors in the complex plane.

clearly when $\omega t = 0$ since the $\bar{A}$ projection is the real ($\cos \phi$) part and the OC projection is the imaginary ($\sin \phi$) part.

The phasor's time derivative is important in vibration work. From Eq. (2.2.1), the first and second derivatives become

$$\frac{d\bar{A}}{dt} = j\omega A \, e^{j\omega t} = \omega A \, e^{j(\omega t + \pi/2)}$$

$$\frac{d^2\bar{A}}{dt^2} = j^2\omega^2 A \, e^{j\omega t} = -\omega^2 A \, e^{j\omega t} = \omega^2 A \, e^{j(\omega t + \pi)} \tag{2.2.4}$$

from which it is evident that a time derivative is equivalent to multiplying the original phasor by $j\omega$. Multiplication by $j\omega$ is equivalent to multiplying by ω and phase shifting by $\pi/2$ radians or ninety degrees, as shown in Fig. 2.2.2. It is evident that integration is the same as dividing the original vector by $j\omega$.

From now on in this book, the bold notation is dropped, since the vector quantity (phasor) is complex with both magnitude and phase information and is simply treated as such.

2.2.2 The Phasor and Real Valued Sinusoids

The sinusoidal function is used extensively in describing vibration responses. This basic time history plays an important role in understanding

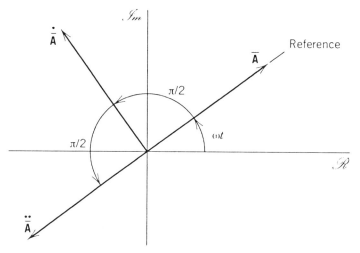

Fig. 2.2.2. A phasor and its derivatives. (From J. W. Dally, W. F. Riley, and K. G. McConnell, *Instrumentation for Engineering Measurements*, 2nd ed., Copyright © 1993 by John Wiley & Sons, New York. Reprinted by permission.)

signal analysis. Consider a sinusoidal time history that is defined by

$$x(t) = B\cos(\omega t + \phi) = B\cos(\theta) \qquad (2.2.5)$$

where B is the signal's peak amplitude.
  ω is the signal's circular frequency (rad/s).
  ϕ is the signal's phase angle; dependent on when $t = 0$.
  $\theta = \omega t + \phi$ is the combined argument.
Dependent on the value of ϕ, Eq. (2.2.5) represents both sine and cosine functions. For example, $\phi = 0$, gives a cosine function while $\phi = \pi/2$ gives a sine function.

The Euler formula in Eq. (2.2.2) can be used to relate the sine and cosine functions to the complex exponential function, giving

$$\cos(\theta) = \frac{e^{j\theta} + e^{-j\theta}}{2}$$
$$\sin(\theta) = \frac{e^{j\theta} - e^{-j\theta}}{2j} \qquad (2.2.6)$$

Combining Eqs. (2.2.5) and (2.2.6) gives a real valued time history that has no imaginary parts. This time history can be expressed as

$$x(t) = B\cos(\omega t + \phi) = \underbrace{\left(\frac{B}{2}e^{j\theta}\right)}_{\text{CCW}} + \underbrace{\left(\frac{B}{2}e^{-j\theta}\right)}_{\text{CW}}$$

$$= \underbrace{\left(\frac{B}{2}e^{j\phi}\right)}_{\text{CCW}} e^{j\omega t} + \underbrace{\left(\frac{B}{2}e^{-j\phi}\right)}_{\text{CW}} e^{-j\omega t} = \underbrace{X e^{j\omega t}}_{\text{CCW}} + \underbrace{X^* e^{-j\omega t}}_{\text{CW}}$$

(2.2.7)

where CW means a clockwise rotating phasor and CCW means counterclockwise rotating phasor. Equation (2.2.7) indicates that any single frequency sinusoid can be described in terms of two counterrotating vectors (X and X^*), each with magnitude $B/2$. These counterrotating vectors are shown in Fig. 2.2.3. The idea of counterrotating vectors introduces the concept of a negative frequency; that is, vector X rotates in the counterclockwise direction with a positive frequency (the $e^{j\theta}$ term), while X^* rotates in the clockwise direction with a negative frequency (the $e^{-j\theta}$ term). The real axis projections (point P_1) in Fig. 2.2.3 add to give the cosine response (point P) that has a magnitude of B. Similarly, the imaginary axis projections (points P_2) are seen to cancel. The real axis is the reference line for measuring all angles.

The exponential form of Fourier series is often used in frequency analysis. Consequently, a vibration data analyst needs to be aware of these concepts in order to read current literature and to understand operation manuals.

The counterrotating vectors in Eq. (2.2.7) can be expressed as

$$X = a + jb = \sqrt{a^2 + b^2}\, e^{j\phi} = \frac{B}{2} e^{j\phi} \quad \text{(counterclockwise)}$$

$$X^* = a - jb = \sqrt{a^2 + b^2}\, e^{-j\phi} = \frac{B}{2} e^{-j\phi} \quad \text{(clockwise)}$$

(2.2.8)

These vectors are *complex conjugates*; that is, they have equal magnitude ($B/2$) but opposite phase ($+jb$ vs. $-jb$ for the same value of a). Equations (2.2.7) and (2.2.8) contain trigonometric relationships of

$$a = \frac{B}{2}\cos(\phi) \qquad b = \frac{B}{2}\sin(\phi)$$

$$\tan(\phi) = \frac{b}{a}$$

(2.2.9)

These relationships are also evident from Fig. 2.2.3 when ωt is zero.

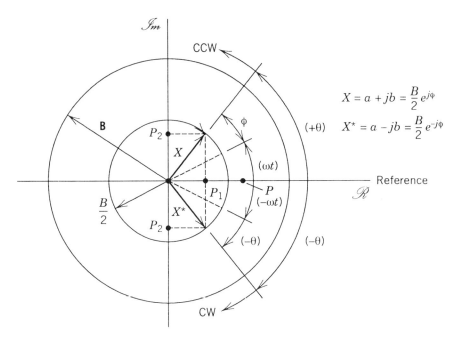

Fig. 2.2.3. Counterrotating complex vectors X and X^* that generate a real valued sinusoid of $B\cos(\omega t + \phi)$. (From J. W. Dally, W. F. Riley, and K. G. McConnell, *Instrumentation for Engineering Measurements*, 2nd ed., Copyright © 1993 by John Wiley & Sons, New York. Reprinted by permission.)

The notation used in Eqs. (2.2.7) through (2.2.9) is used extensively in this book. Generally, the time domain expression uses the lower case like $x(t)$ while the corresponding frequency domain expression uses the upper case like X or X_p or $X(\omega)$ as is appropriate. Only in special cases will this notation scheme be violated where standard time domain notation uses a capital letter. In these cases, a different symbol is used for the frequency domain quantity.

2.3 PERIODIC TIME HISTORIES

Many vibration responses can be classified and adequately described as stationary periodic time histories. Periodic time histories repeat themselves every T seconds; that is, $x(t + T) = x(t)$, independent of time t. These functions have a fundamental circular frequency ω_0 (rad/s) or frequency f_0 (Hz) that is related to the repeat period T by

$$\omega_0 = 2\pi\left(\frac{1}{T}\right) = 2\pi f_0 \qquad (2.3.1)$$

This fundamental frequency is the signal's lowest frequency.

2.3.1 Periodic Fourier Series

Any periodic and continuous real valued time history with a finite number of discontinuities can be written in terms of a Fourier series expressed in terms of sines and cosines. In view of Eq. (2.2.6), the sine and cosine Fourier series can also be expressed in terms of an exponential summation given by

$$x(t) = \sum_{p=-\infty}^{+\infty} X_p \, e^{jp\omega_0 t} \qquad (2.3.2)$$

where the complex Fourier coefficient X_p (with real and imaginary parts) is given by

$$X = \frac{1}{T} \int_{t}^{t+T} x(\tau) \, e^{-jp\omega_0 \tau} \, d\tau \qquad (2.3.3)$$

Equations (2.3.2) and (2.3.3) are called the *periodic Fourier transform pair*. Equation (2.3.2) transforms from a frequency domain description (in terms of X_p) to a time domain description, while Eq. (2.3.3) transforms from a time domain description of $x(t)$ to its frequency domain description. Equation (2.3.2) sums over all discrete frequencies that are multiples of fundamental frequency ω_0.

Equation (2.3.3) shows that the complex *Fourier coefficients* X_p are obtained by integrating $x(\tau)$ over one fundamental time period T. The integration can start at any convenient time t where either 0 or $-T/2$ is often used in theoretical analysis. Coefficient X_p is a measure of how well $x(\tau)$ correlates with the exponential function $(e^{-jp\omega_0 t})$ on a temporal basis over one fundamental period T. These coefficients occur in complex conjugate pairs X and X_p^* that are described by Eqs. (2.2.8) and (2.2.9), that is,

$$\begin{aligned} X_p &= a_p + jb_p \\ X_{-p} &= a_p - jb_p = X_p^* \quad \text{complex conjugate} \end{aligned} \qquad (2.3.4)$$

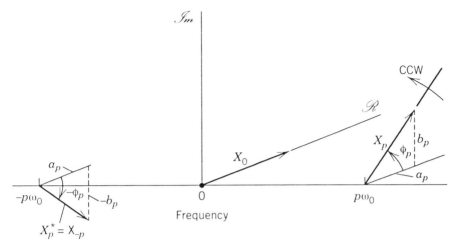

Fig. 2.3.1. A three-dimensional plot of X_0, X_p, and X_p^*.

Figure 2.3.1 shows the pth frequency components (also called the pth *harmonic*) plotted on three axes consisting of a real axis R, an imaginary axis *Im*, and a frequency axis that uses integer multiples of ω_0. Coefficient X_p is a vector that rotates in a counterclockwise direction with frequency $(p\omega_0)$ while its conjugate vector X_p^* rotates in the clockwise direction with frequency $(-p\omega_0)$. Equation (2.3.4) and Fig. 2.3.1 show that frequency component X_p has magnitude and phase given by

$$|X_p| = \sqrt{a_p^2 + b_p^2} = \sqrt{X_p X_p^*} \quad \text{magnitude} \tag{2.3.5}$$

$$\tan \phi_p = \frac{b_p}{a_p} \quad \text{phase}$$

The complex conjugate phase is the negative of ϕ_p as calculated from Eq. (2.3.5) and shown in Fig. 2.3.1.

The complex Fourier series has additional useful properties.

1. Only half of the coefficients need be determined since X_p^* is the complex conjugate of X_p.
2. When $x(t)$ is an even function where $x(-t) = x(t)$, only real coefficients a_p result because the cosine function is also an even time function that correlates with the even signal.
3. When $x(t)$ is an odd function where $x(-t) = -x(t)$, only imaginary coefficients b_p result since the sine function is also an odd time function that correlates with the odd signal.

Knowledge of these properties can simplify analytical work at times. However, when working with experimental data, the instant when $t = 0$ is usually quite arbitrary. Consequently, the Fourier coefficients X_p are usually complex quantities that are expressed in terms of their real and imaginary parts (or magnitude and phase characteristics).

Frequency Spectrum A periodic signal's frequency spectrum consists of a plot of X_p as function of discrete frequencies $p\omega_0$. This is a plot of discrete vectors along the frequency axis, as shown in Fig. 2.3.1. Since X_p and X_p^* are complex conjugates, it is common practice to plot only the positive frequency components. However, due to the complex nature of X_p, two different ways can be used to display these discrete frequency spectra. The first display consists of plotting a_p and b_p as functions of frequency. The second consists of plotting the magnitude $|X_p|$ and phase ϕ_p as functions of frequency. For a person examining a frequency spectrum, the magnitude plot is usually most informative, since phase is dependent on when time is set equal to zero while the magnitude indicates those frequencies of most concern. For mathematical recreation of a signal using the periodic Fourier transform, both magnitude and phase (or real and imaginary parts) are required.

2.3.2 The Mean, Mean Square, and Parseval's Formula

Mean Value Equation (2.3.3) shows that the temporal mean value results when summation index p is zero. It is possible to use a single period T and obtain a good mean value estimate since all harmonics will average to zero over period T. Also, the higher harmonics will drop out rapidly independent of this fortunate feature that $p\omega_0 T$ is a multiple of π. Thus X_0 is the *mean value of a periodic signal*. It will be seen in Chapter 5 that this statement is only approximately true when dealing with a frequency analyzer.

Mean Square A periodic signal's mean square is obtained by substituting Eq. (2.3.2) into Eq. (2.1.2) and integrating to obtain

$$A_{\text{RMS}}^2 = \frac{1}{T}\int_0^T x(\tau)^2\, d\tau = \sum_{p=-\infty}^{\infty} |X_p|^2 = X_0^2 + 2\sum_{p=1}^{\infty} |X_p|^2 \quad (2.3.6)$$

Equation (2.3.6) is known as *Parseval's formula* and shows that the mean square is the sum of the squares of the absolute values (magnitudes) of the Fourier coefficients.

The first summation in Eq. (2.3.6) can be divided into two terms since X_p and X_p^* are complex conjugates with equal magnitudes. The first term is X_0^2 and represents the square of the mean. The second term is twice the summation of $|X_p|^2$ over all positive frequencies. Equation (2.2.8) shows that the magnitude of X_p is half of the sinusoidal magnitude B_p so that $|X_p| = B_p/2$. Then Eq. (2.3.6) becomes

$$A_{\text{RMS}}^2 = X_0^2 + \sum_{p=1}^{\infty} \frac{B_p^2}{2} = X_0^2 + \sum_{p=1}^{\infty} B_{\text{RMS}}^2 \qquad (2.3.7)$$

Equation (2.3.7) is an alternative form of Parseval's formula. A comparison of Eqs. (2.3.7) and (2.1.7) shows that Eq. (2.1.7) is a special case that deals with a single frequency component of magnitude B and a mean value of magnitude D.

The importance of Eqs. (2.3.6) and (2.3.7) is that RMS frequency components add in a power-like manner, that is, on a squared basis, to give the signal's mean square. Many different periodic signals will give the same A_{RMS} value. Thus no unique relationship exists between a periodic signal and A_{RMS} as there is for a single sinusoid with zero mean.

2.3.3 Analysis of a Square Wave

An informative signal for analysis purposes to illustrate some of these concepts is a periodic rectangular square wave, as shown in Fig. 2.3.2a. This signal's Fourier coefficients are obtained directly from Eq. (2.3.3) by integrating from $-T/2$ to $T/2$. This integration gives

$$X_P = \frac{A}{2}\left[\frac{\sin(p\pi/2)}{(p\pi/2)}\right] = \frac{A}{2}\text{sinc}(z) \qquad (2.3.8)$$

where $z = p\pi/2$.

sinc(z) is the *sinc function* that is defined by

$$\text{sinc}(z) = \frac{\sin(z)}{z} \qquad (2.3.9)$$

The sinc function has a value of unity in the limit as z goes to zero and decreases within an envelope bounded by $1/z$. It is also an even function in that $\text{sinc}(-z) = \text{sinc}(z)$.

Since $x(t)$ is an even function as defined in Fig. 2.3.2a, the coefficients are real and appear as plotted in Fig. 2.3.2b. Figures 2.3.2b and 2.3.2c show two different types of frequency spectra. Figure 2.3.2b shows an amplitude spectrum where the magnitudes and signs are shown while Fig. 2.3.2c shows only amplitude magnitudes. It is seen in Fig. 2.3.2b that the mean value is $A/2$, the even coefficients are zero, the odd coefficients

28 DYNAMIC SIGNAL ANALYSIS

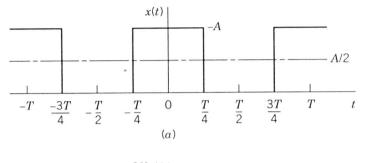

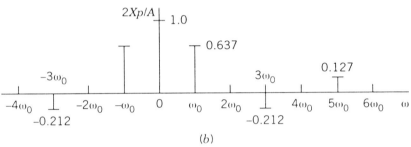

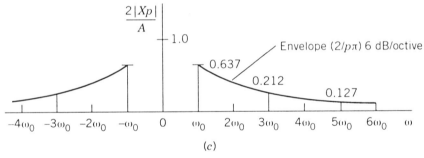

Fig. 2.3.2. Periodic rectangular pulses and Fourier frequency components. (*a*) Rectangular pulse of amplitude A and period T. (*b*) Fourier series coefficients for periodic rectangular pulses. (*c*) Magnitude of Fourier coefficients for periodic rectangular pulses.

alternate in sign, and both frequency spectra are even functions. Both Figs. 2.3.2*b* and 2.3.2*c* show that the magnitudes decrease within the envelope described by $2/p\pi$ in accordance with the sinc function. This envelope decreases at a rate of 6 dB/octave.

It should be evident that either plot is informative to one interested in knowing the signal's frequency components. However, only Fig. 2.3.2*b* shows the component signs (phase information) that are required to mathematically recreate the time history. Thus there can be two distinct needs, one for analysis information through personal observation and one for mathematically recreating the signal.

Parseval's formula in Eq. (2.3.6) can be checked by calculation in this case. The direct integration process and the summation process of Eq. (2.3.6) give the mean square as

$$\text{MS} = \frac{A^2}{2} = \frac{A^2}{4}\left[1 + 2\sum_{p=1}^{\infty} \text{sinc}^2\left(\frac{p\pi}{2}\right)\right] \quad (2.3.10)$$

which reduces to

$$1 = 2\sum_{p=1}^{\infty} \text{sinc}^2\left(\frac{p\pi}{2}\right) \quad (2.3.11)$$

Calculation of the first 200 summation terms in Eq. (2.3.11) gives a value of 0.996 compared to unity. Thus the area under the square of the sinc function is 0.5.

2.4 TRANSIENT SIGNAL ANALYSIS

Figure 2.4.1a shows a transient signal $x(t)$. This signal is characterized by being constant for long periods of time with significant amplitude changes occurring over a short time duration T_d. Application of Fourier series concepts to transient signals causes interesting things to happen.

2.4.1 Difference Between Periodic and Transient Frequency Analysis

The difference between periodic and transient signal behavior is illustrated using the rectangular pulse of amplitude A, duration T, and repeat period T_0, as shown in Fig. 2.4.1b. Let the pulse repeat period T_0 be related to the pulse duration T by

$$T_0 = \beta T \quad (2.4.1)$$

so that the corresponding fundamental frequency becomes

$$\omega_0 = \frac{2\pi}{T_0} = \frac{2\pi}{\beta T} \quad (2.4.2)$$

It is seen in Fig. 2.4.1b that each rectangular pulse becomes more isolated

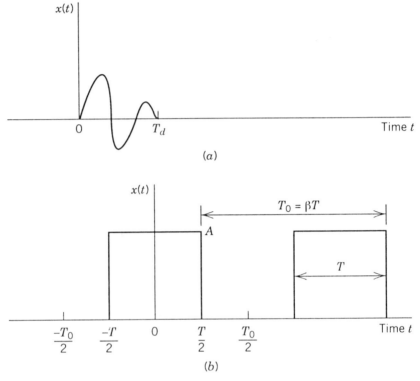

Fig. 2.4.1. Transient time histories. (*a*) Definition. (*b*) Rectangular pulses of duration T and repeat period T_0. (From J. W. Dally, W. F. Riley, and K. G. McConnell, *Instrumentation for Engineering Measurements*, 2nd ed., Copyright © 1993 by John Wiley & Sons, New York. Reprinted by permission.)

in time with increasing values of β, becoming a single pulse of duration T in the limit as β goes to infinity.

For this time history, the Fourier frequency components obtained from Eq. (2.3.3) are real-valued (phase is zero) and are given by

$$X_P = \frac{AT}{T_0}\left[\frac{\sin(p\pi/\beta)}{(p\pi/\beta)}\right] = \frac{AT}{T_0}\operatorname{sinc}(z) \qquad (2.4.3)$$

where $z = p\pi/\beta$ for this case. Equation (2.4.3) gives the same results as Eq. (2.3.8) when $\beta = 2$. The mean value X_0 is AT/T_0 for this case and this is consistent with the results from Eq. (2.1.1).

The values of X_p for $\beta = 2$ and 10 are plotted in Fig. 2.4.2. This figure shows three significant characteristics.

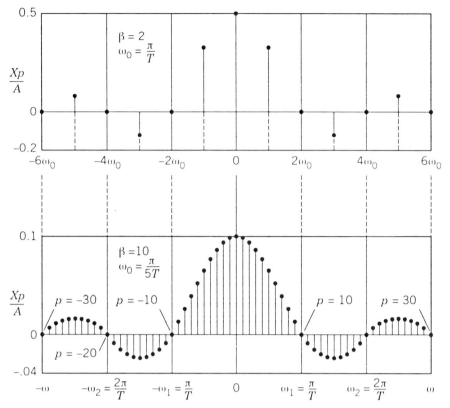

Fig. 2.4.2. Sinc function Fourier coefficients for $\beta = 2$ and $\beta = 10$. (From J. W. Dally, W. F. Riley, and K. G. McConnell, *Instrumentation for Engineering Measurements*, 2nd ed., Copyright © 1993 by John Wiley & Sons, New York. Reprinted by permission.)

1. When $\beta = 10$, the magnitudes are one-fifth those corresponding to $\beta = 2$.
2. Eleven frequency components lie between $\omega = 0$ and $\omega_1 = \pi/T$, showing that the frequency components are much closer together.
3. When argument z in Eq. (2.4.3) is a multiple of π, zero amplitude frequency components occur at fixed frequencies.

Thus it is evident that the Fourier series approach is inadequate to handle transform directly without some modification.

2.4.2 The Transient Fourier Transform

The transient Fourier transform is obtained from the periodic Fourier transform by defining a new Fourier coefficient in the limit as T_0 goes to infinity. Let $X(\omega) = X_p T_0$ be the new Fourier coefficient so that Eqs. (2.3.2) and (2.3.3) become

$$X(\omega) = \int_{-\infty}^{\infty} x(t) e^{-j\omega t} dt \qquad (2.4.4)$$

and

$$x(t) = \frac{1}{2\pi} \int_{-\infty}^{\infty} X(\omega) e^{j\omega t} d\omega \qquad (2.4.5)$$

These equations form the *transient Fourier transform* pair. When $x(t)$ is a real-valued time history, $X(-\omega)$ is the complex conjugate of $X(\omega)$. Equations (2.4.4) and (2.4.5) are valid when $x(t)$ is bounded, defined over the time region of $t = -\infty$ to $t = \infty$, and satisfies the relationship that

$$\int_{-\infty}^{\infty} |x(t)| dt < \infty$$

Most real physical system signals satisfy these generous conditions.

Equations (2.4.4) and (2.4.5) indicate that all frequencies are present in transient time histories. However, each frequency makes an infinitesimal contribution to the signal. This contribution is defined by

$$X(\omega) \frac{d\omega}{2\pi} = X(f) df \qquad (2.4.6)$$

where $d\omega$ or df are infinitesimal quantities. It is evident from Eqs. (2.4.4) through (2.4.6) that $X(\omega)$ is a *spectral density* that is expressed in terms of (units of $x(t)$ per hertz). For example, the spectral density units are lb/Hz when $x(t)$ has force units of pounds or g/Hz when $x(t)$ is an acceleration signal with units of g's, and so on. The spectral density is a continuous complex variable function of parameter ω that has real and imaginary parts (magnitude and phase), just as its discrete cousin X_p is a discrete vector (complex number) that is a function of discrete frequencies.

There is a subtle difference between a *frequency spectrum* and a *spectral density*. The frequency spectrum is a display of discrete amplitudes as a function of discrete frequencies of $p\omega_0$; that is X_p versus $p\omega_0$. The spectral density is a continuous display of amplitude density (units of $x(t)$ per

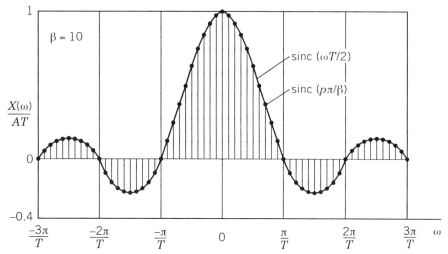

Fig. 2.4.3. Comparison of spectral density $X(\omega)$ and $X_p T = X(p\omega_0)$ for a rectangular pulse of amplitude A and duration T. (From J. W. Dally, W. F. Riley, and K. G. McConnell, *Instrumentation for Engineering Measurements*, 2nd ed., Copyright © 1993 by John Wiley & Sons, New York. Reprinted by permission.)

hertz) as a function of continuous frequencies; that is, $X(\omega)$ versus ω. Clearly, frequency spectrum and spectral density are similar but not the same. A major question is, can the spectral density be calculated using a periodic Fourier transform calculation scheme if the results are properly interpreted?

Consider the limit of the rectangular pulse of amplitude A and duration T, as shown in Fig. 2.4.1b, when T_0 goes to infinity. When Eq. (2.4.4) is integrated over $-T/2$ to $T/2$ (note that $x(t)$ is zero outside of this time range), the result is

$$X(\omega) = AT \frac{\sin(\omega T/2)}{(\omega T/2)} = AT \operatorname{sinc}(\omega T/2) \qquad (2.4.7)$$

A comparison of Eqs. (2.4.3) and (2.4.7) show they are the same when Eq. (2.4.3) is multiplied by T_0 in order to convert its discrete frequency spectrum into a discrete spectral density; that is $X_p T_0 = X(\omega) = X(p\omega_0)$. These two functions are plotted in Fig. 2.4.3 for the case when $\beta = 10$. The zero crossing occurs when either $\omega T/2$ or $p\pi/\beta$ are multiples of π.

This theoretical exercise shows several important points.

1. A periodic Fourier transform can be used to calculate a transient

Fourier transform spectral density if the transient signal occupies 10 percent or less of repeat period T_0; that is, $\beta \geq 10$.

2. The spectral density is obtained by multiplying the discrete frequency components X_p by repeat period T_0.
3. The value of $X_0 T_0$ and $X(0)$ is AT, that is, the area under the transient curve, as is seen from Eq. (2.4.4) when $\omega = 0$. AT is the multiplicative factor in both frequency spectrum displays and represents an impulse type of term.

2.4.3 Transient Mean, Mean Square, and Parseval's Formula

The definitions for mean and mean square as given by Eqs. (2.1.1) and (2.1.2) must be changed when dealing with impulsive functions, since both quantities go to zero in the limit as T goes to infinity. In order to overcome this difficulty the original definitions are multiplied by T so that the *transient signal mean* is defined as

$$\text{mean} = \int_{-\infty}^{\infty} x(t)\, dt \qquad (2.4.8)$$

In mechanics terms, Eq. (2.4.8) can be thought to represent the signal's impulse. Similarly, the *transient signal mean square* is defined as

$$\text{mean square} = \int_{-\infty}^{\infty} x^2(t)\, dt \qquad (2.4.9)$$

Substituting $x(t)$ from Eq. (2.4.5) into Eq. (2.4.9) and taking advantage of Eq. (2.4.4) and the complex conjugate nature of $X(\omega)$ (namely, $X(-\omega)$ is the complex conjugate of $X(\omega)$ so that $X(-\omega)$ times $X(\omega)$ equals $|X(\omega)|^2$), Eq. (2.4.9) reduces to

$$\text{mean square} = \int_{-\infty}^{\infty} x^2(t)\, dt = \frac{1}{2\pi} \int_{-\infty}^{\infty} |X(\omega)|^2\, d\omega \qquad (2.4.10)$$

which is known as *Parseval's formula for integrals*. A comparison of Eq. (2.4.10) with Eq. (2.3.6) shows that they are identical in form, one for a discrete frequency component spectrum and the other for a continuous spectral density. Further, it is seen that the integral in Eq. (2.4.10) is proportional to the area under the square of the spectral density curve. These results are additional evidence of the difference between periodic and transient signals and their respective frequency spectra.

Example: Consider the rectangular step pulse shown in Fig. 2.4.1b when β is infinite so that the pulse stands alone in time. For this case, the transient mean value from Eq. (2.4.8) becomes

$$\text{mean} = AT \qquad (2.4.11)$$

from which it is seen that the transient mean is an impulse-like quantity. The transient mean square from Eq. (2.4.9) gives

$$\text{mean square} = A^2 T \qquad 2.4.12)$$

However, if the frequency spectrum from Eq. (2.4.7) is substituted into Eq. (2.4.10), the result is

$$\text{mean square} = \frac{A^2 T}{\pi} \int_{-\infty}^{\infty} \text{sinc}^2(z)\, dz \qquad (2.4.13)$$

where $z = (\omega T/2)$. Equation (2.4.13) gives the same result as Eq. (2.4.12) so that the integral is equal to π. Note that the periodic mean square is proportional to A^2 while the transient mean square is proportional to $A^2 T$. The extra T term comes from the transient definition where the integral is not averaged by time as is done in the periodic definition.

2.5 CORRELATION CONCEPTS—A STATISTICAL POINT OF VIEW

Correlation is an important means of analyzing dynamic signals. The correlation function can be directly related to a signal's Fourier coefficients. There are three different, but similar, definitions of correlation, depending on whether the signal is periodic, transient, or random in nature. Within each signal classification, there are two subclassifications called auto-correlation and cross-correlation. In this section, statistical definitions of correlation are reviewed while the periodic, transient, and random type time history correlation relationships are developed in the following sections.

Consider the xy data shown in Fig. 2.5.1a. We are interested in determining the relationship between these two sets of data, which are plotted against one another. This task is simplified if the data is transformed by using the linear relationships of

$$\begin{aligned} x_1 &= x - \bar{x} \\ y_1 &= y - \bar{y} \end{aligned} \qquad (2.5.1)$$

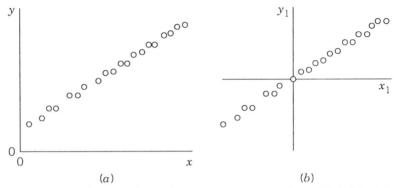

Fig. 2.5.1. Linear data transformation to remove mean values. (*a*) Original data. (*b*) Transformed data.

where $\bar{x}$ and $\bar{y}$ are the mean values of x and y, respectively. Then, x_1 and y_1 represent a set of coordinates that pass through the "center of gravity" of the data, as shown in Fig. 2.5.1*b*.

The best straight line estimate of the data is given by

$$y_p = mx_1 \tag{2.5.2}$$

where y_p is the predicted value of y_1, and m is a straight line slope. The vertical[2] deviation of the data from its best straight estimate is given by

$$\Delta = y_1 - y_p = y_1 - mx_1$$

The idea is to minimize the value of Δ^2 in a least squares sense. This is done by calculating the *mathematical expectation* of Δ^2 (the standard symbol for *mathematical expectation*[3] $E[x]$) so that we have

$$\begin{aligned} E[\Delta^2] &= E[(y_1 - mx_1)^2] \\ &= E[y_1^2] + m^2 E[x_1^2] - 2mE[x_1 y_1] \end{aligned} \tag{2.5.3}$$

The minimum value of m corresponds to the derivative of Eq. (2.5.3) becoming zero. This gives

---

[2] A horizontal deviation can also be used.
[3] The mathematical expectation is defined in terms of the data's probability density $p(x)$ (see Section 2.8 for a detailed definition of probability density). Thus the mean value of data x is estimated from $E[x] = \int_{-\infty}^{\infty} xp(x)\,dx$.

CORRELATION CONCEPTS – A STATISTICAL POINT OF VIEW

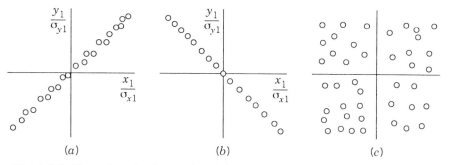

Fig. 2.5.2. Plots of perfectly correlated data and perfectly uncorrelated data. (a) $\rho_{xy} = 1$. (b) $\rho_{xy} = -1$. (c) $\rho_{xy} = 0$.

$$m = \frac{E[x_1 y_1]}{E[x_1^2]} \qquad (2.5.4)$$

Note that the variance of the x_1 and y_1 data is given by

$$\sigma_{x_1}^2 = E[x_1^2] \quad \text{and} \quad \sigma_{y_1}^2 = E[y_1^2] \qquad (2.5.5)$$

Substitution of Eqs. (2.5.1), (2.5.4) and (2.5.5) into Eq. (2.5.2) gives the best straight line estimate relationship between the data to be

$$\frac{y - \bar{y}}{\sigma_y} = \left\{ \frac{E[(x - \bar{x})(y - \bar{y})]}{\sigma_x \sigma_y} \right\} \frac{x - \bar{x}}{\sigma_x} \qquad (2.5.6)$$

The bracketed term is the *normalized correlation coefficient* ρ_{xy} given by

$$\rho_{xy} = \frac{E[(x - \bar{x})(y - \bar{y})]}{\sigma_x \sigma_y} \qquad (2.5.7)$$

When $\rho_{xy} = \pm 1$, the data is perfectly correlated, as shown in Fig. 2.5.2; it is completely uncorrelated when $\rho_{xy} = 0$. Equation (2.5.7) shows that the mathematical expectation $E[(x - \bar{x})(y - \bar{y})]$ represents the correlation of x and y. However, this number has more meaning when normalized by the standard deviation of each parameter from its mean value, in which case, the correlation coefficient ranges from -1 to $+1$ and a universal meaning can be attached to its value. The normalized correlation coefficient is a measure of the "likeness" or "similarity" of the data x and y. In our signal analysis, the likeness or similarity is compared on a time basis.

2.6 CORRELATION CONCEPTS—PERIODIC TIME HISTORIES

In this section, we are interested in analyzing the kinds of correlation relationships that exist between two periodic time histories. In this case, time is the common parameter so that the x and y data of the previous section is compared on a temporal basis. A major assumption is made in working with periodic time histories and correlation; namely, all time histories have the *same fundamental frequency* ω_0. If this assumption is not true, then there is no correlation between the signals over a long period of time. Both cross-correlation and auto-correlation are presented in this section.

2.6.1 Cross-Correlation

Cross-correlation $R_{12}(\tau)$ is defined by the time average integral

$$R_{12}(\tau) = \frac{1}{T} \int_{-T/2}^{T/2} x_1(t) x_2(t + \tau) \, dt \tag{2.6.1}$$

where τ is the time shift.
 T is the periodic function's fundamental period.
 x_1 and x_2 are the time functions being compared.
Standard notation is that the second subscript (2 in this case) corresponds to the time-shifted time history, while the first subscript corresponds to the reference (unshifted) time history.

The process for calculating the cross-correlation function is shown in Fig. 2.6.1. This process consists of several steps.

1. The values of $x_1(t)$ (point A) are multiplied by the values of $x_2(t + \tau)$ (point B) to form a third curve for a given value of time shift τ (see Fig. 2.6.1a and b).
2. The area under this product function curve is then obtained and averaged over one period T (see Fig. 2.6.1b).
3. The average value is plotted as a single point C (see Fig. 2.6.1c).
4. This process is repeated for each value of τ that is of interest to complete the cross-correlation function.

The cross-correlation function indicates how well the time shifted events of $x_2(t + \tau)$ are similar to the events of $x_1(t)$. For example, consider the cross-correlation function peak at τ_0 (point D in Fig. 2.6.1c). This peak can give information about the time delay between two events on a structure, such as between a vehicle's wheel axle and steering wheel accelerations. This time delay is related to the structural path that the vibration

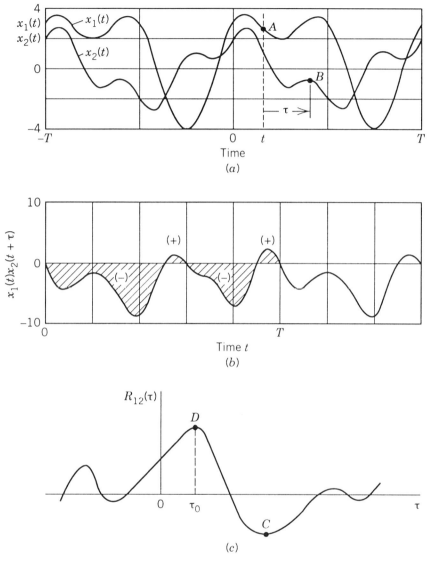

Fig. 2.6.1. Cross-correlation calculation processes. (*a*) Relative shift between time histories. (*b*) Product of time histories for $\tau = T/5$. (*c*) Cross-correlation function.

must travel from the axle to steering wheel. This time delay may give a clue as to where effective corrective action may be taken to control the vibration.

It is desirable to relate the cross-correlation function to the Fourier series coefficients that represent $x_1(t)$ and $x_2(t)$. The corresponding periodic Fourier transforms are given by

$$x_1(t) = \sum_{q=-\infty}^{+\infty} X_{1q} e^{jq\omega_0 t}$$

$$X_{1q} = \frac{1}{T} \int_t^{t+T} x_1(t) e^{-jq\omega_0 t} dt \qquad (2.6.2)$$

for $x_1(t)$ and

$$x_2(t) = \sum_{p=-\infty}^{+\infty} X_{2p} e^{jp\omega_0 t}$$

$$X_{2p} = \frac{1}{T} \int_t^{t+T} x_2(t) e^{-jp\omega_0 t} dt \qquad (2.6.3)$$

for $x_2(t)$. Substituting the expression for $x_2(t + \tau)$ from Eq. (2.6.3) into Eq. (2.6.1) gives

$$R_{12}(\tau) = \frac{1}{T} \int_{-T/2}^{T/2} \left\{ x_1(t) \sum_{p=-\infty}^{+\infty} X_{2q} e^{jp\omega_0 t} e^{jp\omega_0 \tau} \right\} dt \qquad (2.6.4)$$

Then, interchanging integration and summation processes in Eq. (2.6.4), we obtain

$$R_{12}(\tau) = \sum_{p=-\infty}^{+\infty} X_{2p} e^{jp\omega_0 \tau} \left\{ \frac{1}{T} \int_{-T/2}^{T/2} x_1(t) e^{jp\omega_0 t} dt \right\} \qquad (2.6.5)$$

A comparison of the bracketed term in Eq. (2.6.5) with Eq. (2.6.2) shows that the bracketed term is the complex conjugate of X_{1p}. Thus Eq. (2.6.5) reduces to

$$R_{12}(\tau) = \sum_{p=-\infty}^{+\infty} \{X_{1p}^* X_{2p}\} e^{jp\omega_0 t} = \sum_{p=-\infty}^{+\infty} C_{12_p} e^{jp\omega_0 \tau} \qquad (2.6.6)$$

where

$$C_{12_p} = X_{1p}^* X_{2p} \qquad (2.6.7)$$

is the pth cross-correlation *periodic Fourier coefficient*. In general, these cross-correlation Fourier coefficients are complex with real and imaginary parts.

CORRELATION CONCEPTS – PERIODIC TIME HISTORIES   41

Equation (2.6.6) indicates that the cross-correlation is periodic and can be expanded in terms of sinusoids. Thus C_{12_p} can be obtained from

$$C_{12_p} = \frac{1}{T} \int_\tau^{\tau+T} R_{12}(\tau) e^{-jp\omega_0 \tau} \, d\tau \qquad (2.6.8)$$

Equations (2.6.6) and (2.6.8) form a periodic Fourier transform pair relating the cross-correlation function to its frequency components between the time shift domain and the frequency domain. Equation (2.6.7) shows how these frequency components are related to the frequency components of the original functions.

Now, consider the cross-correlation function $R_{21}(\tau)$ that is defined by

$$R_{21}(\tau) = \frac{1}{T} \int_{-T/2}^{T/2} x_2(t) x_1(t+\tau) \, dt \qquad (2.6.9)$$

where $x_1(t)$ is the time shifted function. Then, repeating the steps used in formulating Eqs. (2.6.4) through (2.6.7) with $x_1(t)$ being replaced by its Fourier series expansion in Eq. (2.6.2), the results are

$$R_{21}(\tau) = \sum_{p=-\infty}^{+\infty} \{X_{1p} X_{2p}^*\} e^{jp\omega_0 \tau} = \sum_{p=-\infty}^{+\infty} C_{21_p} e^{jp\omega_0 \tau} \qquad (2.6.10)$$

where

$$C_{21_p} = X_{1p} X_{2p}^* \qquad (2.6.11)$$

and

$$C_{21_p} = \frac{1}{T} \int_\tau^{\tau+T} R_{21}(\tau) e^{-jp\omega_0 \tau} \, d\tau \qquad (2.6.12)$$

It is obvious from comparing Eqs. (2.6.7) and (2.6.11) that the cross-correlation frequency components are complex conjugates of one another; that is,

$$\begin{aligned} C_{12_p} &= C_{21_p}^* \\ C_{21_p} &= C_{12_p}^* \end{aligned} \qquad (2.6.13)$$

The implication of Eq. (2.6.13) is that $R_{12}(\tau) = R_{21}(-\tau)$, that is, the functions appear as though they are rotated about the vertical ($\tau = 0$) axis.

2.6.2 Auto-Correlation

Auto-correlation occurs when the two time histories are the same, so that a time history is compared with itself. Thus, Eqs. (2.6.1) and (2.6.6) through (2.6.8) become

$$R_{11}(\tau) = \frac{1}{T}\int_{-T/2}^{T/2} x_1(t)x_1(t+\tau)\,dt = \sum_{p=-\infty}^{+\infty} C_{11p} e^{jp\omega_0\tau} \qquad (2.6.14)$$

where

$$C_{11p} = X_{1p}^* X_{1p} = |X_{1p}|^2 \qquad (2.6.15)$$

and

$$C_{11p} = \frac{1}{T}\int_{\tau}^{\tau+T} R_{11}(\tau) e^{-jp\omega_0\tau}\,d\tau \qquad (2.6.16)$$

The auto-correlation function has several important features:

1. When $\tau = 0$, Eq. (2.6.14) becomes the mean square definition for periodic functions. Thus $R_{11}(0)$ is the mean square of the signal.
2. Equation (2.6.14) reduces to Parseval's formula for periodic functions when $\tau = 0$; that is, Eq. (2.6.14) becomes

$$R_{11}(0) = \frac{1}{T}\int_{-T/2}^{T/2} x_1(t)^2\,dt = \sum_{p=-\infty}^{+\infty} C_{11p} = \sum_{p=-\infty}^{+\infty} |X_{1p}|^2 \qquad (2.6.17)$$

3. All Fourier series coefficients in Eq. (2.6.15) are seen to be real and positive. This means that the auto-correlation function is an even function in τ; that is, $R_{11}(-\tau) = R_{11}(\tau)$, and it contains only cosine-type terms, since cosine functions are even while sine functions are odd.

Example 1: Determine the cross- and auto-correlation functions for two periodic functions that are described by $x_1 = D_1 + B_1 \cos(\omega t)$ and $x_2 = D_2 + B_2 \cos(\omega t)$. In this case, $C_{-1} = B_1 B_2/4$, $C_0 = D_1 D_2$, and $C_1 = B_1 B_2/4$. All other Fourier coefficients are zero. Thus according to Eqs. (2.6.6) and (2.6.7), the cross-correlation becomes

$$R_{12}(\tau) = D_1 D_2 + \frac{B_1 B_2}{2}\cos(\omega\tau) \qquad (2.6.18)$$

CORRELATION CONCEPTS—PERIODIC TIME HISTORIES   43

Equation (2.6.18) shows $R_{12}(\tau)$ is a sinusoidal function with the same frequency as the original function and has a DC offset. The DC offset comes from the individual function's DC offset. It is seen that if either function has zero offset, then the cross-correlation also has zero offset. The reader should verify that the results of Eq. (2.6.18) can be obtained by direct integration of Eq. (2.6.1) and that the same result occurs if the functions $x_1(t)$ and $x_2(t)$ are sine functions instead of cosine functions.

For the auto-correlation case, Eq. (2.6.18) becomes

$$R_{11}(\tau) = D^2 + \frac{B^2}{2}\cos(\omega\tau) \qquad (2.6.19)$$

It is evident that this result gives the mean square when $\tau = 0$ and is identical to the results of both Eq. (2.1.6) and Parseval's formula (Eqs. (2.3.6) or 2.3.7)).

Example 2: For this example, consider the square wave shown in Fig. 2.3.3a. Determine the auto-correlation of this time history. Substitution of the square wave properties into the integral part of Eq. (2.6.14) gives

$$R_{11}(\tau) = \frac{A^2}{T}\left[\frac{T}{2} - \tau\right] \quad \text{for } 0 < \tau < T/2$$

$$R_{11}(\tau) = \frac{A^2}{T}\left[\frac{T}{2} + \tau\right] \quad \text{for } -T/2 < \tau < 0 \qquad (2.6.20)$$

Equation (2.6.20) repeats itself over each period T so that a triangular auto-correlation function is generated, as shown in Fig. 2.6.2. The largest value of $R_{11}(\tau)$ is A^2.

A Fourier analysis of this triangular waveform gives a Fourier expression of

$$R_{11}(\tau) = \frac{A^2}{4}\left[1 + \frac{8}{\pi^2}\sum_{p=1}^{\infty}\frac{\cos\{(2p-1)\omega_0\tau\}}{(2p-1)^2}\right] \qquad (2.6.21)$$

The Fourier series coefficients for the square wave are given by Eq. (2.3.8). Substituting these coefficients into Eq. (2.6.14) and noting that

44 DYNAMIC SIGNAL ANALYSIS

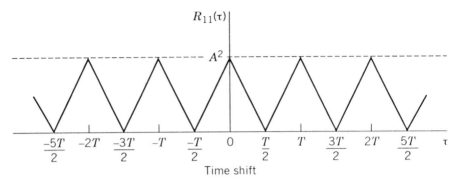

Fig. 2.6.2. Auto-correlation function $R_{11}(\tau)$ for a periodic square wave of duration T and amplitude A (see Fig. 2.3.2a).

auto-correlation functions are even functions so that only cosine terms result, one also obtains the results of Eq. (2.6.21).

2.7 CORRELATION CONCEPTS—TRANSIENT TIME HISTORIES

Recall that a transient time history requires a different formulation of the periodic Fourier transform than a periodic time history. Similar problems exist if the cross-correlation definition of Eq. (2.6.1) is applied to transient time histories, since the time factor T needs to go to infinity. Thus multiplying the previous correlation definition by T gives the corresponding transient definitions.

2.7.1 Cross-Correlation

The *transient cross-correlation* function is defined by

$$R_{12}^t(\tau) = \int_{-\infty}^{+\infty} x_1(t) x_2(t+\tau)\, dt \qquad (2.7.1)$$

where superscript t indicates that this is the transient cross-correlation function. The transient time functions have transient Fourier transform pairs given by

$$\begin{aligned} x_1(t) &= \frac{1}{2\pi} \int_{-\infty}^{+\infty} X_1(\omega) e^{j\omega t}\, d\omega \\ X_1(\omega) &= \int_{-\infty}^{+\infty} x_1(t) e^{-j\omega t}\, dt \end{aligned} \qquad (2.7.2)$$

for $x_1(t)$ and

$$x_2(t) = \frac{1}{2\pi} \int_{-\infty}^{+\infty} X_2(\omega) e^{j\omega t} d\omega$$

$$X_2(\omega) = \int_{-\infty}^{+\infty} x_2(t) e^{-j\omega t} dt$$
(2.7.3)

for $x_2(t)$. Then it is easy to show, by using the process utilized in Section 2.6, that

$$R^t_{12}(\tau) = \frac{1}{2\pi} \int_{-\infty}^{+\infty} \{X_1^*(\omega) X_2(\omega)\} e^{j\omega \tau} d\omega = \frac{1}{2\pi} \int_{-\infty}^{+\infty} C_{12}(\omega) e^{j\omega \tau} d\omega$$
(2.7.4)

where

$$C_{12}(\omega) = X_1^*(\omega) X_2(\omega)$$
(2.7.5)

is the cross-correlation frequency spectrum. This frequency spectrum is a continuous function of frequency that is usually complex. The units for $C_{12}(\omega)$ are (units of $x_1(t)$/Hz)(units of $x_2(t)$/Hz). Note that the asterisk (*) indicates that the $X_1^*(\omega)$ is a complex conjugate. The frequency spectrum in Eqs. (2.7.4) or (2.7.5) is related to $R^t_{12}(\tau)$ by its Fourier transform, given by

$$C_{12}(\omega) = \int_{-\infty}^{+\infty} R^t_{12}(\tau) e^{-j\omega \tau} d\tau$$
(2.7.6)

so that Eqs. (2.7.4) and (2.7.6) create a transient Fourier transform pair.

Similar to the periodic cross-correlation relationships, the transient cross-correlation frequency spectrum of $R^t_{21}(\tau)$ is the complex conjugate of the frequency spectrum of $R^t_{12}(\tau)$. This is easily determined by substituting the transient Fourier transform expression of $x_1(t)$ from Eq. (2.7.2) into

$$R^t_{12}(\tau) = \int_{-\infty}^{+\infty} x_2(t) x_1(t + \tau) dt$$
(2.7.7)

interchanging the integration order and recognizing $X_2^*(\omega)$ as one of the terms. The results are

$$R'_{12}(\tau) = \frac{1}{2\pi} \int_{-\infty}^{+\infty} \{X_1(\omega)X_2^*(\omega)\} e^{j\omega\tau} d\omega \quad (2.7.8)$$

$$= \frac{1}{2\pi} \int_{-\infty}^{+\infty} C_{21}(\omega) e^{j\omega\tau} d\omega$$

where

$$C_{21}(\omega) = X_1(\omega)X_2^*(\omega) \quad (2.7.9)$$

and

$$C_{21}(\omega) = \int_{-\infty}^{+\infty} R'_{21}(\tau) e^{-j\omega\tau} d\tau \quad (2.7.10)$$

It is obvious that Eqs. (2.7.8) and (2.7.10) are a Fourier transform pair. A comparison of Eqs. (2.7.5) and (2.7.9) shows that $C_{12}(\omega)$ and $C_{21}(\omega)$ are complex conjugates of one another, that is,

$$\begin{aligned} C_{21}(\omega) &= C_{12}^*(\omega) \\ C_{12}(\omega) &= C_{21}^*(\omega) \end{aligned} \quad (2.7.11)$$

This result is characteristic of cross-correlation frequency spectra that are obtained from periodic and transient signals. This frequency domain characteristic implies that $R'_{21}(\tau) = R'_{12}(-\tau)$.

2.7.2 Auto-Correlation

The *transient auto-correlation function* is obtained when both functions are the same. Then, Eqs. (2.7.1), and (2.7.4) through (2.7.6) become

$$R'_{11}(\tau) = \int_{-\infty}^{+\infty} x_1(t)x_1(t+\tau) dt = \frac{1}{2\pi} \int_{-\infty}^{+\infty} C_{11}(\omega) e^{j\omega\tau} d\omega \quad (2.7.12)$$

where

$$C_{11}(\omega) = X_1^*(\omega)X_1(\omega) = |X_1(\omega)|^2 \quad (2.7.13)$$

and

$$C_{11}(\omega) = \int_{-\infty}^{+\infty} R'_{11}(\tau) e^{-j\omega\tau} d\tau \quad (2.7.14)$$

CORRELATION CONCEPTS—TRANSIENT TIME HISTORIES 47

It should be clear that the transient auto-correlation function is an even function just like the periodic auto-correlation function since $C_{11}(\omega)$ is real. Thus the auto-correlation is made up of only cosine type terms. Note that the units are (units of $x_1(t)/\text{Hz})^2$.

Example: Consider the rectangular pulse of duration T and amplitude A shown in Fig. 2.4.1b when β is very large. The corresponding auto-correlation function is obtained from Eq. (2.7.12) by direct integration to be

$$R_{11}^t(\tau) = A^2(T - \tau) \quad \text{for } 0 < \tau < T$$
$$R_{11}^t(\tau) = A^2(T + \tau) \quad \text{for } -T < \tau < 0 \quad (2.7.15)$$
$$R_{11}^t(\tau) = 0 \quad \text{for } \tau < -T \text{ and } \tau > T$$

This is a single triangular pulse of duration $2T$ with a maximum value of A^2T when $\tau = 0$. A transient Fourier transform analysis of the transient triangular auto-correlation function in Eq. (2.7.15) gives a frequency spectrum of

$$C(\omega) = 2A^2T^2 \left[\frac{1 - \cos(\omega T)}{(\omega T)^2} \right] \quad (2.7.16)$$

However, a transient Fourier analysis of time-history $x(t)$ shows that

$$X(\omega) = AT \, \text{sinc}\left(\frac{\omega T}{2}\right) \quad (2.7.17)$$

so that substitution of Eq. (2.7.17) into Eq. (2.7.13) gives

$$C(\omega) = X^*(\omega)X(\omega) = A^2T^2 \left[\frac{\sin^2(\omega T/2)}{(\omega T/2)^2} \right] \quad (2.7.18)$$

A comparison of Eqs. (2.7.16) and (2.7.18) shows that the double angle trigonometric identity is satisfied. The frequency spectrum described by these equations shows that the auto-correlation function is an even function with real frequency components. These component's magnitudes decrease with increasing frequency and have a magnitude of A^2T^2 at zero frequency.

This zero frequency value is the transient signal's mean square value multiplied by T, a result that is consistent with the results that are obtained directly from Eq.(2.4.12).

2.8 CORRELATION CONCEPTS—RANDOM TIME HISTORIES

Random signals are much more troublesome to describe than either periodic or transient signals. In order to use transient Fourier transforms on any signal the integral relationship of

$$\int_{-\infty}^{\infty} |x(t)|\, dt < \infty \tag{2.8.1}$$

must be satisfied. However, Eq. (2.8.1) is not satisfied when $x(t)$ is a random signal. Thus transient Fourier transforms cannot be used and another way must be found to analyze random signals.

A more pragmatic approach to analyzing random signals is used, one that sidesteps some critical mathematical problems that plague the transient Fourier transform approach. This approach utilizes the concepts developed for auto- and cross-correlation functions where the signal's frequency spectra are related to the auto-correlation's frequency spectra. It will be seen that auto-correlation, mean square concepts, and Parseval's formula hold the key to working with random signals. In this approach, the periodic frequency spectra are transformed into the random frequency spectra through an averaging procedure that is analogous to how the periodic frequency spectra can estimate the transient frequency spectra.

In this section, it is assumed that the random processes are stationary and ergodic; that is, both ensemble (averaging across many time history records at a given instant in time) and temporal (averaging over time using one time history) calculations give the same mean, mean square, and statistical distributions (probability densities). This is a common assumption that we are forced to make in most practical cases.

Four major topics are addressed in this section. These topics are the auto-correlation relationships, the cross-correlation relationships, the implications of dealing with several random processes added together, and finally statistical probability density functions for common time histories.

2.8.1 Auto-Correlation and Auto-Spectral Density

The *random auto-correlation function* for random signals is defined by

$$R_{xx}(\tau) = \lim_{T \to \infty} \frac{1}{T} \int_{-T/2}^{T/2} x(t)x(t + \tau)\, dt \tag{2.8.2}$$

Note that the random auto-correlation function symbol $R_{xx}(\tau)$ uses the variable's symbol (x in this case) as a subscript. This is done to distinguish the random auto-correlation function from that belonging to the periodic and transient functions ($R_{11}(\tau)$ and $R_{11}^t(\tau)$, respectively). The definition in Eq. (2.8.2) is different from that used for periodic time histories (see Eq. (2.6.1)) in that time T must become large. However, the similarity with periodic time histories, other than the length of time T, provides a convenient practical method for estimating random behavior through use of periodic analysis methods. More on this point later.

It is known that $R_{xx}(\tau)$ has the following general properties.

1. $R_{xx}(\tau)$ must go to zero as τ becomes large (both positively and negatively) since the random concept implies no correlation between current events and either future or past events that are far away.
2. Auto-correlation function is a real valued function.
3. Auto-correlation is an even function so that $R_{xx}(\tau) = R_{xx}(-\tau)$.
4. The Fourier transform requirement of Eq. (2.8.1) is satisfied for the auto-correlation function; that is,

$$\int_{-\infty}^{\infty} |R_{xx}(\tau)|\, d\tau < \infty \qquad (2.8.3)$$

is satisfied.

Granting that all of these statements are true, then the well-known Weiner–Khintchine[4] Fourier transform pair can be used; namely,

$$R_{xx}(\tau) = \frac{1}{2\pi} \int_{-\infty}^{\infty} S_{xx}(\omega) \cos(\omega\tau)\, d\omega$$

$$S_{xx}(\omega) = \int_{-\infty}^{\infty} R_{xx}(\tau) \cos(\omega\tau)\, d\tau \qquad (2.8.4)$$

Since it is known that $R_{xx}(\tau)$ is a real valued even function, then it follows that $S_{xx}(\omega)$ must also be a real valued even function; that is, $S_{xx}(\omega) = S_{xx}(-\omega)$. Further, it can be shown that $S_{xx}(-\omega)$ can only be positive. Thus $\cos(\omega\tau)$ is used in place of $e^{\pm j\omega\tau}$ in this Fourier transform pair. A typical plot of $S_{xx}(\omega)$ is shown in Fig. 2.8.1.

The physical significance of $S_{xx}(\omega)$ can be obtained by setting $\tau = 0$. In

---

[4]N. Wiener, "Generalized Harmonic Analysis," *Acta Mathematica*, Vol. 55, 1930, pp. 11–258. A. Khintchine, "Korrelationstheorie der Stationaren Stochastischen Prozesse," *Mathematische Annalen*, Vol. 109, 1934, pp. 604–615.

50 DYNAMIC SIGNAL ANALYSIS

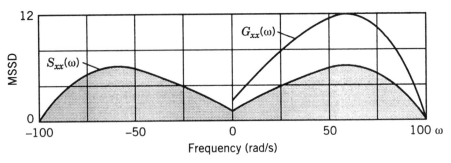

Fig. 2.8.1. Single and double sided auto-spectral densities.

this case, Eqs. (2.8.2) and (2.8.4) combine to give

$$R_{xx}(0) = \lim_{T \to \infty} \frac{1}{T} \int_{-T/2}^{T/2} x^2(t)\, dt = \frac{1}{2\pi} \int_{-\infty}^{\infty} S_{xx}(\omega)\, d\omega \qquad (2.8.5)$$

The first integral is clearly the definition of the signal's mean square, while the second integral is the area under the complete plot of $S_{xx}(\omega)$, as shown in Fig. 2.8.1. Thus the preferred name for $S_{xx}(\omega)$ is either the *mean square spectral density* (MSSD) or the *auto-spectral density* (ASD). Unfortunately most frequency analysis concepts were originally developed in either acoustics or communication theory so that the $S_{xx}(\omega)$ is commonly called the *power spectral density* (PSD). The PSD misnomer has caused endless confusion to many mechanical engineers and is not used in this book.

The units of $S_{xx}(\omega)$ are dependent on the location of $(1/2\pi)$ in Eq. (2.8.4). This term can be assigned to either the $R_{xx}(\tau)$ expression or the $S_{xx}(\omega)$ expression without affecting the mathematics.[5] However, the assignment of $(1/2\pi)$ has a big effect on the units. The definition used in Eq. (2.8.4) is preferred because $S_{xx}(\omega)$ will then have units of x^2/Hz, which are the same as used in frequency analyzers. The units are most clearly seen from Eq. (2.8.5), where the left-hand side has units of variable x squared while $S_{xx}(\omega)$ is multiplied by $[d\omega/(2\pi)]$ or df in hertz. Hence, its units are those of x^2/Hz.

Theoretically, ASD is a double-sided frequency spectrum. Experimentally, only a single-sided frequency spectrum is used. These frequency spectra are related by

$$G_{xx}(\omega) = 2S_{xx}(\omega) \qquad \text{for } 0 < \omega < \infty$$
$$G_{xx}(0) = S_{xx}(0) \qquad \text{for } \omega = 0 \qquad (2.8.6)$$

---

[5]Some mathematicians will use $(1/\sqrt{2\pi})$ with each term in Eq. (2.8.4).

where it is noted that the zero frequency value is related to the mean value, a result that was found previously for periodic and transient autocorrelation functions. Equation (2.8.6) is shown graphically in Fig. 2.8.1. The area under the $G_{xx}(\omega)$ curve must be the same as that under the $S_{xx}(\omega)$ curve.

Conceptually, Eq. (2.8.5) is the same as Parseval's formula for periodic functions given by Eq. (2.3.6); namely,

$$A_{RMS}^2 = \frac{1}{T}\int_0^T f(t)^2 \, dt = \sum_{p=-\infty}^{\infty} |X_p|^2 = X_0^2 + 2\sum_{p=1}^{\infty} |X_p|^2 \quad (2.8.7)$$

The major difference between these two equations is the averaging time T. This time difference is overcome by performing the periodic analysis of Eq. (2.8.7) many times, averaging the resulting discrete frequency spectra of X_p^2, and scaling the results to give an estimate of either $S_{xx}(\omega)$ or $G_{xx}(\omega)$. Thus the mean square is the connecting link between a random signal's frequency spectra and its ASD. This concept is the basis for constructing a digital frequency analyzer to analyze a random signal and is developed further in Chapter 5.

In order to grasp the significance of Eq. (2.8.4), consider the simple ASD density curve shown in Fig. 2.8.2a. Here, the spectral density is constant over the frequency range from $-\omega_2$ to $-\omega_1$ and from ω_1 to ω_2. If this spectral density is substituted into Eq. (2.8.4) and integrated, the auto-correlation function becomes

$$R_{xx}(\tau) = \frac{S_0}{\pi \tau} [\sin(\omega_2 \tau) - \sin(\omega_1 \tau)]$$

or

$$= \left\{\frac{2S_0 \Delta \omega}{\pi}\right\} \cos(\omega_0 \tau) \, \text{sinc}(\Delta \omega \tau) \quad (2.8.8)$$

where

$$\omega_0 = \left(\frac{\omega_1 + \omega_2}{2}\right) \quad (2.8.9)$$

is the average frequency and

$$\Delta \omega = \left(\frac{\omega_2 - \omega_1}{2}\right) \quad (2.8.10)$$

is half of the ASD curve's bandwidth. Equations (2.8.8) through (2.8.10) are used to gain insight into two limiting types of random processes, one being a narrow-band process and the other a wide-band process.

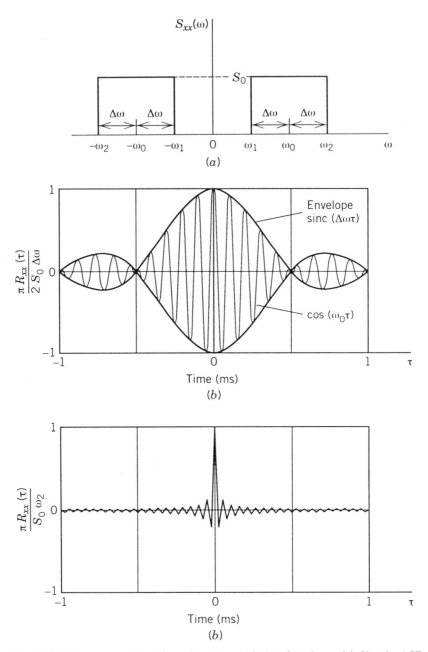

Fig. 2.8.2. Narrow- and broad-band auto-correlation functions. (*a*) Simple ASD curve for theoretical calculation. (*b*) Typical narrow-band auto-correlations function $\omega_0 = 62.8$ rad/s and $\Delta\omega = 6.28$ rad/s. (*c*) Typical broad-band auto-correlation function $\omega_2 = 62.8$ rad/s.

CORRELATION CONCEPTS—RANDOM TIME HISTORIES   53

Narrow-Band Random Process   A narrow-band random process is described as one where $\Delta\omega$ is small compared to ω_0. In this case, Eq. (2.8.8) shows that the sinc($\Delta\omega\tau$) function is the envelope function that decays slowly ($1/\Delta\omega\tau$) while the $\cos(\omega_0\tau)$ term oscillates a number of times per envelope half period ($\Delta\omega\tau = \pi$), as shown in Fig. 2.8.2b. The auto-correlation function is seen to go to zero for large values of τ. A frequency ratio of ten to one (ω_0 is ten times $\Delta\omega$) is used in Fig. 2.8.2b for illustrative purposes. This ratio should be more like 30 to 100 to one in order to be considered a narrow-band process, a condition that is more difficult to illustrate. A narrow-band random process, $x(t)$, looks like a sinusoid with center frequency ω_0 while it has a variable amplitude. The auto-correlation function shows this fundamental frequency oscillation in the $\cos(\omega_0\tau)$ term. As τ becomes larger, the correlation is lost, eventually going to zero as τ becomes large. The narrower the bandwidth $\Delta\omega$, the larger τ must be before correlation is lost.

Broad-Band Random Process   A broad-band random process is characterized by ω_1 being zero or close to zero while ω_2 is allowed to become large. In this case, Eqs. (2.8.9) and (2.8.10) show that ω_0 and $\Delta\omega$ become equal to $\omega_2/2$. Then the auto-correlation function begins to look like that shown in Fig. 2.8.2c. In this case, Eq. (2.8.8) becomes

$$R_{xx}(\tau) = \left\{\frac{S_0\omega_2}{\pi}\right\} \text{sinc}(\omega_2\tau) \qquad (2.8.11)$$

and is plotted in Fig. 2.8.2c for the case where ω_2 has the same value as used in Fig. 2.8.2b so that the range of τ remains the same. It is apparent that the function decays very quickly in this case so that there is little correlation within a broad-band signal.

However, an examination of Eq. (2.8.11) shows that the magnitude at $\tau = 0$ increases directly with the value of ω_2. In the limit as ω_2 becomes larger and larger, Eq. (2.8.11) begins to look more and more like a *Dirac delta function*, so that in the limit Eq. (2.8.11) can be written as

$$R_{xx}(\tau) = S_0\, \delta(\tau) \qquad (2.8.12)$$

Then, the corresponding ASD from Eq. (2.8.4) becomes

$$S_{xx}(\omega) = \int_{-\infty}^{\infty} R_{xx}(\tau)\cos(\omega\tau)\, d\tau = \int_{-\infty}^{\infty} S_0\, \delta(\tau)\cos(\omega\tau)\, d\tau = S_0$$

$$(2.8.13)$$

54 DYNAMIC SIGNAL ANALYSIS

This result indicates that a Dirac delta function auto correlation corresponds to an ASD that is constant over all frequencies. Such a random process is called *white noise*. It is noted that this is a limiting ideal process that is physically unattainable. However, it does point out that a broadband random process behaves as though it is uncorrelated except at very small values τ.

2.8.2 Cross-Correlation and Cross-Spectral Density

There are two cross-correlation functions of interest, $R_{xy}(\tau)$ and $R_{yx}(\tau)$. They are defined in terms of their temporal averages and corresponding Fourier transforms as

$$R_{xy}(\tau) = \lim_{T \to \infty} \frac{1}{T} \int_{-T/2}^{T/2} x(t)y(t+\tau)\,dt = \frac{1}{2\pi}\int_{-\infty}^{\infty} S_{xy}(\omega)\,e^{j\omega\tau}\,d\omega$$

$$S_{xy}(\omega) = \int_{-\infty}^{\infty} R_{xy}(\tau)\,e^{-j\omega\tau}\,d\tau$$

(2.8.14)

and

$$R_{yx}(\tau) = \lim_{T \to \infty} \frac{1}{T} \int_{-T/2}^{T/2} y(t)x(t+\tau)\,dt = \frac{1}{2\pi}\int_{-\infty}^{\infty} S_{yx}(\omega)\,e^{j\omega\tau}\,d\omega$$

$$S_{yx}(\omega) = \int_{-\infty}^{\infty} R_{yx}(\tau)\,e^{-j\omega\tau}\,d\tau$$

(2.8.15)

where $S_{xy}(\omega)$ and $S_{yx}(\omega)$ are the *cross-spectral densities* (CSDs). The CSDs are complex frequency spectra with real and imaginary parts (magnitude and phase). Note that auto-correlation ASDs contain only magnitudes and have no phase information, while CSDs contain phase information. CSDs have units of xy/Hz since the $(1/2\pi)$ is used with the cross-correlation integral. Equations (2.8.14) and (2.8.15) will hold as long as the functions x and y are uncorrelated for large values of τ and one of the functions has a zero mean. Then Eq. (2.8.1) is satisfied when it is applied to the cross-correlation function.

Similar to periodic and transient cross-correlation, the CSDs are complex conjugates of one another so that

$$S_{xy}(\omega) = S_{yx}^{*}(\omega)$$
$$S_{yx}(\omega) = S_{xy}^{*}(\omega)$$

(2.8.16)

This implies that $R_{xy}(\tau) = R_{yx}(-\tau)$, the same result as obtained for periodic and transient cross-correlation functions.

It is noted that the cross-spectral densities can be estimated from periodic frequency spectra through an averaging process to accomplish the long time required in Eqs. (2.8.14) and (2.8.15). The spectral density units are obtained by multiplying by an appropriate scale factor. This process is described in detail in Chapter 5.

2.8.3 Correlation and Spectral Densities of Multiple Random Processes

Random signals may contain significant noise. Consider that $x(t)$ and $y(t)$ are the processes of interest while $n(t)$ and $m(t)$ are signal noise. Then, let us define two new random processes as

$$u(t) = x(t) + n(t)$$
$$v(t) = y(t) + m(t) \quad (2.8.17)$$

It is of interest to determine the corresponding cross-correlation function between $u(t)$ and $v(t)$. This gives

$$R_{uv}(\tau) = R_{xy}(\tau) + R_{ny}(\tau) + R_{xm}(\tau) + R_{nm}(\tau) \quad (2.8.18)$$

Now, it is common for the noises to be statistically independent, so assume that $x(t)$ and $m(t)$ are statistically independent and that $y(t)$ and $n(t)$ are also statistically independent. Then, Eq. (2.8.18) becomes

$$R_{uv}(\tau) = R_{xy}(\tau) + R_{nm}(\tau) \quad (2.8.19)$$

Equation (2.8.19) shows that the output cross-correlation function is the sum of the signal's cross-correlation function plus that of the noise. Substitution of Eq. (2.8.19) into the Fourier transform of Eq. (2.8.14) or (2.8.15) gives

$$S_{uv}(\omega) = S_{xy}(\omega) + S_{nm}(\omega) \quad (2.8.20)$$

which shows that the noise CSD adds to the signal's CSD. If the noises are statistcally independent of one another, then $R_{nm}(\tau)$ and $S_{nm}(\omega)$ are zero so that

$$R_{uv}(\tau) = R_{xy}(\tau) \quad (2.8.21)$$

and

$$S_{uv}(\omega) = S_{xy}(\omega) \quad (2.8.22)$$

Equations (2.8.21) and (2.8.22) show that cross-correlation information is noise-free, provided the noise sources are statistically independent. Unfortunately, there have been cases where there is cross talk between data channels, so that not only are the noises statistically dependent but there is significant signal interdependence as well. Such situations must be carefully examined before any data is obtained. A test procedure for this situation is outlined in Chapter 5.

Now if we assume that $x(t) = y(t)$ and $n(t) = m(t)$, then these cross-correlation and CSD relationships become auto-correlation and ASD relationships described by

$$R_{uu}(\tau) = R_{xx}(\tau) + R_{xn}(\tau) + R_{nx}(\tau) + R_{nn}(\tau) \quad (2.8.23)$$

and

$$S_{uu}(\omega) = S_{xx}(\omega) + S_{xn}(\omega) + S_{nx}(\omega) + S_{nn}(\omega) \quad (2.8.24)$$

It is clear from Eqs. (2.8.23) and (2.8.24) that the presence of any noise will contaminate both the auto-correlation $R_{xx}(\tau)$ and the ASD $S_{xx}(\omega)$ functions. Note that the cross-spectral terms drop out if x and n are uncorrelated. These ideas are developed further in our discussion of dual channel frequency analyzers. The following example shows how broad-band and narrow-band noise affects the measurement of a sinusoid.

Example: The broad-band white noise has a spectral density given by Eq. (2.8.13) as $G_{nn}(\omega) = 2S_0$ so that $R_{nn}(\tau) = 2S_0 \delta(\tau)$. The auto-correlation function for a sinusoid is given by Eq. (2.6.19) so that $R_{uu}(\tau)$ becomes

$$R_{uu}(\tau) = \frac{B^2}{2} \cos(\omega_0 \tau) + 2S_0 \delta(\tau) \quad (2.8.25)$$

The auto-correlation function shows that the broad-band noise drops out immediately. This result indicates that a sinusoid buried in broad-band noise can be found by examining the auto-correlation function.

Now, let us examine the case where the noise is band-limited with the same center frequency as the sinusoid. Then the auto-correlation function is given by

$$R_{uu}(\tau) = \left[\frac{B^2}{2} + \left\{ \frac{2S_0 \Delta\omega}{\pi} \right\} \operatorname{sinc}(\Delta\omega \tau) \right] \cos(\omega_0 \tau) \quad (2.8.26)$$

It is evident from Eq. (2.8.26) that the larger the value of $\Delta\omega$, the faster the sinc function envelope decays, leaving the sine wave exposed. The smaller the value of $\Delta\omega$, the more trouble it is to find the sine wave. Thus these two cases show that broad-band random noise is preferred over narrow-band random noise in terms of measuring a sinusoidal signal that is buried in noise.

2.8.4 Statistical Distributions

It was mentioned in Section 2.1 that the mean and the mean square have no meaning unless some other information is provided, like statistical distributions. A common method for describing a statistical distribution is the *probability density*. The probability density is defined by a limiting process

$$p(x) = \lim_{\Delta x \to 0} \frac{P(x + \Delta x) - P(x)}{\Delta x} \qquad (2.8.27)$$

where $p(x)$ is the probability density function.
$P(x)$ is the probability that a number lies between $-\infty$ and x.
$P(x + \Delta x)$ is the probability that a number lies between $-\infty$ and $x + \Delta x$.
Δx is the increment or range of x.
The probability that x is in the region bounded by x_1 and x_2 is given by

$$P(x_1 < x < x_2) = \int_{x_1}^{x_2} p(x)\, dx \qquad (2.8.28)$$

Similarly, the probability that $-\infty < x < \infty$ is unity, since every number must lie in this region. In view of Eq. (2.8.28), this statement is equivalent to

$$P(-\infty < x < \infty) = \int_{-\infty}^{\infty} p(x)\, dx = 1 \qquad (2.8.29)$$

The central limit theorem shows that many random processes have a Gaussian probability density distribution since they are the result of many independent random sources that do not have to be Gaussian random processes themselves. Thus it is desirable to be able to describe Gaussian random processes.

It is common to use two standard variables when describing random probability distributions. These are

$$z = \frac{x - \bar{x}}{\sigma_x} \tag{2.8.30}$$

where $\bar{x}$ is the data's mean value and σ_x is the data's standard deviation, which is defined as the square root of the variance or

$$\sigma_x = \sqrt{E(x - \bar{x})^2} \tag{2.8.31}$$

and $E[(x - \bar{x})^2]$ is the mathematical expectation of $(x - \bar{x})^2$ or variance.

The Gaussian probability density is expressed as

$$p(z) = \frac{1}{\sqrt{2\pi}} e^{-(z^2/2)} \tag{2.8.32}$$

which is plotted in Fig. 2.8.3a. It is evident from Eq. (2.8.28) that the area under the curve is the probability that a number lies within given limits. For example, 68.3 percent of the readings lie within $\pm\sigma_x$, while 99.7 percent of the readings lie within $\pm 3\sigma_x$. This fact, that 99.7 percent of the values lie in the $\pm 3\sigma_x$ range, is used in specifying a true RMS voltmeter. The *crest factor* (CF) is defined as

$$\text{CF} = \frac{x_{\max}}{\sigma_x} \tag{2.8.33}$$

and is usually specified to be in the range of 3 to 5. It is obvious that the minimum crest factor should exceed 3.

The Rayleigh probability density describes the peak values of a narrowband random process as the bandwidth $\Delta\omega$ goes to zero. It is given by

$$p(z) = z \, e^{-(z^2/2)} \tag{2.8.34}$$

and is plotted in Fig. 2.8.3b. In this case, 39.5 percent of the readings lie below one σ_x while 60.5 percent of the peak values are above σ_x.

The *chi-square* probability density is given by

$$p(\chi^2) = [2^{(n/2)} \Gamma(n/2)]^{-1} (\chi^2)^{((n-2)/2)} e^{-(\chi^2/2)} \tag{2.8.35}$$

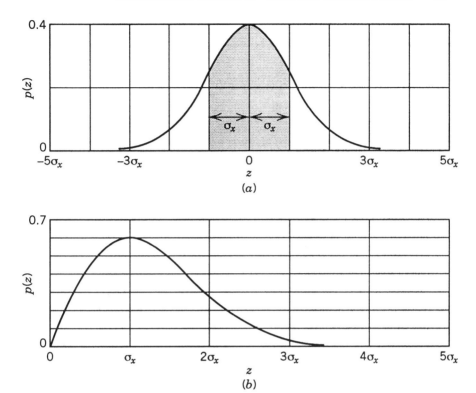

Fig. 2.8.3. Two common random probability densities. (*a*) Gaussian probability density. (*b*) Rayleigh probability density.

where χ^2 is the chi-square variable.
$\Gamma(n/2)$ is the Gamma Function.
n is the integer *degrees of freedom*.

Equation (2.8.35) is important in estimating the measurement errors that occur in experimentally determined frequency spectra.

It is interesting to compare these random probability distributions with those of a sinusoid, a square wave, and a triangular wave, since these simple deterministic waveforms are often used as reference waveforms for theoretical calculations. The results are:

$$p(x) = \frac{1}{\pi\sqrt{B^2 - x^2}} \quad \text{for } |x| < B$$
$$p(x) = 0 \quad \text{for } x < -B \text{ and } x > B \quad (2.8.36)$$

for the sinusoid of amplitude B:

$$p(x) = \frac{\delta(x-B) + \delta(x+B)}{2} \qquad (2.8.37)$$

for a square wave with amplitudes of $\pm B$; and

$$\begin{aligned} p(x) &= B^{-1} & 0 < x < B \\ &= 0 & x > B \end{aligned} \qquad (2.8.38)$$

for a triangular wave described by

$$\begin{aligned} x(t) &= \frac{2B}{T}\left(t + \frac{T}{2}\right) & -\frac{T}{2} < t < 0 \\ x(t) &= \frac{2B}{T}\left(\frac{T}{2} - t\right) & 0 < t < \frac{T}{2} \end{aligned} \qquad (2.8.39)$$

2.9 SUMMARY

This chapter is concerned with describing the types of signals encountered in dynamic measurements. The first step is to identify the different signal classes that are encountered, such as deterministic (including both periodic and transient) and random signals. Chaotic processes represent the newest classification wherein a deterministic process generates a random appearing signal. Obviously, this is a new area for research. Currently, we are forced to analyze chaotic signals as random due to a lack of understanding.

The temporal mean and mean square parameters are considered as single numbers that can be used to describe a signal. However, they are extremely limited without further information about the signal's process, particularly its probability density. The frequency spectrum is seen as a method to visualize the frequencies of a signal.

Periodic time histories are analyzed using the periodic Fourier series (periodic Fourier transform), which allows the transformation from a signal's time domain description to its frequency domain description. The signal's mean square calculation leads to Parseval's formula. The analysis of a square wave generates the commonly occurring sinc function that shows up in later sections as well.

A transient time history is seen to come from a periodic time history where the repeat period is allowed to increase to infinite length. This concept of extending the repeat period shows the connection between the periodic Fourier transform and the transient Fourier transform. This connection is the multiplication of the periodic Fourier transform frequency spectrum by the repeat period time T_0, a concept that works so long as the transient signal occupies less than 10 percent of repeat period T_0. It is also seen that the concept of mean, mean square, and Parseval's formula required a change in definition when working with transient signals.

Correlation is a statistical concept. This concept is reviewed in its statistical context and is then translated into a temporal context. A different temporal definition is required for each general class of signal; namely, periodic, transient, and random. In addition, for each signal class, there are cross-correlation and auto-correlation types of analysis.

Cross-correlation is a measure of how well the events of one signal correlate with another signal that is time shifted with respect to the first signal. The cross-correlation function and the cross spectral density are Fourier transforms of one another. The exact form of Fourier transform depends on the signal type being processed. Generally, cross-spectral densities are complex (with real and imaginary parts) and can be related to the frequency spectra of each signal that is being analyzed.

Auto-correlation is a measure of how well the events in a signal correlate with each other when time shifted with respect to one another. When the time shift is zero, the auto-correlation function becomes the signal's mean square, which leads to Parseval's formula relating the mean square to the signal's frequency components. It is not surprising that auto-correlation functions are even functions in time shift τ so that the corresponding frequency spectrum is real valued.

The ideas and concepts developed here impact on how we approach vibration analysis, instrumentation requirements, and signal processing. These topics are reviewed in depth in the following chapters.

REFERENCES

1. Bendat, J. S. and A. G. Piersol, *Measurement and Analysis of Random Data*, John Wiley & Sons, New York, 1966.
2. Bendat, J. S. and A. G. Piersol, *Random Data: Analysis and Measurement Procedures*, John Wiley & Sons, New York, 1971.
3. Blackman, R. B. and J. W. Tukey: *The Measurement of Power Spectra*, Dover, New York, 1959.
4. Bruel & Kjaer Instruments, *Technical Review*, A quarterly publication available from Bruel & Kjaer Instruments, Inc., Marlborough, MA.
    (a) "On the Measurement of Frequency Response Functions," No. 4, 1975.
    (b) "Digital Filters and FFT Technique," No. 1, 1978.
    (c) "Discrete Fourier Transform and FFT Analyzers," No. 1, 1979.
    (d) "Zoom-FFT," No. 2, 1980.

(e) "Cepstrum Analysis," No. 3, 1981.
(f) "System Analysis and Time Delay Spectrometry" (Part 1), No. 1, 1983.
(g) "System Analysis and Time Delay Spectrometry" (Part II), No. 2, 1983.
(h) "Dual Channel FFT Analysis" (Part I), No. 1, 1984.
(i) "Dual Channel FFT Analysis" (Part II), No. 2, 1984.
5. Crandall, Stephen H. and William D. Mark, *Random Vibration in Mechanical Systems*, Academic Press, New York, 1963.
6. Ewins, David J., *Modal Testing: Theory and Practice*, Research Studies Press Ltd., Lecthworth, Hertfordshire, UK, 1984.
7. Lange, F. H., *Correlation Techniques*, D. Van Nostrand Co., Princeton, NJ, 1967.
8. Lee, Y. W., *Statistical Theory of Communication*, John Wiley & Sons, New York, 1960.
9. Lyon, Richard H., *Machinery Noise and Diagnostics*, Butterworth, Boston, MA, 1987.
10. Moon, F. C., *Chaotic Vibrations: An Introduction of Chaotic Dynamics for Applied Scientists and Engineers*, John Wiley & Sons, New York, 1987.
11. Newland, D. E., *An Introduction to Random Vibrations and Spectral Analysis*, Longman Group Limited, London, UK, 1975.
12. Randall, R. B., *Frequency Analysis*, Available from Bruel & Kjaer Instruments, Inc., Marlborough, MA, 1987.
13. Rao, Singiresu S., *Mechanical Vibrations*, Addison-Wesley, Reading, MA, 1990.
14. Stroud, K. A., *Fourier Series and Harmonic Analysis*, Stanley Thornes Ltd., Cheltenham, UK, 1984.
15. Thomson, William T., *Theory of Vibration with Applications*, 3rd ed., Prentice Hall, Englewood Cliffs, NJ, 1988.

3 Vibration Concepts

Engineers use a multichannel measurement system to measure the dynamic response characteristics of panels on a Titan 4 launch vehicle. The understanding of the measured frequency response functions and the corresponding mode shapes requires a clear understanding of the fundamental principles of vibration behavior. (Photo Courtesy of *Sound and Vibration*).

64 VIBRATION CONCEPTS

3.1 INTRODUCTION

The goal of this chapter is to review those concepts that are most often involved in vibration testing. There are many texts and papers on the theory of vibrations. What is often lacking in these sources is a detailed look at how the resulting theory applies to vibration testing. Those concepts that seem to be fundamental to vibration testing are reviewed and interpreted from a testing point of view.

The single degree of freedom (DOF) transient and steady-state forced response is fundamental to most vibration behavior, including impulse response. The frequency response function approach is emphasized in these developments. Some nonlinear effects are also considered, since these effects often show up in testing.

The two DOF system is introduced to contrast its behavior to that of a single DOF system in terms of transient and forced steady-state response. Then the modal model is developed and the direct steady-state and modal steady state responses are compared.

The two DOF model is followed by second order continuous systems such as taut string, axial, and torsional vibrations. In these models, space and time variables are separated. This separation leads to mode shapes and modal analysis for continuous systems. Such ideas as mode participation factor, modal mass, modal stiffness, and modal damping follow naturally. These concepts help to graphically illustrate why certain testing response occurs.

Finally, the fourth order continuous system is reviewed. Both a beam and a beam under tension are examined in detail. The separation of variables and modal approach are again used to illustrate the type of phenomenon we are dealing with when vibration testing is attempted. The importance of boundary conditions in vibration testing is explored in depth.

The concepts of this chapter will be used in Chapter 4 on instrumentation, Chapter 6 on vibration exciter characteristics, Chapter 7 on vibration testing concepts, and finally in Chapter 8 in developing a general vibration testing model.

3.2 THE SINGLE DOF MODEL

The single DOF system is the starting point in studying vibration response. In this section, the fundamental equation of motion and free vibration responses of a single DOF system are considered.

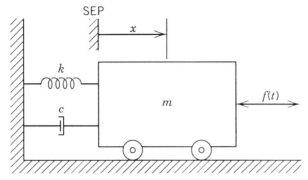

Fig. 3.2.1. Elements of the simplest single DOF system.

3.2.1 Equations of Motion

The linear single DOF system consists of a mass m, linear viscous damper c, linear spring k, and excitation force $f(t)$, as shown in Fig. 3.2.1. Let x be the coordinate that locates mass m from its static equilibrium position (SEP). Then Newton's second law of motion gives

$$\underset{\text{inertia}}{m\ddot{x}} + \underset{\text{damping}}{c\dot{x}} + \underset{\text{spring}}{kx} = \underset{\text{excitation}}{f(t)} \quad (3.2.1)$$

where[1] $m\ddot{x}$ is the inertia force.
$c\dot{x}$ is the damping force (energy dissipator).
kx is the spring force (restoring force).
$f(t)$ is the excitation force.

Equation (3.1.1) is a second order linear differential equation of motion with constant coefficients. The significance of Eq. (3.2.1) is that it shows that a vibration consists of a balance between inertia, damping, spring, and excitation forces at each instant of time. Ideally, this is a linear equation as shown. In reality, it is only approximately linear since damping and restoring forces tend to have nonlinear characteristics, particularly for large vibration amplitudes.

The principle of superposition can be used for linear systems described by a differential equation such as Eq. (3.2.1). This principle means that

---

[1] One dot over a variable indicates the first time derivative, while two dots indicates the second time derivative.

if $x_1(t)$ and $x_2(t)$ are solutions to Eq. (3.2.1), then any linear combination is also a solution.

3.2.2 Free Undamped Vibration

The *free undamped* vibration response occurs when $c = f(t) = 0$ so that Eq. (3.2.1) becomes

$$m\ddot{x} + kx = 0 \qquad (3.2.2)$$

This vibration is characterized by the inertia force and spring force canceling one another. Substitution of $x = Ae^{st}$ gives the characteristic frequency equation as

$$ms^2 + k = 0$$

from which

$$s = \pm\sqrt{-\frac{k}{m}} = \pm j\omega_n$$

where

$$\omega_n = \sqrt{\frac{k}{m}} \qquad (3.2.3)$$

is the *natural frequency* in rad/s. This natural frequency is a property of the system and is independent of coordinates selected to describe the system as well as of units employed in formulating the equations.

The *complementary* solution of Eq. (3.2.2) is given by

$$x_c = A\cos(\omega_n t) + B\sin(\omega_n t) \qquad (3.2.4)$$

Substitution of Eq. (3.2.4) into Eq. (3.2.2) gives

$$A(k - m\omega_n^2) + B(k - m\omega_n^2) = 0 \qquad (3.2.5)$$

Note how the spring force (kA or kB) is canceled by its corresponding inertia force ($m\omega_n^2 A$ or $m\omega_n^2 B$), independent of amplitude. This shows that these two forces completely cancel one another when the system vibrates at its natural frequency. As will be seen, this force cancellation causes resonance in forced vibrations.

3.2.3 Free Damped Vibration

The damped vibration model is more realistic so let $f(t) = 0$ in Eq. (3.2.1) and substitute $x = Ae^{st}$ as before. The result is the characteristic frequency equation given by

$$ms^2 + cs + k = 0 \qquad (3.2.6)$$

This equation has either two real negative roots or two complex conjugate roots for s, depending on the value of c relative to its *critical value* given by

$$c_c = 2\sqrt{km} \qquad (3.2.7)$$

In vibration work, the underdamped case that corresponds to $c < c_c$ is of most interest. For convenience, we define the *damping ratio* ζ as

$$\zeta = \frac{c}{c_c} = \frac{c}{2\sqrt{km}} \qquad (3.2.8)$$

($\zeta < 1$ for the underdamped case), and the *damped frequency* ω_d is given by

$$\omega_d = \omega_n \sqrt{1 - \zeta^2} \qquad (3.2.9)$$

The corresponding complementary (transient) solution is

$$x_c = e^{-\zeta \omega_n t}[A \cos(\omega_d t) + B \sin(\omega_d t)] \qquad (3.2.10)$$

If the particular solution is zero (recall this requires $f(t) = 0$), then *initial conditions* of x_0 and v_0 at $t = 0$ lead to a complementary solution of

$$x_c = e^{-\zeta \omega_n t}\left[x_0 \cos(\omega_d t) + \frac{v_0 + \zeta \omega_n x_0}{\omega_d} \sin(\omega_d t)\right] \qquad (3.2.11)$$

The importance of Eq. (3.2.11) is illustrated in Fig. 3.2.2. It is seen that this vibration decays within the exponential envelope function of $e^{-\zeta \omega_n t}$, while it oscillates at the damped frequency ω_d. Both the natural

68 VIBRATION CONCEPTS

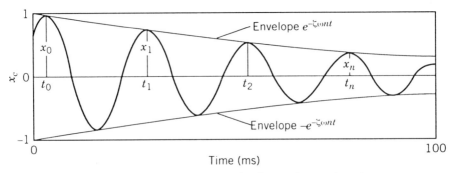

Fig. 3.2.2. Transient response for linear viscous damping.

frequency and the damping ratio can be obtained experimentally from this damped oscillation curve.

The *logarithmic decrement* δ is derived from Eq. (3.2.11) and Fig. 3.2.2 as

$$\delta = \frac{2\pi\zeta}{\sqrt{1-\zeta^2}} = \frac{1}{n}\ln\left[\frac{x_0}{x_n}\right] \qquad (3.2.12)$$

where n is the number of cycles used.
x_0 is the cycle amplitude at time $t = t_0$.
x_n is the cycle amplitude at time $t = t_n$.

Unfortunately, damping is often amplitude dependent since it comes from many different sources or mechanisms. Joints or connections are an important source of amplitude dependent damping in structures. At low amplitudes of vibration joints are rigid, while at higher amplitudes of vibration slippage often occurs. This slippage is a Coulomb friction type of phenomenon rather than a linear viscous damping phenomenon. Let us look at how we may be able to tell these two damping mechanisms from one another.

Equation (3.2.12) can be manipulated to show that the exponential envelope function is given by

$$\log(x_n) = \log(x_0) - (0.4343\delta)n \qquad (3.2.13)$$

so that a plot of peak amplitudes versus cycle number n gives a straight line on log paper with a slope of (-0.4343δ). Similarly, it can be shown that *Coulomb damping* causes a linear decrease in amplitude with cycles given by

$$x_n = x_0 - (4F_d/k)n \qquad (3.2.14)$$

THE SINGLE DOF MODEL   69

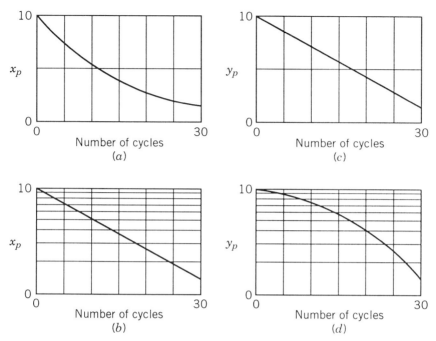

Fig. 3.2.3. Linear and log plots of viscous and Coulomb damping models. $x_0 = 10$ and $x_{30} = 1.52$ in all cases. (*a*) Linear plot for viscous damping. (*b*) Log plot for viscous damping. (*c*) Linear plot of Coulomb damping. (*d*) Log plot of Coulomb damping.

where F_d is the Coulomb damping friction force. Figure 3.2.3 is a plot of cycle amplitude x_n as a function of cycle number n on a linear and logarithmic vertical scale for both linear viscous damping and Coulomb damping. Figure 3.2.3a shows the exponential envelope function on a linear vertical scale, while Fig. 3.2.3b shows the straight line relationship of Eq. (3.2.13) when using a vertical log scale. Figure 3.2.3c shows the linear amplitude relationship of Eq. (3.2.14) when using a linear vertical scale, while Fig. 3.2.3d shows the log plot for the Coulomb damping case. It is evident that significant deviation from one model or the other is revealed in these plots. This deviation should help to identify the damping type. Note that $x_0 = 10$ and $x_{30} = 1.52$ for each case illustrated so that the beginning and end points are the same; hence the same amount of energy is removed from the system during the 30 cycles of oscillation.

The above discussion indicates that this is one means of determining damping mechanisms. How should such a system be excited to obtain the initial conditions? There are three distinct ways to excite this system to measure its natural frequency and damping characteristics. First, a static

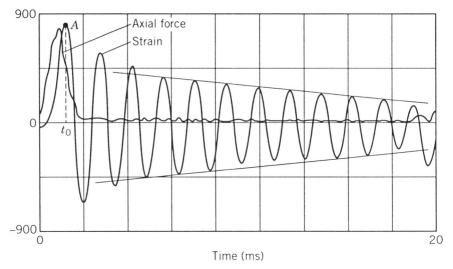

Fig. 3.2.4. Measured axial force and flexural strain in a machine element showing Coulomb friction as the controlling phenomenon.

load can be applied, and then quickly released to give the initial condition combination of $x_0 = \delta$ (static deflection) and $v_0 = 0$. In this case, the motion is dominated by the cosine term in Eq. (3.2.11). Second, an impulse is applied to the mass to cause a velocity v_0 with $x_0 = 0$. In this case, the response motion is dominated by the sine term in Eq. (3.2.11) and is often referred to as *impulse loading*. Third, both a static deflection and an initial velocity are present at time $t = 0$. This condition can be obtained by any force combination (or means) that produces a set of initial conditions at a certain time after which the force $f(t)$ remains zero while the vibration decays to zero. No one method is superior to the others. The test objective is to create an excitation that results in a clean decay signal.

It is important to establish the basic damping mechanism through these different plots. The data shown in Fig. 3.2.4 comes from strain data obtained from a dynamic machine element. The engineers assigned to this problem were unsuccessful in building a finite element model to predict the element's behavior because they missed the data's telltale message: Coulomb friction is dominant.

The driving mechanism is Coulomb friction combined with the axial load and a machine geometry change. The friction force became so large that the element's ends acted as though they were welded to the mating parts. This caused significant flexural strain to build up and then be released at point A as the axial load began to decrease. The resulting flexural vibration had an essentially straight line envelope as shown. Then the cycle repeated itself all over again. Needless to say, many hours of effort were wasted without recognizing the role of this controlling mechanism in formulating a proper dynamic model. It turned out that the entire design problem could have been adequately worked out on a couple of sheets of paper without finite element elegance being employed at all.

3.2.4 Structure Orientation and Natural Frequency

Consider the simple pendulous rigid body structure shown in Fig. 3.2.5. It pivots about a horizontal axis at O and is attached to a spring at distance a from the origin O. The body has a mass center at G, at distance b from origin O. It is rather easy to show that the governing differential equation of motion for the free undamped vibration of this structure is given by

$$I_0\ddot{\theta} + \{ka^2 - Wb\cos(\theta_0)\}\theta = Wb\sin(\theta_0) - ka\delta = 0 \quad (3.2.15)$$
$$\text{static equilibrium}$$

where I_0 is the mass moment of inertia with respect to the axis at O.
$\quad\theta$ describes motion from the SEP.
$\quad k$ is the linear spring.
$\quad W$ is the weight of the mass.
$\quad\theta_0$ is the initial static equilibrium position.
$\quad\delta$ is the initial stretch in the spring to hold the body in its static equilibrium position.

The right-hand side of Eq. (3.2.15) is the static equilibrium condition that tells how large δ must be to keep the body in static equilibrium.

The interesting aspect of this simple problem is the natural frequency that is given by

$$\omega_n = \sqrt{\frac{ka^2}{I_0}\left[1 - \frac{Wb\cos(\theta_0)}{ka^2}\right]} \quad (3.2.16)$$

This equation clearly shows that the natural frequency is dependent on its initial orientation; namely, angle θ_0. The amount that the natural frequency changes is given by the bracketed terms. The smaller the ratio of Wb/ka^2, the less effect this phenomenon will have. However, for low natural frequency applications, this type of term can cause considerable confusion in experimental results. For example, an engineer was measuring the natural frequencies of a very flexible beam in different orientations. It was found that the fundamental natural frequency changed with the

72 VIBRATION CONCEPTS

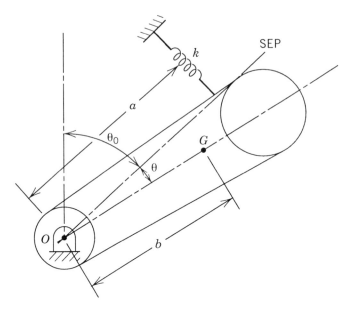

Fig. 3.2.5. Simple pendulous type body to show natural frequency being dependent on orientation relative to gravity.

beam's orientation in the vertical plane. The above analysis suggested that significant gravity forces were at work. A theoretical analysis of the beam's dynamics that included gravity forces verified that these forces were significant in this case and accounted for the measured variation.

3.3 SINGLE DOF FORCED RESPONSE

This section describes the forced response of a single DOF linear system. Understanding this response gives insight into measured dynamic response as well as different ways to display and interpret this response. A crucial element is the damping model that is employed. For many structures, it is found that traditional viscous damping is inadequate. The characteristics of viscous, Coulomb, and structural damping models are compared for use in describing and interpreting forced vibration response. Of these three, viscous and structural damping are found to be most useful. The common graphical schemes for displaying the three commonly used frequency response functions of receptance, mobility, and accelerance are presented.

3.3.1 The Viscous Damping Case

The steady-state solution is obtained by assuming that the excitation force $f(t)$ in Eq. (3.2.1) is a phasor described by

$$f(t) = F_0 e^{j\omega t}$$

where F_0 is the excitation vector. Then Eq. (3.2.1) becomes

$$m\ddot{x} + c\dot{x} + kx = F_0 e^{j\omega t} \qquad (3.3.1)$$

The response phasor is assumed to have the same frequency ω so that

$$x(t) = X_0 e^{j\omega t} \qquad (3.3.2)$$

where X_0 is the response vector. Substitution of Eq. (3.3.2) into Eq. (3.3.1) gives

$$(k - m\omega^2 + jc\omega)X_0 e^{j\omega t} = F_0 e^{j\omega t}$$

Canceling the $e^{j\omega t}$ term and solving for the response vector X_0 gives

$$X_0 = \frac{F_0}{\underbrace{k - m\omega^2}_{\text{real}} + \underbrace{jc\omega}_{\text{imaginary}}} = \frac{F_0}{k[\underbrace{1 - r^2}_{\text{real}} + \underbrace{j2\zeta r}_{\text{imaginary}}]} = H(\omega)F_0 \qquad (3.3.3)$$

where $r = \omega/\omega_n$ is the dimensionless frequency ratio. $H(\omega)$ is called the *frequency response function* (FRF), and it relates the output (displacement in this case) per unit of input (excitation force in this case) at each frequency ω. This FRF is called the *receptance* (as well as *admittance*, *dynamic compliance*, and *dynamic flexibility*) when x is structural displacement and F is an excitation force. It is seen in Eq. (3.3.3) that $H(\omega)$ is complex with real and imaginary parts. Consequently, it has magnitude $|H(\omega)|$ and phase ϕ so that the steady-state output can be written as

$$x = H(\omega)F_0 e^{j\omega t} = |H(\omega)|e^{-j\phi}F_0 e^{j\omega t} = |H(\omega)|F_0 e^{j(\omega t - \phi)} \qquad (3.3.4)$$

Equation (3.3.4) indicates that the response phasor is phase shifted by ϕ radians relative to the excitation phasor. The phase angle ϕ is obtained from the real and imaginary parts as

$$\tan \phi = \frac{\text{imaginary}}{\text{real}} = \frac{c\omega}{k - m\omega^2} = \frac{2\zeta r}{1 - r^2} \qquad (3.3.5)$$

The minus sign in Eq. (3.3.4) is due to the real and imaginary parts being

74 VIBRATION CONCEPTS

in the denominator in Eq. (3.3.3). It is noted that the process of using phasors to solve Eq. (3.3.1) is an efficient means to convert from the time domain to the frequency domain where we have only complex (real and imaginary) algebraic equations to work with.

3.3.2 Common FRFs

Equation (3.3.4) shows the input–output relationship for force and displacement. Often we measure either velocity or acceleration instead of displacement. For velocity, we have

$$v = \dot{x} = j\omega H(\omega) F_0 e^{j\omega t} = Y(\omega) F_0 e^{j\omega t} \tag{3.3.6}$$

so that *mobility* $Y(\omega)$ and receptance $H(\omega)$ are related by

$$Y(\omega) = j\omega H(\omega) = \omega |H(\omega)| e^{j\theta} \tag{3.3.7}$$

The phase angle θ is given by

$$\theta = \phi + \pi/2 \tag{3.3.8}$$

since $j = e^{j\pi/2}$. Equations (3.3.7) and (3.3.8) clearly show that mobility is simply ω times the magnitude of the receptance with a 90 degree phase shift.

The acceleration is related to the input force by taking the derivative of the velocity so that

$$a = \dot{v} = j\omega Y(\omega) F_0 e^{j\omega t} = (j\omega)^2 H(\omega) F_0 e^{j\omega t} = A(\omega) F_0 e^{j\omega t} \tag{3.3.9}$$

so that *accelerance* $A(\omega)$ is related to mobility $Y(\omega)$ and receptance $H(\omega)$ by

$$A(\omega) = j\omega Y(\omega) = \omega |Y(\omega)| e^{j\Theta} = -\omega^2 H(\omega) \tag{3.3.10}$$

where angle θ is given by

$$\Theta = \theta + \pi/2 = \phi + \pi \tag{3.3.11}$$

These results show that receptance, mobility, and accelerance are easily

TABLE 3.3.1. Definitions of Frequency Response Functions

Response	Definition	Name[a]
Displacement	$x/F = H(\omega)$	**Receptance**
		Admittance
		Dynamic compliance
Velocity	$v/F = Y(\omega) = j\omega H(\omega)$	**Mobility**
Acceleration	$a/F = A(\omega) = j\omega Y(\omega)$	**Accelerance**
		Inertance

[a]Preferred names are boldface.

obtained from one another. The names and definition of these ratios are shown in Table 3.3.1.

From these relationships we can construct three other ratios, called *apparent stiffness* (force per unit of displacement), *mechanical impedance* (force per unit of velocity), and *apparent mass* (force per unit of acceleration). It is noted that the name of mechanical impedance is also given to the ratios of F/x and F/a in the literature. Thus we need to be careful when reading any papers that use the name of mechanical impedance. The names shown in Table 3.1.1 are most commonly used and are recommended.

3.3.3 Damping Models in Forced Response

Experience has shown that viscous damping is inadequate for describing damping in many real structures since damping is found to be either independent or weakly dependent on frequency. In this subsection, three forced vibration damping models (viscous, Coulomb, and structural) and their interrelationships are considered. The energy dissipated per cycle of oscillation is a convenient way to compare these damping models.

It is assumed that the motion is sinusoidal for all models, that is, damping forces are either close to the viscous model or small enough to cause only minor motion distortion. Then the energy dissipated per cycle for the viscous case is calculated from

$$W_d = \int F_d \, dx = \pi c \omega A^2 \qquad (3.3.12)$$

where A is the sinusoidal amplitude of motion and F_d is the damping force. This damping force can be written in terms of displacement x:

$$F_d = \pm c\omega \sqrt{A^2 - x^2} \qquad (3.3.13)$$

76 VIBRATION CONCEPTS

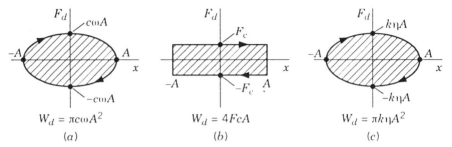

Fig. 3.3.1. Three steady-state damping models. Cross-hatched areas indicate the energy dissipated per cycle. (*a*) Viscous. (*b*) Coulomb. (*c*) Structural.

Equation (3.3.13) is plotted in Fig. 3.3.1*a*, where it is seen that the path is elliptical with maximum values of $\pm c\omega A$ at $x = 0$. The enclosed area is the energy dissipated per cycle, as given by Eq. (3.3.12), and is proportional to amplitude squared and frequency ω. For steady state response conditions, the energy dissipated per cycle is equal to the work done per cycle on mass m by the exciting force. Thus the net work done per cycle by the exciter force is equal to the energy dissipated per cycle and is the reason that steady state motion occurs.

The Coulomb model is shown in Fig. 3.3.1*b* where F_c is the Coulomb damping force. The enclosed rectangular area is the energy dissipated per cycle and can be expressed by

$$W_d = 4F_c A \tag{3.3.14}$$

This energy dissipation model varies only linearly with amplitude. The sinusoidal waveform is not distorted in this case.

Experience shows that energy dissipation in many structures is proportional to amplitude squared so that

$$W_d = \alpha A^2 \tag{3.3.15}$$

The problem is to translate this damping concept into the systems response equations. The *loss factor* η is defined as the energy lost per cycle divided by 2π times the maximum potential energy U ($= kA^2/2$ for a single DOF system). Thus

$$\eta = \frac{W_d}{2\pi U} = \frac{\alpha}{\pi k} \tag{3.3.16}$$

Then, equating the expressions for energy lost per cycle gives the results shown in Table 3.3.2. It is evident from Table 3.3.2 that

SINGLE DOF FORCED RESPONSE

TABLE 3.3.2. Steady State Response Damping Model Relationships

Type of Damping	$W_d{}^a$	$C_{eq}{}^b$	$\eta_{eq}{}^b$
Viscous (c)	$\pi c \omega A^2$	c	$(c\omega)/k$
Coulomb (F_c)	$4F_c A$	$(4F_c)/\pi \omega A$	$(4F_c)/\pi k A$
Loss factor (η)	$\pi k \eta A^2$	$(k\eta)/\omega$	η

[a] Energy lost per cycle is W_d.
[b] Equivalent damping expression. Note that $\alpha = \pi k \eta$ and that structural or hysteresis damping $h = k\eta$.

and that
$$c\omega = k\eta$$
$$\eta = 2\zeta \quad (3.3.17)$$

at resonance so that either the $c\omega$ term in our equations can be replaced by $k\eta$ or ζ by $\eta/2$. Then the structural energy loss per cycle is given by

$$W_d = \pi k \eta A^2 \quad (3.3.18)$$

and the corresponding damping force versus position plot is as shown in Fig. 3.3.1c. The difference in the curves shown in Figs. 3.3.1a and c is that the maximum damping force (when $x = 0$) is independent of frequency in the structural (loss factor) damping model.

A better feel for the differences between the viscous and loss factor approach is illustrated in Fig. 3.3.2, where η is plotted as a function of ω for the ideal and actual cases. The viscous damping curve is also plotted as a straight line. When the viscous and η curves intersect, they represent the same damping. This intersection should occur at resonance, the frequency where damping is most important. Transient damping occurs at essentially the natural frequency so that the transient envelope function $e^{-\zeta \omega_n t}$ becomes $e^{-\eta \omega_n t/2}$ when ζ is replaced by $\eta/2$. This result implies that structural damping has the same transient behavior as viscous damping and that the transient techniques can be used to measure the values of η.

3.3.4 The Structural Damping Response

For sinusoidal motion, the damping and spring terms in Eq. (3.3.3) can be written as

$$c\dot{x} + kx = (k + jc\omega)X_0 e^{j\omega t} = (k + jk\eta)X_0 e^{j\omega t}$$

when the $c\omega = k\eta$ relationship of Eq. (3.3.17) is used. Thus we define the

78   VIBRATION CONCEPTS

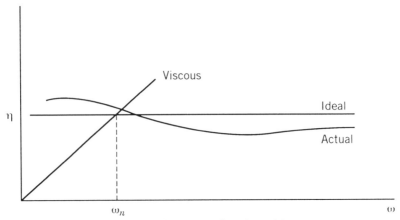

Fig. 3.3.2. Loss factor as a function of frequency.

complex structural stiffness

$$k^* = k(1 + j\eta) \qquad (3.3.19)$$

where k^* is seen to represent both the standard spring k and the damping term. When the results of Eqs. (3.3.17) and (3.3.19) are used in Eqs. (3.3.1) and (3.3.2), Eqs. (3.3.3) and (3.3.5) become

$$X_0 = \underbrace{\frac{F_0}{k - m\omega^2 + jk\eta}}_{\text{real} \quad \text{imaginary}} = \underbrace{\frac{F_0}{k[1 - r^2 + j\eta]}}_{\text{real} \quad \text{imaginary}} = H(\omega)F_0 \qquad (3.3.20)$$

$$\tan \phi = \frac{\text{imaginary}}{\text{real}} = \frac{k\eta}{k - m\omega^2} = \frac{\eta}{1 - r^2} \qquad (3.3.21)$$

A comparison of Eqs. (3.3.3) and (3.3.5) with Eqs. (3.3.20) and (3.3.21) shows that they are the same form and have nearly identical characteristics. Thus the same FRF concepts of receptance, mobility, and accelerance also apply to the structural damping case.

Next, we need to address how to graphically display typical FRFs. One thing is evident: these FRFs are complex with real and imaginary parts and they can cover a wide range of both magnitudes and frequencies. Three methods are used to display these results: (1) *Bode* diagrams, (2) plots of real and imaginary parts versus frequency, and (3) *Nyquist* diagrams. In the following subsections, each of these plots is used to display receptance, mobility, and accelerance curves. The case displayed corresponds to values $k = 1000$ lb/in, $m = 0.1$ lb-s²/in., and $c = 20$ lb-s/in. so that $\omega_n = 100$ rad/s, $\zeta = 0.10$, and the equivalent value of η is 0.20 or $k\eta = 200$.

3.3.5 The Bode Diagram

Receptance A *Bode diagram* plots the FRF's magnitude and phase as functions of frequency. Figure 3.3.3a is a receptance plot for the viscous damping case and Fig. 3.3.3b is the receptance plot for the structural damping case. The two curves are nearly the same in magnitude. However, the phase angle is seen to be different at frequencies well below the natural frequency.

Vertical log scales and horizontal log scales are commonly used in Bode plots for several reasons. First, the responses' dynamic range can be of the order of 10,000 to 1. This large range requires some way to maintain detail at various levels. The log scale accomplishes this goal. Second, the use of log scales for both magnitude and frequency allow data that plot as curves on linear scales to become asymptotic straight lines on log scales. Third, it is common practice to display results in terms of the dB log scale given by

$$dB = 20 \log(x/x_r) \tag{3.3.22}$$

where x_r is the reference value that must be known before dB have any meaning.

The magnitude curve in Fig. 3.3.3 can be used to estimate system parameters. When $\omega \ll \omega_n$, the dominant term in either Eq. (3.3.3) or Eq. (3.3.20) is spring constant k, so that we have a horizontal *spring line* that can be used to obtain the spring constant as shown. The downward pointing arrow on the left side indicates the direction of increasing spring constants. On the other hand, when $\omega \gg \omega_n$, the dominant denominator term in either Eq. (3.3.3) or Eq. (3.3.20) is $m\omega^2$. This magnitude becomes a straight line on log paper since

$$\log\left(\frac{1}{m\omega^2}\right) = -\log(m) - 2\log(\omega) \tag{3.3.23}$$

This line is called the *mass line* since its position is controlled by mass m while its slope is controlled by ω. The mass line has a slope that is 12 dB/octave (40 dB/decade), as shown in Fig. 3.3.3, the slope being calculated directly from Eq. (3.3.23). The direction of increasing mass is shown by the downward pointing arrow on the right side of the plot. The spring line and the mass line intersect at point A, the system's natural frequency. Recall that the spring force and inertia force cancel when the system

80 VIBRATION CONCEPTS

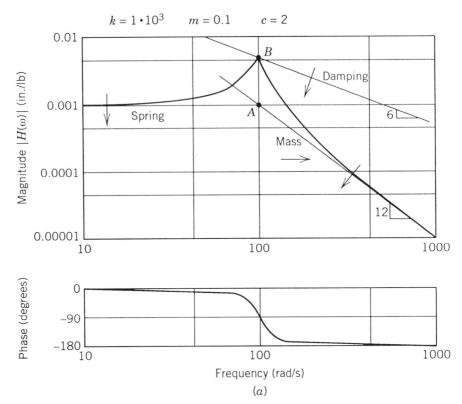

Fig. 3.3.3. Receptance Bode plots. (*a*) Viscous damping model. (*b*) Structural damping model.

oscillates at its natural frequency. For frequencies around ω_n, the only significant force left in Eq. (3.3.3) to counter the applied force is damping. Thus near resonance the response, is controlled by $1/c\omega$, so that a *damping line* is defined by

$$\log\left(\frac{1}{c\omega}\right) = -\log(c) - \log(\omega) \qquad (3.3.24)$$

Equation (3.3.24) indicates that this line has a slope of 6 dB/octave (20 dB/decade), as shown in Fig. 3.3.3*a*. The peak response of $H(\omega)$ is determined by damping c when $\omega = \omega_n$, point B in Fig. 3.3.3*a*. Point B is the intersection of the vertical natural frequency line and the damping line. However, from Eq. (3.3.20), it is seen that the peak is controlled by $1/k\eta$ for frequencies close to ω_n, so that the damping curves in Fig. 3.3.3*b* are described by

SINGLE DOF FORCED RESPONSE

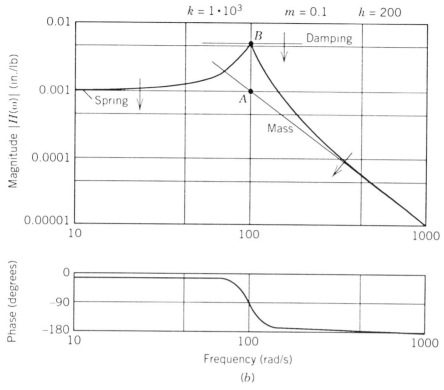

Fig. 3.3.3. (*Continued*).

$$\log\left(\frac{1}{k\eta}\right) = -\log(k) - \log(\eta) \qquad (3.3.25)$$

so that the loss factor type of damping is given by the distance from the spring line to point *B*. The direction of increasing damping is shown by the arrow. The vertical height between points *A* and *B* is the system *Q*.

Phase angle ϕ is seen to vary from zero to -180 degrees as the frequency passes through resonance in Fig. 3.3.3a, while it varies from $\phi \cong -\eta$ well below resonance to -180 degrees above resonance, as shown in Fig. 3.3.3b. The phase angle is -90 degrees in both models at resonance for this type of physical system. However, a 90 degree phase shift may not occur for other systems at resonance, contrary to a common opinion. It is only true for systems that follow this type of vibration model. The fact that $\tan(\phi) \cong -\eta$ for the structural damping model when well below resonance has been used to measure the loss factor of rubbery and plastic type materials.

82 VIBRATION CONCEPTS

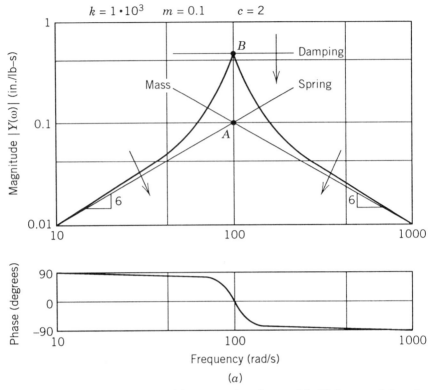

Fig. 3.3.4. Mobility Bode plot. (*a*) Viscous damping model. (*b*) Structural damping model.

Mobility The mobility FRF is plotted in the Bode diagram format, as shown in Fig. 3.3.4, for both viscous and structural damping. The multiplication of receptance by $j\omega$ rotates the receptance diagram by 6 dB/octave so that the spring line has a positive slope of 6 dB/octave, the viscous damping lines are horizontal, and the mass line slope becomes minus 6 dB/octave, as shown in Fig. 3.3.4a. The structural damping lines now become inclined with a slope of 6 dB/octave, as shown in Fig. 3.3.4b. The natural frequency occurs at the spring line–mass line intersection, point A in both Fig. 3.3.4a and b. Point B locates the peak value from which the respective damping can be estimated.

The phase angle is shifted by 90 degrees in both cases due to multiplying the receptance by $j\omega$. The phase angle is 0 degrees at resonance for this case. Standard calibrated mobility paper is available to plot data so that a systems mass, stiffness, and viscous damping can be estimated from

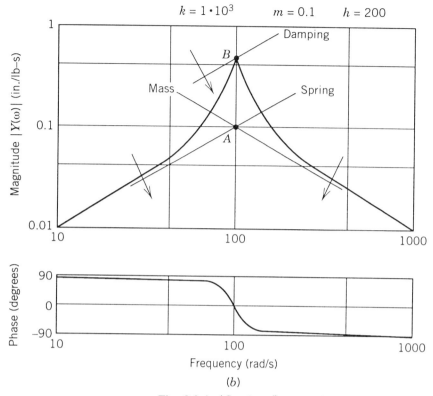

Fig. 3.3.4. (*Continued*).

the plots. The loss factor can be estimated from the measured viscous damping.

Accelerance The accelerance FRF is plotted in the Bode diagram format as shown in Fig. 3.3.5a (viscous damping) and 3.3.5b (structural damping). The multiplication of mobility by $j\omega$ causes the mobility diagram to be rotated by 6 dB/octave so that the spring line has a positive slope of 12 dB/octave, the viscous damping a positive slope of 6 dB/octave, and the mass lines are horizontal, as shown in Fig. 3.3.5a. In the structural damping case, the damping line remains parallel to the spring line and has a slope of 12 dB/octave, as shown in Fig. 3.3.5b. Point A is the natural frequency corresponding to the intersection of the mass and spring lines. The phase angle is shifted by 90 degrees again due to the $j\omega$ multiplier. In both cases the phase angle is a positive 90 degrees at resonance. Remember that this 90 degree phase shift is true only for this type of system.

84 VIBRATION CONCEPTS

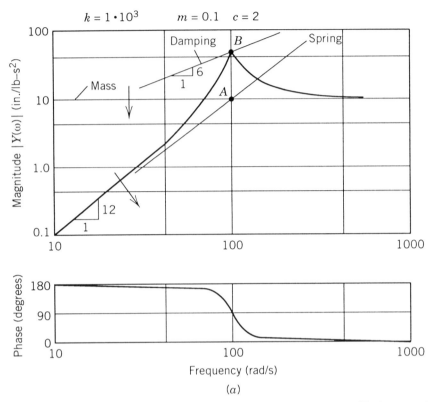

Fig. 3.3.5. Accelerance Bode plot. (*a*) Viscous damping model. (*b*) Structural damping model.

3.3.6 Real and Imaginary Plots and Nyquist Diagrams

The real and imaginary plots and the Nyquist plots are considered together, since they are closely related and reinforce one another.

Receptance The receptance real and imaginary plots for the viscous case are shown in Figs. 3.3.6*a* and *b*, respectively. In the real plot, three points, *A*, *B*, and *C*, are marked. The real part is zero at point *A*, corresponding to the system's natural frequency, that is, to the point where the inertia force and the spring force cancel. Point *D* is the corresponding imaginary component at resonance and allows the calculation of the effective damping. Points *B* and *C* are special points on the Nyquist diagram that correspond to the largest difference on the real axis. Also note that the imaginary plot starts at zero for frequencies well below resonance and goes to zero for frequencies well above resonance.

The Nyquist plot in Fig. 3.3.6*c* shows the magnitude of *H*(ω) and its

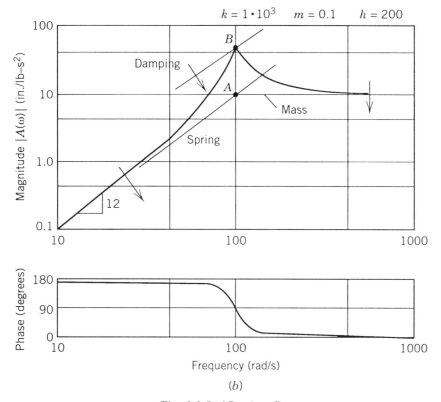

Fig. 3.3.5. (*Continued*).

phase angle ϕ without any frequency information. Points A, B, C, and D in the real and imaginary plots are also shown so that the connection is straightforward. The Nyquist plot is nearly a circle with the center at O, becoming more circle like with center O approaching the imaginary axis as the system damping decreases.

The corresponding structural damping case is shown in Fig. 3.3.7. In this case, the Nyquist diagram can be shown to be a perfect circle[2] with a diameter that is equal to $1/k\eta$. The fact that this curve is nearly a circle for small amounts of viscous damping and is a circle for structural damping has led to the concept of circle fitting as used in modal analysis. Note that the imaginary part starts at $-\eta$ for low frequencies below resonance and goes to zero at high frequencies above resonance.

---

[2]For example, see D. J. Ewins, *Modal Testing: Theory and Practice*, Research Studies Press, Ltd., Letchworth, Hertfordshire, UK, pp. 36–40.

86 VIBRATION CONCEPTS

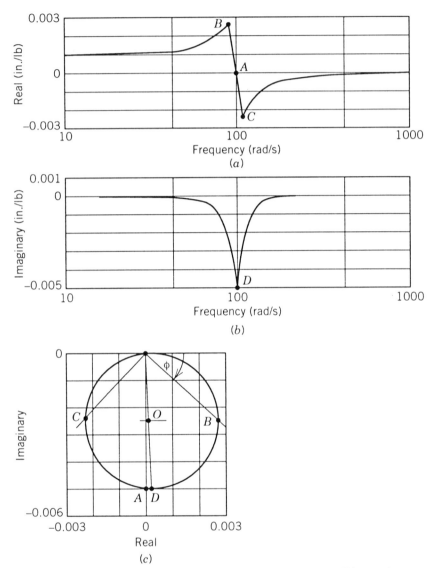

Fig. 3.3.6. Viscous damping model. (*a*) Real receptance plot. (*b*) Imaginary receptance plot. (*c*) Nyquist receptance plot.

Mobility The mobility case for viscous and structural damping is shown in Figs. 3.3.8 and 3.3.9. In the viscous damping case, the mobility is also a perfect circle with a center at O, the diameter of which is $1/c$, as shown in Fig. 3.3.8c. In the structural damping case, the circle is distorted and its center O is not on the real axis. Thus a circle fit is most appropriate when the damping is viscous or extremely light, in which case the structural

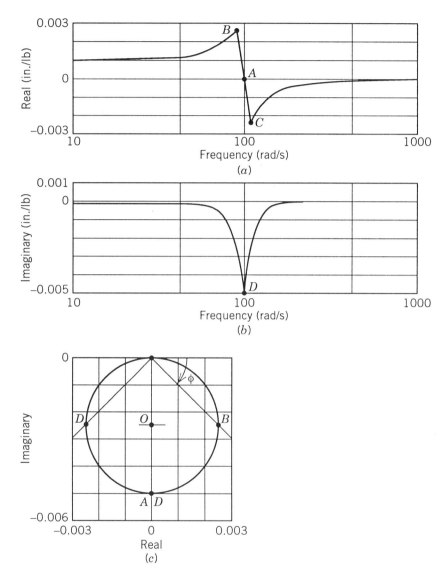

Fig. 3.3.7. Structural damping model. (*a*) Real receptance plot. (*b*) Imaginary receptance plot. (*c*) Nyquist receptance plot.

damping is nearly a circle. Note that the phase angle θ is measured from the real axis, starts near 90 degrees well below resonance, goes to zero at resonance, and then to -90 degrees when significantly above resonance.

Accelerance The accelerance for both viscous and structural damping is shown in Figs. 3.3.10 and 3.3.11. In this case the Nyquist diagram is

88 VIBRATION CONCEPTS

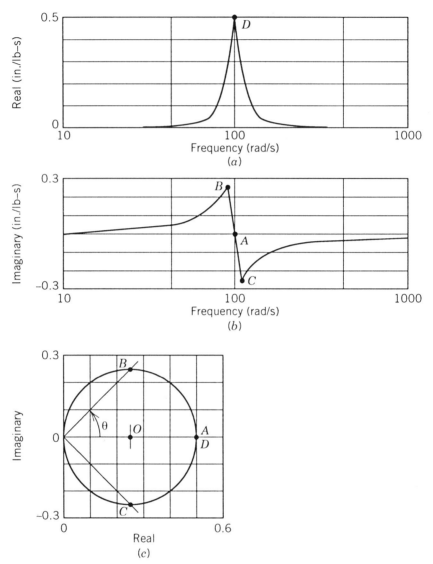

Fig. 3.3.8. Viscous damping model. (*a*) Real mobility plot. (*b*) Imaginary mobility plot. (*c*) Nyquist mobility plot.

noncircular in both cases, since the circle center O is not on the vertical axis in both cases. The phase angle θ is seen to start at 180 for small values of ω and goes to 90 degrees at resonance, and then toward zero for frequencies significantly above resonance.

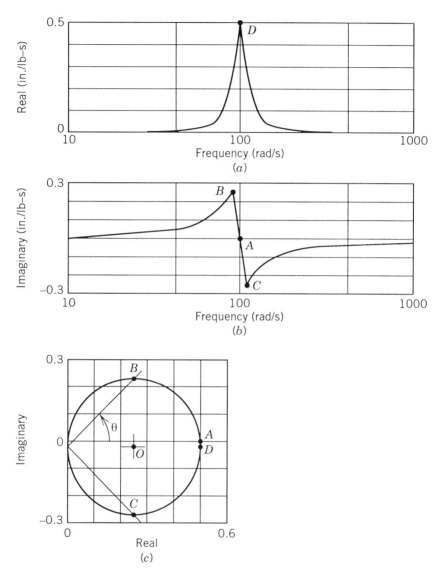

Fig. 3.3.9. Structural damping model. (*a*) Real mobility point. (*b*) Imaginary mobility point. (*c*) Nyquist mobility plot.

Of the three methods for displaying the FRF data, the Bode plot is most informative in estimating the system parameters.[3]

---

[3] J. P. Salter, *Steady State Vibration*, Kenneth Mason Press, Homewell Havant Hampshire, UK, 1969, pp. 13–14. This is an excellent book on interpreting physical vibration models from Bode plots.

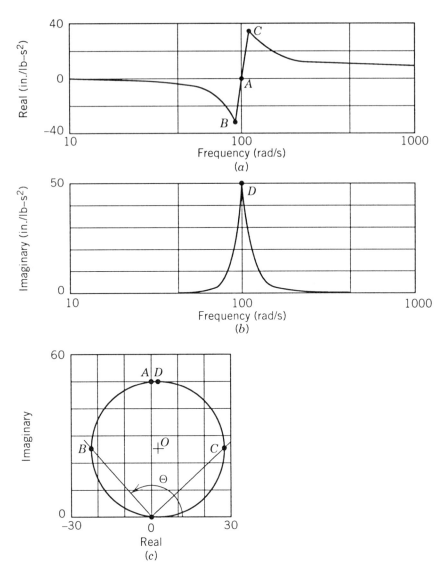

Fig. 3.3.10. Viscous damping model. (*a*) Real accelerance plot. (*b*) Imaginary accelerance plot. (*c*) Nyquist accelerance plot.

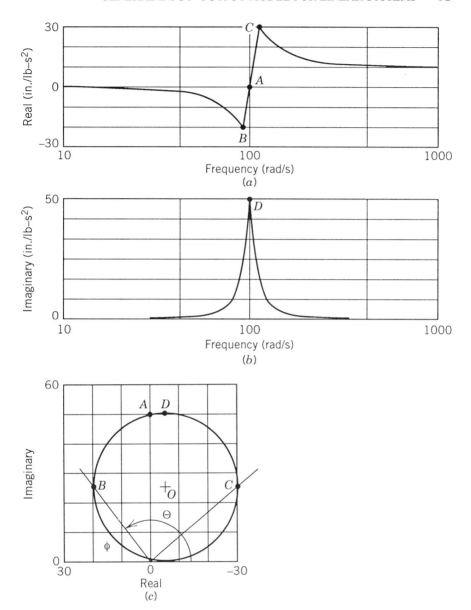

Fig. 3.3.11. Structural damping model. (*a*) Real accelerance plot. (*b*) Imaginary accelerance plot. (*c*) Nyquist accelerance plot.

3.4 GENERAL INPUT–OUTPUT MODEL FOR LINEAR SYSTEMS

General input–output relationships for linear systems are desirable since, as was shown in Chapter 2, most real signals contain many frequency

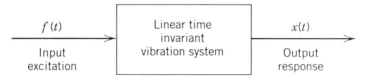

Fig. 3.4.1. Block diagram of general linear system. (from J. W. Dally, W. F. Riley, and K. G. McConnell, *Instrumentation for Engineering Measurements*, 2nd ed., copyright © 1993 by John Wiley & Sons, New York. Reprinted by permission.)

components, while in Section 3.3, particularly Eqs. (3.3.4), (3.3.6), (3.3.9), and (3.3.20), it is shown that each frequency component is modified by a structure's FRF. The question is: How do time and frequency response concepts tie together for arbitrary input excitation?

3.4.1 The Frequency-Domain (Fourier Transform) Approach

The block diagram in Fig. 3.4.1 shows a linear time invariant system. The input excitation time history is $f(t)$, while the output response time history is $x(t)$. If a deterministic sinusoidal excitation phasor

$$f(t) = F_0 e^{j\omega t} \tag{3.4.1}$$

is applied to a linear system, a sinusoidal response phasor occurs at the same frequency such that

$$x(t) = X_0 e^{j\omega t} = H(\omega) F_0 e^{j\omega t} \tag{3.4.2}$$

where X_0 is the output vector.

$H(\omega)$ is the complex frequency response function that alters both magnitude and phase of the output vector X_0 relative to the input vector F_0 as a function of frequency ω.

We can reduce Eq. (3.4.2) to

$$X_0 = H(\omega) F_0 \tag{3.4.3}$$

as a general relationship for each frequency ω.

In Chapter 2 it was shown that periodic inputs are composed of discrete frequency components described by a discrete frequency spectrum while transient inputs are composed of all frequencies that are described by a continuous frequency spectrum. Now Eq. (3.4.3) is valid for each and every frequency, so we can write it as

GENERAL INPUT–OUTPUT MODEL FOR LINEAR SYSTEMS

and as
$$X_p = H(p\omega_0) F_p \quad \text{for discrete frequency spectra}$$
$$X(\omega) = H(\omega) F(\omega) \quad \text{for continuous spectral density} \quad (3.4.4)$$

Equations (3.4.4) suggest that the functional form of $H(\omega)$ can be obtained by measuring both input and output frequency spectra. These measurements can be done on either a discrete ($p\omega_0$) or a continuous frequency basis so long as magnitude and phase information is preserved.

From Eq. (3.4.4) we see that the units of $H(\omega)$ are those of $x(t)$ divided by those of $f(t)$ and that they can be determined from either X_p/F_p or $X(\omega)/F(\omega)$. For example, if $x(t)$ has units of inches and $f(t)$ has units of pounds, then X_p has units of inches and F_p has units of pounds, while $X(\omega)$ has units of inches per hertz and $F(\omega)$ has units of pounds per hertz, so that $H(\omega)$ has units of inches per pounds, independent of whether a discrete or a continuous frequency spectrum is used.

Recall that an input frequency spectrum is determined by using either periodic Fourier transform (PFT) or a transient Fourier transform (TFT) so that

$$F_p = \frac{1}{T} \int_t^{t+T} f(t) e^{-jp\omega_0 t} dt$$

or
$$F(\omega) = \int_{-\infty}^{\infty} f(t) e^{-j\omega t} dt \quad (3.4.5)$$

Either the periodic inverse Fourier transform (PIFT) or the transient inverse Fourier transform (TIFT) and the frequency domain input–output relationship of Eq. (3.3.4) are used to give the output $x(t)$ as

$$x(t) = \sum_{p=-\infty}^{\infty} X_p e^{jp\omega_0 t} = \sum_{p=-\infty}^{\infty} H(p\omega_0) F_0 e^{jp\omega_0 t}$$

$$x(t) = \frac{1}{2\pi} \int_{-\infty}^{\infty} X(\omega) e^{j\omega t} d\omega = \frac{1}{2\pi} \int_{-\infty}^{\infty} H(\omega) F(\omega) e^{j\omega t} d\omega \quad (3.4.6)$$

Equations (3.4.4) through (3.4.6) indicate a very desirable general input–output process. It consists of determining the input frequency spectrum from Eq. (3.4.5), then using Eq. (3.4.4) to obtain the output frequency

3.4.2 The Time-Domain Impulse Response Approach

A general time-domain method for estimating a system's response to an arbitrary input can be developed by using the *Dirac delta function*. Assume that the input is given by

$$f(t) = \delta(t - \tau) \tag{3.4.7}$$

The Dirac delta function $\delta(t - \tau)$ is zero everywhere except at $t = \tau$ where the ordinate is infinity with zero time duration but has unit area. Conceptually, the Dirac delta function can be represented in the limit as Δt goes to zero by a rectangular area of width Δt and height of $1/\Delta t$, as shown in Fig. 3.4.2a. The impulse function has units of $f(t)$ times time t. Input excitation signals with very short time durations are commonly called *impulsive* signals regardless of the variable involved. Only when $f(t)$ is a force is the result an impulse according to classical engineering definitions.

System response $x(t)$ due to $\delta(t - \tau)$ is called the *impulse-response function* $h(t - \tau)$, as shown in Fig. 3.4.2b where the relative time shift $\epsilon = t - \tau$ is measured relative to dummy time variable τ. We see that the impulse response function is zero before the impulse occurs so that $h(\epsilon) = 0$ for $\epsilon < 0$. Also, $h(\epsilon)$ decays with time in real systems. The units of $h(\epsilon)$ are those of response $x(t)$ divided by units of excitation $f(t)$ multiplied by time.

The response $x(t)$ due to an arbitrary input $f(t)$ is obtained by using the impulse response function shown in Fig. 3.4.2c. In this case, the impulse at time τ is $\{f(\tau) d\tau\}$ and the response at time t (where $t > \tau$) due to this impulse is

$$x_i(t) = [f(\tau) d\tau] h(t - \tau) \tag{3.4.8}$$

Since the system is linear, these x_i responses can be *superimposed* (summed) in the *time-domain* at time t to give

$$x(t) = \int_{-\infty}^{\infty} f(\tau) h(t - \tau) d\tau \tag{3.4.9}$$

since a summation becomes an integration in the limit as $\tau \to 0$. Equation (3.4.9) has a number of different names including *convolution*, *Faltung*, and *Duhamel*. Convolution and Duhamel are used most often in vibration

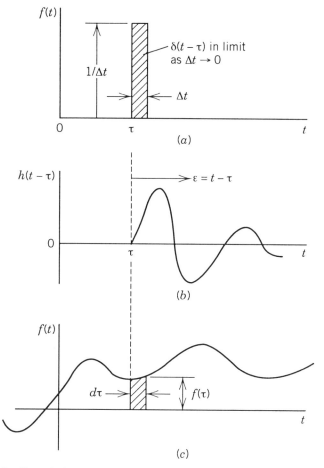

Fig. 3.4.2. Convolution concepts. (*a*) Dirac delta functions. (*b*) Characteristic impulse response function. (*c*) Arbitrary input time history showing impulse at time τ. (From J. W. Dally, W. F. Riley, and K. G. McConnell, *Instrumentation for Engineering Measurements*, 2nd ed., copyright © 1993 by John Wiley & Sons, New York. Reprinted by permission.)

testing. The upper integration limit can be either t or ∞ since $h(t - \tau)$ is zero when its argument ($\epsilon = t - \tau$) is negative; that is, $\tau > t$. Equation (3.4.9) is a general input–output relationship that gives the same result for $x(t)$ as Eq. (3.4.6) even though the processes are completely different. One process involves frequency domain multiplication before integration over frequency using Fourier transform concepts, while the other process involves integration (summation) in the time domain while using the impulse-response function.

3.4.3 Receptance FRF Versus Impulse Response Function

The receptance frequency response function $H(\omega)$ and the impulse-response function $h(t - \tau)$ are related to each other. To show this relationship, let the Dirac delta function be the excitation term in Eq. (3.4.9) when $\tau = 0$. Then, Eq. (3.4.9) shows that $x(t) = h(t)$, while Eq. (3.4.5) shows that $F(\omega) = 1$ for all frequencies. Then Eq. (3.4.6) becomes

$$h(t) = \frac{1}{2\pi} \int_{-\infty}^{\infty} H(\omega) e^{j\omega t} d\omega$$

$$H(\omega) = \int_{-\infty}^{\infty} h(t) e^{-j\omega t} dt$$
(3.4.10)

Equation (3.4.10) shows that the impulse-response function and the frequency response function form a transient Fourier transform pair.

The time domain convolution process described by Eq. (3.4.9) to give output $x(t)$ from input time history $f(t)$ is represented by the left-hand column of boxes in Fig. 3.4.3. The right-hand column of boxes represent the frequency-domain multiplication process described by Eq. (3.4.4). Either the periodic Fourier transform or a transient Fourier transform is used to transform from time domain to frequency domain for their respective type of signal. The process of transferring from the frequency domain to the time domain is either the periodic inverse Fourier transform or transient inverse Fourier transform.

The process of converting from impulse-response function $h(t)$ to the receptance FRF $H(\omega)$ is the transient Fourier transform while the process of converting from the receptance FRF $H(\omega)$ to the impulse-response function $h(t)$ is the transient inverse Fourier transform as shown. Note that transient Fourier transforms are required in this definition.

Example: The second-order differential equation of motion that occurs throughout vibration formulations is given as

$$m\ddot{x} + c\dot{x} + kx = f(t) \qquad (3.4.11)$$

When $f(t)$ is replaced by the Dirac delta function, the initial conditions

GENERAL INPUT–OUTPUT MODEL FOR LINEAR SYSTEMS

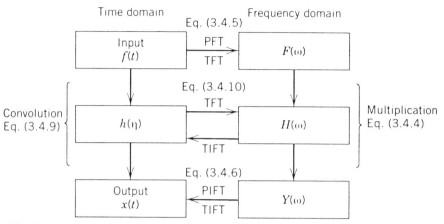

Fig. 3.4.3. Illustration of linear system input-output relationships showing Fourier transform as bridge between convolution and frequency multiplication processes. (From J. W. Dally, W. F. Riley, and K. G. McConnell, *Instrumentation for Engineering Measurements*, 2nd ed., copyright © 1993 by John Wiley & Sons, New York. Reprinted by permission.)

for this unit impulse are $x(0) = 0$ and $\dot{x}(0) = 1/m$. Then the impulse-response function becomes

$$h(t) = \frac{e^{-\zeta\omega_n t}}{m\omega_d}\sin(\omega_d t) \quad (3.4.12)$$

where ω_n is the natural frequency given by $\sqrt{k/m}$, ζ is the damping ratio given by $c/(2\sqrt{km})$, and ω_d is the damped frequency given by $\omega_n\sqrt{1-\zeta^2}$. Equation (3.4.12) shows that the impulse-response function decays within an exponential envelope. This is characteristic behavior of impulsive response for stable linear dynamic systems. The receptance FRF for Eq. (3.4.11) is given by

$$H(\omega) = \frac{1}{k - m\omega^2 + jc\omega} = \frac{1}{m(\omega_n^2 - \omega^2 + j2\zeta\omega_n\omega)} \quad (3.4.13)$$

Note that $H(-\omega)$ is the complex conjugate of $H(\omega)$. It is left as an exercise for the reader to show that Eqs. (3.4.12) and (3.4.13) are transient Fourier transforms of one another.

3.4.4 Random Input–Output Relationships

Bendat and Piersol[4] show that the output (auto-spectral density) (ASD) is related to the input ASD by

$$S_{xx}(\omega) = |H(\omega)|^2 S_{ff}(\omega)$$

or (3.4.14)

$$G_{xx}(\omega) = |H(\omega)|^2 G_{ff}(\omega)$$

where $S_{xx}(\omega)$ and $S_{ff}(\omega)$ are the output and input double-sided ASD.
$G_{xx}(\omega)$ and $G_{ff}(\omega)$ are the output and input single-sided ASD.
Similarly, the cross-spectral density (CSD) is related to the input ASD by

$$S_{fx}(\omega) = H(\omega) S_{ff}(\omega)$$
$$G_{fx}(\omega) = H(\omega) G_{ff}(\omega)$$
(3.4.15)

where $S_{fx}(\omega)$ is the dual sided output CSD.
$G_{fx}(\omega)$ is the output single sided CSD.
It is evident that there is no FRF phase information preserved in Eq. (3.4.14), while Eq. (3.4.15) preserves both magnitude and phase information. These equations are used a great deal in doing frequency analysis of random signals.

3.4.5 Shock Response Spectra

Design engineers need to estimate the stress that occurs in a structure when exposed to *shock loads*. A shock load is one with large magnitudes and relatively short time duration t_0, as shown in Fig. 3.4.4b. The concept of *shock response spectra* (SRS) is usually based on the response of an undamped single DOF system. The shock spectra concept was suggested by Biot in 1932 and was used by him 10 years later to study building stresses caused by seismic events.[5] The designer's concern is: What is the maximum response (stress or strain) that will occur in the structure when subjected to a particular shock load? The undamped response is usually used in estimating SRS since this is the maximum possible response for a given shock load.

The impulse response function $h(t)$ for the undamped single DOF system in Fig. 3.4.4a is given by Eq. (3.4.12) when $\zeta = 0$. When this $h(t)$

---

[4] J. S. Bendat and A. G. Piersol, *Measurement and Analysis of Random Data*, John Wiley & Sons, New York, 1966, pp. 98–99.
[5] M. A. Biot, "Analytical and Experimental Methods in Engineering Seismology," *Proceedings of the ASCE*, Jan. 1942.

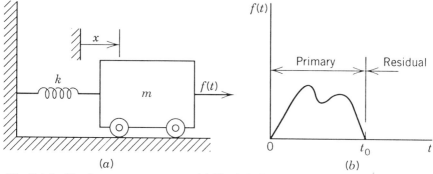

Fig. 3.4.4. Shock response concepts. (*a*) Single DOF mechanical system. (*b*) Shock load.

is substituted into Eq.(3.4.9) and the limits of integration for a shock load as shown in Fig. 3.4.4*b* are used, the response becomes

$$x(t) = \frac{\omega_n}{k} \int_0^t f(\tau) \sin\{\omega_n(t - \tau)\} \, d\tau \qquad (3.4.16)$$

since there is no input before time $\tau = 0$. The maximum response that occurs for a given shock load $f(\tau)$ depends on the value of natural frequency ω_n, stiffness k, pulse duration t_0, and time t used to evaluate the maximum response. For a given stiffness k, Eq. (3.4.16) indicates that the maximum response is a function of $\omega_n t_0$. Note that both t_0/T (where T is the natural period) and $f_n t_0$ are used for graphing SRS. A SRS consists of plotting $x_{\max}$ versus the dimensionless parameter $\omega_n t_0$ or $f_n t_0$. We find that the three possible shock spectra that can be plotted are:

1. *Primary SRS*: The maximum response that occurs within the time window of $0 < t < t_0$ (see Fig. 3.4.4*b*)
2. *Residual SRS*: The maximum response that occurs after time t_0, that is, $t > t_0$
3. *Maxi-Max SRS*: The maximum response that occurs at any time

The maxi-max SRS is most useful since it gives the worst possible response. The SRS interpretation is straightforward for a single DOF system. It displays the maximum response that occurs for a given shock load as function of the system's natural frequency. Note that this definition is for a single DOF system.

Example: Consider the ramp hold excitation shown in Fig. 3.4.5a. For $t < t_0$, the systems response is given by[6]

$$x(t) = \frac{F_0}{k}\left(\frac{t}{t_0} - \frac{\sin \omega_n t}{\omega_n t_0}\right) \tag{3.4.17}$$

Similarly, for $t > t_0$, the system's response is given by

$$x(t) = \frac{F_0}{k}\left(1 - \frac{\sin \omega_n t}{\omega_n t_0} + \frac{\sin \omega_n(t - t_0)}{\omega_n t_0}\right) \tag{3.4.18}$$

We see that the primary SRS from Eq. (3.4.17) occurs when $t = t_0$ so that

$$x_{\max} = \frac{F_0}{k}\left(1 - \frac{\sin \omega_n t_0}{\omega_n t_0}\right) \tag{3.4.19}$$

The residual SRS is obtained from Eq. (3.4.18) by differentiating Eq. (3.4.18) with respect to time, setting the derivative to zero in order to find the time of maximum response. This procedure gives

$$x_{\max} = \frac{F_0}{k}\left(1 + \frac{1}{\omega_n t_0}\sqrt{2(1 - \cos \omega_n t_0)}\right) \tag{3.4.20}$$

The primary SRS from Eq. (3.4.19) and the residual SRS from Eq. (3.4.20) are plotted in Fig. 3.4.5b, along with the maxi-max SRS as functions of $f_n t_0$. The maxi-max SRS must take on the largest value of either the primary SRS or the residual SRS. In this case, the maxi-max SRS and the residual SRS are the same as shown.

What is the physical meaning SRS of these curves? First, consider the primary SRS shown in Fig. 3.4.5b. When the $f_n t_0$ term is small, the system cannot respond adequately within time t_0, since the system is too soft. Thus the primary SRS must start near zero. The maximum primary response occurs around $f_n t_0 = 0.7$ when its value is nearly 1.22. The primary response SRS is seen to oscillate about the unity line for values of $f_n t_0 > 1$. This means that the response is nearly equal to the static deflection at the end of the ramp.

The residual SRS is seen to start at values of 2 for small values of $f_n t_0$. This is due to the input looking like a step input to a low natural frequency

---

[6]See W. T. Thomson, *Theory of Vibrations with Applications*, 4th ed., Prentice Hall, Englewood Cliffs, NJ, 1988.

GENERAL INPUT–OUTPUT MODEL FOR LINEAR SYSTEMS   101

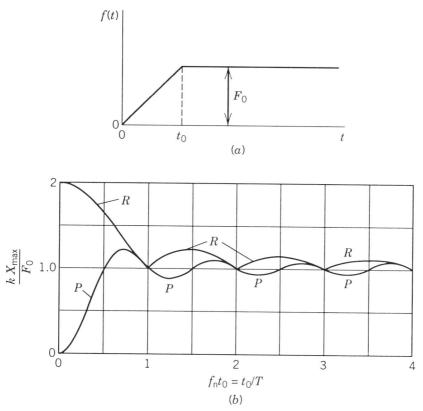

Fig. 3.4.5. Shock response spectra for ramp hold loading. (*a*) Ramp hold input. (*b*) Shock response spectra.

system, since the step input response is proportional to $\{1 - \cos(\omega_n t)\}$, which always has a maximum value of 2. As the values of $f_n t_0$ increase beyond unity, the maximum residual SRS values are gradually approaching a value of unity. There are special points where the SRS is unity; namely, when $f_n t_0$ has integer values. It is obvious that the residual and maxi-max SRS are the same since the residual SRS is greater than or equal to the primary SRS.

Two time histories have the same time period (10 ms) and amplitude (100 force units), as shown in Fig. 3.4.6a. One pulse is a half-sine while the other is a single sinusoidal cycle. The corresponding maxi-max SRS are shown in Fig. 3.4.6b when the structure has 5 percent damping. It is evident that the half-sine is significantly more severe (by a factor of 10) than the single sinusoidal cycle at low frequencies below 10 Hz. Between 70 and about 500 Hz the single sinusoidal cycle is more severe than the

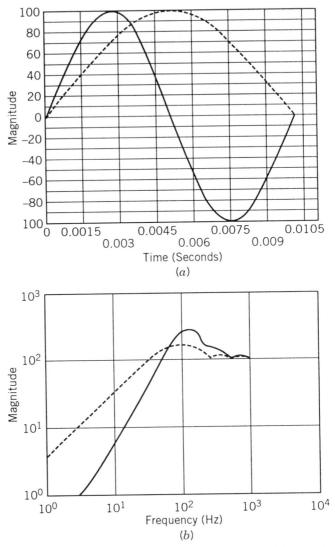

Fig. 3.4.6. (*a*) Plot of half sine and single sine shock loads. (*b*) Plot of corresponding shock spectra. (Courtesy Dr. J. Rogers, Sandia National Laboratory, Albuquerque, NM.)

half-sine. If the structure has a natural frequency around 10 Hz, it is overtested by using the half-sine input if the real input is the single sinusoidal cycle. Similarly, it is undertested if the structure's natural frequency is 100 Hz. This simple example shows the importance of waveform on SRS for a single DOF system. The situation is more complex for multi DOF systems.

The SRS is one of the most misunderstood and abused concepts in vibration testing. The concept is based on a single DOF linear system. Most real structures that are exposed to shock are multi DOF systems and often exhibit nonlinear behavior. In addition, it is assumed that any two time-histories with the same shock response spectra have the same potential to damage the structure. This assumption is not true, except in very special circumstances. Thus caution must be exercised when trying to employ SRS concepts.

3.5 NONLINEAR BEHAVIOR

Generally engineers and scientists are trained to think in terms of linear systems and most likely believe that any response can be calculated using Newton's laws of motion in a deterministic fashion with a large enough computer. However, we are often puzzled by strange behavior caused by nonlinear phenomena. For many, nonlinear behavior is a convenient scapegoat when an unusual response occurs, especially since there are few who can challenge vague claims of nonlinear behavior. The objective of this section is to introduce the reader to some basic nonlinear response concepts, and hopefully, to encourage that reader to study both nonlinear behavior and the new field of chaotic motion. It is noted that chaotic motion can occur only in nonlinear systems.

Some sources of nonlinear behavior are:

- Nonlinear springs and damping elements
- Nonlinear boundary conditions
- Fluid structure interaction
- Mechanical backlash in mating parts
- Nonlinear electric and magnetic forces
- Servo controllers and electronic elements

This section addresses some fundamental nonlinear concepts. First, nonlinear behavior requires different ways of looking at vibration responses, such as the phase plane. The Duffing equation is a good model for explaining the commonly encountered jump phenomenon that occurs during forced vibration tests. Concepts of self-excited vibrations and limit cycle behavior are introduced using the van der Pol equation. The Mathieu equation is used to introduce the concept of time variable parametric excitation. Finally, this section ends with a few words about chaotic behavior and how chaotic behavior may be identified in some instances.

3.5.1 The Phase Plane

The *phase plane* is a plot of velocity $\dot{x}$ versus displacement x. We know that the undamped oscillation of a linear system can be written as

$$x = A \sin(\omega_n t) \qquad (3.5.1)$$

so that the corresponding velocity is given by

$$\dot{x} = \omega_n A \cos(\omega_n t) \qquad (3.5.2)$$

A plot of $\dot{x}$ versus x is shown in Fig. 3.5.1a. It is evident that point A corresponds to $\omega_n t = 0$, B to $\omega_n t = \pi/2$, C to $\omega_n t = \pi$, and D to $\omega_n t = 3\pi/2$, so that a clockwise moving point P traces out an enclosed and

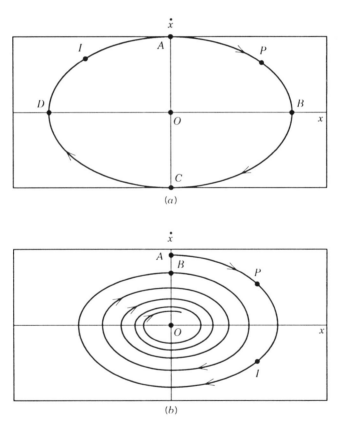

Fig. 3.5.1. (*a*) Phase plane plot for undamped sinusoidal motion. (*b*) Phase plane plot for damped sinusoidal motion. Both curves start with the initial conditions at point A.

repeating elliptical path. This phase plane plot represents a stable center point O that corresponds to a static equilibrium point of minimum potential energy.

Similarly, we know that the damped oscillation is described by

$$x = A e^{-\zeta \omega_n t} \sin(\omega_d t) \qquad (3.5.3)$$

so that

$$\dot{x} = A e^{-\zeta \omega_n t}[-\zeta \omega_n \sin(\omega_d t) + \omega_d \cos(\omega_d t)] \qquad (3.5.4)$$

Equations (3.5.3) and (3.5.4) give the decaying spiral path shown in Fig. 3.5.1b, where point P moves in the clockwise direction as before. The spiral path is due to energy being dissipated during each cycle so that when the process starts at point A, it is seen to decay to point B after one cycle. It is clear that Fig. 3.5.1b represents a stable system so that point O is a stable *focus point* and is the static equilibrium position with minimum potential energy. If the phase plane plot shows point P to move around point O along a continuously changing path, then the system may be unstable. The mark of a stable system is for point P to eventually enclose point O with a repeatable path.

3.5.2 The Simple Pendulum

The simple pendulum in Fig. 3.5.2a has a mass moment of inertia I about its pivot axis at O and a mass center at G that is located a distance b from O. θ is the coordinate describing the pendulum's position relative to the vertical line. The equation of motion for this system is given by

$$I\ddot{\theta} + mgb \sin(\theta) = 0 \qquad (3.5.5)$$

For small angles $\sin(\theta) \cong \theta$, giving a linear system response. Equation (3.5.5) can be written as

$$\ddot{\theta} + \omega_0^2 \sin(\theta) \cong \ddot{\theta} + \omega_0^2 \left(\theta - \frac{\theta^3}{6}\right) = 0 \qquad (3.5.6)$$

where ω_0 is the system's small angle natural frequency and the sine function is approximated by the first two terms of its series expansion. Equation (3.5.6) can be written in a more general form as

$$\ddot{\theta} + \omega_0^2 \theta + \alpha \theta^3 = 0 \qquad (3.5.7)$$

where $\alpha = -\omega_0^2/6$ in this case. Equation (3.5.7) is known as the *free*

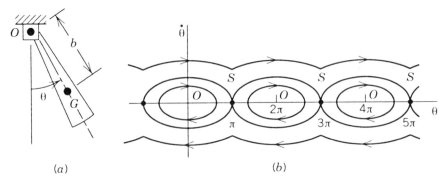

Fig. 3.5.2. (a) Simple pendulum. (b) Its phase plane for small, medium, and large amplitude motions.

undamped Duffing equation. When α is negative, Eq. (3.5.7) represents a softening nonlinear spring, while a positive α represents a hardening spring.

Equation (3.5.7) is often solved approximately by using Lindstedt's perturbation method.[7] For initial conditions of $\theta(0) = \theta_0$ and $\dot{\theta}(0) = 0$, the first perturbation gives

$$\theta = \theta_0 \cos(\omega t) + \frac{\omega_0^2 \theta_0^3}{192\omega^2}(\cos(\omega t) - \cos(3\omega t)) \qquad (3.5.8)$$

for motion and

$$\omega^2 = \omega_0^2\left(1 - \frac{\theta_0^2}{6}\right) \qquad (3.5.9)$$

for natural frequency ω. This approximate perturbation solution shows that a free oscillation has at least two frequency components, ω and 3ω in Eq. (3.5.8). Frequency ω is dependent on motion amplitude; its value decreases with increasing amplitude of motion as shown in Eq. (3.5.9). The generation of higher frequency harmonics (called *super harmonics*) is a characteristic of nonlinear systems. These super harmonics usually are integer multiples of the fundamental frequency as obtained in this example. Nonlinear systems are also known to generate *subharmonics*; namely, ω/n where n is an integer.

---

[7]See S. S. Rao, *Mechanical Vibrations*, 2nd ed., Addison-Wesley, Reading, MA, 1990, Chapter 13.

The simple pendulum described by Eq. (3.5.5) has an interesting phase plane, as shown in Fig. 3.5.2b. It has multiple equilibrium points at O, located at multiples of 2π as shown. However, it also has a separation point S that occurs at odd multiples of π. The separation points correspond to the pendulum being at the top of its swing. The curve that passes through points S is called the *separatrix*, for it separates oscillatory motion from a continuous motion where θ increases or decreases with time as the pendulum swings around and around and around. It is clear that the phase plane can give valuable information about a system's behavior.

3.5.3 The Duffing Equation of Forced Vibration

Duffing[8] studied the equation of motion of a mass on a nonlinear cubic hardening or softening spring subjected to harmonic excitation; the equation he devised is now known as the *Duffing equation*. This equation is described by

$$m\ddot{x} + c\dot{x} + kx \pm \mu x^3 = F\cos(\omega t) \qquad (3.5.10)$$

The undamped normalized form of Eq. (3.5.10) is most often solved so that it becomes

$$\ddot{x} + \omega_0^2 x \pm \alpha x^3 = P\cos(\omega t) \qquad (3.5.11)$$

It is obvious that Eq. (3.5.11) is Eq. (3.5.10) divided by mass m, ω_0 is the linear system natural frequency, and $P = F/m$. Equation (3.5.11) is usually solved using the *method of iteration*. The first solution is assumed to be of the form

$$x_0 = A\cos(\omega t) \qquad (3.5.12)$$

where A must be determined. When this solution is substituted into Eq. (3.5.11) and the second derivative expression is integrated twice, the first iterative approximation becomes

$$x_1 = \frac{1}{\omega^2}(\omega_0^2 A \pm 0.75\alpha A^3 - P)\cos(\omega t) \pm \cdots \qquad (3.5.13)$$

when higher order terms are neglected. Duffing argued that the amplitudes

---

[8]G. Duffing, *Erwugene Schwingungen bei veranderlicher Eigenfrequenz*, F. Vieweg u. Sohn, Braunschweig, Germany, 1918.

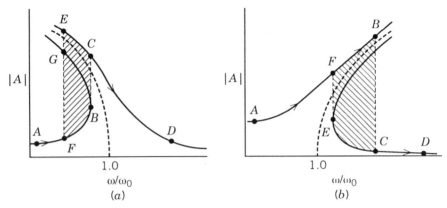

Fig. 3.5.3. Forced response showing jump phenomena. (a) For softening spring. (b) For hardening spring.

in Eqs. (3.5.12) and (3.5.13) must be the same. Thus amplitude A must satisfy the relationship of

$$y = \mp \frac{0.75\alpha A^3}{\omega_0^2} = \left(1 - \frac{\omega^2}{\omega_0^2}\right) A - \frac{P}{\omega_0^2} \qquad (3.5.14)$$

for a given values of ω_0, ω, α, and P. Amplitudes A can be determined by plotting the cubic curve as a function of A for given values of α and reference frequency ω_0. Then plotting the linear function for a given value of P and various values ω/ω_0 shows that the intersection of these two curves gives the values of A that satisfy Eq. (3.5.14). This procedure produces the amplitude of response curves shown in Fig. 3.5.3 for both softening and hardening springs. The central dashed line corresponds to the system's natural frequency and is obtained from Eq. (3.5.14) with $P = 0$. Thus

$$\frac{\omega^2}{\omega_0^2} = 1 \pm \frac{0.75\alpha A^2}{\omega_0^2} \qquad (3.5.15)$$

which agrees with Eq. (3.5.9).

The response of this simple system shows an interesting behavior called the *jump phenomenon*. For the softening spring in Fig. 3.5.3a, we start at point A. As the frequency increases, the amplitude increases up to point

B, where a sudden jump occurs up to point C. Then, as the frequency increases, we move along the curve toward point D. When the frequency decreases, we move from point D up along the curve past point C to point E where, due to damping, the response drops out to point F. The curve's lower branch starting at point B to point G and beyond does not exist. This sudden jump phenomenon is a tip-off that nonlinear spring behavior is taking place. The hardening spring has a similar phenomenon with increasing frequency starting at point A, moving to point B where damping causes the dropping out to point C and then on to small amplitudes at point D. For decreasing frequencies, the jump occurs at point E to point F and then the response follows along the curve to point A.

The sudden dropout or growth in vibration at the jump phenomenon is totally surprising when observed on a time history record. The signal can drop almost instantly by over an order of magnitude (a factor of 10) within a one or two cycles. One's first impulse is to think that the test equipment has failed.

Example: A demonstration of typical nonlinear behavior is illustrated in Fig. 3.5.4. The test item can be modeled as mass m_1 attached to a light mass m_2 by a coulomb friction mechanism. Mass m_2 is attached to spring k, which is attached to the device base as shown in Fig. 3.5.4a. The device's base was attached to the armature of a vibration exciter for test purposes. The exciter was swept through a frequency range of 20 Hz to 1 kHz with a sinusoidal input at a level of 1 g pk while the resulting output acceleration level of mass m_1 was measured and is shown in Fig. 3.5.4b, where a peak output acceleration of 14.1 g_{RMS} occurred at 600 Hz. A careful examination of this response curve shows a rather sharp increase in amplitude with frequency below the peak and a slightly shallow slope above resonance compared to the usual near symmetry of a classical resonance peak. Then the input level was increased to 5.0 g pk and the frequency was swept up through the 400 to 800 Hz frequency range and then back down again with the results shown in Fig. 3.5.4c. A sudden jump in response occurred at about 559 Hz with a peak amplitude of 26.98 g_{RMS} on the increasing frequency sweep. The sudden jump occurred at about 522 Hz at a level of 16.97 g_{RMS} on the decreasing frequency sweep. Clearly, the coulomb friction played very important part in this nonlinear behavior. Continual testing caused the friction to be reduced due to wear so that the test could not be repeated. The final result was that the designers had to reevaluate their design and its environment.

110   VIBRATION CONCEPTS

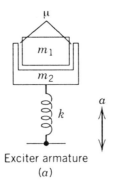

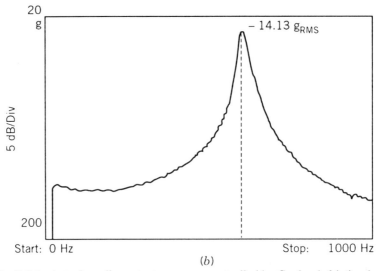

Fig. 3.5.4. Actual nonlinear test response controlled by Coulomb friction between masses. (*a*) Model of test system. (*b*) 1 g pk sinusoidal input showing slight softening characteristics. (*c*) 5 g pk sinusoidal input with frequencies sweeping up and down.

3.5.4 The van der Pol Equation and Limit Cycles

Van der Pol[9] studied the effect of negative damping with amplitude modification. The van der Pol equation is described by

$$\ddot{x} - \alpha(1 - x^2)\dot{x} + x = 0 \qquad (3.5.16)$$

---

[9]B. van der Pol, "Relaxation Oscillations," *Philosophical Magazine*, Vol. 2, 1926, pp. 978–992.

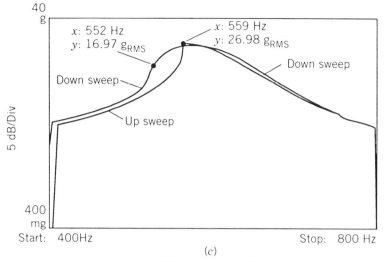

Fig. 3.5.4. (*Continued*).

where parameter α controls the velocity term's significance. When α is small ($\alpha < 0.1$), the solution to Eq. (3.5.16) is approximately

$$x \cong 2\cos(t) + \cdots \qquad (3.5.17)$$

when very small higher harmonic terms are neglected. This means that the phase plane plot is an ellipse, as shown in Fig. 3.5.5a. Equation (3.5.16) has an interesting characteristic behavior called a *limit cycle*. When amplitude A is less than 2, the velocity term pumps energy into the system. When amplitude A is greater than 2, the velocity term removes energy from the system. Only when amplitude A is 2 will it remain constant. Thus for any initial condition that starts inside the limit cycle, the motion spirals out to the limit cycle, while for any initial condition that starts outside the limit cycle, the motion decays until it reaches the limit cycle. The van der Pol equation is the underlying principle for building electronic oscillators.

When the value of α is larger than 10, a different kind of motion called *relaxation oscillation* occurs, as shown in Fig. 3.5.5b. In this case (starting at point A), the velocity remains a nearly constant positive value over most of the displacement until point B, where the velocity changes very quickly to a negative value at point C, remains at a nearly constant negative value until point D is reached, and then changes quickly to the positive value at point A. The motion quickly becomes that of the limit cycle, depending on initial conditions being inside or outside of the limit cycle.

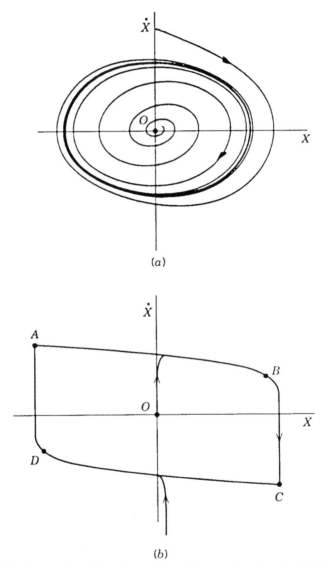

Fig. 3.5.5. Phase plane plots for van der Pol equation showing limit cycles. (*a*) For $\alpha = 0.1$, which produces sinusoidal motion. (*b*) For $\alpha = 10$, which produces relaxation motion.

3.5.5 The Mathieu Equation

Mathieu type equations occur in a number of simple physical systems where the differential equation of motion coefficients are time dependent under external control. For example, consider the simple inverted pendulum of length l and mass m shown in Fig. 3.5.6a where pivot O has vertical sinusoidal motion (source of external control). The corresponding differential equation of motion is

$$\ddot{\theta} + \frac{g}{l}\left(-1 + \frac{\omega^2 Y}{g}\cos(\omega t)\right)\theta = 0 \qquad (3.5.18)$$

where it is evident that θ's "spring rate parameter" is time dependent. Obviously, when $\omega = 0$, the pendulum falls down but it has been demonstrated[10] both theoretically and experimentally that the pendulum will oscillate about the vertical position in a stable manner for certain values of ω.

Equation (3.5.18) can be written in the form of

$$\frac{d^2\theta}{d\lambda^2} + \{a - 2b\cos(2\lambda)\}\theta = 0 \qquad (3.5.19)$$

which is known as the Mathieu equation. The regions of stability and instability of Eq. (3.5.19) are shown in Fig. 3.5.6b.

3.5.6 Chaotic Vibrations

An entire new field of dynamic research has developed over the past 25 years that is generally referred to as chaotic motion or chaotic vibrations. Nonlinear system relationships are required to generate chaotic motion. Past terminology often spoke of a system as being "unstable" under certain conditions. Often the case is that under certain conditions a region of operation is reached where a nonlinear phenomenon can cause strange "unstable" behavior. This unstable behavior is often a deterministic process, even though the resulting motion may appear to be random.

How can chaotic vibrations be distinguished from random vibrations? Moon[11] suggests a number of items that help identify when nonperiodic or chaotic motions may be present. First, can a nonlinear element be identified in the system? This is important for a linear system is not

---

[10] J. P. den Hartog, *Mechanical Vibrations*, 4th ed., McGraw-Hill, New York, 1956.
[11] See F. C. Moon, *Chaotic Vibrations: An Introduction to Chaotic Dynamics for Applied Scientists and Engineers*, John Wiley and Sons, New York, 1987, chapter 2.

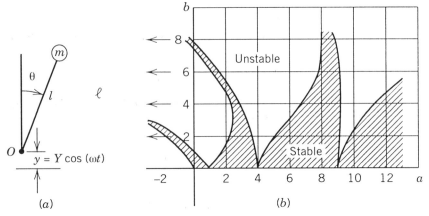

Fig. 3.5.6. Mathieu equation. (*a*) Inverted pendulum, the system that generates time dependent parameters. (*b*) Plot showing stable and unstable regions for parameters *a* and *b*.

chaotic; only nonlinear systems are. Second, are there sources of random input into the system? If there are no known sources of random input, but the output appears to be random by having a broad frequency spectrum, then there is a reasonable chance that some kind of chaotic behavior is taking place. Third, look at the phase plane history. The interpretation of these plots is beyond this book, but generally chaotic phase plane plots traverse a wide region of phase space instead of a fairly well defined curve. A further indication of chaotic motion is the presence of many frequency components below the excitation frequency. The upshot of this subsection is that we all need to study chaotic concepts and phenomenon more carefully in order to clearly identify when we are dealing with either random or chaotic processes. Moon's book on chaotic vibration is a good place to start.

3.6 THE TWO DOF VIBRATION MODEL

In Sections 3.2 and 3.3, a single DOF vibration model was introduced to describe a structure's dynamic behavior in the simplest possible terms in order to establish concepts of natural frequency, damping, resonance, and so on. The more complicated two DOF model is explored in this section in order to establish additional important vibration response concepts. Such concepts as multiple natural frequencies and natural mode shapes that make up modal analysis as well as multiple FRF functions are helpful in understanding vibration testing. Both driving point and transfer FRF models are developed in two ways; a direct solution method and a modal

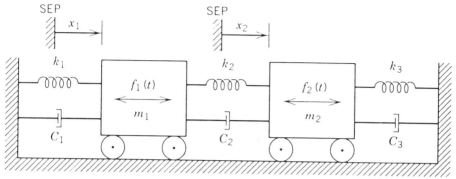

Fig. 3.6.1. Two DOF model showing system parameters. (From J. W. Dally, W. F. Riley, and K. G. McConnell, *Instrumentation for Engineering Measurements*, 2nd ed., copyright © 1993 by John Wiley & Sons, New York. Reprinted by permission.)

synthesis method. Finally, these two formulation methods are compared in order to establish their equivalence in describing dynamic behavior.

3.6.1 Equations of Motion

A two DOF model consisting of two masses (m_1 and m_2), three dampers (c_1, c_2, and c_3), three springs (k_1, k_2, and k_3), and two external excitation forces ($f_1(t)$ and $f_2(t)$) is shown in Fig. 3.6.1. Two coordinates (x_1 and x_2) are required to describe the position of the two masses relative to their respective static equilibrium positions (SEP). The minimum components required to have a two DOF system are two masses, one spring k_2, and one damper c_2, so that all other parameters in Fig. 3.6.1 can be set equal to zero. A special case, called *semi-definite system*, occurs when springs k_1 and k_3 and dampers c_1 and c_3 are zero. In this situation, the masses can have a common rigid body motion superimposed upon any vibration that may occur. Common examples are rotating machinery, a truck and trailer, a train (many degrees of freedom), a force transducer attached to a structure, and so on.

Each mass in Fig. 3.6.1 has its own differential equation of motion based on Newton's second law of motion. These equations are[12]

$$(m_1)\ddot{x}_1 + (c_1 + c_2)\dot{x}_1 + (-c_2)\dot{x}_2 + (k_1 + k_2)x_1 + (-k_2)x_2 = f_1(t)$$
$$(m_2)\ddot{x}_2 + (-c_2)\dot{x}_1 + (c_2 + c_3)\dot{x}_2 + (-k_2)x_1 + (k_2 + k_3)x_2 = f_2(t)$$

(3.6.1)

---

[12] For example, see W. T. Thomson, *Theory of Vibration with Applications*, 4th ed., Prentice Hall, Englewood Cliffs, NJ, 1988.

116 VIBRATION CONCEPTS

Each equation represents a balance of inertia, damping, stiffness, and excitation forces that act on a given mass at each instant in time. We can write these equations in a more general form as

$$m_{11}\ddot{x}_1 + c_{11}\dot{x}_1 + c_{12}\dot{x}_2 + k_{11}x_1 + k_{12}x_2 = f_1(t)$$
$$m_{22}\ddot{x}_2 + c_{21}\dot{x}_1 + c_{22}\dot{x}_2 + k_{21}x_1 + k_{22}x_2 = f_2(t)$$
(3.6.2)

The double-subscripted terms represent the mass, damping, and stiffness constants that are defined in Eq. (3.6.1) by the bracketed coefficients on x_1 and x_2 and their time derivatives. Equation (3.6.2) can be written in matrix form as

$$[m]\{\ddot{x}\} + [c]\{\dot{x}\} + [k]\{x\} = \{f\}$$
(3.6.3)

Mass matrix $[m]$, damping matrix $[c]$, and stiffness matrix $[k]$ are defined by

$$[m] = \begin{bmatrix} m_{11} & m_{12} \\ m_{21} & m_{22} \end{bmatrix} = \begin{bmatrix} m_1 & 0 \\ 0 & m_2 \end{bmatrix}$$

$$[c] = \begin{bmatrix} c_{11} & c_{12} \\ c_{21} & c_{22} \end{bmatrix} = \begin{bmatrix} (c_1 + c_2) & (-c_2) \\ (-c_2) & (c_2 + c_3) \end{bmatrix}$$
(3.6.4)

$$[k] = \begin{bmatrix} k_{11} & k_{12} \\ k_{21} & k_{22} \end{bmatrix} = \begin{bmatrix} (k_1 + k_2) & (-k_2) \\ (-k_2) & (k_2 + k_3) \end{bmatrix}$$

The column matrices in Eq. (3.6.3) are called *displacement vector* $\{x\}$, *velocity vector* $\{\dot{x}\}$, *acceleration vector* $\{\ddot{x}\}$, and *force excitation vector* $\{f\}$, respectively. The $[m]$, $[c]$, and $[k]$ matrices are square with dimensions $N \times N$, while the vectors have dimensions of $1 \times N$ for an N DOF model. In this book, only a two DOF model is manipulated; but when appropriate, the results are generalized for an N DOF system.

We are interested in both free undamped vibration and damped steady-state forced vibration responses. An efficient method to determine these responses is to assume that both excitation and response have a phasor form defined by $f_i(t) = F_i e^{j\omega t}$ and $x_i(t) = X_i e^{j\omega t}$. Then Eqs. (3.6.2) reduce to

$$D_{11}X_1 + D_{12}X_2 = F_1$$
$$D_{21}X_1 + D_{22}X_2 = F_2$$
(3.6.5)

when $e^{j\omega t}$ terms are canceled. The *dynamic stiffness* is defined by

$$D_{ip} = k_{ip} - m_{ip}\omega^2 + jC_{ip}\omega \qquad (3.6.6)$$

where it is understood that D_{ip} is a function of frequency ω. The viscous damping term can be replaced by the structural damping equivalent in all equations of this section. We use the viscous damping model for developmental purposes. Equation (3.6.5) is an algebraic equation containing the system's frequency dependent dynamic characteristics. Equation (3.6.6) shows how stiffness, mass, damping, and frequency affect each dynamic stiffness term.

3.6.2 Undamped Natural Frequencies and Mode Shapes

We are interested in how the system vibrates naturally; that is, in the undamped natural frequencies and in the corresponding natural motion that goes with each natural frequency. For the undamped case, we let both the damping and excitation terms be zero in Eq. (3.6.5) so that these equations become

$$D'_{11}X'_1 + D'_{21}X'_2 = 0$$
$$D'_{21}X'_1 + D'_{22}X'_2 = 0 \qquad (3.6.7)$$

Undamped free vibration amplitudes (X'_1 and X'_2 in this case) are nonzero when the undamped dynamic stiffness determinant Δ' is zero; that is,

$$\Delta' = D'_{11}D'_{22} - D'_{12}D'_{21} = m_1 m_2(\omega_1^2 - \omega^2)(\omega_2^2 - \omega^2) \qquad (3.6.8)$$

The roots of Eq. (3.6.8) are called *natural frequencies* where ω_1 is the *fundamental natural frequency*, ω_2 is the second natural frequency, and so on. Equation (3.6.8) is called the *characteristic frequency equation* and can be written in terms of its roots ω_1 and ω_2 as shown. It is seen that this equation is fourth order in ω and second order quadratic in ω^2. Generally, ω has an order of $2N$ for an N DOF system. The values of ω are real with equal $\pm$ values for each natural frequency so that the solutions are real sinusoidal functions in accordance with the assumed phasor format.

The values of X'_1 and X'_2 in Eq. (3.6.7) must satisfy a ratio for each natural frequency so that

$$\frac{X'_2}{X'_1} = -\frac{D'_{11}}{D'_{12}} = -\frac{D'_{21}}{D'_{22}} = u_{2i} \quad \text{for } i = 1 \text{ and } 2 \qquad (3.6.9)$$

where u_{2i} is the *modal parameter* for the second coordinate and the *i*th natural frequency relative to amplitude X'_1. Equation (3.6.9) shows how the motion of X_2' is related to the value of X_1' for each natural frequency. These amplitude ratios describe the *mode shape* (shape of motion) that must occur at that natural frequency.

We assumed that $x_i = X_i e^{j\omega t}$ in formulating Eqs. (3.6.5) and (3.6.7). We also found that there are two values of ω (since we are dealing with a two DOF system) that satisfy the characteristic frequency equation and there are two corresponding modal parameters for each coordinate X_i. Thus the undamped free vibration response can be written as

$$\begin{Bmatrix} x_1 \\ x_2 \end{Bmatrix} = B_1 \begin{Bmatrix} u_{11} \\ u_{21} \end{Bmatrix} e^{j\omega_1 t} + B_2 \begin{Bmatrix} u_{12} \\ u_{22} \end{Bmatrix} e^{j\omega_2 t} \qquad (3.6.10)$$

↑ first mode   ↑ second mode

The $\{u\}$ vectors are called *modal vectors* and define the natural vibration's mode shape, one for each natural frequency. The mode shape is not affected by multiplying a vector by a constant value. Thus constants B_1 and B_2 are the *modal amplitudes* of motion: their values are dependent on the initial conditions (values of x_1, $\dot{x}_1$, x_2, and $\dot{x}_2$ at time equal to zero) as well as on the scale factor used with each modal vector. Many different schemes are used to normalize modal vectors since multiplication by a constant does not change the *mode shape*. Thus we can change the numerical value of a modal vector by multiplying it by a number, but we will not change the mode's actual shape.

Modal vectors can be used to form a *modal matrix* such that

$$[u] = \begin{bmatrix} \begin{Bmatrix} u_{11} \\ u_{21} \end{Bmatrix} \begin{Bmatrix} u_{12} \\ u_{22} \end{Bmatrix} \end{bmatrix} = \begin{bmatrix} u_{11} & u_{12} \\ u_{21} & u_{22} \end{bmatrix} \qquad (3.6.11)$$

↑ second mode
↑ first mode

Modal parameter u_{ip} is the mode shape value for coordinate i at natural frequency ω_p; that is, the first subscript i refers to the coordinate while the second subscript refers to the pth natural frequency. Thus all modal parameters in a given column have the same second subscript since these parameters refer to the same natural frequency. Note that some authors write the modal parameter as

$$u_{ip} = u_i^p \qquad (3.6.12)$$

where superscript p refers to the pth natural frequency. For the undamped case, the modal vectors are real valued so that all mass motions are either in phase ($+u$'s) or out of phase ($-u$'s).

It can be shown that the modal matrix $[u]$ and its transpose $[u]^T$ have important orthogonality properties that *diagonalize* both mass and stiffness matrices; that is,

$$[u]^T[m][u] = \begin{bmatrix} \ddots & \\ & m \\ & & \ddots \end{bmatrix} \quad \text{and} \quad [u]^T[k][u] = \begin{bmatrix} \ddots & \\ & k \\ & & \ddots \end{bmatrix} \quad (3.6.13)$$

so that only diagonal values are nonzero. This orthogonality property is important in developing a modal response model.

3.6.3 Steady-State Forced Vibration Response (Direct Method)

The forced vibration response can be obtained in a number of ways. The direct method is to use Eq. (3.6.5) where X_1 and X_2 are obtained directly by using Cramer's rule. The result is

$$x_1 = \left(\frac{D_{22}}{\Delta}\right) F_1 e^{j\omega t} + \left(\frac{-D_{12}}{\Delta}\right) F_2 e^{j\omega t}$$
$$x_2 = \left(\frac{-D_{21}}{\Delta}\right) F_1 e^{j\omega t} + \left(\frac{D_{11}}{\Delta}\right) F_2 e^{j\omega t} \quad (3.6.14)$$

where determinant Δ is defined by

$$\Delta = D_{11}D_{11} - D_{12}D_{21} \quad (3.6.15)$$

Determinant Δ has both real and imaginary parts and is a minimum whenever ω is a natural frequency, thus causing the real part to be zero that creates a peak or resonant response. Equation (3.6.14) can be written in terms of receptance FRFs as

$$x_1 = H_{11}F_1 e^{j\omega t} + H_{12}F_2 e^{j\omega t}$$
$$x_2 = H_{21}F_1 e^{j\omega t} + H_{22}F_2 e^{j\omega t} \quad (3.6.16)$$

where again it is understood that the H_{pq}'s are functions of frequency ω and have special names. H_{pp} is called the *driving point receptance* for mass p (where p can be either 1 or 2 in this case) and H_{pq} is called the *transfer receptance* and represents the motion at location p due to excitation at location q. Obviously, driving point receptance is a special case of transfer receptance when the subscripts are the same.

3.6.4 Steady-State Forced Response (Modal Method)

The mode shapes can be used to develop a modal response to a general excitation. This method assumes that a forced vibration response can be obtained by summing modal responses according to a coordinate transformation from the actual space coordinates x_i to *generalized coordinates* $q_p(t)$ such that

$$\{x(t)\} = [u]\{q(t)\} \tag{3.6.17}$$

This is done by inserting Eq. (3.6.17) into Eq. (3.6.3) and multiplying by $[u]^T$ to obtain

$$[u]^T[m][u]\{\ddot{q}\} + [u]^T[c][u]\{\dot{q}\} + [u]^T[k][u]\{q\} = [u]^T\{F\} \tag{3.6.18}$$

If it is assumed that damping is proportional to either mass and/or stiffness such that[13]

$$[c] = \alpha[m] + \beta[k] \tag{3.6.19}$$

where α and β are constants, then, by virtue of orthogonality as defined in Eq. (3.6.13), Eq. (3.6.18) becomes a set of N ($N = 2$ in this case) uncoupled differential equations described by

$$\begin{bmatrix} m \\ \end{bmatrix}\{\ddot{q}\} + \begin{bmatrix} c \\ \end{bmatrix}\{\dot{q}\} + \begin{bmatrix} k \\ \end{bmatrix}\{q\} = \{Q\} \tag{3.6.20}$$

so that the pth natural mode is governed by

$$m_p\ddot{q}_p + c_p\dot{q}_p + k_pq_p = Q_p \tag{3.6.21}$$

where k_p is the pth modal stiffness.
 m_p is the pth modal mass.
 c_p is the pth modal damping.
The pth *generalized excitation force* Q_p is given by

$$Q_p = u_{1p}F_1 + u_{2p}F_2 = \sum_{k=1}^{N} u_{kp}F_k \tag{3.6.22}$$

---

[13]When Eq. (3.6.19) is not valid one obtains complex mode shapes. See: L. D. Mitchell, "Complex Modes: A Review," *Proceedings of the 8th International Modal Analysis Conference*, Kissimmee, FL, 1990, pp. 891–899.

Equation (3.6.22) shows that force F_k cannot excite the pth mode if modal parameter u_k is zero, a result that is valid for static as well as dynamic loads. The generalized excitation force concept is important when exciting structures at one or more locations since either a single load or a combination of loads can cause Q_p to be zero. Regardless of the cause, when Q_p is zero, the pth mode will be absent from any vibration data. The particular combination of loads is masked by modal parameters u_{kp} that are unknown in an unknown structure until after the tests are completed. Consequently, significant information is often missed unless more than one excitation point is used, one point at a time.

Equation (3.6.21) is a second order differential equation with constant coefficients. It has a steady-state solution given by

$$q_p = \frac{Q_p}{D_p} e^{j\omega t} = \sum_{k=1}^{N} \left(\frac{u_{kp}}{D_p}\right) F_k e^{j\omega t} \tag{3.6.23}$$

The *modal dynamic stiffness* D_p in Eq. (3.6.23) is given by

$$D_p = k_p - m_p \omega^2 + jc_p \omega = k_p(1 - r_p^2 + j2\zeta_p r_p) \tag{3.6.24}$$

where r_p is the pth dimensionless modal frequency ratio (ω/ω_p).
ζ_p is the pth dimensionless modal damping ratio.
Equation (3.6.24) shows that the pth modal natural frequency is

$$\omega_p = \sqrt{\frac{k_p}{m_p}} \tag{3.6.25}$$

and the pth *modal damping ratio* is

$$\zeta_p = \frac{c_p}{2\sqrt{m_p k_p}} = \frac{\eta_p}{2} \tag{3.6.26}$$

where η_p is the pth mode loss factor. Recall that natural frequency is an invariant dynamic property of the system, a property that is independent of coordinate system and/or units employed. A closer examination of Eq. (3.6.13) shows that both the modal stiffness k_p and modal mass m_p are proportional to the product of the modal vector terms. Thus the absolute values for k_p and m_p change with the way the mode shape is scaled, but the natural frequency remains unchanged since both m_p and k_p have the same scaling factors. It is common practice to normalize both theoretical and experimental mode shapes so that the modal masses are unity. Then, performance of the calculations in Eq. (3.6.13) can be used to judge the adequacy of the mode shapes since the off diagonal values will not be zero unless precise modal vectors are used. Usually, we do not have precise modal vectors, only estimates. It has been found that the stiffness matrix is highly sensitive to errors in the modal vectors compared to the mass matrix.

If we insert Eq. (3.6.23) into Eq. (3.6.17), we obtain the steady-state modal responses as

$$x_1 = \left[u_{11}\left(\frac{u_{11}}{D_1}\right) + u_{12}\left(\frac{u_{12}}{D_2}\right)\right]F_1 e^{j\omega t} + \left[u_{11}\left(\frac{u_{21}}{D_1}\right) + u_{12}\left(\frac{u_{22}}{D_2}\right)\right]F_2 e^{j\omega t}$$

$$x_2 = \left[u_{21}\left(\frac{u_{11}}{D_1}\right) + u_{22}\left(\frac{u_{12}}{D_2}\right)\right]F_1 e^{j\omega t} + \left[u_{21}\left(\frac{u_{21}}{D_1}\right) + u_{22}\left(\frac{u_{22}}{D_2}\right)\right]F_2 e^{j\omega t}$$

(3.6.27)

It is evident in Eq. (3.6.27) that the bracketed terms represent the input-output FRF's, i.e., H_{ir}. It is also seen that the dominant term in each bracket is controlled by the denominator term (modal dynamic stiffness), D_k being its smallest at modal resonance (see Eq. (3.6.24)).

3.6.5 Comparison of Direct and Modal Response FRFs

We have obtained the input output relationships for the two DOF system in Fig. 3.6.1 by two methods, direct and modal. We know that both methods should give the same results. Comparing the results in Eqs. (3.6.14), (3.6.16), and (3.6.27) shows that the FRFs can be expressed as

$$H_{11} = \left(\frac{D_{22}}{\Delta}\right) = \left[u_{11}\left(\frac{u_{11}}{D_1}\right) + u_{12}\left(\frac{u_{12}}{D_2}\right)\right] \qquad (3.6.28)$$

$$H_{12} = \left(\frac{-D_{12}}{\Delta}\right) = \left[u_{11}\left(\frac{u_{21}}{D_1}\right) + u_{12}\left(\frac{u_{22}}{D_2}\right)\right]$$

$$H_{21} = \left(\frac{-D_{21}}{\Delta}\right) = \left[u_{21}\left(\frac{u_{11}}{D_1}\right) + u_{22}\left(\frac{u_{12}}{D_2}\right)\right]$$

(3.6.29)

$$H_{22} = \left(\frac{D_{11}}{\Delta}\right) = \left[u_{21}\left(\frac{u_{21}}{D_1}\right) + u_{22}\left(\frac{u_{22}}{D_2}\right)\right] \qquad (3.6.30)$$

The first bracketed term in each of Eqs. (3.6.28) and (3.6.29), and (3.6.30) is the direct steady-state solution from Eq. (3.6.14), while the second bracketed term is the modal steady-state solution from Eq. (3.6.27). Note that $H_{12} = H_{21}$ in Eq. (3.6.29).

Equations (3.6.14) and (3.6.16), as well as (3.6.27) through (3.6.30), indicate that the response measured at one point is dependent on all excitation forces, modal parameters, and modal dynamic stiffness D_r. This explains why significantly different experimental results are obtained when a transducer is located at different points in the structure, since the modal parameter associated with the transducer location changes. Similarly, with a fixed transducer location, the output changes when the excitation forces change location. This result is both a curse and a blessing. The curse part is associated with obtaining only limited information when a single transducer is mounted on the structure. The blessing part comes from controlled experiments that allow us to measure the mode shapes, natural frequencies, and damping if we obtain sufficient information in an orderly manner.

In order to measure mode shapes, it is obvious from Eqs. (3.6.28) through (3.6.30) that a single excitation source is preferred; otherwise the output is dependent on several sets of modal parameters instead of simply one set. Under multiple excitation, cross coupling between responses, mode shapes, and excitation forces occurs. It will be shown later that the vibration exciter interacts significantly with the structure so this additional interaction must be accounted for as well. Otherwise, the experimental results can include the exciter as part of the structure.

Now let us look at how we might go about measuring a mode shape. Let us apply a single force excitation F_1 so that F_2 is zero. In addition, let the excitation frequency be ω_1. Then responses x_1 and x_2 (see H_{11} and H_{21} terms) reflect modal parameters u_{11} and u_{21}, since the ratio of modal parameter u_{11} divided by D_1 is common to both terms that multiply F_1 at frequency ω_1. This suggests that keeping an excitation force at location 1 and moving a motion transducer from location 1 to location 2, and so on, will reveal the first mode shape since modal parameters u_{i1} vary from location to location and describe the first modal vector. A similar result occurs when ω_2 is the excitation frequency, giving modal parameters u_{i2}. Similarly, F_1 can be zero while F_2 is the non-zero exciter force. For this case, it is seen that the same modal vectors are obtained when ω is equal to first ω_1 and then ω_2. However, no information is obtained if the excitation force is located at a node point, a point of zero response so that u_{ii} is zero. Thus we should repeat the measurement set on an unknown structure, since we do not know which points may be node points until the experiment is completed and the data analyzed.

For an N DOF system, the above ideas can be extended so that the

124 VIBRATION CONCEPTS

frequency response function H_{pr} becomes

$$H_{pr} = \sum_{k=1}^{N} u_{pk}\left(\frac{u_{rk}}{D_k}\right) = u_{p1}\left(\frac{u_{r1}}{D_1}\right) + u_{p2}\left(\frac{u_{r2}}{D_2}\right) + u_{p3}\left(\frac{u_{r3}}{D_3}\right) + \cdots \quad (3.6.31)$$

$$\text{first} \qquad \text{second} \qquad \text{third}$$
$$\text{mode} \qquad \text{mode} \qquad \text{mode}$$

where u_{pk} are the modal parameters and D_k is the modal dynamic stiffness. The ratio $[u_{rk}/D_k]$ reflects the efficiency of excitation force F_r at location r in exciting the kth natural mode. Modal parameter u_{pk} refers to the response efficiency at location p for the kth natural mode. Equation (3.6.31) is the basis of experimental modal analysis and shows that peak values at any natural frequency are contaminated by adjacent modal responses, due to the summation process. Ewins[14] studied the significance of leaving out some of the modes in Eq. (3.6.31) in computing a check on measurement adequacy. He showed that the modes measured are satisfactory but that other FRFs cannot be predicted or used as a quality check.

Example: In order to have a feel for how these equations and modal models are related to one another, let us consider a system with the following values. The mass, damping, and stiffness matrices are:

$$[m] = \begin{bmatrix} 1 & 0 \\ 0 & 2 \end{bmatrix} \quad [c] = \begin{bmatrix} 2 & 0 \\ 0 & 4 \end{bmatrix} \quad [k] = \begin{bmatrix} 8100 & -7200 \\ -7200 & 7200 \end{bmatrix} \quad (3.6.32)$$

where m, c, and k are expressed in a consistent set of units. The corresponding natural frequencies and modal matrix become

$$\omega_1 = 16.85 \text{ rad/s} \qquad \omega_2 = 106.9 \text{ rad/s} \qquad [u] = \begin{bmatrix} 1 & 1 \\ 1.086 & -0.461 \end{bmatrix} \quad (3.6.33)$$

Then, Eq. (3.6.13) gives the modal mass, damping, and stiffness matrices as

---

[14] D. J. Ewins, "On Predicting Point Mobility Plots from Measurements of Other Mobility Parameters," *Journal of Sound and Vibration*, Vol. 70, No. 1, 1980, pp. 69–75.

$$\begin{bmatrix} \ddots & & \\ & m & \\ & & \ddots \end{bmatrix} = \begin{bmatrix} 3.357 & 0 \\ 0 & 1.424 \end{bmatrix} \quad \begin{bmatrix} \ddots & & \\ & c & \\ & & \ddots \end{bmatrix} = \begin{bmatrix} 6.71 & 0 \\ 0 & 2.85 \end{bmatrix}$$

$$\begin{bmatrix} \ddots & & \\ & k & \\ & & \ddots \end{bmatrix} = \begin{bmatrix} 953 & 0 \\ 0 & 16300 \end{bmatrix}$$

(3.6.34)

Note that the values of $[c]$ are proportional to mass $[m]$ so that the modal damping matrix is diagonal and the modal vectors are real. These parameter values are used in Eqs. (3.6.28) and (3.6.29) to calculate typical driving point H_{11} and transfer H_{12} FRFs, as shown in Fig. 3.6.2.

The magnitude of the driving point FRF H_{11} is shown in Fig. 3.6.2a. H_{11} is seen to be the sum of two modes that form the combined response curve according to Eq. (3.6.31) for $N = 2$. These modes are of the same sign in the region below resonance ω_1, of opposite sign so they subtract in the region between resonances ω_1 and ω_2 since mode one is now out of phase, and then add negatively in a region above resonance ω_2 where both modes are now out of phase. The signs of each curve are shown in Fig. 3.6.2a.

We see that the peak at ω_1 is controlled by mode one while the peak at ω_2 is controlled by mode two. The modal damping ratios for these plots are $\zeta_1 = 0.059$ and $\zeta_2 = 0.009$, respectively. The corresponding dynamic range is about 2000 from valley to peak. A deep valley occurs in the region between ω_1 and ω_2 at point A. This valley corresponds to a test structure acting like a "dynamic absorber" to the excitation force; that is, the structure is vibrating in such a way as to force the driven point to have zero response. Driving point receptance always shows these valleys when the structure is grounded. This valley is located by the real part of D_{22} in Eq. (3.6.14) being zero and should occur at $\omega_A = 60$ rad/s for this example. The minimum value may not be shown in this graph since it is calculated digitally at distinct points, none of which may necessarily correspond precisely to ω_A. Digital calculations always have this problem of missing peaks and valleys when working with highly variable FRF graphs on both a theoretical or experimental basis.

The wide dynamic range of these FRFs from valleys to peaks can present measurement problems. At a valley, the motion signal is essentially zero while the force signal is large on a relative basis. Thus the motion signal may be full of noise since the instrument system is set up to measure the large peak motion amplitudes. On the other hand, we know the motion is large for a given amount of force at the peaks so that the force signal may be full of noise relative to the motion signal. The resonant peaks contain the most significant system information in terms

126 VIBRATION CONCEPTS

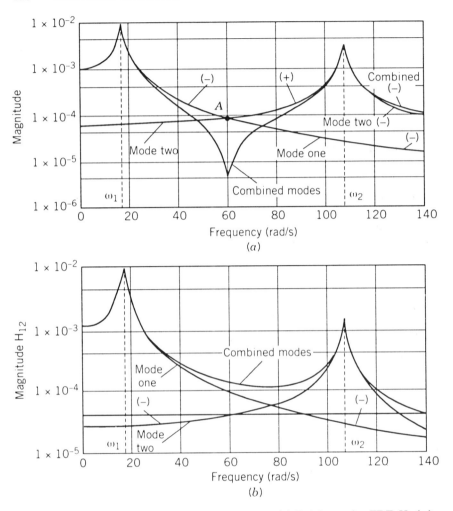

Fig. 3.6.2. Typical FRFs for a two DOF system. (*a*) Driving point FRF $H_{11}(\omega)$. (*b*) Transfer FRF $H_{12}(\omega)$.

of natural frequencies, mode shapes, and damping. Consequently, it is the force signal that is most easily corrupted when making these FRF measurements. In addition, we find in Chapter 4 that the force transducer may significantly interact with the test structure to cause additional measurement error around structural resonances, while in Chapters 6 and 7 we find that vibration exciters suffer a force drop-out phenomenon at test system resonance. All of these error sources contribute to measurement problems under the best of conditions.

The magnitude of the transfer FRF H_{12} is plotted in Fig. 3.6.2*b*. This

FRF does not have a deep valley in the combined curve where the first and second modes are added together. This absence of a valley is due to the fact that both modal contributions have the same sign in this region between ω_1 and ω_2.

The driving point FRF H_{11} can be calculated by either the direct method (using D_{22} and Δ; see Eq. (3.6.28)) or by the modal method (using u_{ip} and D_p terms). A comparison of these two calculation methods shows that the percent error is bounded by $\pm 3 \times 10^{-12}$. This close agreement is a measure of the computer's numerical accuracy since the theoretical error should be zero. Thus the two solution methods give the same results theoretically. In addition we see that the modal model is an effective way to view the dynamic behavior of a given structural system. However, we should keep in mind that errors can result when we truncate Eq. (3.6.31) and use only the first few modes and ignore the higher modes.[15]

It should be clear that successful vibration experiments rely on accurate experimental data to generate adequate FRFs from which the mode shapes, natural frequencies, and modal damping may be extracted. The most serious experimental problems occur at resonance.

3.7 THE SECOND ORDER CONTINUOUS VIBRATION MODEL

So far we have progressed from a single DOF system in Sections 3.3 and 3.4 to a multi DOF dynamic system in Section 3.6. The multi DOF system showed increasing complexity of motion, FRFs, and interaction. In this section, we examine the second order continuous system for additional insight concerning vibration response and how this response influences our testing procedures and results. In fact, the continuous system dynamic model allows us to see the influence of using a finite number of masses and coordinates to represent a real structure that is always a continuous system. Again, we start by looking at the governing differential equation of motion. Then, we look at the free undamped modes of vibration in order to see how the system wants to move naturally at each natural frequency. The modal model of forced vibration response is developed along with the response to different kinds of excitation force systems.

3.7.1 The Fundamental Equation of Motion

The three classical linear systems shown in Fig. 3.7.1 have a common partial differential equation of motion that describes their behavior. These systems are the taut string, axial rod (also linear springs come under this

---

[15]See previous reference to D. J. Ewins reference in footnote 12.

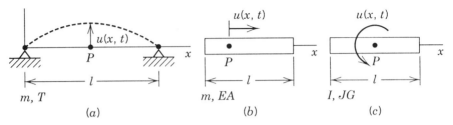

Fig. 3.7.1. Three simple second order continuous vibrating systems. (a) String. (b) Axial rod. (c) Torsion rod.

model), and torsion rod. The string model of length l is shown in Fig. 3.7.1a where $u(x, t)$ is the lateral string motion of point P at location x and time t. The string has mass per unit length m and constant tension force T. The axial rod model of length l is shown in Fig. 3.7.1b where $u(x, t)$ is the axial motion of point P at location x and time t. The rod has mass per unit length m and axial force of EA. The linear spring is the same as shown in Fig. 3.7.1b except that kl is its axial force. The torsional model of length l is shown in Fig. 3.7.1c where $u(x, t)$ is the cross-sectional rotation of point P at location x and time t. The rod has inertia per unit length of m and torsional moment JG.

The partial differential equation of motion for these systems can be written as[16]

$$m \frac{\partial^2 u}{\partial t^2} + C \frac{\partial u}{\partial t} - \frac{\partial}{\partial x}\left[K \frac{\partial u}{\partial x}\right] = f(x, t) \qquad (3.7.1)$$

inertia damping spring   Excitation

where C is the damping per unit length.

$f(x, t)$ is the excitation force or couple per unit length.

Equation (3.7.1) represents the dynamic force or torque balance that must exist at each point in these simple structures at each instant of time. It is seen that this dynamic force (torque) balance consists of inertia, damping (energy dissipation), spring (restoring force or moment), and excitation terms. These are the same types of terms that are present in single and multi DOF systems as well. It is the balance of these force or moment terms at any frequency that determines the structure's behavior even though Eq. (3.7.1) applies to every point in the structure.

When $f(x, t)$ is zero, the response described by Eq. (3.7.1) is that of a decaying transient vibration. When there is an excitation, the solution to

---

[16] See S. S. Rao, *Mechanical Vibrations*, 2nd ed., Addison-Wesley, Reading, MA, 1990, chapter 8.

THE SECOND ORDER CONTINUOUS VIBRATION MODEL

TABLE 3.7.1. Second Order System Parameters[a]

System	Stiffness K	Inertia m	Damping C
Taut string	Tension force T	Mass per unit length ρA	Damping force per unit length
Axial rod	Axial force EA	Mass per unit length ρA	Damping force per unit length
Linear spring	Axial force kl	Mass per unit length m_s/l	Damping force per unit length
Torsion rod	Torque couple JG	Inertial per unit length ρJ	Damping couple unit length

[a]E—Young's modulus; G—shear modulus; A—area; J—polar moment of inertia of area; k—spring constant, m_s—spring mass.

Eq. (3.7.1) gives the forced vibration response. The different physical parameters used for m, C, and k are summarized in Table 3.7.1.

The membrane's normal displacement $u(x, y, t)$ has an equation of motion similar to Eq. (3.7.1), given by

$$m\frac{\partial^2 u}{\partial t^2} + C\frac{\partial u}{\partial t} - T\nabla^2 u = f(x, y, t) \qquad (3.7.2)$$

inertia   damping   spring   excitation

where the Laplace operator ∇^2 is given by

$$\nabla^2 = \frac{\partial^2}{\partial x^2} + \frac{\partial^2}{\partial y^2} \qquad (3.7.3)$$

As written, Eq. (3.7.2) requires that tension T be the same in both the x and y directions. In Eq. (3.7.2), m is mass per unit area, C is the damping per unit area, and $f(x, y, t)$ is the excitation force per unit area. Equation (3.7.2) represents the force balance at each point on a per unit area basis. It is clear that a membrane has two spatial coordinates while the classical problems described by Eq. (3.7.1) depends on only one spatial coordinate. The following techniques can be applied to Eq. (3.7.2) as well as to Eq. (3.7.1). It turns out that there is a significant increase in complexity in the two dimensional case compared to the single dimensional case. However, we learn little more about basic response characteristics other than that multidimensional problems are a lot more work.

3.7.2 Separation of Space and Time Variables

A common method to determine the free undamped response of systems described by Eqs. (3.7.1) and (3.7.2) is to use the separation of variables approach. This is done by assuming that $C = f(x, t) = 0$ in Eq. (3.7.1) and that

$$u(x, t) = U(x)\eta(t) \qquad (3.7.4)$$

where $U(x)$ is the motion's spatial shape while $\eta(t)$ is the motion's time variation. When Eq. (3.7.4) is inserted into Eq. (3.7.1), two uncoupled second order differential equations result, which are given by

$$\ddot{\eta} + \left(\lambda^2 \frac{K}{m}\right)\eta = \ddot{\eta} + \omega^2 \eta = 0 \qquad (3.7.5)$$

for time function $\eta(t)$ where ω is the frequency of oscillation and

$$U'' + \lambda^2 U = 0 \qquad (3.7.6)$$

for space function $U(x)$ where the primes indicate differentiation with respect to x. The *constant of separation* λ^2 must be determined from the problem's boundary conditions. Frequency ω and separation constant λ are related to the system parameters m and K in Eq. (3.7.5) by

$$\omega = \lambda \sqrt{K/m} = \lambda c \qquad (3.7.7)$$

where c is the *wave speed*.

The solution to both Eqs. (3.7.5) and (3.7.6) is composed of sines and cosines so that Eq. (3.7.4) becomes

$$u(x, t) = \underbrace{[A_1 \cos(\lambda x) + A_2 \sin(\lambda x)]}_{\text{boundary conditions}} \underbrace{[B_1 \cos(\omega t) + B_2 \sin(\omega t)]}_{\text{initial conditions}} \qquad (3.7.8)$$

The problem's boundary conditions affect the spatial function and allow us to determine the values of λ for which these solutions exist. The initial conditions, along with the steady-state solution, determine constants B_1 and B_2.

Example: Now consider the simple example of a taut string, as shown in Fig. 3.7.1a. The boundary conditions are $u(0, t) = 0$ and $u(l, t) = 0$. Since these conditions must hold for all time, they are equivalent to $U(0) = 0$

and $U(l) = 0$. The first boundary condition gives $A_1 = 0$ while the second boundary condition requires that

$$A_2 \sin(\lambda l) = 0$$

If $A_2 = 0$, there is no vibration problem to consider, and we become unemployed. Thus the sine function must be zero. This happens whenever λl is a multiple of π so that we have an *eigenvalue* of

$$\lambda_p = \frac{p\pi}{l} \tag{3.7.9}$$

which gives the corresponding *natural frequencies* from Eq. (3.7.7) of

$$\omega_p = \lambda_p \sqrt{\frac{K}{m}} = \frac{p\pi}{l}\sqrt{\frac{K}{m}} \tag{3.7.10}$$

The corresponding spatial function $U(x)$ (*eigenfunction*) becomes

$$U_p(x) = \sin(\lambda_p x) = \sin\left(\frac{p\pi}{l}x\right) \tag{3.7.11}$$

$U_p(x)$ is the pth *natural mode shape* or simply *mode shape* and is the shape of motion for each natural frequency ω_p.

These mode shapes are shown in Fig. 3.7.2a for $p = 1, 2, 3,$ and 4. Note that all modes satisfy the boundary conditions and that the higher modes ($p > 1$) have *node points*, points of zero motion, located between the ends. The locations of these node points play a significant role in forced vibration response when the point excitation force is applied at a node point. In addition, the modal vectors for a multi DOF system correspond to evaluating Eq. (3.7.11) at certain values of x; say, at every $0.2l$ as shown by circles in Fig. 3.7.2a.

Problems begin to arise in describing the higher natural frequency mode shapes in a multi DOF model since there are too few points to adequately describe the mode shape as seen in Fig. 3.7.2a. This shortage of points is called *spatial aliasing* and is a consequence of using a finite number of masses and coordinates to describe a system that has nearly an infinite number of point masses, mode shapes, and natural frequencies. Thus a

132 VIBRATION CONCEPTS

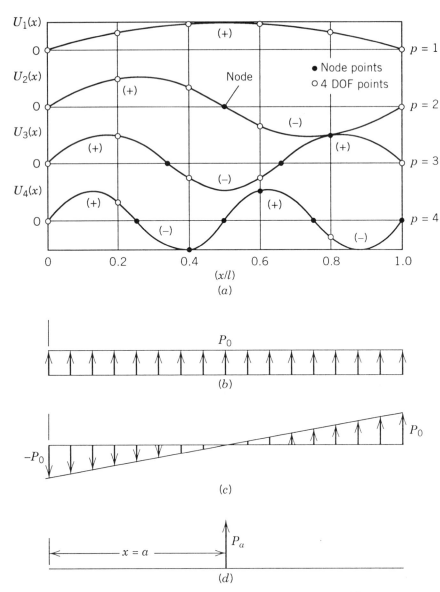

Fig. 3.7.2. Mode shapes and typical distributed excitation forces. (*a*) Mode shapes one through four. (*b*) Uniformly distributed excitation force. (*c*) Triangularly distributed excitation force. (*d*) Concentrated (point) excitation force.

multi DOF model is limited as to the number of modes it can adequately represent spatially.

Theoretically, the complementary solution for free vibration must include all possible solutions as p goes to infinity. However, due to the fact that assumptions are made in formulating these simple models, the equations begin to have significant errors at higher modes. Thus we write Eq. (3.7.8) as

$$u(x,t) = \sum_{p=1}^{N} U_p(x)\eta_p(t) = \sum_{p=1}^{N} \sin(\lambda_p x)[B_1 \cos(\omega_p t) + B_2 \sin(\omega_p t)] \quad (3.7.12)$$

where N is used to remind us of an upper limit. The initial condition at time $t = 0$ gives the constants B_1 and B_2. Equation (3.7.12) plays the same role for this continuous system that Eq. (3.6.10) does for the two DOF system.

Table 3.7.2 shows three common boundary conditions that occur for axial vibration and torsional vibration of uniform shafts that are called free-free, fixed-free, and fixed-fixed. The frequency equation, the normal mode shape, and natural frequencies are shown.

3.7.3 Orthogonality Conditions

The natural mode shapes possess an orthogonality property so that modal mass and modal stiffness can be calculated for continuous systems just as they are for multi DOF systems. This orthogonality property is contained in the Sturm–Liouville theorems[17] for second order systems under rather broad conditions. The mass orthogonality condition requires integration over the system's length l so that

$$\int_0^l m(x) U_n(x) U_p(x)\, dx = \begin{cases} 0 & \text{for } n \neq p \\ m_p & \text{for } n = p \end{cases} \quad (3.7.13)$$

where m_p is the pth mode's modal mass. Similarly, stiffness orthogonality requires integration over the system's length l so that

$$-\int_0^l \frac{d}{dx}\left[K \frac{dU_p}{dx}\right] U_n\, dx = \begin{cases} 0 & \text{for } n \neq p \\ k_p & \text{for } n = p \end{cases} \quad (3.7.14)$$

---

[17]See any applied mathematics book such as: L. A. Pipes and L. R. Harvill, *Applied Mathematics for Engineers and Physicists*, 3rd ed., McGraw-Hill, New York, 1970.

TABLE 3.7.2. Common Boundary Conditions, Natural Frequencies, and Mode Shapes for Second Order Continuous Systems

End Conditions	Boundary Conditions	Frequency Equation	Mode Shape or Normal Mode	Natural Frequencies[a]
Free-free	$U'(0) = 0$ $U'(l) = 0$	$\sin(\lambda l) = 0$	$U_p = \cos\left(\dfrac{p\pi x}{l}\right)$	$\omega_p = \dfrac{p\pi c}{l}$ $p = 0, 1, 2, 3$
Fixed-free	$U(0) = 0$ $U'(l) = 0$	$\cos(\lambda l) = 0$	$U_p = \sin\left(\dfrac{(2p+1)\pi x}{l}\right)$	$\omega_p = \dfrac{(2p+1)\pi c}{2l}$ $p = 1, 2, 3$
Fixed-fixed	$U(0) = 0$ $U(l) = 0$	$\sin(\lambda l) = 0$	$U_p = \sin\left(\dfrac{p\pi x}{l}\right)$	$\omega_p = \dfrac{p\pi c}{l}$ $p = 1, 2, 3$

[a] $c = \sqrt{K/m}$ speed of sound in the rod, string, and so on. See Table 3.7.1 for definitions of K and m for various systems.

where k_p is the pth mode's modal stiffness. Equations (3.7.13) and (3.7.14) play the same role for continuous system modeling that Eq. (3.6.13) does for multi DOF system modeling. Orthogonality is a fundamental characteristic of vibrating systems in that mode shapes possess orthogonality properties when weighted with the mass and stiffness distribution.

When Eq. (3.7.12) is substituted into Eq. (3.7.1) for the undamped free vibration case, the equation is multiplied by $U_n(x)\,dx$ and integrated over length l, and the orthogonality conditions of Eqs. (3.7.13) and (3.7.14) are used, one obtains the pth natural frequency as

$$\omega_p = \sqrt{\frac{k_p}{m_p}} \qquad (3.7.15)$$

The value of ω_p is independent of the mode shape function's magnitude. For example, when one mode shape is increased by a factor of 10 relative to the others, both the modal mass and stiffness from Eqs. (3.7.13) and (3.7.14) will be increased by a factor of 100 for that mode so that there is no effect on ω. Thus natural frequency is a system property that is independent of units and coordinate systems employed.

3.7.4 The Modal Model and Forced Vibrations

We are interested in finding the particular solution to Eq. (3.7.1) for a general excitation force or couple per unit length. First, we assume the excitation force is given by

$$f(x, t) = P(x)f(t) \qquad (3.7.16)$$

where $P(x)$ is the load's spatial distribution.
$f(t)$ is the input time history.
Second, we assume that the particular solution is given by

$$u(x, t) = \sum_{p=1}^{N} U_p(x) q_p(t) \qquad (3.7.17)$$

where $q_p(t)$ is a *generalized coordinate* in modal space that is to be determined. Note that Eq. (3.7.17) is identical in purpose and general form to Eq. (3.6.17). Insertion of Eqs. (3.7.16) and (3.7.17) into Eq. (3.7.1) gives

$$\sum_{p=1}^{N} \left\{ m U_p \ddot{q}_p + C U_p \dot{q}_p - \frac{d}{dx}\left[K \frac{dU_p}{dx} \right] q_p \right\} = P(x)f(t) \qquad (3.7.18)$$

The complexity of Eq. (3.7.18) is reduced by multiplying it by $U_n(x)\,dx$, integrating on length l, and employing the orthogonality conditions of Eqs. (3.7.13) and (3.7.14), along with assuming that damping is proportional to the mass and stiffness distributions (see Eq. (3.6.19)). The result is

$$m_p \ddot{q}_p + C_p \dot{q}_p + k_p q_p = Q_p f(t) \qquad (3.7.19)$$

where m_p is the pth modal mass.
C_p is the pth modal damping.
k_p is the pth modal stiffness.
Q_p is the pth *modal* or *generalized excitation* force.
The generalized excitation force Q_p is given by

$$Q_p = \int^l P(x) U_p(x)\,dx \qquad (3.7.20)$$

Equation (3.7.20) is extremely important to vibration testing for it tells us how much a given mode will be excited by a given distribution of excitation force $P(x)$. We explore the importance of this equation in detail in the following subsection.

When the excitation time-history is given by phasor $f(t) = e^{j\omega t}$, the steady-state solution to Eq. (3.7.19) becomes

$$u(x,t) = \sum_{p=1}^{N} \frac{U_p(x) Q_p}{D_p} e^{j\omega t} = \sum_{p=1}^{N} U_p(x) H_p Q_p\, e^{j\omega t} \qquad (3.7.21)$$

where the *modal FRF* H_p is defined as

$$H_p = \frac{1}{D_p} \qquad (3.7.22)$$

where D_p is the *modal dynamic stiffness* that is given by Eq. (3.6.24):

$$D_p = k_p - m_p \omega^2 + j C_p \omega = k_p(1 - r_p^2 + j 2 \zeta_p r_p) \qquad (3.6.24)$$

The modal FRF in Eq. (3.7.22) relates the pth mode's output per unit of modal excitation Q_p. This modal output is then modified by the pth mode shape according to Eq. (3.7.21). Clearly, Eq. (3.6.24) shows that resonance occurs when the real part is zero.

Equation (3.7.21) shows that $H_p Q_p$ controls the amount of response that is available for the pth mode, while $U_p(x)$ tells us how much of the pth mode's response will occur at location x. When $U_p(x)$ is zero, the pth modal responses are unmeasurable. Thus Eq. (3.7.21) represents the input-output FRF for a given modal excitation force Q_p. Now we shall turn our attention to the importance of Eq. (3.7.20) in exciting structures.

3.7.5 The Generalized Excitation Force for Distributed Loads

Equation (3.7.20) is equivalent to Eq. (3.6.22) for the multi DOF system, and it describes how effective a given excitation force distribution $P(x)$ is in exciting the pth natural frequency and mode shape. We consider three useful distributed loads of uniform, triangular, and concentrated (point) loads, as shown in Figs. 3.7.2b, 3.7.2c, and 3.7.2d, respectively. These load distributions are often encountered in vibration testing in both field and laboratory environments.

Example: Continuous Loads: When the uniform continuous loading (P_0 force units/unit length) as shown in Fig. 3.7.2b is used in Eq. (3.7.20), the resulting modal excitation force becomes

$$Q_p = \frac{P_0 l}{p\pi}(1 - \cos(p\pi)) = \begin{cases} 0 & \text{for } p \text{ even} \\ 2P_0 l/p\pi & \text{for } p \text{ odd} \end{cases} \quad (3.7.23)$$

First, Eq. (3.7.23) shows that the even natural frequencies are absent from any experimental results since modal excitation force Q_p is zero. Second, we see that all odd natural frequencies are excited, but with diminishing amplitudes due to the $1/p$ term. This indicates that the higher modes are excited less efficiently than the lower modes. A physical explanation for this behavior can be seen in Fig. 3.7.2, where the product of mode shape and the excitation force is seen to have a net positive area for the first and third modes and a zero net area for the even (second and fourth) modes. Similarly, the triangular load distribution in Fig. 3.7.2c will excite only the even natural modes while it will not excite the odd natural modes, a result that the reader can easily verify using Eq. (3.7.20).

Such startling results indicate that simply attaching a structure to an exciter and shaking it will not necessarily give all the natural frequencies. In fact, it is seen that we could miss a lot of natural frequencies from the excitation mechanism alone. These concepts hold for shock loading as well, since modal loads Q_p are independent of $f(t)$ and depend only on load distribution $P(x)$ and mode shape $U_p(x)$. A similar set of statements apply to the field excitation as well. We shall address these topics further in Chapter 7.

Example: Concentrated Load: Now let us look at the implications of using a concentrated excitation force located at $x = a$ as shown in Fig. 3.7.2d.

138 VIBRATION CONCEPTS

This force is described by

$$P(x) = P_a \delta(x - a) \tag{3.7.24}$$

where $\delta(x - a)$ is the Dirac delta function. Substitution of Eq. (3.7.24) into Eq. (3.7.20) gives

$$Q_p = P_a U_p(a) \tag{3.7.25}$$

Equation (3.7.25) clearly shows that any time $U_p(a)$ is zero, the pth natural frequency and mode shape are not excited. Thus Fig. 3.7.2a shows that applying excitation at the midpoint will result in suppression of all even natural frequencies and mode shapes, while applying excitation at a quarter point will result in the suppression of the fourth natural frequency as well as all multiples of the fourth natural frequency.

This modal suppression due to excitation location is independent of time history $f(t)$ and excitation force magnitude. These results also apply to an initial condition problem. The initial condition is generated through a time history where a concentrated load builds slowly, and then is suddenly released. The reader can work out the theoretical details or demonstrate this result experimentally using a guitar or taut wire, a microphone, and frequency analyzer. For example, apply the concentrated load at the midpoint. The resulting frequency analysis should show the fundamental frequency and a number of odd multiples but no even multiples of the fundamental natural frequency. Similarly, plucking at the third point will show the third natural frequency and its multiples are missing, while plucking at the quarter point will show the absence of fourth natural frequency and its multiples. These last few paragraphs demonstrate the importance and effect of excitation force distribution on dynamic and transient experiments started with initial conditions.

3.7.6 Continuous Model FRFs

There are two kinds of FRFs that are generated with continuous systems. The first type corresponds to a distributed force, while the second corresponds to a concentrated excitation force or forces.

Distributed Excitation When the excitation load is a distributed function $P(x)$, then the modal load Q_p will have a particular value for each mode so that the response at location b is given by Eq. (3.7.21) as

THE SECOND ORDER CONTINUOUS VIBRATION MODEL 139

$$u(b,t) = \sum_{p=1}^{N} U_p(b)H_pQ_p \, e^{j\omega t} = H_b \, e^{j\omega t} \qquad (3.7.26)$$

where the point response due to distributed excitation FRF is given by

$$H_b = \sum_{p=1}^{N} U_p(b)H_pQ_p \qquad (3.7.27)$$

Equation (3.2.27) clearly shows us that this output FRF contains the pth mode shapes as defined by $U_p(b)$, the modal FRFs H_p, and the modal excitation force Q_p. Clearly, we can determine the pth mode shape by measuring the responses at a number of points (by changing the value of b) provided that Q_p is nonzero. This means that we can measure mode shapes of smaller structures by attaching them to a vibration exciter to shake (excite) them with an inertial load. However, the restrictions on loading discussed above apply. Also, we note that the FRF is designated to have only one subscript since the input is distributed and no specific input location can be determined except under specific circumstances so that the results are application dependent.

Concentrated Force Excitation Loading Similarly, substitution of Eq. (3.7.25) into Eq. (3.7.21) gives the response at location b due to a concentrated excitation force applied at location a as

$$u(b,t) = \sum_{p=1}^{N} U_p(b)H_pU_p(a)P_a \, e^{j\omega t} = H_{ba}P_a \, e^{j\omega t} \qquad (3.7.28)$$

where the point to point input-output FRF H_{ba} is given by

$$H_{ba} = \sum_{p=1}^{N} U_p(b)H_pU_p(a) \qquad (3.7.29)$$

Again, we see that the measured FRF is composed of the product of two mode shapes and modal FRF functions H_p. In this case, Eq. (3.7.29) indicates that we can measure the pth mode shape either by changing the location of the motion measurement (change value of b) and keeping the excitation location fixed or by changing the load location (change the value of a) and keeping the motion measurement location fixed, and since the pth mode shape is involved with both locations a or b. It is clear that if either $U_p(b)$ or $U_p(a)$ is zero, then the corresponding pth modal data is missing from experimental data. This result suggests that one cannot use a single input or output location in conducting a test without prior knowledge of a structure's mode shapes. Otherwise, there is high risk that a significant mode is missing from the experimental data.

The point to point FRF in Eq. (3.7.29) does not include the excitation force's magnitude P_a like the previous FRF given by Eq. (3.7.27) where Q_p cannot be removed from the summation except in special cases. This simple modal model shows us that considerable skill may be required to adequately plan a successful vibration test in either the field or the laboratory environment.

3.8 FOURTH ORDER CONTINUOUS VIBRATION SYSTEM— THE BEAM

This is the last theoretical model that is developed to give us further insight into vibration response that is important in understanding vibration testing. The governing differential equation of motion is found to be a fourth order partial differential equation that describes the motion of a simple beam-type structure. In this section, the natural frequencies and their corresponding mode shape are developed along with the modal model. The effects of boundary conditions on the fundamental natural frequency are explored in depth. A beam under axial tension is also considered to gain insight into the effects of this tension on the structure's behavior.

3.8.1 The Fundamental Equation of Motion

Consider the beam shown in Fig. 3.8.1. The beam's parameters are described by lateral motion $u(x, t)$, by mass per unit length m, by bending stiffness EI (E = Young's modulus and I = area moment of inertia of beam cross section about its centroidal axis), by axial tension force T, by damping per unit length C, and by excitation force per unit length $f(x, t)$. For these parameters, the differential equation for lateral motion is given by[18]

$$m\frac{\partial^2 u}{\partial t^2} + C\frac{\partial u}{\partial t} - \underbrace{\frac{\partial}{\partial x}\left[T\frac{\partial u}{\partial x}\right]}_{\text{tension}} + \underbrace{\frac{\partial^2}{\partial x^2}\left[EI\frac{\partial^2 u}{\partial x^2}\right]}_{\text{bending}} = f(x, t) \qquad (3.8.1)$$

$$\underbrace{\phantom{m\frac{\partial^2 u}{\partial t^2}}}_{\text{inertia}} \underbrace{\phantom{C\frac{\partial u}{\partial t}}}_{\text{damping}} \underbrace{\phantom{\frac{\partial}{\partial x}\left[T\frac{\partial u}{\partial x}\right]}}_{\text{stiffness}} \underbrace{\phantom{\frac{\partial^2}{\partial x^2}\left[EI\frac{\partial^2 u}{\partial x^2}\right]}}_{\text{excitation}}$$

---

[18]See, for example, S. S. Rao, *Mechanical Vibrations*, 2nd ed., Addison-Wesley, Reading, MA, 1990, chapter 8.

FOURTH ORDER CONTINUOUS VIBRATION SYSTEM—THE BEAM

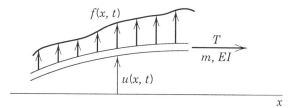

Fig. 3.8.1. Definition of a beam element with displacement $u(x,t)$, mass per unit length m, external load $f(x,t)$, axial tension T, bending stiffness EI, and spatial coordinate x.

Equation (3.8.1) is a fourth order partial differential equation and represents the force balance on a per unit length basis that must occur at each point in the beam at each instant of time. This balance consists of inertia, damping, spring, and excitation forces. The spring is seen to consist of two terms: one is the beam's tension force T and the other is the beam's bending stiffness EI. Equation (3.8.1) is sometimes referred to as either a *thin beam* or an *Euler–Bernoulli* beam. When shear deformation and rotary inertia are included, the model is referred to as the *Timoshenko beam*. The thin beam model of Eq. (3.8.1) is sufficient for our purposes.

3.8.2 Natural Frequencies and Mode Shapes

The undamped natural frequencies and mode shapes for a uniform cross sectioned (m and EI are constant) thin beam without axial tension force T are considered here. The separation of variables approach is used as in Section 3.7 so that

$$u(x,t) = U(x)\eta(t) \tag{3.8.2}$$

Two differential equations are obtained by insertion of Eq. (3.8.2) into Eq. (3.8.1). These are:

$$\ddot{\eta} + \left(\lambda^4 \frac{EI}{m}\right)\eta = \ddot{\eta} + \omega^2 \eta = 0 \tag{3.8.3}$$

for time function $\eta(t)$ and

$$\frac{d^4 U}{dx^4} - \lambda^4 U = 0 \tag{3.8.4}$$

for the spatial function $U(x)$ where λ^4 is the *separation constant*. Constant

142 VIBRATION CONCEPTS

λ and frequency ω are related by

$$\omega^2 = \lambda^4 \frac{EI}{m} \tag{3.8.5}$$

The solutions to Eqs. (3.8.3) and (3.8.4) are:

$$\eta(t) = B_1 \cos(\omega t) + B_2 \sin(\omega t) \tag{3.8.6}$$

and

$$U(x) = A_1 \cos(\lambda x) + A_2 \sin(\lambda x) + A_3 \cosh(\lambda x) + A_4 \sinh(\lambda x) \tag{3.8.7}$$

Equation (3.8.7) is seen to contain both sinusoidal and hyperbolic functions due to the fourth order character of Eq. (3.8.4), and it represents natural mode shapes for each eigenvalue of λ.

The Sturm–Liouville theorem[19] shows that mode shapes are orthogonal when weighted with the mass and stiffness distributions. These orthogonality conditions are:

$$\int_0^l m(x) U_p(x) U_r(x)\, dx = \begin{cases} 0 & \text{for } r \neq p \\ m_p & \text{for } r = p \end{cases} \tag{3.8.8}$$

for the pth modal mass m_p and

$$\int_0^l \frac{d^2}{dx^2}\left[EI \frac{d^2 U_p(x)}{dx^2}\right] U_r(x)\, dx = \begin{cases} 0 & \text{for } r \neq p \\ k_p & \text{for } r = p \end{cases} \tag{3.8.9}$$

for the pth modal stiffness.

Example: A uniform simply supported (pinned-pinned) beam of length l is shown in Fig. 3.8.2 along with its boundary conditions. It is evident that both ends are fixed to have no translation motion but are free to rotate and have no bending moments. The Euler–Bernoulli bending moment-displacement equation (from elementary mechanics of materials) is given by

$$M(x,t) = EI \frac{\partial^2 u(x,t)}{\partial x^2}$$

---

[19]See: L. A. Pipes and L. R. Harvill, *Applied Mathematics for Engineers and Physicists*, 3rd ed., McGraw-Hill, New York, 1970.

FOURTH ORDER CONTINUOUS VIBRATION SYSTEM – THE BEAM

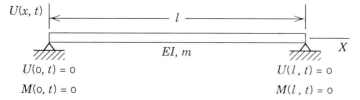

Fig. 3.8.2. Uniform simply supported beam of length l.

and is used to convert boundary condition moments into displacement boundary conditions. Since the boundary conditions apply for all time, they can be reduced to conditions that $U(x)$ in Eq. (3.8.7) must satisfy. These conditions are: $U(0) = U(l) = 0$ and $U''(0) = U''(l) = 0$. Application of these boundary conditions to Eq. (3.8.7) give $A_1 = A_3 = A_4 = 0$ and a *characteristic frequency equation* of

$$\sin(\lambda l) = 0$$

that is satisfied when λl is a multiple of π. Thus the *eigenvalues* become

$$\lambda_p = \frac{p\pi}{l} \qquad (3.8.10)$$

so that mode shapes (*eigenfunctions*) from Eq. (3.8.7) become

$$U_p(x) = \sin(\lambda_p x) = \sin\left(\frac{p\pi x}{l}\right) \qquad (3.8.11)$$

and natural frequencies from Eq. (3.8.5) become

$$\omega_p = \lambda_p^2 \sqrt{\frac{EI}{m}} = \left(\frac{p\pi}{l}\right)^2 \sqrt{\frac{EI}{m}} \qquad (3.8.12)$$

for each value of p. For this case, the mode shapes in Eq. (3.8.11) are the same as those for the taut string. The natural frequencies in Eq. (3.8.12) are seen to increase according to p^2, while in the taut string's case they increased linearly with p. When the mode shape (as scaled in Eq. (3.8.11)) is substituted into Eqs. (3.8.8) and (3.8.9), one finds that the orthogonality terms are satisfied and the modal mass becomes

$$m_p = ml/2$$

144 VIBRATION CONCEPTS

(a constant) for all values of p, while the modal stiffness becomes

$$k_p = p^4 \frac{\pi^4}{2} \frac{EI}{l^3} = p^4(1.015)k_s$$

where k_s is the static spring constant for the beam's midpoint; that is, $k_s = 48EI/l^3$. When $p = 1$, the values of m_p and k_p are the same as those estimated by using the Rayleigh energy method to obtain the beam's fundamental natural frequency. When these values are substituted into the natural frequency expression (modal stiffness divided by modal mass), we obtain the same pth natural frequency as given by Eq. (3.8.12).

3.8.3 Natural Frequencies and Boundary Conditions

The first five values of $\lambda_p l$ for uniform beams under six cases of common boundary conditions are given in Table 3.8.1. The cases shown are: (1) pinned-pinned (simply-supported), (2) free-free, (3) fixed-fixed, (4) fixed-free (cantilever), (5) fixed-pinned, and (6) pinned-free. There are several important observations that we can draw from Table 3.8.1 relative to the $\lambda_p l$ values.

First, the free-free (case 2) and pinned-free (case 6) can have rigid body motion so that the first $\lambda_p l$ value is zero. The zero $\lambda_p l$ value produces a fundamental natural frequency that is also zero. Such a system is called *semidefinite* and is capable of rigid body motion. All vehicles such as aircraft, automobiles, and so on are semidefinite systems where vibration is superimposed on top of the rigid body motion.

Second, there are two sets of case pairs where the $\lambda_p l$ values are the same even though the boundary conditions are opposites. The first set consists of free-free (case 2) and fixed-fixed (case 3). The second set consist of fixed-pinned (case 5) and pinned-free (case 6). In both sets, one case is semidefinite with rigid body motion even though the frequency equations are the same in each set. In each set, rigid body motion is possible in one case while it is not in the other. Thus we need to be aware from the physical situation that zero is an admissible value for the first eigenvalue.

Third, the higher natural frequencies tend to have the same $\lambda_p l$ values. For example, there are values in the range of 7 to 8 (see $\lambda_2 l$ in cases 2, 3, 5, and 6 and $\lambda_3 l$ in case 4), values in the range of 10 to 11 (see $\lambda_3 l$ in cases 2, 3, 5, and 6 and $\lambda_4 l$ in case 4), and values in the range of 13.3 to 14.2 (see $\lambda_4 l$ in cases 2, 3, 5, and 6 and $\lambda_5 l$ in case 4). So what causes this kind of behavior?

The fundamental natural frequency is highly dependent on the beam's boundary conditions. This dependency is illustrated in Fig. 3.8.3a, where

FOURTH ORDER CONTINUOUS VIBRATION SYSTEM – THE BEAM

TABLE 3.8.1. Common Beam Boundary Conditions, Frequency Equation, and $\lambda_p l$ Values for Beams with Uniform Cross Sections

Case	Boundary Conditions	Frequency Equation	$\lambda_p l$ Values
1	Pinned–pinned	$\sin(\lambda_p l) = 0$	$\lambda_1 l = \pi = 3.142$ $\lambda_2 l = 2\pi = 6.284$ $\lambda_3 l = 3\pi = 9.425$ $\lambda_4 l = 4\pi = 12.566$ $\lambda_5 l = 5\pi = 15.708$
2	Free–free	$\cos(\lambda l)\cosh(\lambda l) = 1$	$\lambda_0 l = 0$ (R.B.)[a] $\lambda_1 l = 4.730$ $\lambda_2 l = 7.853$ $\lambda_3 l = 10.996$ $\lambda_4 l = 14.137$
3	Fixed–fixed	$\cos(\lambda l)\cosh(\lambda l) = 1$	$\lambda_1 l = 4.730$ $\lambda_2 l = 7.853$ $\lambda_3 l = 10.996$ $\lambda_4 l = 14.137$ $\lambda_5 l = 17.278$
4	Fixed–free	$\cos(\lambda l)\cosh(\lambda l) = -1$	$\lambda_1 l = 1.875$ $\lambda_2 l = 4.694$ $\lambda_3 l = 7.855$ $\lambda_4 l = 10.996$ $\lambda_5 l = 14.137$
5	Fixed–pinned	$\tan(\lambda l) = \tanh(\lambda l)$	$\lambda_1 l = 3.926$ $\lambda_2 l = 7.069$ $\lambda_3 l = 10.210$ $\lambda_4 l = 13.352$ $\lambda_5 l = 16.494$
6	Pinned–free	$\tan(\lambda l) = \tanh(\lambda l)$	$\lambda_0 l = 0$ (R.B.)[a] $\lambda_1 l = 3.926$ $\lambda_2 l = 7.069$ $\lambda_3 l = 10.210$ $\lambda_4 l = 13.352$

[a] R.B. means rigid body motion.

the first mode shape for a pinned-pinned and fixed-fixed uniform beam is plotted. The fixed beam is constrained (due to zero slope at its boundaries) to have higher strain energy per unit of midpoint deflection. The effects of this constraint are distributed over the beam's entire length. Consequently, the pinned-pinned beam has less strain energy per unit of midpoint deflection than the fixed-fixed beam so that ω_1 (case 1) is less than ω_1 (case 3); that is, $\lambda_1 l = 3.142$ (case 1) while $\lambda_1 l = 4.730$ (case 3).

The mode shapes for the fifth natural frequency are shown in Fig. 3.8.3b, where the fixed boundary condition's effects are confined to the first and last quarter of the beam. The beam's center half behaves nearly

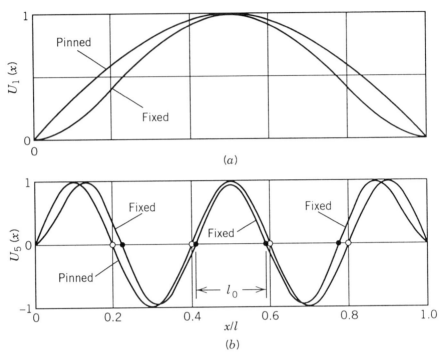

Fig. 3.8.3. First and fifth normal modes for pinned-pinned and fixed-fixed beams. (*a*) First mode comparison. (*b*) Fifth mode comparison.

the same as the pinned-pinned beam. If we use the wave length l_0 (distance between node points in Fig. 3.8.3*b* for the fixed-fixed mode shape) in the pinned-pinned beam's fundamental natural frequency equation, we can predict the corresponding natural frequency of the fixed-fixed beam. In this case, $l_0 \cong 0.183l$. Insertion of this value into Eq. (3.8.12) for the fundamental natural frequency gives

$$\omega = \pi^2 \sqrt{\frac{EI}{ml_0^4}} = \left(\frac{\pi^2}{0.183}\right) \sqrt{\frac{EI}{ml^4}} = (17.136)^2 \sqrt{\frac{EI}{ml^4}}$$

The value of $\lambda_5 l$ for the fixed-fixed beam (case 3) is 17.278, compared to 17.136, that we have predicted here, an error of less than one percent. The reason this works so well is that, in the middle portion of the beam, the section of beam between node points behaves exactly like a pinned-pinned beam of length l_0. This type of approach can be used to check out beam-like behavior if ω and l_0 can be measured so long as l_0 is less than about one-fourth of the beam. When l_0 is greater than one-fourth of the beam's length, boundary conditions can significantly affect the results. These results are the dynamic equivalent of Saint Venant's principle used in mechanics of materials.

The fundamental natural frequency of the fixed-free (cantilever) beam in case 4 is extremely sensitive to the rotational boundary condition of zero slope. The model shown in Fig. 3.8.4a has a torsional spring K that allows some rotation to occur around support point O. Point O is assumed to a pinned boundary condition. For this case, we can show that the frequency equation involves λl and a stiffness ratio β defined by

$$\beta = \frac{EI}{Kl} \tag{3.8.13}$$

A fixed boundary condition corresponds to K being infinite and $\beta = 0$. When the frequency equation is solved for various values of β, we can compute the system's fundamental natural frequency and form a dimensionless frequency ratio of ω/ω_R where ω_R is the reference natural frequency that corresponds to $\beta = 0$, the perfect cantilever beam. Figures 3.8.4b and 3.8.4c show the functional relationship between ω/ω_R and β.

When β is large ($\beta > 2$), the solution approaches values that are equivalent to

$$\omega = \sqrt{K/I_0} \tag{3.8.14}$$

where I_0 is the beam's rigid body mass moment of inertia about pivot point O in Fig. 3.8.4a. This case shows that rigid body behavior is built into the beam model and is the limiting condition when the support natural frequencies are low enough compared to the fundamental free-free natural frequency.

The ratio ω/ω_R is extremely sensitive to changes in β for small values of β as shown in Fig. 3.8.4c. The initial portion of the curve can be estimated by

$$\frac{\omega}{\omega_R} = 1 - 1.6\beta \qquad \text{for } \beta < 0.1$$

This high sensitivity turns out to be important in understanding a discrepancy between field and simulated structural responses.

Practical Example: When an aircraft carrier came out of overhaul, the island-mast vibrations when it launched aircraft from its waist catapults

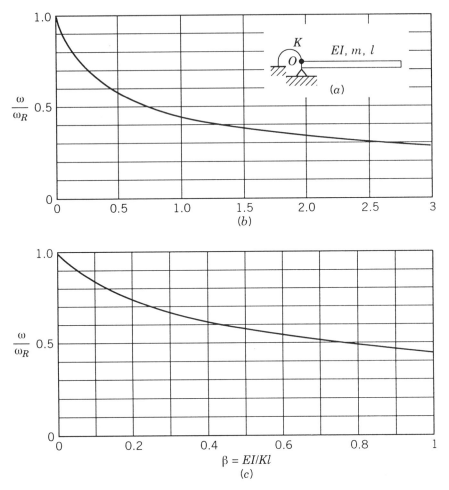

Fig. 3.8.4. (*a*) Pinned–torsionally supported system. (*b*) Fundamental ratio β (= EI/Kl). Note $\beta = 0$ gives the reference cantilever beam's natural frequency ω_R. (*c*) An expansion of the plot in Fig. 3.8.4*b*.

were excessive compared to its preoverhaul operating behavior.[20] One area of concern was the low natural frequency of the island-mast structure compared to that of similar ships. The measured natural frequency was about 3.2 Hz while a simulated model of the island structure predicted values greater than 4.7 Hz, giving a frequency ratio around 0.68. Once

---

[20]K. G. McConnell, "Tracking the Cause of Large Shipboard Vibrations During Catapult Launching," *Proceedings, SEM Fall Conference*, Savannah, GA, Nov. 1987.

Fig. 3.8.4c was developed, it became evident that the supporting structure and stiffness K played a significant role in controlling this natural frequency and that β should have a value around 0.3. Subsequent experimental tests on the ship as well as investigations of the island's rotational structural stiffness gave values of K that were consistent with these values. Once this behavior was understood, it was possible to eliminate the structure's response as the cause of the excessive vibration and to judge the structure as being sound. It was found that a critical part was left out of the launch controller during overhaul, which, when replaced, returned the ship to normal operation.

This practical example shows the importance of a simple theoretical model that included a key system parameter to understanding the discrepancy between field measurements and predicted fundamental natural frequency. Once the significance of base stiffness was understood, informed progress could be made. The amount of sensitivity to base stiffness was somewhat surprising since the unrealistic fixed boundary condition regularly appears in papers and books without mentioning the consequences of the zero slope assumption.

The purpose of this lengthy discussion is to introduce the following general statements concerning testing and boundary conditions.

1. Boundary conditions are often the source of error between field and either laboratory measurements or a computer simulation.
2. Attempts to place force transducers in a structure at certain key points may significantly alter the structure's boundary conditions. Strain gages on existing structural members are preferred to inserting force transducers since strain gages have little effect on most structures.
3. When measuring dynamic characteristics of a structure for model verification, the free-free boundary condition is preferred over the pinned boundary condition, which is preferred over the fixed boundary condition. The free state can be approximated by using soft suspension springs so that the rigid body natural frequencies are well below the first free boundary natural frequency. Usually, a frequency separation ratio of 5 or more is sufficient.

3.8.4 The Modal Model

It is evident that orthogonality properties of Eqs. (3.8.8) and (3.8.9) are similar to those used in Section 3.7 and these properties can be used to obtain the modal model's differential equation of motion for generalized

150 VIBRATION CONCEPTS

coordinate $q_p(t)$, a differential equation identical to Eq. (3.7.19), so that

$$m_p \ddot{q}_p + C_p \dot{q}_p + k_p q_p = Q_p f(t) \qquad (3.7.19)$$

where the generalized excitation force Q_p is given by Eq. (3.7.20):

$$Q_p = \int_0^l P(x) U_p(x)\, dx \qquad (3.7.20)$$

and the particular solution is given by Eq. (3.7.21):

$$u(x,t) = \sum_{p=1}^{N} U_p(x) H_p Q_p\, e^{j\omega t} \qquad (3.7.21)$$

where H_p is the pth modal FRF, that is, $H_p = 1/D_p$.

Conceptually, the second and fourth order modal models are the same. The only significant difference is the mode shape function $U_p(x)$, which is usually more complicated in the case of a beam than in that of a string or rod. This means that the discussion of modal response in Section 3.7 under different kinds of distributed excitation loads is applicable to beams as well as general structures. Hence, the case of a beam under axial tension is more informative in gaining insight into real structural behavior than repeating the previous discussion on distributed excitation loads.

3.8.5 The Beam Under Tension

It is informative to examine the response of a simply supported beam subjected to a tension load T. We use the uniform beam shown in Fig. 3.8.2 for the purposes of this discussion. In addition, let us neglect both the damping and excitation terms in Eq. (3.8.1) and look only at the free vibration response. In this case, we assume the motion is sinusoidal at frequency ω as described by Eq. (3.8.6). Then, substituting Eq. (3.8.6) into Eq. (3.8.2), we obtain

$$u(x,t) = U(x)[B_1 \cos(\omega t) + B_2 \sin(\omega t)] \qquad (3.8.15)$$

Insertion of Eq. (3.8.15) into Eq. (3.8.1) with $C = f(x,t) = 0$ gives

$$EI\frac{d^4 U}{dx^4} - T\frac{d^2 U}{dx^2} - m\omega^2 U = 0 \qquad (3.8.16)$$

as the fourth order differential equation that $U(x)$ must satisfy. If we let

FOURTH ORDER CONTINUOUS VIBRATION SYSTEM—THE BEAM

$U(x) = Ae^{sx}$, then Eq. (3.8.16) gives

$$s^4 - \frac{T}{EI}s^2 - \frac{m\omega^2}{EI} = 0$$

which has two roots in s^2 that are given by

$$s_1^2, s_2^2 = \left(\frac{T}{2EI}\right) \pm \left(\left(\frac{T}{2EI}\right)^2 + \frac{m\omega^2}{EI}\right)^{1/2} \tag{3.8.17}$$

where it is noted that s_1^2 is positive and s_2^2 is negative. If either value of s_1 or s_2 is known, the corresponding natural frequency ω can be determined. These two roots give a solution for $U(x)$ as

$$U(x) = A_1 \cosh(s_1 x) + A_2 \sinh(s_1 x) + A_3 \cos(s_2 x) + A_4 \sin(s_2 x) \tag{3.8.18}$$

where the hyperbolic functions and trigonometric functions have different arguments. The values of s_1 and s_2 are determined by applying the boundary conditions shown in Fig. 3.8.2. This process gives $A_1 = A_3 = 0$ and the requirement that

$$\sinh(s_1 l) \sin(s_2 l) = 0 \tag{3.8.19}$$

must be satisfied. Since $\sinh(s_1 l) \neq 0$ for any nonzero value of s_1, it is clear that Eq. (3.8.19) is satisfied only when $s_2 l$ is a multiple of π ($s_{2p} = p\pi/l$). Then, Eq. (3.8.17) gives the pth natural frequency as

$$\omega_p = \frac{\pi^2}{l^2}\sqrt{\frac{EI}{m}\left(p^4 + p^2 \frac{T}{T_{\text{crit}}}\right)} \tag{3.8.20}$$

where T_{crit} is the *critical Euler buckling* load given by

$$T_{\text{crit}} = \frac{\pi^2 EI}{l^2} \tag{3.8.21}$$

The corresponding values of s_{1p} are obtained by substituting Eq. (3.8.20) into Eq. (3.8.17). The resulting pth mode shape becomes

$$U_p(x) = \sin(s_{2p} x) = \sin\left(\frac{p\pi x}{l}\right) \tag{3.8.22}$$

which is the same mode shape result we obtained for the beam without tension T. Thus tension (or compression for that matter) force has no significant effect on the mode shape according to this result. In the case of dynamic compressive loads, a more advanced analysis by McIvor and Bernard[21] shows significantly different and more complicated behavior.

A number of conclusions can be made concerning the natural frequencies given by Eq. (3.8.20). These are:

1. When $T = 0$, Eq. (3.8.20) gives the same natural frequencies as Eq. (3.8.12).
2. When T is very large compared to T_{crit}, the lower ($p = 1, 2, \ldots$) natural frequencies are close to those of a taut string. However, the high mode ($p \gg 3$) natural frequencies are controlled by bending; namely, the EI/m term controls since p^4 becomes dominant over $p^2 T/T_{crit}$.
3. When $EI = 0$, the natural frequencies correspond to those of a taut string.
4. If T is positive, the natural frequencies are higher than they are for bending alone since both tension and bending contribute to the effective spring force.
5. If T is negative (compressive like in a column), the natural frequencies are lower than predicted by conventional beam theory since the compressive force reduces the effective spring rate due to bending alone. As T approaches T_{crit}, the natural frequency also approaches zero.

This simple example shows that axial tension or compression forces can significantly alter what we would normally anticipate to be natural frequencies while having little or no effect on mode shapes. Thus we need to be aware of axial tension effects in structural responses. Frequencies may not be distributed as we would anticipate or may change from experiment to experiment when tension is allowed to vary considerably during test.

---

[21] I. K. McIvor and J. E. Bernard, "The Dynamic Response of Columns Under Short Duration Axial Loads," *Transactions of the ASME, Journal of Applied Mechanics*, Sept. 1973, pp. 688–692.

3.9 SUMMARY

The objective of this chapter has been to review well known vibration relationships from a vibration testing point of view. We started with the transient vibration of a single DOF system and found the governing second order differential equation of motion represents a balance of inertia, damping, and spring forces at each instant of time. In the free undamped vibrations, the spring and inertia forces cancel one another when the system is oscillating at its natural frequency. It is this cancellation of inertia and spring forces that leads to the resonance phenomenon when this system is excited at its natural frequency. The damped transient response is bounded by an exponential envelope for viscous damping and a straight line envelope for Coulomb damping. It is also shown that low natural frequencies may be sensitive to a structure's orientation relative to gravity.

In the single DOF forced response case, three useful output/input FRFs of receptance (displacement/unit force), mobility (velocity/unit force), and accelerance (acceleration/unit force) are introduced. These FRF relationships are found to be related to one another by either multiplying or dividing one FRF by either $j\omega$ or $(j\omega)^2$ to obtain one of the other FRFs. The viscous, Coulomb, and structural (hysteresis) damping models are explored for use with forced vibration response. Either the viscous or structural damping model is usually best suited to use in predicting forced vibration response. The various FRFs are presented in the Bode diagram (magnitude and phase versus frequency plots), real and imaginary as a function of frequency plots, and Nyquist (imaginary versus real) plot. The Bode diagram is the most commonly employed plot and uses log or dB scales to accommodate large dynamic ranges. The Nyquist receptance plot with structural damping is a circle centered on the vertical imaginary axis, while the Nyquist mobility plot with viscous damping is a circle centered on the horizontal axis as the system goes through resonance, leading to the circle fit concept.

The general input-output model for linear systems shows that two distinct approaches can be used. One approach uses Fourier series concepts and multiplication in the frequency domain, while the other approach uses the impulse response function and superposition in the time domain. It is shown that the impulse response function $h(t)$ and receptance FRF $H(\omega)$ are transient Fourier transforms of one another. The shock response spectra SRS is shown to be a single DOF concept that converts a time history into a plot of maximum response as a function of the single DOF oscillator's natural frequency. The SRS is an often misused concept.

Nonlinear effects are caused by many different mechanisms in a given system such as springs, damping, electromagnetic effects, and so on. The phase plane (velocity versus displacement) plot is introduced as one means to examine nonlinear behavior. The linear free undamped system follows a closed elliptical path, while a linear free damped oscillation follows a spiral path, as opposed to the limit cycle behavior of a self excited nonlinear system. The simple pendulum generates odd multiples of the fundamental frequency and has a natural frequency that decreases with increasing amplitude of motion due to the softening spring. The Duffing equation is used to examine nonlinear forced vibration response with its associated shift in resonant frequency with amplitude of motion and sudden jump response. The van der Pol equation is used to show limit cycle behavior and self excited behavior. The Mathieu equation is used to illustrate systems that exhibit parametric excitation that can be either stable or unstable, depending on the system parameter values. The idea that chaotic motion can occur only in nonlinear systems is important; particularly since these motions can be misinterpreted as being random. In such cases the engineer may miss the most significant experimental result.

The two DOF model is used to introduce multi DOF vibration response concepts. The free undamped response is examined for natural frequencies and the natural motion that occurs at each natural frequency. This natural motion is described by the modal vector. Modal vectors are the key to orthogonality conditions that involve the mass and stiffness matrices. The steady-state forced vibration response is obtained by using both a direct matrix method and a modal analysis approach where each natural mode makes its own contribution to the response. These response modes are added together to make up the input-output FRF. It is shown that the direct and modal models give the same FRF if a sufficient number of modal responses are included. It is recognized in real systems that a limited number of degrees of freedom are used to describe the system, and hence, limit the accuracy of any experimentally determined modal model.

The second order continuous model is presented in terms of a taut string, axial rod (and spring), and torsional rod vibrations. The separation of variables method of analysis is used to develop the modal model. Boundary conditions are imposed on the spatial solution $U(x)$ to generate natural frequencies and mode shapes. These resulting natural frequencies and mode shapes are then employed through use of orthogonality to generate the modal differential equation of motion utilizing a generalized modal space coordinate $q_p(t)$. The generalized modal space excitation force Q_p is found to depend on the excitation force distribution. The effects of excitation force distribution are explored in depth. It is found that certain modes may not be excited due to excitation distribution. An example showed that half of the potential modes of vibration could be missed. Finally, the FRF functions that can be generated in such a case are examined in terms of the influence of the distributed excitation force. It is found that only for a concentrated force excitation can a point to point input-output FRF be defined. When the excitation force is distributed, a point to point FRF cannot be defined. Each case needs to be treated on a separate basis.

The fourth order beam is the last system to be explored. A beam under tension has two spring type terms, one due to tension and one due to the beam's bending stiffness. The undamped free vibration model shows the mode shapes are composed of hyperbolic and sinusoidal functions. The effect of boundary conditions on natural frequencies is explored in depth, and it is seen that carelessness in handling boundary conditions can cause serious testing problems. The modal model gives identical results as the second order continuous system except for mode shape complexity. The beam under tension shows that tension increases the beam's natural frequencies while compressive forces lower the beam's natural frequencies relative to those of an untensioned beam.

Finally, it is extremely important to have a theoretical model to explain basic dynamic behavior so that the implications of measured data are understood and appropriate informed action is taken. Otherwise, the meaning of experimental data can be misinterpreted, an important engineering design change is missed, and resources are wasted. For example, power line cables hang in a shape called the catenary and generally have sag/span ratios in the 2 to 5 percent range. Traditionally, cable vibration observations of natural frequency and mode shape are compared to those of the taut string model.

A more advanced dynamic model has been developed by Nariboli and McConnell[22] that takes the cable's curvature and axial elasticity into account. This model shows a dramatically different behavior relative to the taut string model in terms of natural frequencies and mode shapes. The natural frequencies and mode shapes are dependent on a cable parameter $EA/\rho gAL_0$ (where E = Young's modulus, A = cross-sectional area, ρg = cable weight per unit volume, L_0 = half span length) and the sag/span ratio. A typical plot of dimensionless frequency ($\omega = \omega_n/\omega_R$) versus sag/span ratio ($\delta/2L_0$) for a typical cable parameter of 14,000 is shown in Fig. 3.9.1. When the sag/span ratio is less than 0.001, the natural frequencies are close to those predicted by conventional taut string theory, that is, integer multiples of the fundamental string frequency ω_R.

However, as the sag/span ratio increases, many interesting things happen. First, the odd natural frequencies cross over the even natural frequencies along the line $\omega = a$ in Fig. 3.9.1. This means that for a sag/span ratio of about 2.25 percent, the first and second cable frequencies become the same but there are two different mode shapes, as shown in

---

[22]G. A. Nariboli and K. G. McConnell, "Curvature Coupling of Catenary Cable Equations," *International Journal of Analytical and Experimental Modal Analysis*, Vol. 3, No. 2, Apr. 1988, pp. 49–56.

156 VIBRATION CONCEPTS

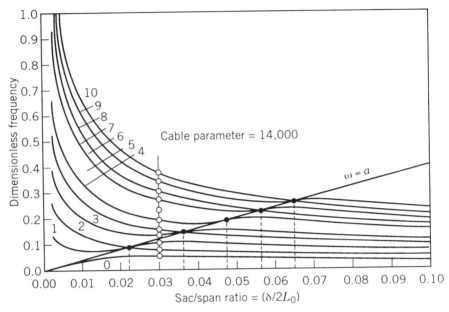

Fig. 3.9.1. Dimensionless cable natural frequencies versus sag/span ratio for a cable when cable parameter $(EA/\rho AgL_0)$ is 14,000.

Figs. 3.9.2b and 3.9.2c. Second, we see that a new natural frequency and mode shape begin to occur as the sag/span ratio increases from zero; this is called the zero root mode in Fig. 3.9.1 and has the same mode shape as the first string mode when the sag/span ratio is zero (compare Figs. 3.9.2a and 3.9.2b). Third, we see that the first cable mode shape becomes the same as a third taut string mode shape for sag/span ratios above five percent (see Fig. 3.9.2b).

In order to get a handle on the implications of these curves, let us assume that the sag/span ratio is 3 percent. Then, as we move up the vertical line at 3 percent sag/span ratio in Fig. 3.9.1, we find that the fundamental frequency that we should observe corresponds to the zero root mode frequency and has the zero root mode shape shown in Fig. 3.9.2a. The second frequency we should observe corresponds to the second cable frequency line in Fig. 3.9.1 and has the root 2 mode shape shown in Fig. 3.9.2c. The third frequency that we should observe corresponds to the first taut string frequency line (curve 1 in Fig. 3.9.1) and has the mode shape of root 1 in Fig. 3.9.2b that corresponds to the 0.03 sag/span ratio. In addition, as we move up along the vertical line at 3 percent sag/span ratio, we see that the natural frequencies are not integer multiples of the fundamental natural frequency.

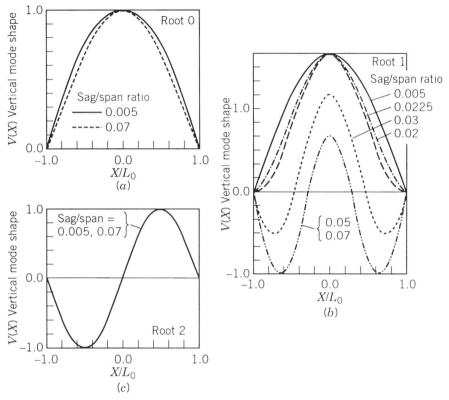

Fig. 3.9.2. Vertical mode shapes for the elastic caternary cable. (*a*) Root 0 vertical mode shape. (*b*) Root 1 mode shape that changes from mode 1 to mode 3. (*c*) Root 2 mode shape that remains unchanged with sag/span ratio.

In this example, mode shapes are similar between models and are hard to measure accurately. On the other hand, natural frequencies do not match the taut string model where these frequencies are integer multiples of the fundamental. Also, on a hot day, these measured natural frequencies change, since the sag/span ratio changes with temperature. While this more advanced theoretical model is not perfect, it shows that the sag/span ratio and the cable parameter significantly alter system response characteristics so we cannot use the taut string model without being significantly misled.

Hence, it is seen that considerable confusion and frustration occur when we try to interpret data with ideas from an inadequate model, as the data appears to be inconsistent and does not fit behavior predicted by these ideas.

REFERENCES

1. Bandstra, J. P., "Comparison of Equivalent Viscous Damping and Nonlinear Damping in Discrete and Continuous Vibrating Systems," *Journal of Vibration, Acoustics, Stress, and Reliability in Design*, Vol. 105, 1983, p. 382–392.
2. Bendat, J. S. and A. G. Piersol, *Measurement and Analysis of Random Data*, John Wiley & Sons, New York, 1966.
3. Bert, C. W., "Material Damping: An Introductory Review of Mathematical Models, Measures, and Experimental Techniques," *Journal of Sound and Vibration*, Vol. 29, No. 2, 1973, pp. 129–153.
4. Biot, M. A., "Analytical and Experimental Methods in Engineering Seismology," *Proceedings of the ASCE*, Jan. 1942.
5. Blevins, R. D., *Formulas for Natural Frequency and Mode Shape*, Van Nostrand Reinhold, New York, 1979.
6. Bogoliubov, N. N. and Y. A. Mitropolsky, *Asymptotic Methods in the Theory of Nonlinear Oscillations*, Hindustan Publishing, Delhi, India, 1961.
7. Broch, J. T., *Mechanical Vibration and Shock Measurements*, 2nd ed., available from Bruel & Kjaer Instruments, 1980.
8. Caughey, T. K., "Classical Normal Modes in Damped Linear Dynamics Systems," *Journal of Applied Mechanics*, Vol. 27, 1960, pp. 269–271.
9. Crandall, S. H., "The Role of Damping in Vibration Theory," *Journal of Sound and Vibration*, Vol. 11, No. 1, 1970, pp. 3–18.
10. Crandall, S. H. and W. D. Mark, *Random Vibration in Mechanical Systems*, Academic Press, New York, 1963.
11. Den Hartog, J. P., *Mechanical Vibrations*, 4th ed. McGraw-Hill, New York, 1956.
12. Ewins, D. J., "Measurement and Application of Mechanical Impedance Data, Part 1—Introduction and Ground Rules, Part 2—Measurement Techniques, Part 3—Interpretation and Application of Measured Data," *Journal of the Society of Environmental Engineers*, Dec. 1975–June 1976.
13. Ewins, D. J., *Modal Testing: Theory and Practice*, Research Studies Press, Ltd., Letchworth, Hertfordshire, UK (available from Bruel & Kjaer Instruments, Inc.), 1984.
14. Harris, C. M. (Ed), *Shock and Vibration Handbook*, 3rd ed. McGraw-Hill, New York, 1988.
15. Hayashi, C., *Nonlinear Oscillations in Physical Systems*, McGraw-Hill, New York, 1964.
16. Hurty, W. C. and M. R. Rubinstein, *Dynamics of Structures*, Prentice-Hall, Englewood Cliffs, NJ, 1964.
17. Klein, L., "Transverse Vibrations of Non-Uniform Beams," *Journal of Sound and Vibration*, Vol. 37, 1974, pp. 491–505.
18. Kreyszig, E., *Advanced Engineering Mathematics*, 4th ed. John Wiley, New York, 1979.

19. Meirovitch, L., *Analytical Methods in Vibration Analysis*, Macmillan, New York, 1967.
20. Mickens, R. E., *An Introduction to Nonlinear Oscillations*, Cambridge University Press, Cambridge, UK, 1981.
21. Minorsky, N., *Nonlinear Oscillations*, D. Van Nostrand, Princeton, NJ, 1962.
22. Mitchell, L. D., "Complex Modes: A Review," *Proceedings of the 8th International Model Analaysis Conference*, Kissimmee, FL, 1990.
23. Moon, F. C., *Chaotic Vibrations: An Introduction to Chaotic Dynamics for Applied Scientists and Engineers*, John Wiley & Sons, New York, 1987.
24. Nayfeh, A. H. and D. T. Mook, *Nonlinear Oscillations*, John Wiley and Sons, New York, 1979.
25. Otts, J. V., "Force-Controlled Vibration Tests; A Step Toward Practical Application of Mechanical Impedance," *Shock and Vibration Bulletin*, Vol. 34, No. 5, Feb. 1965, pp. 45–53.
26. Perkins, K. A. R., "The Effect of Support Flexibility on the Natural Frequencies of a Uniform Cantilever Beam," *Journal of Sound and Vibration*, Vol. 4, 1966, pp. 1–8.
27. Pipes, L. A. and L. R. Harvill, *Applied Mathematics for Engineers and Physicists*, 3rd Ed. McGraw-Hill, New York, 1970.
28. Rao, S. S., *Mechanical Vibrations*, 2nd ed., Addison-Wesley, Reading, MA, 1990.
29. Salter, J. P., *Steady-State Vibration*, Kenneth Mason Press, Homewell Hovant, Hampshire, UK, 1969.
30. Thomson, W. T., *Theory of Vibration with Applications*, 4th ed., Prentice Hall, Englewood Cliffs, NJ, 1988.
31. Timoshenko, T., D. H. Young, and W. Weaver, Jr., *Vibration Problems in Engineering*, 4th ed., John Wiley & Sons, New York, 1974.
32. Tomlinson, G. R., "A Simple Theoretical and Experimental Study of the Force Characteristics from Electrodynamic Exciters on Linear and Non-Linear Systems," *Proceedings of the 5th International Modal Analysis Conference*, 1987, pp. 1479–1486.

4 Transducer Measurement Considerations

Typical piezoelectric accelerometer transducers used in vibration testing. (Photo Courtesy of PCB Piezotronics, Inc.)

4.1 INTRODUCTION

Vibration testing requires using transducers to measure motion as well as forces. Many different methods have been developed to measure vibratory motion. These methods include full field video schemes, laser beam scanning schemes, optical fibers, magnetic sensors, and so on. These methods are not addressed in this book due to their high cost and limited use up to the present time. Some optical methods are becoming more popular as their costs continue to decline. There are certain problems where optical techniques are the only ones that can be used.

Force, torque, and moment transducers are often constructed by utilizing strain gages bonded to the structure. One engineer confided to me his great dislike for using strain gages because the results were so difficult to understand due to all of those structural resonances. Unfortunately, this test engineer did not understand the information conveyed by the strain gages, the actual forces and/or moments at a particular point in the structure. Unfortunately, this engineer was applying static mechanics of materials concepts to a dynamic mass-elastic system. It is the author's opinion that carefully located strain gages can be most informative as to the forces and/or moments that exist in a structure at a given point at each instant in time. The application of strain gages to structural testing is well documented in many excellent sources,[1] so these sensors are not discussed in any detail.

Piezoelectric sensing transducers are commonly used in accelerometers and force gages. Consequently, we explore the characteristics of these transducers in depth. The popularity of piezoelectric sensors is due to their small size, high stiffness, and high output. However, piezoelectric sensors are charge generators that require special amplifiers, called charge amplifiers, and built-in voltage followers. These amplifiers are explored in depth since their improper use can significantly affect our measurements.

The accelerometer can be represented by a single degree of freedom (DOF) mechanical device. This device's response to sinusoidal and transient motions are explored. The force transducer is a special device that interacts significantly with its environment. A unique two DOF model is developed that describes a force transducer when attached between the ground and a structure, when attached to an impact hammer to measure impulse forces, and when attached between a structure and a vibration exciter. In the latter case of exciter and structure, it is shown that the force transducer does not have a natural frequency, a most surprising result.

It is found that both accelerometers and force transducers are careless with the truth under certain conditions that cause errors in our measured frequency response functions (FRFs). Methods are described to correct for these FRF errors after the measurements are completed. These correction schemes are extremely helpful in reducing confusion in the measured vibration data and the resulting experimental vibration models.

This chapter ends with a look at calibration considerations. Some calibration schemes that can be used to check out a transducer under various simple test conditions are presented.

---

[1]J. W. Dally, W. F. Riley, and K. G. McConnell, *Instrumentation for Engineering Measurements*, 2nd ed., John Wiley and Sons, New York, 1993.

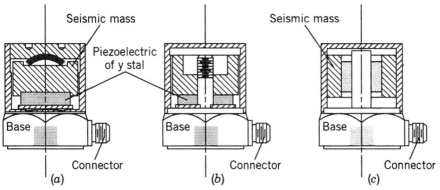

Fig. 4.2.1. Three common piezoelectric accelerometer designs. (*a*) Isolated compression. (*b*) Single-ended compression. (*c*) Shear.

4.2 MECHANICAL MODEL OF SEISMIC TRANSDUCERS— THE ACCELEROMETER

The accelerometer is the most commonly used vibration measurement transducer due to its small size, wide range of sensitivities, and large usable frequency range. Other transducers, such as displacement and velocity, have various shortcomings that the accelerometer does not have, such as size, weight, and usable frequency range. Two commonly used accelerometer sensing elements are strain gages and piezoelectric crystals. Of these two sensors, the piezoelectric is most popular since lightweight transducers with high sensitivity and natural frequencies are more easily obtained, due to the fact that the sensor also provides the transducer's spring and damping. The instrument scene is rapidly changing, due to micro machining techniques that allow us to produce new instruments.

Figure 4.2.1 shows three common accelerometer designs that are called isolated compression, single-ended compression, and shear. Each transducer design has a base, and a piezoelectric crystal and a *seismic mass*, which are contained within the protective case. The piezoelectric crystal and structural case (in some instances) combine to give an effective spring rate to support seismic mass m. The shear design reduces the transducer's sensitivity to base bending that occurs when an accelerometer is attached to structures with high bending strains. It is common to have mounting threads in the transducer's base and to use a special coaxial cable to connect the transducer to its interface electronics. Section 4.3 will discuss the characteristics of piezoelectric circuits, while this section concentrates on the mechanical characteristics of accelerometers.

4.2.1 The Basic Mechanical Model

The designs in Fig. 4.2.1 suggest that the accelerometer can be modeled as shown in Fig. 4.2.2. In Fig. 4.2.2a, we see that seismic mass m is attached to base mass m_b by a spring k and damping element c. There is an external force $f(t)$ acting on seismic mass m while force $f_b(t)$ acts on base mass m_b. $f_b(t)$ is the force required to hold the accelerometer to the test structure. The seismic mass motion is described by coordinate y, while base mass motion is described by coordinate x. If we assume $x > y$, then the free body diagram (FBD) of Fig. 4.2.2b results. Applying Newton's second law of motion and using the relative motion definition of $z = y - x$, we obtain

$$m\ddot{z} + c\dot{z} + kz = f(t) - m\ddot{x} \qquad (4.2.1)$$

as the fundamental differential equation of motion that describes the accelerometer's mechanical behavior. Equation (4.2.1) is a force balance that must occur on seismic mass m. It is composed of inertia, damping, and spring forces on the left-hand side and has two excitation terms on the right-hand side. The first excitation term $f(t)$ is constant in most situations since the transducer case is usually sealed so that only gravity forces can act on seismic mass m. The second excitation term is an inertial force due to base acceleration; that is, $m\ddot{x}$. The transducer is designed to measure this inertial force. The fact that there are two potential excitation terms should remind us of the possibility that we have a transducer that is sensitive to external force as well as base acceleration. This dual sensitivity is explored separately later in this section. Let us first see how we can measure acceleration.

Assume that base motion is a sinusoidal phasor given by $x = X_0 e^{j\omega t}$ and the corresponding response phasor is $z = Z_0 e^{j\omega t}$. Then, inserting these two phasors into Eq. (4.2.1) gives

$$Z_0 = \frac{-m\omega^2 X_0}{k - m\omega^2 + jc\omega} = \frac{-m\omega^2 X_0}{k(1 - r^2 + j2\zeta r)} \qquad (4.2.2)$$

which is the standard forced vibration response we observed in Chapter 3 for a single DOF system. We recognize that $-\omega^2 X_0 = a_0$ is the base acceleration's magnitude. Consequently, we can write Eq. (4.2.2) as

$$Z_0 = \frac{ma_0}{k - m\omega^2 + jc\omega} = \frac{ma_0}{k(1 - r^2 + j2\zeta r)} = H(\omega)a_0 \qquad (4.2.3)$$

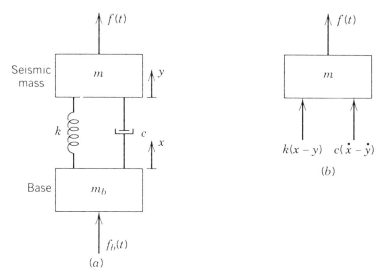

Fig. 4.2.2. Mechanical model of an accelerometer. (*a*) Schematic sketch. (*b*) Free body diagram (FBD). (From J. W. Dally, W. F. Riley, and K. G. McConnell, *Instrumentation for Engineering Measurements*, 2nd ed., copyright © 1993 by John Wiley & Sons, New York. Reprinted by permission.)

where $H(\omega)$ is the accelerometer's mechanical FRF that is given by

$$H(\omega) = \frac{m}{k - m\omega^2 + jc\omega} = \frac{m}{k(1 - r^2 + j2\zeta r)} \qquad (4.2.4)$$

where r is the dimensionless frequency ratio.
  ζ is the dimensionless damping ratio.
When the frequency is well below the transducer's natural frequency, Z_0 has a magnitude of

$$Z_0 = \frac{ma_0}{k} \qquad (4.2.5)$$

since the inertial ($m\omega^2$) and damping ($c\omega$) terms are negligible compared to stiffness k. Equation (4.2.5) shows that the desired behavior is that of relative motion of the spring element, which counteracts the inertial force of ma_0. This desired behavior is dependent on the accelerometer's FRF characteristic response given by Eq. (4.2.4). This is the same as the receptance plot shown in Fig. 3.3.3a, where it has a constant value of m/k when ω is very small compared to the natural frequency, passes through resonance when $\omega \cong \omega_n$ where damping controls the response, and then falls off as $1/\omega^2$ for frequencies above resonance. This means that the mechanical response is significantly altered with excitation frequency relative to the natural frequency since both magnitude and phase change with frequency ratio r.

We see from Fig. 3.3.3a and Eq. (4.2.4) that Eq. (4.2.5) is valid with a maximum error of 5 percent when $r < 0.2$. This means that an accelerometer is driven by inertial force ma_0 and this inertial force is resisted by transducer stiffness k so long as we are below 20 percent of the transducer's natural frequency. For a 10 percent error, the value of r increases to 0.32. Thus a convenient rule of thumb is that *a fifth gets you 5 percent error while a third gets you 10 percent error*.

The transducer's natural frequency is an important variable in predicting a usable frequency range. We are interested in what happens to this natural frequency if the transducer is suspended in midair and is not attached to any structure, giving a semidefinite two DOF system. The natural frequency of a two DOF semidefinite system is given by

$$\omega_n^* = \sqrt{\frac{k}{m}\left(1 + \frac{m}{m_b}\right)} = \omega_n \sqrt{1 + \frac{m}{m_b}} \qquad (4.2.6)$$

where it is evident that the natural frequency mounted on a very large mass ($m_b >>> m$) is the lowest possible transducer natural frequency. Hence, we only quote $\omega_n = \sqrt{k/m}$ as the natural frequency, since this is the lowest possible natural frequency that can occur and our errors will be less than those predicted using ω_n.

4.2.2 Gravity Forces and Acceleration Measurements

In the previous section we considered the external force $f(t)$ in Eq. (4.2.1) to be constant so that it can be zeroed out. The implication is that external gravity forces do not change with time. This assumption is not true. Gravity forces can change with time and can have significant effects. We use an example to illustrate this situation.

Consider the pendulum shown in Fig. 4.2.3a. An accelerometer is mounted to the bar at point D, which is located a distance l from the pendulum's pivot axis at O. The pendulum's mass center is at G at a distance of b from pivot axis O. Angle ϕ defines the pendulum's position relative to the vertical line through O, as shown.

We would like to measure the pendulum's acceleration at point D where the accelerometer is attached. If we draw a FBD of the seismic mass, as shown in Fig. 4.2.3b, we see that an external gravity force acts

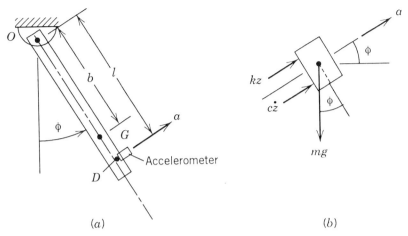

Fig. 4.2.3. Accelerometer mounted on a swinging pendulum. (*a*) Sketch of system. (*b*) FBD of seismic mass. (Fig. 4.2.3a from J. W. Dally, W. F. Riley, and K. G. McConnell, *Instrumentation for Engineering Measurements*, 2nd ed., copyright © 1993 by John Wiley & Sons, New York. Reprinted by permission.)

on this mass so that the external force becomes

$$f(t) = -mg\sin(\phi) \qquad (4.2.7)$$

when we use the direction of acceleration a as positive. Similarly, we know that the differential equation of motion for the pendulum is

$$I_0\ddot{\phi} + m_p g b \sin(\phi) = 0 \qquad (4.2.8)$$

where I_0 is the pendulum's mass moment of inertia about point O and m_p is the pendulum's mass. The acceleration of point D in the direction of a is given by the kinematic relationship of

$$\ddot{x} = l\ddot{\phi} \qquad (4.2.9)$$

Now, if we combine Eqs. (4.2.8) and (4.2.9), we obtain an expression for $\ddot{x}$. Then, using this $\ddot{x}$ expression and Eq. (4.2.7) in Eq. (4.2.1), we find that Eq. (4.2.1) becomes

$$m\ddot{z} + c\dot{z} + kz = m\left[\frac{m_p b l}{I_0} - 1\right]g\sin(\phi) \qquad (4.2.10)$$

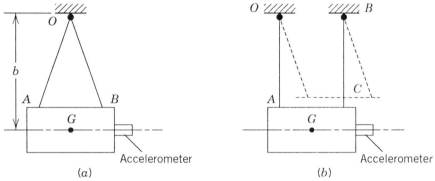

Fig. 4.2.4. (a) Accelerometer mounted on a rotating pendulum mass. (b) Accelerometer mounted on a translating mass with parallel bar support.

Now, assuming that the frequency of oscillation is well below the transducer's natural frequency, Eq. (4.2.10) effectively reduces to

$$z = \frac{m}{k}\left[\frac{m_p b l}{I_0} - 1\right] g \sin(\phi) \qquad (4.2.11)$$
$$\phantom{z = \frac{m}{k}\,}\text{expect}\quad\text{error}$$

The first bracketed term in Eq. (4.2.11) is the expected measurement, while the second term is the error due to gravity. However, we see that the local gravity force alters this measurement, sometimes significantly, since the bracketed term can be zero for the correct value of b in some cases.

Consider the following simple but dramatic example. Construct the pendulum shown in Fig. 4.2.4a where both support strings pass through point O. In this case, $b = l$ and $I_0 = m_p b^2$ so that Eq. (4.2.11) gives a zero value. Then construct the support system as shown in Fig. 4.2.4b. In this case, the mass moves with a translating motion so that its angle is always constant, and the accelerometer measures the pendulum's true acceleration. Why the difference? In Fig. 4.2.4a, the accelerometer's axis rotates with the body and has the same time function as the body. If angle ϕ is constant, this problem does not occur. These experiments are an interesting experience for my students who want to believe the transducer's output regardless of the evidence before them.

This problem points out a very subtle fact of dealing with accelerometers. If the body rotates so that the accelerometer's axis changes direction, then the gravity body force can generate significant errors. This gravity body force problem is particularly troublesome when working with large low frequency structures where significant rotations can take place at the test frequency and the acceleration levels are low. We can rewrite Eq. (4.2.11) as

$$z = -\frac{m}{k}[\ddot{x} + g\sin(\phi)] \qquad (4.2.12)$$

which gives us more perspective of the relative magnitudes that we are talking about. For example, suppose that our accelerometer is mounted at a structural node point for natural frequency ω_1. Then, $\ddot{x}$ is zero but we have $\phi = \phi_0 \sin(\omega_1 t)$ as the rotation angle for this resonance condition. Equation (4.2.12) shows that we measure an acceleration of

$$g_0 \phi_0 \sin(\omega_1 t)$$

that is not zero. Suppose that the excitation is broad-band random or an impulse. Then, Eq. (4.2.12) shows that the accelerometer output is a mixture of linear and rotational-gravity response terms. This effect is particularly troublesome at low frequencies where accelerations are relatively small so that $\ddot{x}$ and $g\phi_0$ have the same order of magnitude.

The problem with this type of measurement error is that there is no simple way to detect its presence unless it is so dramatic that there is no signal. Thus we should have a rule of thumb that states: *Changing orientation of an accelerometer's sensing axis with time generates signals due to gravity.* So beware: Accelerometers do not lie; they are just careless with the truth under certain circumstances.

The force transducer is usually presented along with the accelerometer as having the same mechanical model. Recent research into force transducer structure interaction has lead to a more refined model where the above model is a limiting case. Other cases, where a force transducer is attached to either an impact hammer or between a structure and vibration exciter, require a more advanced model to describe their behavior, as developed in Section 4.7.

4.3 PIEZOELECTRIC SENSOR CHARACTERISTICS

Accelerometers and force transducers often use piezoelectric sensing elements to achieve certain design natural frequencies, weight, and sensitivity characteristics. Piezoelectric sensing elements (crystals) are charge generating devices that require high input impedance signal conditioning instruments to interface between them and a recording device. Charge-amplifiers and built-in voltage-followers convert the generated charge into an output voltage that can be measured and recorded. We review the

4.3.1 Basic Circuits and Operational Amplifiers

There are three basic electrical circuit elements that are called *capacitance*, *resistance*, and *inductance*. In electrical circuits, the driving force or potential is called *voltage* while either *charge* or *current* represents the quantity moved. The relationship between voltage, capacitance, and charge is

$$E = \frac{q}{C} \tag{4.3.1}$$

where E is the voltage drop (with units of volts) across capacitance C (with units of farads) when a charge q (with units of coulombs) exists on capacitance C. Charge q and current I (with units of amperes) are related by

$$I = \dot{q} \tag{4.3.2}$$

Voltage, resistance, and current (or charge) are related by

$$E = RI = R\dot{q} \tag{4.3.3}$$

where R is resistance with units of ohms. Voltage, inductance, and current (or charge) are related by

$$E = L\dot{I} = L\ddot{q} \tag{4.3.4}$$

where L is inductance with units in henrys.

The *operational amplifier*, often simply called an op-amp, is an extremely useful element in modern instrumentation electronic circuits. It is useful in explaining how more complicated circuits work as well as used in such circuits. Figure 4.3.1a shows the symbol used for an op-amp. We see that it has two inputs, one labeled + for noninverting operation and the other labelled − for inverting operation. The output voltage is related to the input voltages by

$$E_0 = G(E_1 - E_2) \tag{4.3.5}$$

where G is the amplifier's *open loop gain*. G is very large, usually

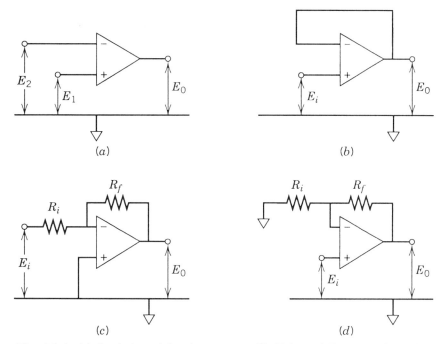

Fig. 4.3.1. (*a*) Symbol used for the op-amp. (*b*) Voltage follower configuration. (*c*) Inverting with gain configuration. (*d*) Noninverting with gain configuration. (Figs. 4.3.1*b* through 4.3.1*d* from J. W. Dally, W. F. Riley, and K. G. McConnell, *Instrumentation for Engineering Measurements*, 2nd ed., copyright © 1993 by John Wiley & Sons, New York. Reprinted by permission.)

$G > 100{,}000$. The large open loop gain is required to obtain the op-amp's useful performance characteristics.

The voltage follower configuration is connected as shown in Fig. 4.3.1*b*, for which the output voltage is related to the input voltage by

$$E_0 = E_i \tag{4.3.6}$$

The voltage follower's advantages are that it has a nearly infinite input resistance (to draw essentially zero current from the voltage source), that it has zero output impedance, and that it has adequate power to drive long signal lines. Consequently, it is nearly an ideal isolation amplifier between a sensor and a recorder device.

An op-amp connected as shown in Fig. 4.3.1c gives us an inverting amplifier with gain. The output voltage is

$$E_0 = -\frac{R_f}{R_i} E_i \qquad (4.3.7)$$

where R_f is the feedback resistance.

R_i is the input resistance.

This arrangement changes the signal's sign and gives it a voltage gain of R_f/R_i. When multiple voltages are inputted through multiple equal resistances R_i, the result is a voltage summing amplifier.

The noninverting amplifier with gain is shown in Fig. 4.3.1d. In this case, the output voltage is related to the input voltage by

$$E_0 = \left(1 + \frac{R_f}{R_i}\right) E_i \qquad (4.3.8)$$

where the terms in the brackets control the voltage gain. It is obvious that Eq. (4.3.8) becomes the same as Eq. (4.3.6) when R_f is zero.

We note that all of these input-output relationships are independent of open loop gain G so long as G is large, such as $>100{,}000$. This property of performance being independent of G makes the op-amp an attractive electronic circuit element. We use these properties in explaining piezoelectric circuit characteristics. Unfortunately, the above equations are limited by the amplifier's useful frequency range, usually expressed in terms of the gain bandpass product.

4.3.2 Charge Sensitivity Model

The piezoelectric crystal generates a charge when it is subjected to either normal or shearing stresses, depending on how the crystal is cut relative to its crystallographic planes. Our concern, as users, is to describe the instrument's overall performance, so we are not interested in how the crystal is cut. Experimentally, we have found that the charge generated is proportional to the crystal's deformation so that

$$q = S_z z \qquad (4.3.9)$$

where q is the charge generated by the crystal in *picocoulombs* (pC).

S_z is the *displacement charge sensitivity* (pC/unit of z).

z is the relative motion that the crystal experiences from its zero equilibrium position.

The value of S_z depends on the crystal material utilized as well as cross-sectional area and length. While Eq. (4.3.9) is helpful, we are interested in relating the charge generated to the unit being measured such as g's of acceleration or pounds or newtons of force. We have seen in the previous section that relative motion is proportional to the quantity being measured so that $z = Ka$. Then, Eq. (4.3.9) becomes

$$q = KS_z a = S_q a \qquad (4.3.10)$$

where S_q is the transducer's *charge sensitivity* (pC/unit of a such as pC/g or pC/lb or pC/newton, etc.).

K is the constant of proportionality, $1/k$ for force transducer or m/k for an accelerometer where k is the transducer spring rate and m is its seismic mass.

We use the linear charge-measurand relationship of Eq. (4.3.10) throughout this book.

4.3.3 The Charge Amplifier

The charge amplifier's operating characteristics can be explained in terms of two op-amps that are connected in series, as shown in Fig. 4.3.2. The first op-amp serves as the charge amplifier that converts charge q into voltage E_2 and has both resistive R_f and capacitive C_f feedback components. The calibration capacitor C_{cal} is often connected as shown. The calibration capacitance has no effect on normal operation when its input end is shielded but floating relative to ground. It is used to generate a calibration charge when a known voltage is applied.

The second op-amp is used to standardize the transducer's voltage sensitivity. It has an input resistance $R_1 = bR$ where b is the potentiometer position that ranges from zero to unity; that is, $0 < b < 1$.

The piezoelectric sensor consists of charge generator q and a capacitance C_t connected in parallel. The transducer has a very large resistance R_t that is in parallel with C_t, but this R_t has no effect on the system's performance so it is not shown. *Note: We must keep all connections clean to ensure there are no low resistance paths in parallel with these capacitors. Failure to maintain clean connections will cause serious measurement errors.* The cable is characterized by capacitance C_c, while the op-amp's input capacitance is characterized by C_a. Since the transducer, cable, and amplifier capacitances are in parallel, they can be replaced by a single input capacitance C that is given by

$$C = C_t + C_c + C_a \qquad (4.3.11)$$

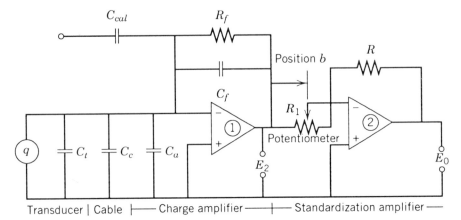

Fig. 4.3.2. Schematic diagram of a charge amplifier connected to a piezoelectric transducer. (From J. W. Dally, W. F. Riley, and K. G. McConnell, *Instrumentation for Engineering Measurements*, 2nd ed., copyright © 1993 by John Wiley & Sons, New York. Reprinted by permission.)

The differential equation that governs the charge-amp's output voltage E can be shown to be given by[2]

$$b\left[\frac{C}{G_1} + C_f\left(1 + \frac{1}{G_1}\right)\right]\dot{E} + b\left(1 + \frac{1}{G_1}\right)\frac{E}{R_f} = S_q \dot{a} \qquad (4.3.12)$$

where G_1 is the open-loop gain of the first op-amp.
 b is the second op-amp's resistance gain.
Equation (4.3.12) can be reduced to

$$\dot{E} + \frac{E}{R_f C_{eq}} = \left[\frac{S_q}{bC_{eq}}\right]\dot{a} \qquad (4.3.13)$$

where the equivalent capacitance is given by

$$C_{eq} = \frac{C}{G_1} + C_f = C_f\left[1 + \frac{C}{C_f G_1}\right] \qquad (4.3.14)$$

We see from Eq. (4.3.14) that input (source) capacitance C has little effect on this measurement system; this is because $C/C_f G_1$ is very small compared to unity in most cases, since G_1 is greater than 100,000 for most

---

[2]See J. W. Dally, W. F. Riley, and K. G. McConnell, *Instrumentation for Engineering Measurement*, 2nd ed., John Wiley and Sons, New York, 1993, Chapter 9.

op-amps. This term's effect on charge amplifier performance is often specified in terms of a maximum input capacitance C that is permitted for each feedback capacitance C_f. When the input capacitance limitation is satisfied, we find the equivalent capacitance C_{eq} is the feedback capacitance C_f. Then the charge-amplifier's governing differential equation becomes

$$\dot{E} + \frac{E}{R_f C_f} = \left[\frac{S_q}{bC_f}\right] \dot{a} \qquad (4.3.15)$$

The bracket quantity is the measurement system's voltage sensitivity S_v (units of volts/unit of variable a so that volts/g, etc., are common) as

$$S_v = \left[\frac{S_q}{b}\right]\left[\frac{1}{C_f}\right] = \frac{S_q^*}{C_f} \qquad (4.3.16)$$

where S_q^* is the standardized charge sensitivity. It is evident from Eq. (4.3.16) that two system parameters are available to control voltage sensitivity, b and C_f. Parameter b is used to convert a charge sensitivity S_q into a standard charge sensitivity S_q^* by making b numerically equal to transducer charge sensitivity. In this way, S_q^* has standard values of 1 or 10 or 100 pC/unit of a dependent on S_q being in the range of either 0.1 to 1 or 1 to 10 or 10 to 100 pC/unit of a. Once the potentiometer position parameter b is set, we can use feedback capacitor C_f to provide a wide range of standard voltage sensitivities. These sensitivities are usually printed on a range dial. Most charge amplifiers provide feedback capacitances that range from 10 to 50,000 pF in a 1–2–5–10 sequence.

The input-output FRF for the transducer is obtained by assuming that the input is described by phasor $a = a_0 e^{j\omega t}$ and output is described by phasor $E = E_0 e^{j\omega t}$. Insertion of these two phasors and Eq. (4.3.16) into Eq. (4.3.15) gives

$$E_0 = \left[\frac{jR_f C_f \omega}{1 + jR_f C_f \omega}\right] S_v a_0 \qquad (4.3.17)$$

where $R_f C_f$ is called the circuit's *RC time constant*. The corresponding FRF becomes

$$H(\omega) = \frac{E_0}{S_v a_0} = \frac{jR_f C_f \omega}{1 + jR_f C_f \omega} = \frac{\omega T}{\sqrt{1 + (\omega T)^2}} e^{j\phi}$$

$$T = R_f C_f \qquad (4.3.18)$$

$$\phi = \frac{\pi}{2} - \tan^{-1}(\omega T)$$

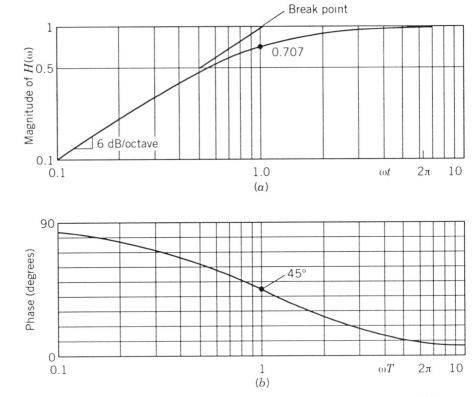

Fig. 4.3.3. Bode diagram of piezoelectric low frequency characteristics. (*a*) Log log magnitude plot. (*b*) Linear log phase plot. (From J. W. Dally, W. F. Riley, and K. G. McConnell, *Instrumentation for Engineering Measurements*, 2nd ed., copyright © 1993 by John Wiley & Sons, New York. Reprinted by permission.)

where T is the circuit time constant (seconds).

ϕ is the phase angle of the output phasor relative to the input phasor (radians).

The FRF's low frequency characteristics are shown in Fig. 4.3.3 in Bode diagrams using a log log magnitude plot (Fig. 4.3.3*a*) and linear log phase plot (Fig. 4.3.3*b*). We see that the magnitude curve starts out as a straight line with a slope of 6 dB/octave. This straight line intersects the unity curve at $\omega T = 1$ while the actual curve is low by 3 dB at $\omega T = 1$, an error of 30 percent since its value is 0.707 at this point. Then, the curve asymptotically approaches a value of unity. The straight line intersection point is called the *break point*, while the corresponding frequency is called the *break frequency*. The break frequency is most often quoted in specifications as the 3 dB point. The phase angle starts with a leading angle of

90 degrees and drops off to 45 degrees when $\omega T = 1$, eventually approaching zero for $\omega T > 100$.

The magnitude error is about 1 percent when $\omega T = 2\pi$ or $fT = 1$, so that the magnitude is 0.9876 and the phase angle is 9.04 degrees. These curves show that the low frequency FRF characteristic has significant effect on both magnitude and phase information. We shall find that, in some measurements, it is very important that this low frequency FRF be the same for all measurement channels; otherwise significant measurement errors can occur between data channels.

The charge amplifier has several distinct advantages. First, the system's performance is controlled by the internal feedback capacitance and resistance independent of source capacitance from transducer and cable so long as the source capacitance is less than the maximum allowed. Second, the output voltage sensitivity can be standardized by using the standardization potentiometer position b. Third, a wide range of voltage sensitivities are available by changing the feedback capacitance. Now we look at the characteristics of the built-in voltage follower circuit that has become popular.

4.3.4 Built-In Voltage Followers

Solid state electronics have progressed to the point where miniature unity gain voltage follower amplifiers are built inside the transducer's housing. A typical circuit schematic diagram is shown in Fig. 4.3.4a. The power supply is connected to the amplifier and transducer by a two wire shielded cable, while the output signal is connected to the recording instrument by a two wire shielded cable as well.

Cable capacitance C_c is eliminated from having any effect on system performance since this capacitance is on the output (low impedance) side of the built-in voltage follower. Both sensor capacitance C and the amplifier input resistance R are unaffected by environmental conditions other than temperature because these components are inside the transducer housing. Resistance R and capacitance C are selected to give a nominal voltage sensitivity, draw the correct current from the power supply, and provide a reasonable internal time constant $T = RC$.

The power supply in Fig. 4.3.4a consists of a current regulating diode (CRD), a dc supply voltage E_i, a meter M, and a coupling capacitance C_1. The CRD maintains a nominal +11 volts at the transistor source (S) when there is no input or transducer signal. Coupling capacitance C_1 shields the recording instrument from the 11 volt dc voltage. Meter (M) monitors the transducer cable connection. If the meter reads zero, a short exists in the transducer cable connections, while if the meter reads the supply voltage E_i, the cable connection is broken.

The operating characteristics of the circuit in Fig. 4.3.4a can be ex-

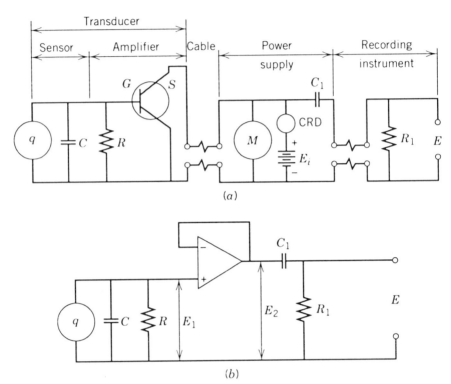

Fig. 4.3.4. The built-in voltage follower transducer system. (*a*) Schematic circuit. (*b*) Equivalent circuit. (From J. W. Dally, W. F. Riley, and K. G. McConnell, *Instrumentation for Engineering Measurements*, 2nd ed., copyright © 1993 by John Wiley & Sons, New York. Reprinted by permission.)

plained by using the equivalent circuit shown in Fig. 4.3.4b. The MOSFET amplifier is replaced with a unity gain voltage follower op-amp. The input side of the voltage follower amplifier yields one differential equation, while the output side yields a second differential equation. The input side gives

$$\dot{E}_1 + \frac{E_1}{RC} = \left[\frac{S_q}{C}\right]\dot{a} = S_v\dot{a} \qquad (4.3.19)$$

while the output side gives

$$\dot{E} + \frac{E}{R_1 C_1} = \dot{E}_2 = \dot{E}_1 \qquad (4.3.20)$$

The steady-state input-output FRF is obtained from Eqs. (4.3.19) and (4.3.20) as

$$H(\omega) = \frac{CE_0}{S_q a_0} = \left[\frac{jRC_1\omega}{1 + jRC_1\omega}\right]\left[\frac{jRC\omega}{1 + jRC\omega}\right] \quad (4.3.21)$$

Equation (4.3.21) has two time constants of

$$T = RC \quad \text{and} \quad T_1 = R_1 C_1 \quad (4.3.22)$$

$T = RC$ is the internal time constant that is determined by the manufacturer. T should be very large; however, amplifier current requirements limit the value of R and voltage sensitivity requirements (see Eq. (4.3.19)) limit the value of C. T usually ranges from 0.5 to 2000 seconds.

The external time constant $T_1 = R_1 C_1$ is controlled by the instrument user. While the blocking capacitor C_1 is fixed by power supply manufacturer, the readout instrument input resistance R_1 is selected by the user. Typically C_1 is 10 microfarads while readout instruments have input resistances that range from 0.01 MΩ to 1 MΩ. This gives a range of 0.1 to 10 seconds for external time constant T_1 so that it can easily become the controlling time constant. Magnitude and phase angle characteristics for the dual time constant instrument described by Eq. (4.3.21) are shown in Fig. 4.3.5. We see from Eq. (4.3.21) that we can either express T_1 as a multiple of T or T as a multiple of T_1. We have expressed T_1 as a multiple of T, using values of 1, 10, and 1000.

Figure 4.3.5 is a Bode diagram using log log magnitude plots and linear log phase angle plots. These plots show the effects of time constants and different time constant ratios on instrument response. The curve corresponding to $T_1 = 1000T$ starts with a slope of 6 dB/octave and follows the response for a single time constant instrument over the range shown. Thus we see that the second time constant has no significant effect on instrument behavior. However, when $T_1 = T$, low-frequency response drops off at a 12 dB/octave rate and the phase-angle rapidly approaches 180 degrees with decreasing ωT. For increasing ωT, we see that ωT must be close to 10 before the response approaches the line for $T_1 = 1000T$. When $\omega T = 1$ at break point (B), the amplitude is attenuated 6 dB (down 50 percent) and a 90 degree phase shift exits for this case. When $T_1 = 10T$, the curve has an additional break point at A ($\omega T = 0.1$) below which the attenuation drops off at 12 dB/octave and above which the attenuation rises at 6 dB/octave. This curve is seen to follow the single time constant instrument closely for values of ωT greater than 2. Hence, for all practical purposes, we see that the minimum time constant ratio is about 10 while an ideal ratio is over 100 to one. The important point of all this is that

180 TRANSDUCER MEASUREMENT CONSIDERATIONS

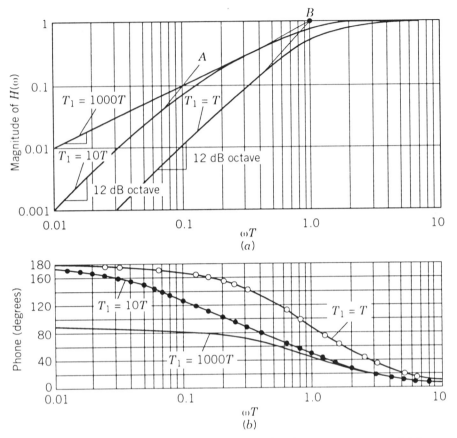

Fig. 4.3.5. Bode diagram for dual time constant low frequency response. (*a*) Log log magnitude plot. (*b*) Linear log phase plot. (From J. W. Dally, W. F. Riley, and K. G. McConnell, *Instrumentation for Engineering Measurements*, 2nd ed., copyright © 1993 by John Wiley & Sons, New York. Reprinted by permission.)

we users can seriously affect the instrument's low frequency performance by connecting the power supply output through capacitance C_1 to a recording instrument with too small of an input impedance. In this way, we seriously alter this low frequency response.

The built-in voltage follower system has several advantages. First, the manufacturer fixes the instrument's voltage sensitivity S_v so that we users have no voltage gains or range switches to adjust. Second, the cable capacitance and length have no significant effect on the output. Third, we obtain high-level output voltage signals along with low-level noise. Fourth, battery power supplies can be easily used in field applications since power requirements are modest. Fifth, it connects directly with most readout instruments. Sixth, it requires little technician training to set up compared to the more versatile charge-amp.

4.3.5 The Overall Accelerometer FRF

We are interested in examining the accelerometer's overall FRF. This is done by determining the output voltage when we take both mechanical effects from Eq. (4.2.4) and electrical effects from Eq. (4.3.18) into account to obtain

$$E_0 = \left(\frac{S_z}{C}\right)\left(\frac{m}{k(1 - r^2 + j2\zeta r)}\right)\left(\frac{jRC\omega}{1 + jRC\omega}\right) a_0 \qquad (4.3.23)$$

which can be written as

$$\frac{E_0}{S_v a_0} = \underbrace{\left(\frac{jRC\omega}{1 + jRC\omega}\right)}_{\text{electrical}} \underbrace{\left(\frac{1}{(1 - r^2 + j2\zeta r)}\right)}_{\text{mechanical}} = H_e(\omega) H_m(\omega) = H_a(\omega) \qquad (4.3.24)$$

since the voltage sensitivity is given by

$$S_v = \frac{S_q}{C} = \left(\frac{mS_z}{k}\right)\left(\frac{1}{C}\right) \qquad (4.3.25)$$

Equation (4.3.24) shows that the accelerometer's overall FRF is the product of the electrical and mechanical FRFs. If the accelerometer is using a dual time constant amplifier system, then the electrical FRF is given by Eq. (4.3.21) instead of Eq. (4.3.18).

A generic magnitude plot of Eq. (4.3.24) is shown in Fig. 4.3.6. In this plot, we see that the piezoelectric circuit controls the low frequency response that has a slope of 6 dB/octave. Between frequencies ω_1 and ω_2, we find ideal instrument behavior where there is minimum phase shift and amplitude distortion. For frequencies above ω_2, the mechanical FRF controls the response with a resonant peak that is followed by the 12 dB/octave drop-off. If the sensing element is a strain gage, the low frequency response remains unity down to zero frequency. We shall ex-

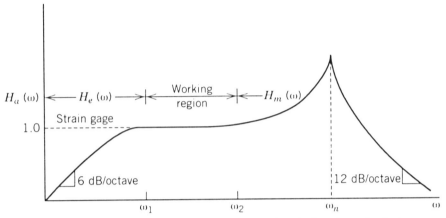

Fig. 4.3.6. Overall accelerometer FRF showing electrical and mechanical error regions as well as working region. (From J. W. Dally, W. F. Riley, and K. G. McConnell, *Instrumentation for Engineering Measurements*, 2nd ed., copyright © 1993 by John Wiley & Sons, New York. Reprinted by permission.)

plore the effects of low frequency roll-off on the instrument's ability to measure transient events in Section 4.5.

4.4 COMBINED LINEAR AND ANGULAR ACCELEROMETERS

There are many situations where we need to measure both linear and angular motions at a point on either a structure or a rigid body. We explore two techniques in this section. In the first, we use two accelerometers that are mounted apart from one another either directly on the structure or on a special mounting body. In the second, we use a recently developed transducer that measures both linear acceleration and angular acceleration in one lightweight unit that is attached to a single point. We describe the principles of operation of this new transducer.

4.4.1 Using Multiple Accelerometers to Measure Combined Motions

The two accelerometer method of measuring both linear and angular accelerations is depicted in Fig. 4.4.1. In this method, two accelerometers are mounted on a "rigid bar" with parallel sensing axis and are spaced a distance of $2l$ apart, as shown in Fig. 4.4.1a. The rigid bar is attached to the structure of interest at the rigid bar's midpoint.

The kinematic motion of each accelerometer is shown in Fig. 4.4.1b.

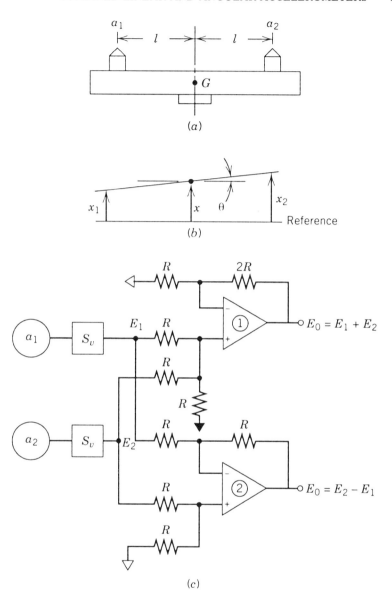

Fig. 4.4.1. Experimental setup to measure the combined linear and angular acceleration. (*a*) Mounting of two accelerometers on a rigid bar. (*b*) Kinematic diagram. (*c*) Circuit used to add and subtract acceleration signals.

184 TRANSDUCER MEASUREMENT CONSIDERATIONS

From this diagram, we can express the average midpoint motion described by x as

$$x = \frac{x_2 + x_1}{2} \tag{4.4.1}$$

where x_1 is the left end motion.
x_2 is the right end motion.

The average acceleration is obtained by differentiating Eq. (4.4.1) twice to give

$$\ddot{x} = \frac{a_1 + a_2}{2} \tag{4.4.2}$$

where a_1 and a_2 are the accelerations of the left and right ends, respectively. Similarly, angle θ is seen to be given by

$$\theta = \frac{x_2 - x_1}{2l} \tag{4.4.3}$$

so that the angular acceleration becomes

$$\ddot{\theta} = \frac{a_2 - a_1}{2l} \tag{4.4.4}$$

In Eq. (4.4.2), the two accelerations add together and are averaged. In Eq. (4.4.4), the two accelerations are subtracted from one another. In either case, we can have significant measurement errors for one quantity and not for the other. It turns out that the smaller signal has the larger error. For example, if the motion is mostly translation (x is large and θ is small), then the angle measurement has a larger error. However, if the motion is mostly rotational (θ is large and x is small), then the linear acceleration signal has the larger error.

The output voltage for each transducer is given by

$$E_1 = S_{v1} a_1 \tag{4.4.5}$$

and

$$E_2 = S_{v2} a_2 \tag{4.4.6}$$

where S_{vi} is the ith transducer's voltage sensitivity. When Eqs. (4.4.5) and (4.4.6) are substituted into Eqs. (4.4.2) and (4.4.4), we obtain

COMBINED LINEAR AND ANGULAR ACCELEROMETERS

$$\ddot{x} = \frac{S_{a2}E_2 + S_{a1}E_1}{2} = \frac{S_a}{2}(E_2 + E_1) \qquad (4.4.7)$$

for the average acceleration and

$$\ddot{\theta} = \frac{S_{a2}E_2 - S_{a1}E_1}{2l} = \frac{S_a}{2l}(E_2 - E_1) \qquad (4.4.8)$$

for the angular acceleration where S_{ai} is the *unit sensitivity* of the ith transducer. The unit sensitivity is related to the voltage sensitivity by

$$S_a = \frac{1}{S_v} \qquad (4.4.9)$$

It is clear from Eqs. (4.4.7) and (4.4.8) that it is convenient to have the same unit sensitivity for each measurement channel.

A summing and difference circuit utilizing op-amps is shown in Fig. 4.4.1c. In this circuit, it is assumed that both acceleration channels have the same voltage sensitivity S_v and that each acceleration measuring channel contains a voltage follower op-amp as its output interface so that it has sufficient power to drive the attached circuits without introducing additional error. Then, op-amp 1 is a noninverting summing amplifier with its gain adjusted so that its output is simply the sum of voltages E_1 and E_2. Op-amp 2 is set up to give the difference of input voltages E_2 and E_1. Other gains can be employed with this circuit so that the output voltages are linear acceleration in g's per volt and angular acceleration in rad/s² per volt.

The physical arrangement shown in Fig. 4.4.1 involves a number of assumptions. First, the "rigid bar" is rigid. This is not true when we go to high frequencies, for the bar becomes a vibrating beam with its own resonances, and so on. Second, the rigid body has no effect on the structure on which it is attached. This is not true, for the bar exhibits rigid body characteristics of mass m and mass moment of inertia about its mass center. These characteristics alter the structure's dynamic behavior, the amount being dependent on the structure's inertial characteristics versus those of the rigid bar. Third, we can mount the accelerometers directly on the structure and avoid all rigid body problems. Unfortunately, the kinematic assumption can be violated except in the structure's rigid body mode of vibration. The best we can hope for is an estimate of the structure's average motion and rotation. This approach is limited to lower frequencies.

A great deal of effort has been expended in trying to use the two accelerometer approach to measure the average linear acceleration and the rotational acceleration at a point in the structure. Recently a new accelerometer has been developed that measures both quantities at a single point. This combined linear and angular accelerometer is described next.

4.4.2 The Combined Linear and Angular Accelerometer

Angular seismic accelerometers are difficult instruments to design so that angular acceleration can be measured independently of linear acceleration. Micro machining and assembly techniques applied to man-made piezoelectric materials make new instruments available that can measure both linear and angular acceleration within the same transducer. We explore how this transducer design works.

Figure 4.4.2a shows that the transducer has a base, an enclosing case, a center post, and two electrically isolated cantilever beams. These beams are constructed from a piezoelectric ceramic material and are called *piezobeams* 1 and 2. The piezobeam's orientation is described by the *xyz* axes. Both *y*-direction linear acceleration and *z*-direction angular acceleration are sensed by this instrument. The piezobeam has four electrical connections that are indicated by symbols A, B, C, and D. Each piezobeam has one electrode on its entire upper surface and another electrode on its lower surface. These electrodes are connected at points A through D to internal amplifiers.

Figure 4.4.2b shows the charge distribution that occurs on the two piezobeams for a positive linear acceleration condition. This charge distribution requires the piezobeam's material to be properly orientated so that each beam's top is positive while its bottom is negative. The electrical charge's polarity is reversed when the beam accelerates downward. A positive angular acceleration condition is shown in Fig. 4.4.2c. For this case, the top of piezobeam 1 is negative while its bottom is positive. Similarly, the top of piezobeam 2 is positive while its bottom is negative.

The circuit in Fig. 4.4.2d consists of two voltage follower amplifiers that are connected to one of the piezobeams as indicated by $A\,C$ and $B\,D$. The voltage follower outputs are connected to the summing and difference amplifiers through a four wire connection cable. We assume that the linear and angular acceleration inputs are at frequencies well below the sensor's natural frequency. We also assume that the charge generated by each motion is proportional to that motion. Then, we can express the charge generated between each beam surface as

$$q_{A/C} = K_1 a_y - K_2 \alpha_z$$
$$q_{B/D} = K_1 a_y + K_2 \alpha_z$$

(4.4.10)

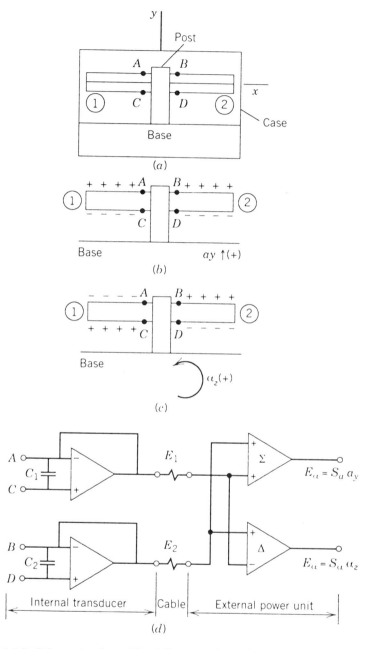

Fig. 4.4.2. Schematic of combined linear and angular accelerometer. (*a*) Dual piezobeam. (*b*) Charge distribution for linear acceleration. (*c*) Charge distribution for angular acceleration. (*d*) Circuit employed to separate linear and angular acceleration signals. (From J. W. Dally, W. F. Riley, and K. G. McConnell, *Instrumentation for Engineering Measurements*, 2nd ed., copyright © 1993 by John Wiley & Sons, New York. Reprinted by permission.)

for positive accelerations as shown. The K_i's are proportionality constants that should be the same for identical beams. We assume that each voltage follower amplifier has an output voltage that is proportional to the charge generated by the corresponding piezobeam so that for piezobeam 1, we have

$$E_1 = \frac{q_{A/C}}{C_1} = \frac{K_1 a_y - K_2 \alpha_z}{C_1} = S_1 a_y - S_2 \alpha_z \qquad (4.4.11)$$

where S_1 and S_2 are beam 1 voltage sensitivities. Similarly, we find the output voltage for piezobeam 2 is

$$E_2 = \frac{q_{B/D}}{C_2} = \frac{K_1 a_y + K_2 \alpha_z}{C_2} = S_3 a_y + S_4 \alpha_z \qquad (4.4.12)$$

where S_3 and S_4 are beam 2 voltage sensitivities. Now, if we add and subtract voltages E_1 and E_2 as shown in Fig. 4.4.2d, we find that

and
$$\begin{aligned} E_a &= E_1 + E_2 = (S_1 + S_3)a_y + (S_4 - S_2)\alpha_z \cong S_a a_y \\ E_\alpha &= E_2 - E_1 = (S_3 - S_1)a_y + (S_2 + S_4)\alpha_z \cong S_\alpha \alpha_z \end{aligned} \qquad (4.4.13)$$

Equation (4.4.13) shows us that we can achieve the desired results if relationships of $S_1 = S_3$ and $S_2 = S_4$ are satisfied. Ideally, we can make the two beams identical, but in reality such adjustments are nearly impossible. It is much simpler to provide a power unit with each transducer where we carefully adjust the gains so that the signals cancel and the instrument functions satisfactorily. Consequently, this accelerometer and its power supply unit are a functioning pair that cannot be separated. Current design has the voltage followers located in the transducer as shown. The summing and difference amplifiers with gain adjustments are located in the power unit. The system requires a four wire connection cable that contains two grounds and two positive supply voltages. The connection of this instrument to a recording device is identical to the built-in voltage follower described in Section 4.3. Thus each channel of this instrument has the same low frequency response characteristic as the dual time constant system.

The piezobeam design has a high voltage sensitivity that is in the order of 1000 mV/g (± 10 g range) for linear accelerations. Several angular acceleration sensitivities are available such as either 0.5, 5, or 50 mV per rad/s^2. The usable frequency range for these transducers is 0.5 to 2000 Hz for an 8 kHz natural frequency transducer. This frequency range is based on a ± 5 percent magnitude error. The major advantage of this lightweight transducer is that both linear and angular motion can be conveniently measured at a single point with minimum mass loading. When these transducers are arranged in a triaxial mounting so that all six degrees of motion at a single location are measured simultaneously, we have a powerful measurement system for some applications. The major disadvantages are the instrument's limited range (in terms of frequency, temperature, and magnitude of acceleration) as well as the requirement that transducer and power unit must be used as a single unit.

4.5 TRANSDUCER RESPONSE TO TRANSIENT INPUTS

So far we have looked at a transducer's steady state response, which generates its input output FRF. In this section, we are concerned with the effects that low frequency piezoelectric and high frequency mechanical resonance characteristics have on the transducer's response to a transient input. A transient input is characterized by rapid changes that occur over a short time period, preceded and followed by a nearly constant value for periods of time that are long in comparison to the measurement system's characteristic response times. More importantly, we need to know what telltale signs are available in the transducer's response that indicate that the transducer's ability to measure a given transient event are being exceeded. Finally, we relate the kinds of effects that a shock loading can have on both electrical and mechanical responses that are unusual and may be troublesome.

4.5.1 Mechanical Response

We need to know how a transducer's mechanical structure responds to transient inputs. Both step and ramp-hold inputs are classic signals that are used to develop theoretical transient transducer response characteristics. Understanding these theoretical responses gives us a means of detecting when a transducer's response capabilities are exceeded.

Step Excitation A step input is shown in Fig. 4.5.1a where the acceleration is zero for negative time, equal to a_0 for positive time, and changes instantaneously at zero time. The response to this transient input is obtained by solving the transducer's differential equation of motion (see Eq. (4.2.1)), which becomes

$$m\ddot{z} + c\dot{z} + kz = -ma_0 \tag{4.5.1}$$

for this case. The solution to Eq. (4.5.1) when initial conditions of $z(0) =$

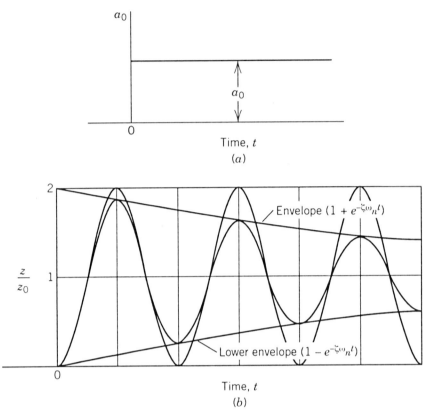

Fig. 4.5.1. Classic step input acceleration. (*b*) Mechanical response for 0 and 5 percent damping.

$0 = \dot{z}(0)$ are applied is given by

$$z = z_0 \left[1 - \frac{e^{-\zeta \omega_n t}}{\sqrt{1 - \zeta^2}} \cos(\omega_d t - \psi) \right] \quad (4.5.2)$$

where $z_0 = -ma_0/k$ is a constant deflection under a load of ma_0.
  $\omega_d = \omega_n \sqrt{1 - \zeta^2}$ is the damped damped frequency.
  $\tan \psi = \zeta/\sqrt{1 - \zeta^2}$ is the response phase angle.
When damping ratio ζ is zero, this response reduces to

$$z = z_0[1 - \cos(\omega_n t)] \quad (4.5.3)$$

The responses described by Eqs. (4.5.2) and (4.5.3) are shown in Fig.

4.5.1b. A positive dimensionless ratio of z/z_0 is used in plotting since accelerometers are designed to give a positive output if $a(t)$ is a positive input. We see that this mechanical response is composed of a damped oscillation at the transducer's damped frequency about the line of unity and is bounded by upper and lower exponential envelopes. The maximum overshoot occurs in the undamped case and is limited to a value of 2. This response does not look at all like the step input, and is called *transducer ringing* since the transducer is "ringing like a bell." This ringing response is a telltale sign that the transient input occurs too fast for the transducer to mechanically respond.

Obviously, a transducer's "ringing response" bears little resemblance to the step input that excited it. This response gives us an indication of the type of measurement errors that can occur when the transducer is forced too rapidly. We consider the ramp-hold time history next since this time history becomes a step input as a limiting case. In this way, we can gain insight into the relationships that govern the basic question of: How fast can a transducer respond to a transient input? It is obvious that a step input is far too rapid.

Ramp-Hold Excitation A typical ramp-hold input function is shown in Fig. 4.5.2a. The input acceleration increases linearly from 0 to a_0 over time period t_0 and remains at a_0 for all time greater than t_0. This input acceleration is expressed mathematically as

$$a(t) = \frac{a_0 t}{t_0} \quad \text{for } 0 \le t \le t_0 \quad (4.5.4)$$

$$a(t) = a_0 \quad \text{for } t > t_0$$

The undamped response to this acceleration input is available in most vibrations textbooks[3] and can be expressed in terms of a reference deflection ($z_0 = ma_0/k$) as

$$\frac{z}{z_0} = \frac{t}{t_0} - \frac{\sin(\omega_n t)}{\omega_n t_0} \quad \text{for } 0 < t < t_0 \quad (4.5.5)$$

$$\frac{z}{z_0} = 1 - \left[\frac{\sin(\omega_n t) - \sin(\omega_n(t - t_0))}{\omega_n t_0}\right] = 1 - \frac{D \sin(\omega_n t - \beta)}{\omega_n t_0} \quad (4.5.6)$$

---

[3] William T. Thomson, *Theory of Vibration with Practice*, 4th ed., Prentice Hall, Englewood Cliffs, NJ, 1988, pp. 100–101.

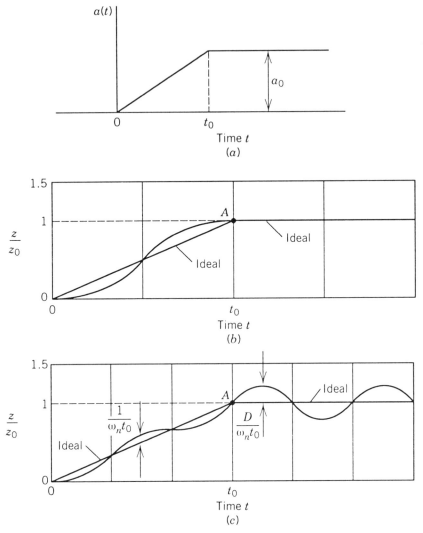

Fig. 4.5.2. (*a*) Ramp type input acceleration. (*b*) Accelerometer response $t_0 = T$. (*c*) Accelerometer response $t_0 = 1.5T$. (Figs. 4.5.2*a* and 4.5.2*c* from J. W. Dally, W. F. Riley, and K. G. McConnell, *Instrumentation for Engineering Measurements*, 2nd ed., copyright © 1993 by John Wiley & Sons, New York. Reprinted by permission.)

where

$$D = \sqrt{2\{1 - \cos(\omega_n t_0)\}} \quad \text{and} \quad \tan\beta = \frac{\sin(\omega_n t_0)}{\{1 - \cos(\omega_n t_0)\}} \quad (4.5.7)$$

The first term in Eq. (4.5.5) is the ideal straight line ramp, while the second term is the transducer's oscillation at its natural frequency about the ramp with an amplitude of $1/(\omega_n t_0)$. This response is illustrated in Figs. 4.5.2b and 4.5.2c. The maximum displacement error at ramp termination (point A) is obtained from Eq. (4.5.5) when $\omega_n t_0$ to is an odd multiple of $\pi/2$ so that

$$\frac{z}{z_0} = 1 \pm \frac{1}{\omega_n t_0} \quad (4.5.8)$$

where the $\pm$ sign is dependent on the sine function being $\pm$ unity.

Both Figs. 4.5.2b and 4.5.2c show cases where $\omega_n t_0$ is a multiple of π. In Fig. 4.5.2b, $\omega_n t_0 = 2\pi$ so that there is no oscillation after point A, but there are large errors along the ramp portion of the input, particularly when t is close to zero. The reason there is no oscillation for $t > t_0$ is that both magnitude and slope match at point A. Figure 4.5.2c shows the case $\omega_n t_0 = 3\pi$ where the amplitude matches at point A while the response has positive slope so that overshoot occurs. We see from Eq. (4.5.6) and Fig. 4.5.2c that the transducer's response after $t > t_0$ is an oscillation at its natural frequency with an amplitude $D/(\omega_n t_0)$. The amplitude of oscillation is seen to vary from zero to $2/(\omega_n t_0)$ depending on value of $\omega_n t_0$. When $\omega_n t_0$ is a multiple of 2π, the value of D is zero. This happens because a complete cycle of oscillation occurs within the ramp time t_0 so that there is no magnitude error and the slope is zero at the end of the ramp. However, when $\omega_n t_0$ is an odd multiple of π, a maximum error of

$$\eta = 2/(\omega_n t_0) \quad (4.5.9)$$

occurs.

The practical result of this analysis is that at least five transducer oscillations must occur during the ramp period if the maximum deviation after time t_0 is limited to 6.4 percent or that at least ten transducer oscillations are required to limit the maximum deviation to less than 3.2 percent. Thus the shortest ramp time duration is $t_0 > 5T$ in order to reduce ringing type of errors. Obviously, transducer damping helps reduce these errors, but remember that damping is small in most transducers so the undamped model gives a slightly conservative estimate in most instances.

194 TRANSDUCER MEASUREMENT CONSIDERATIONS

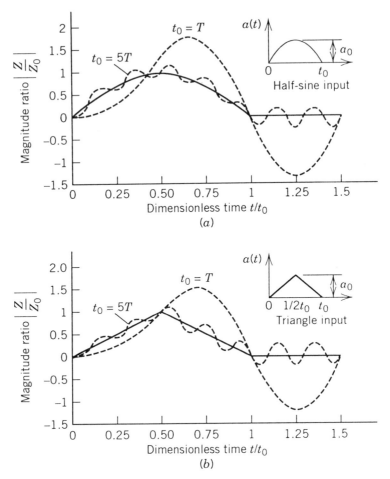

Fig. 4.5.3. Mechanical responses of an accelerometer to transient pulses for pulse durations of $t_0 = T$ and $t_0 = 5T$. (*a*) Half-sine transient. (*b*) Triangular transient. (From J. W. Dally, W. F. Riley, and K. G. McConnell, *Instrumentation for Engineering Measurements*, 2nd ed., copyright © 1993 by John Wiley & Sons, New York. Reprinted by permission.)

Two Common Transient Waveforms and the 5 and 10 Rule The half-sine and triangle transients, as shown in Figs. 4.5.3*a* and 4.5.3*b*, are used as further evidence to support our current observations based on the ramp function. The undamped transducer responses are shown for pulse durations of T and $5T$ where T is the transducer's natural period. These plots clearly indicate that a reasonable transducer output requires that the pulse duration be at least five times the transducer's natural period. Thus we can state a general rule of thumb that can be called the five times rule: Either the transient rise time or fall time or the duration time must be greater than five times the transducer's natural period. The 5 and 10 rule is that: 5 is good but 10 is better. A transducer that is "ringing" at its installed natural frequency is a clear indication that these rules are being violated and an invalid measurement has occurred.

4.5.2 Piezoelectric Circuit Response to Transient Signals

Transient signals contain DC signal components that are blocked by the piezoelectric sensor's AC-coupled response characteristic. The differential equation controlling charge amplifier behavior is given by Eq. (4.3.15) as

$$\dot{E} + \frac{E}{RC} = S_v \dot{a} \tag{4.5.10}$$

where S_v is the voltage sensitivity. Equation (4.5.10) has a particular solution for any transient signal that can be calculated using

$$E = S_v e^{-t/RC} \int_0^t e^{\tau/RC} \dot{a}(\tau) \, d\tau \tag{4.5.11}$$

where τ is a dummy variable of integration.

In order to see the effect of low frequency signal attenuation on measuring transient signals, let us consider this circuit's response to a rectangular pulse of duration t_1 as shown in Fig. 4.5.4a. We know that such a pulse is physically impossible to generate with a mechanical system, but it is a useful limiting case for judging the piezoelectric circuit's low-frequency response characteristics.

Equation (4.5.11) requires an expression for the input's derivative. The derivative for this case is shown in Fig. 4.5.4b. It is composed of two *Dirac impulse functions* $\delta(\tau)$ and can be expressed as

$$\dot{a} = a_0 \, \delta(\tau) - a_0 \, \delta(\tau - t_1) \tag{4.5.12}$$

The Dirac delta function's property of unity area when integrated over the argument's zero region is used when Eq. (4.5.12) is inserted into Eq. (4.5.11) to give

$$E = S_v a_0 [u(t) \, e^{-t/RC} - u(t - t_1) \, e^{-(t-t_1)}] \tag{4.5.13}$$

where $u(t - t_1)$ is the *unit-step function*. Figure 4.5.4c shows the electrical response described by Eq. (4.5.13). This circuit has two significant charac-

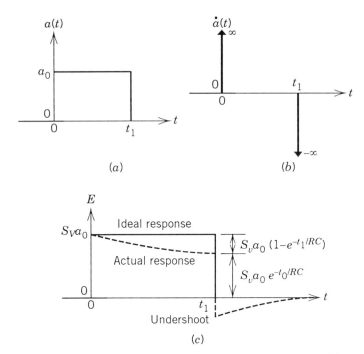

Fig. 4.5.4. Charge amplifier response to rectangular transient pulse. (*a*) Transient pulse. (*b*) Derivative of pulse. (*c*) Output voltage. (From J. W. Dally, W. F. Riley, and K. G. McConnell, *Instrumentation for Engineering Measurements*, 2nd ed., copyright © 1993 by John Wiley & Sons, New York. Reprinted by permission.)

teristics. First, the output signal decays exponentially when $0 < t < t_1$. This exponential decay generates an error of $(1 - e^{-t/RC})$ when compared to the input. Second, an undershoot occurs at the end of the pulse. The undershoot is the maximum error that will occur for this pulse.

We can estimate the maximum error that is associated with Eq. (4.5.13) from a series expansion of $(1 - e^{-t_1/RC})$. This expansion gives

$$\eta_{max} = \frac{t_1}{RC}\left[1 - \frac{1}{2}\frac{t_1}{RC} + \frac{1}{6}\left(\frac{t_1}{RC}\right)^2 - \cdots\right] \qquad (4.5.14)$$

For small values of $t_1/(RC)$, Eq. (4.5.14) reduces to

$$\eta_{max} = \frac{t_1}{RC} \quad \text{or} \quad T = RC = \frac{t_1}{\eta_{max}} \qquad (4.5.15)$$

TABLE 4.5.1. Time Constant (T = RC) Requirements for Various Error Levels

Pulse Shape		2% Error	5% Error	10% Error
Rectangular pulse		$50 t_1$	$20 t_1$	$10 t_1$
Triangular pulse		$25 t_1$	$10 t_1$	$5 t_1$
Half-sine pulse	A	$16 t_1$	$6 t_1$	$3 t_1$
	B	$31 t_1$	$12 t_1$	$6 t_1$

Source: From J. W. Dally, W. F. Riley, and K. G. McConnell, *Instrumentation for Engineering Measurements*, 2nd ed., copyright © 1993 by John Wiley & Sons, New York. Reprinted by permission.)

We can use Eq. (4.5.15) to calculate the time constant versus percent error requirements that are listed in Table 4.5.1 for the rectangular pulse.

We often encounter triangular and half-sine transient pulse shapes while making measurements. Table 4.5.1 gives the required time constants for maintaining specified measurement errors for these pulse shapes as well. We see from this table that for a given error limit, the rectangular pulse has the most severe requirements. Table 4.5.1 can be used to estimate measurement errors for similar transient waveforms. A good indicator of inadequate *RC* time constant in a given transient measurement is the presence of undershoot and its telltale exponential decay.

Transducer systems that use the built-in voltage follower are a special problem since two time constants are involved. Analysis of a step input pulse shows that an effective time constant T_e can be obtained by comparing the initial slope of the dual time constant response to that of a single time constant system.[4] The resulting equivalent time constant is found

---

[4]See: J. W. Dally, W. F. Riley, and K. G. McConnell, *Instrumentation for Engineering Measurements*, 2nd ed., John Wiley & Sons, New York, 1993, p. 318.

to be

$$T_e = \frac{T_1 T}{T_1 + T} \tag{4.5.16}$$

We can use this value of T_e in place of $T(=RC)$ in Table 4.5.1 to estimate the maximum errors associated with a given measurement.

4.5.3 Field Experience with Shock Loading

Vibration test personnel are being requested to test to higher and higher frequencies, particularly when dealing with pyroshock loading. Consequently, considerable effort is being expended to design and understand high frequency transducer and electronic system performance. A paper by Chu[5] explored the effects of high natural frequency "ringing" spikes on a charge amplifier's output signal. He used the circuit shown in Fig. 4.5.5a, where a programmable waveform generator is used to create a repeatable input voltage through the high speed amplifier. The voltage variation used is shown in Fig. 4.5.5b, where it is seen to reach nearly +40 volt peak in about 0.5 μs and a nearly −40 volt peak in slightly over 1 μs. This waveform is typical of pyroshock ringing for shock accelerometers with natural frequencies around 300 kHz. The 1000 pF calibration capacitance is used to generate the charge input. The output voltages ranged from charge amplifier lock-up, as shown in Fig. 4.5.6a, to excellent response, as shown in Fig. 4.5.6b, to distorted response caused by a 10 kHz filter as shown in Fig. 4.5.6c. Note the different time scales are used in displaying each response.

The results of this study showed that some charge amplifiers cannot be used for pyroshock applications due to inadequate response, while others can be used satisfactorily. The category that a given charge amplifier falls into cannot be predicted ahead of time by looking at its specifications. Generally, it appears that the higher the maximum charge that is acceptable and the higher the upper 3 dB bandpass frequency that a charge amplifier has, the better are its chances of performing in an acceptable manner. However, the built-in filters were found to cause significant distortion where the transient pulse shape is removed by the filter and replaced by a totally false hump, as shown in Fig. 4.5.6c. This hump can cause serious distortion of any shock response spectra (SRS) that are calculated from this erroneous response. Chu's paper is recommended reading for anyone doing pyroshock measurements, for the results should

---

[5]A. S. Chu, "A Shock Amplifier Evaluation," *Proceedings Institute of Environmental Sciences*, Apr. 1990, pp. 708–719.

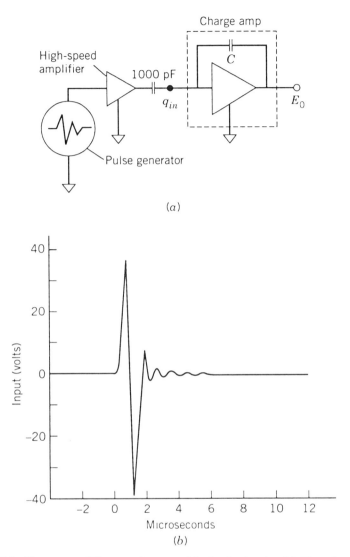

Fig. 4.5.5. Charge amplifier test for use with shock signals. (*a*) Test circuit. (*b*) Input voltage waveform. (Figure adapted from A. S. Chu, "A Shock Amplifier Evaluation," *Proceedings, Institute of Environmental Sciences*, Apr. 1990, pp. 708–719. Reprinted with permission.)

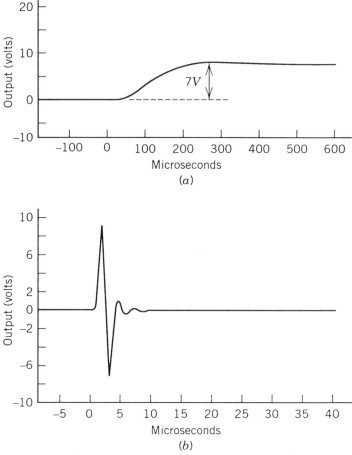

Fig. 4.5.6. Three different types of shock rated charge amplifier outputs when subjected to the same charge pulse shown in Fig. 4.5.5b. (*a*) Lock-up response. (*b*) Tracking response. (*c*) 10 kHz filtered response. (Figure adapted from A. S. Chu, "A Shock Amplifier Evaluation," *Proceedings, Institute of Environmental Sciences*, Apr. 1990, pp. 708–719. Reprinted with permission.)

cause goose bumps in those who have not checked out their own charge amplifiers in a similar fashion.

A somewhat similar war story was personally related to the author by a test engineer some years ago. The engineer and his colleagues had been using specially designed charge amplifiers that had to endure the same shock loading as the structure under test. The test data looked "different" when compared to previous data from similar tests; it exhibited some curious looking signals. After some thought, it was decided to test the

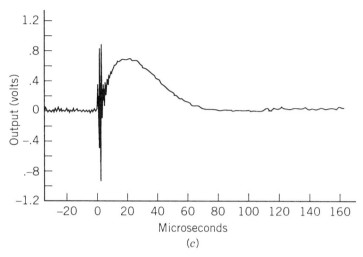

Fig. 4.5.6. (*Continued*).

special charge amplifier with a similar shock loading but with the transducer inactive. The results revealed that the charge amplifier was also functioning as a transducer for these shock loads, generating its own electrical signals. The charge amplifier's development program had not included this type of test, and hence, its sensitivity to shock accelerations was unknown. Apparently, it did not respond to sinusoidal loading in the same way that it responded to shock loading, possibly due to differences in test levels and frequency range.

Today, we are using more transducers with built-in voltage follower amplifiers. We can expect these transducers to function normally in most environments. However, there are situations where they may be susceptible to a phenomenon similar to that of the special charge amplifier; they may give spurious measurements. Thus comparison tests with transducers known to work properly in a given environment are recommended as a check on the suitability of these accelerometers and/or force transducers for shock environments that are considerably different from those encountered in general vibration environments.

4.6 ACCELEROMETER CROSS-AXIS SENSITIVITY

We would like to believe that accelerometers are sensitive to motion in only one direction. Unfortunately, the actual sensing direction is not the direction that we think of as the transducer's primary sensing axis. Consequently, we must accept the premise that accelerometers also sense motion in a plane that is perpendicular to their primary sensing axis. We call this sensitivity to orthogonal acceleration components *cross-axis sensitivity*. Cross-axis sensitivity is primarily due to manufacturing imperfections, even in well designed transducers. Generally, the maximum transverse sensitivity is less than 5 percent, but Han and McConnell[6] have shown that even this small amount of cross-axis sensitivity can seriously contaminate experimental FRFs that are used in experimental modal analysis. Our purpose in making acceleration measurements is to obtain the acceleration vector at a given point in a structure so we can do different kinds of vibration analysis. The significance of accelerometer cross-axis sensitivity is explored in this section.

4.6.1 The Single Accelerometer Cross-Axis Sensitivity Model

McConnell and Han[7] have developed a theoretical model to explain single axis accelerometer cross-axis characteristics and have extended it to include tri-axial accelerometers. In these subsections, we follow the approach they used.

The accelerometer is oriented with the z-axis as its primary sensing axis, while the x and y axes form the cross-axis plane, as shown in Fig. 4.6.1. We represent the accelerometer's voltage sensitivity by vector $\bar{\mathbf{S}}_0$ so that voltage sensitivity has both magnitude and direction and can be written in terms of unit vectors $\mathbf{i}$, $\mathbf{j}$, and $\mathbf{k}$ as

$$\mathbf{S}_0 = \{S_0 \sin(\phi) \cos(\theta)\}\mathbf{i} + \{S_0 \sin(\phi) \sin(\theta)\}\mathbf{j} + \{S_0 \cos(\phi)\}\mathbf{k} \quad (4.6.1)$$

where ϕ is the angle between the sensitivity vector and the z-axis.

θ defines a vertical plane formed by vector $\mathbf{S}_0$ and the z-axis.

The acceleration vector $\mathbf{a}$ that we want to measure can also be expressed in this coordinate system with its components as

$$\mathbf{a} = a_x \mathbf{i} + a_y \mathbf{j} + a_z \quad (4.6.2)$$

It is hypothesized that the output voltage is the dot product of acceleration and voltage sensitivity vectors so that $E_z = \mathbf{S}_0 \cdot \mathbf{a}$ or

$$E_z = \{S_0 \sin(\phi) \cos(\theta)\}a_x + \{S_0 \sin(\phi) \sin(\theta)\}a_y + \{S_0 \cos(\phi)\}a_z \quad (4.6.3)$$

---

[6] S. Han and K. G. McConnell, "The Effects of Transducer Cross-Axis Sensitivity in Modal Analysis," *Proceedings of the 7th International Modal Analysis Conference*, Vol. I, Las Vegas, NV, Jan. 1989, pp. 505–511.

[7] K. G. McConnell and S. Han, "A Theoretical Basis for Cross-Axis Corrections in Tri-Axial Accelerometers," *Proceedings of the 9th International Modal Analysis Conference*, Vol. I, Florence, Italy, Apr. 1991, pp. 171–175.

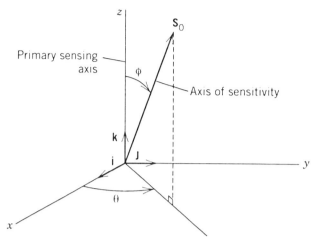

Fig. 4.6.1. Orientation of axis of sensitivity, primary sensing axis, and cross-axis plane for an accelerometer. (Courtesy of the *Proceedings for the 9th International Modal Analysis Conference*, Society for Experimental Mechanics.)

Since angle ϕ is small, we can approximate $\cos(\phi)$ as unity and $\sin(\phi)$ as ϕ so that Eq. (4.6.3) becomes

$$E_z = S_{zx}a_x + S_{zy}a_y + S_{zz}a_z \qquad (4.6.4)$$

where the cross-axis voltage sensitivities are

$$S_{zx} = S_0 \sin(\phi) \cos(\theta) \cong (S_0\phi) \cos(\theta) \qquad (4.6.5)$$

$$S_{zy} = S_0 \sin(\phi) \sin(\theta) \cong (S_0\phi) \sin(\theta) \qquad (4.6.6)$$

and the primary voltage sensitivity is

$$S_{zz} = S_0 \cos(\phi) \cong S_0 \qquad (4.6.7)$$

It is important to realize that the output voltage is a scalar value that is independent of coordinates employed since it comes from a dot product. Now, we apply these equations to a tri-axial accelerometer.

4.6.2 The Tri-Axial Accelerometer Cross-Axis Sensitivity Model

A tri-axial accelerometer is constructed by either mounting three single axis accelerometers on a mounting block's orthogonal surfaces or by a manufacturer's assembling three orthogonal single axis accelerometers inside of a mounting block. In either case, there are three orthogonal axes, one for each accelerometer. The output voltage for each of these accelerometers is described by an equation that is a subscript perturbation of Eq. (4.6.4). Thus we have

$$S_{xx}a_x + S_{xy}a_y + S_{xz}a_z = E_x$$
$$S_{yx}a_x + S_{yy}a_y + S_{yz}a_z = E_y \quad (4.6.8)$$
$$S_{zx}a_x + S_{zy}a_y + S_{zz}a_z = E_z$$

which can be written in matrix form as

$$[S]\{a\} = \{E\} \quad (4.6.9)$$

where $[S]$ is the transducer's *voltage sensitivity matrix*.

The cross-axis voltage sensitivity in a tri-axial accelerometer is determined when the transducer assembled. However, we saw earlier that the output voltage is independent of the coordinate axes used, since the output voltage is the result of a dot product that has no coordinate dependency. This means that it is sufficient to determine all voltage sensitivities by calibration for a given tri-axial accelerometer and that these sensitivities can be used to correctly measure any given acceleration vector. Now we examine how we can correct our voltage readings to obtain a better estimate of the actual acceleration vector.

4.6.3 Correcting Tri-Axial Acceleration Voltage Readings

We need to solve for acceleration vector $\{a\}$ in Eq. (4.6.9). This is done by dividing each equation in Eq. (4.6.8) by its corresponding primary voltage sensitivity S_{ii}. This procedure gives us a set of linear equations that relate the actual acceleration components a_i to the apparent acceleration components b_i such that

$$\epsilon_{xx}a_x + \epsilon_{xy}a_y + \epsilon_{xz}a_z = b_x$$
$$\epsilon_{yx}a_x + \epsilon_{yy}a_y + \epsilon_{yz}a_z = b_y \quad (4.6.10)$$
$$\epsilon_{zx}a_x + \epsilon_{zy}a_y + \epsilon_{zz}a_z = b_z$$

where

$$\epsilon_{pi} = \frac{S_{pi}}{S_{pp}} \quad (4.6.11)$$

is the *cross-axis sensitivity coefficient* that relates the ith direction's acceleration component's contribution to the pth direction's output voltage. Note that $\epsilon_{pp} = 1$ while the pth accelerometer's apparent acceleration is given by

$$b_p = \frac{E_p}{S_{pp}} \qquad (4.6.12)$$

It is clear from Eq. (4.6.10) that a_i and b_i may be essentially equal for some directions and may vary considerably for other directions. Now, we can write Eq. (4.6.10) in matrix notation as

$$[\epsilon]\{a\} = \{b\} \qquad (4.6.13)$$

Since we want the true acceleration vector $\{a\}$, we need to obtain the inverse of $[\epsilon]$. If we let $[C] = [\epsilon]^{-1}$ be the *correction matrix*, we can rewrite Eq. (4.6.13) as

$$\{a\} = [C]\{b\} \qquad (4.6.14)$$

Generally, we find that the values of ϵ are less than 5 to 7 percent so that the correction matrix can be approximated by

$$[C] = \begin{bmatrix} 1 & -\epsilon_{xy} & -\epsilon_{xz} \\ -\epsilon_{yx} & 1 & -\epsilon_{yz} \\ -\epsilon_{zx} & -\epsilon_{zy} & 1 \end{bmatrix} \qquad (4.6.15)$$

and the input accelerations from Eq. (4.6.14) can be estimated from

$$\begin{aligned} a_x &= b_x - \epsilon_{xy}b_y - \epsilon_{xz}b_z \\ a_y &= -\epsilon_{yx}b_x + b_y - \epsilon_{yz}b_z \\ a_z &= -\epsilon_{zx}b_x - \epsilon_{zy}b_y + b_z \end{aligned} \qquad (4.6.16)$$

McConnell and Han performed an error analysis for the linear approximation of $[C]$ when the values ϵ_{pi} are either ± 5 percent or ± 10 percent, as given in Table 4.6.1. It is apparent from this table that the approximation should be acceptable as a correction scheme when the cross-axis coefficients are bounded by 5 to 7 percent. Now, we look at some experimental data to see how cross-axis sensitivity affects experimental

TABLE 4.6.1. Coefficient C_{pi} and Errors from Nominal Values

ϵ_{pi}	$C_{pp}{}^a$ Range of Values	$C_{pi}{}^b$ Range of Values	Nominal Value C_{pi}	C_{pi} Max Error Percent
0.05	1.005–1.0098	0.0478–0.052	0.05	5.8
0.10	1.021–1.04	0.093–0.113	0.10	7.0–11.3

[a]Range of actual values to be compared to nominal value of unity.
[b]Range of actual values to be compared with nominal C_{pi} value.

FRFs and how these FRFs can be improved by using correction matrix $[C]$.

4.6.4 FRF Contamination and Its Removal

Successful experimental modal analysis is highly dependent on obtaining high quality input-output FRFs. Accelerometer cross-axis sensitivity can significantly contaminate these FRFs, even for small amounts of cross-axis sensitivity. Han[8] used a free-free beam with a tri-axial accelerometer attached to its left end, as shown in Fig. 4.6.2, to demonstrate the presence of cross-axis contamination and how it can be successfully removed from the resulting FRF data by using the approximate correction matrix $[C]$.

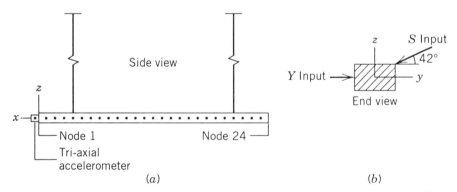

Fig. 4.6.2. Free-free beam. (a) Coordinate directions and transducer location. (b) Impact directions Y and S. (Courtesy of the *Proceedings for the 9th International Modal Analysis Conference*, Society for Experimental Mechanics.)

---

[8]Sangbo Han, "Effects of Transducer Cross-Axis Sensitivity on Modal Analysis," Ph.D. Dissertation, Iowa State University, Ames, IA, 1988.

TABLE 4.6.2. Tri-Axial Accelerometer Sensitivity Matrix

1.000	0.028	−0.051
−0.045	1.000	0.041
0.020	−0.036	1.000

A steel beam with a length of 2337 mm and a 25.4 × 28.6 mm cross-section is used as a known vibration source. The beam was excited by impulses applied at 24 discrete points along the beam's length while standard modal analysis data gathering procedures were employed. These impulses were applied in two directions, called Y and S in Fig. 4.6.2b. The S direction is inclined at an angle of 42 degrees so that the impulse force is directed at the beam's cross-sectional center, thus preventing the excitation of significant torsional accelerations. The S direction input excites both y and z direction vibrations, while the Y input excites only y direction vibrations. In this way, a reference set of uncontaminated experimental data can be obtained against which the contaminated data can be compared, both uncorrected and as corrected by matrix $[C]$. The tri-axial accelerometer used in these experiments has cross-axis coefficients shown in Table 4.6.2. It is seen that all of these coefficients are bounded by ±5.1 percent.

Figure 4.6.3a shows the y direction acceleration FRF that results from excitation impacts in the Y direction. Both the contaminated and corrected data are shown. These two plots are the same since the Y direction input causes little or no motion in either the x or z directions. Consequently, corrections are unnecessary.

S direction impacts excite significant vibratory motion in both the y and z directions so that the y direction acceleration FRF contains significant cross axis contamination, as shown in Fig. 4.6.3b for frequencies around the fundamental natural frequency. This contamination shows up as a radically different response in the region of points A and B where the raw data are significantly different than the data obtained from the Y input case, as shown in Fig. 4.6.3a. The point A contamination peak corresponds to z direction resonance motion, and is measured by the y direction accelerometer's cross-axis sensitivity. When the correction matrix $[C]$ is applied to the raw FRF data, we obtain the corrected or compensated curve shown in Fig. 4.6.3b. We see that this compensation method is effective in removing cross-axis signal contamination in this case. In fact, if one overlays the curve from Fig. 4.6.3a onto that of Fig. 4.6.3b, the compensated curve matches that of Fig. 4.6.3a. Also note that the raw phase angle has an extra shift that is removed by the compensation tech-

208 TRANSDUCER MEASUREMENT CONSIDERATIONS

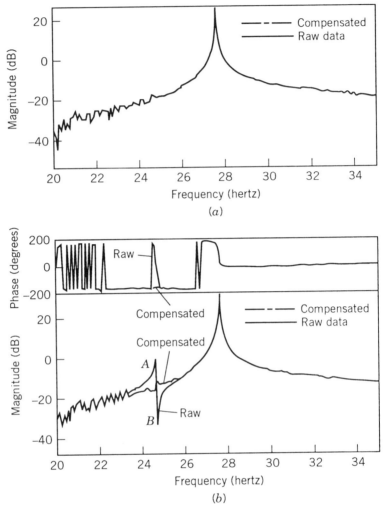

Fig. 4.6.3. y-direction accelerance FRF. (*a*) *Y*-direction input. (*b*) *S*-direction input. (Courtesy of the *Proceedings for the 9th International Modal Analysis Conference*, Society for Experimental Mechanics.)

nique. Now let us look at a broad-band FRF analysis that is presented in a paper by Han and McConnell.[9]

---

[9]S. Han and K. G. McConnell, "Analysis of Frequency Response Functions Affected by the Coupled Modes of the Structure," *The International Journal for Analytical and Experimental Modal Analysis*, Vol. 6, No. 3, Apr. 1991, pp. 147–159.

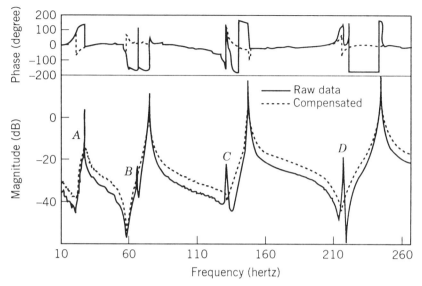

Fig. 4.6.4. Broad-band y-direction accelerance FRF excited by S direction impulses showing cross-axis signal contamination and compensation. (Courtesy of the *Proceedings for the 9th International Modal Analysis Conference*, Society for Experimental Mechanics.)

Node one is located at the beam's left end. Figure 4.6.4 shows node one's broad band y direction acceleration FRF when the input is also at node one and in the S direction. We see that multiple contaminated peaks show up at points B, C, and D. The contamination is not evident at point A, due to the frequency analyzer's frequency resolution being too large, so the extra peak is smeared over. Point A's contamination is shown in Fig. 4.6.3b, where the frequency analyzer's frequency resolution is much smaller. We see that the correction matrix $[C]$ effectively removes these contaminated peaks as well. We are concerned with how this contamination affects the resulting modal analysis that is done by computer software.

The Y direction input data generated four natural frequencies, mode shapes, and damping values from an analysis of the corresponding y direction FRFs over the frequency range shown in Fig. 4.6.4. The natural frequencies and damping values for the first four modes are given in Table 4.6.3. When this software is used on the raw FRF data that includes the contamination peaks, there are seven natural frequencies and damping values, as given in Table 4.6.3. When the software is used to analyze the corresponding compensated FRF, there are four natural frequencies and damping values as given in Table 4.6.3.

TABLE 4.6.3. Natural Frequencies and Modal Damping Values Extracted from Three Different Sets of FRFs

Mode	Y Input	y Output	S Input	y Output	S Input	cy Output[a]
1st	27.58 Hz	0.00560	27.59 Hz	0.00401	27.59 Hz	0.00501
2nd	75.86 Hz	0.00154	67.55 Hz	0.00008	75.91 Hz	0.00143
3rd	148.4 Hz	0.00131	75.91 Hz	0.00142	148.5 Hz	0.00122
4th	244.9 Hz	0.00056	132.3 Hz	0.00038	244.9 Hz	0.00060
5th			148.5 Hz	0.00073		
6th			218.1 Hz	0.00068		
7th			244.9 Hz	0.00070		

[a]The corrected output should be compared to the Y input y output column.

From the natural frequencies and damping values shown in Table 4.6.3, it is evident that the extra (contaminated) peaks caused significant errors to occur in modal damping values as well as the apparent natural frequencies. It is also shown in the paper that the software had to create nonexisting mode shapes and modal parameters to cope with the cross-axis contamination peaks. This means that the experimental data analysis created an erroneous experimental model. Unfortunately, this model is often compared with a theoretical finite element model that does not have these extra mode shapes and natural frequencies. This causes real problems when trying to compare experimental and theoretical modal models because any attempt to correct the theoretical finite element model leads to an incorrect theoretical model as well. Thus we must conclude that accelerometer cross-axis sensitivity can cause significant unwanted problems. The correction matrix $[C]$ effectively removed this contamination in the case cited. Further detailed information is available from the paper cited.

4.6.5 Cross-Axis Resonance

It was assumed in the above experimental investigation that the cross-axis sensitivities were constant over the range of frequencies measured. Unfortunately, accelerometers have cross-axis resonances that are often in the range of 0.5 to 0.8 of its response in the primary sensing direction. This means that cross-axis sensitivities change more rapidly with frequency than the primary axis does and can lead to significant changes in matrix $[C]$ with frequency. Actually, these variations with frequency can be used in the compensation scheme if the variations can be determined through calibration. Experience in calibrating cross-axis sensitivities shows that calibration is extremely difficult to do over a broad range of frequencies, due to the small values encountered for cross-axis directions compared to the primary axis sensitivity. It should be clear from this section that accelerometer cross-axis sensitivity is an area that requires careful consideration in order to avoid contaminated experimental results.

4.7 THE FORCE TRANSDUCER—GENERAL MODEL

The force transducer (also called load cell) is a more complicated instrument than an accelerometer since it can interact directly with the structure under test. Because of this interaction, we need to use a two DOF model to describe a load cell when it is applied to three different environments that were defined by McConnell.[10] The first environment is one where we attach the load cell to a "rigid" foundation. The second environment has the force transducer attached to a "hammer" type of device in order to measure impulse loads. The third environment involves placing the force transducer between a vibration exciter and the structure under test. We assume that the load cell uses a piezoelectric sensor in these developments.

4.7.1 General Electromechanical Model

A typical piezoelectric load cell design is shown in Fig. 4.7.1 where base and seismic masses are nearly equal. The piezoelectric sensing element separates these two masses and measures the relative motion between them. For other designs, the seismic mass is much smaller, like 20 percent of the load cell's total mass, so that with these designs it is important to connect the transducer's seismic mass end to the structure under test. In the equal mass transducer designs, it makes no difference which end is connected to the structure under test. The reason for correctly connecting a load cell becomes apparent later in this section.

The load cell's cross section in Fig. 4.7.1 suggests that a force transducer can be modeled as two masses, consisting of m_1 at the seismic end and m_2 at the base end, a spring k, and a damper c (that are due to the load cell's structure and piezoelectric sensor), and two external excitation forces $f_1(t)$ and $f_2(t)$, as shown in Fig. 4.7.2. External force $f_1(t)$ comes from seismic mass contact with the structure under test, acts on the seismic mass end, and is the force we want to measure with the sensing element. External force $f_2(t)$ is the force required to act on base mass m_2 so that the force transducer moves properly and the desired force $f_1(t)$ acts on the structure under test. A comparison of Fig. 4.7.2 with the two DOF

---

[10]K. G. McConnell, "The Interaction of Force Transducers with Their Test Environment," *Modal Analysis: The International Journal of Analytical and Experimental Modal Analysis*, Vol. 8, No. 2, Apr. 1993, pp. 137–150.

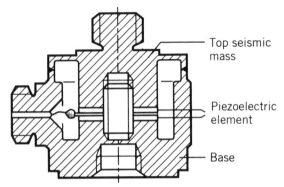

Fig. 4.7.1. Cross-sectional view of piezoelectric force transducer showing nearly equal seismic and base masses. (Courtesy Bruel & Kjaer Instruments, Inc.)

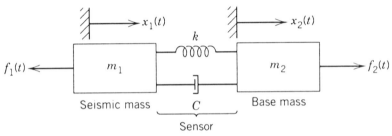

Fig. 4.7.2. Two DOF model of a force transducer. (Courtesy of *Modal Analysis: The International Journal of Analytical and Experimental Modal Analysis*, Society for Experimental Mechanics.)

system shown in Fig. 3.6.1 shows us that $f_1(t) = -f_1(t)$, $k_2 = k$, $c_2 = c$, $k_1 = k_3 = 0$, and $c_1 = c_2 = 0$. Then Eqs. (3.6.1) become

and
$$(m_1)\ddot{x}_1 + (c)\dot{x}_1 + (-c)\dot{x}_2 + (k)x_1 + (-k)x_2 = -f_1(t)$$
$$(m_2)\ddot{x}_2 + (-c)\dot{x}_1 + (c)\dot{x}_2 + (-k)x_1 + (k)x_2 = f_2(t)$$
(4.7.1)

as the differential equations governing a force transducer without consideration of its environment.

We can obtain a steady-state sinusoidal response of Eq. (4.7.1) by assuming that

$$f_1(t) = F_1 e^{j\omega t} \quad \text{and} \quad f_2(t) = F_2 e^{j\omega t}$$

and that

$$x_1 = X_1 e^{j\omega t} \quad \text{and} \quad x_2 = X_2 e^{j\omega t}$$

Inserting these relationships into Eq. (4.7.1), canceling the $e^{j\omega t}$ terms, and letting

$$S = k + jc\omega \tag{4.7.2}$$

be the transducer's apparent stiffness, we can write the solution of Eqs. (4.7.1) as algebraic equations given by

$$\begin{aligned}(S - m_1\omega^2)X_1 - SX_2 &= -F_1 \\ -SX_1 + (S - m_2\omega^2)X_2 &= F_2\end{aligned} \tag{4.7.3}$$

The characteristic frequency equation comes from the determinant of X_1 and X_2 coefficients in Eq. (4.7.3) and can be expressed as

$$\begin{aligned}\Delta(\omega) &= -[(m_1 + m_2)S - m_1 m_2 \omega^2]\omega^2 \\ &= -(m_1 + m_2)k[1 - r^2 + j2\zeta r]\omega^2\end{aligned} \tag{4.7.4}$$

When the real part of $\Delta(\omega)$ in Eq. (4.7.4) is zero, we have the transducer's natural frequency. The first natural frequency is seen to be zero since ω^2 factors out of the equation. This tells us that the transducer is free to move as a rigid body under the influence of forces F_1 and F_2. The second natural frequency and corresponding damping ratio are defined by

$$\begin{aligned}\omega_n &= \sqrt{\frac{k}{m_e}} \\ \zeta &= \frac{c}{2\sqrt{km_e}} \\ m_e &= \frac{m_1 m_2}{m_1 + m_2} \\ r &= \frac{\omega}{\omega_n}\end{aligned} \tag{4.7.5}$$

where m_e is the system's effective mass.
r is a dimensionless frequency ratio.
We see that the effective mass is a combination of masses and not the seismic mass so that changing either mass can alter the transducer's natural frequency and effective damping.

214 TRANSDUCER MEASUREMENT CONSIDERATIONS

The steady state amplitudes of motion can be obtained from Eq. (4.7.3) by using Cramer's rule. This procedure gives

$$X_1 = \frac{SF_2 - (S - m_2\omega^2)F_1}{\Delta(\omega)} \tag{4.7.6}$$

and

$$X_2 = \frac{(S - m_1\omega^2)F_2 - SF_1}{\Delta(\omega)} \tag{4.7.7}$$

We have shown in Eqs. (4.3.9), (4.3.16), and (4.3.17) that the piezoelectric sensor output voltage can be written in terms of relative motion $z = X_2 - X_1$ so that

$$E_f = H_f^e(\omega)\frac{S_z}{C_f}(X_2 - X_1) \tag{4.7.8}$$

where S_z is displacement charge sensitivity (pCb/unit of displacement).
C_f is load cell capacitance.
$H_f^e(\omega)$ is the transducer's electrical low frequency characteristic (indicated by superscript e), which is described by

$$H_f^e(\omega) = \frac{jT_f\omega}{1 + jT_f\omega} \tag{4.7.9}$$

where T_f is the force transducer's electrical time constant; that is, $R_f C_f$.

Inserting X_1 and X_2 from Eqs. (4.7.6) and (4.7.7), as well as the definition of $\Delta(\omega)$ from Eq. (4.7.4), into Eq. (4.7.8) gives the load cell's output voltage as

$$E_f = S_f H_f(\omega)\left[\frac{m_2 F_1}{(m_1 + m_2)} + \frac{m_1 F_2}{(m_1 + m_2)}\right] \tag{4.7.10}$$

where $S_f = S_z/(C_f k) = S_q/C_f$ is the voltage sensitivity (volts/unit of force).
$S_q = S_z/k$ is the transducer charge sensitivity (coulombs/unit of force).

The transducer's FRF is given by

$$H_f(\omega) = \underbrace{\left[\frac{jT_f\omega}{1 + jT_f\omega}\right]}_{\text{electrical}}\underbrace{\left[\frac{1}{1 - r^2 + j2\zeta r}\right]}_{\text{mechanical}} \tag{4.7.11}$$

where r is defined in Eqs. (4.7.5). The reference natural frequency for r is the installed natural frequency and is application dependent. The FRF described in Eq. (4.7.11) is that of a typical electrical and single DOF mechanical response. However, we also see from Eq. (4.7.10) that both forces F_1 and F_2 and masses m_1 and m_2 are involved in the output voltage expression. These variables are application dependent.

We need a relationship between forces F_1 and F_2 before we can use Eqs. (4.7.10) and (4.7.11) to predict force F_1. We now consider the three general application situations of a load cell attached to a fixed foundation, attached to an impulse hammer, and attached between a vibration exciter and a structure under test.

4.7.2 Force Transducer Attached to a Fixed Foundation

Figure 4.7.3 shows the transducer's base attached to an ideal "rigid" foundation. In this case, base motion $x_2(t)$ is zero so that X_2 must be zero as well. Applying this condition of X_2 to Eq. (4.7.7), we obtain an expression of force F_2 in terms of force F_1 that is given by

$$F_2 = \frac{S}{S - m_1\omega^2} F_1 \qquad (4.7.12)$$

Substitution of Eq. (4.7.12) into Eq. (4.7.10) gives

$$E_f = H_f^e(\omega) \frac{S_z}{C_f} \left[\frac{1}{S - m_1\omega^2} \right] F_1 = S_f H_f(\omega) F_1 \qquad (4.7.13)$$

where $H_f(\omega)$ has the same form as given in Eq. (4.7.11) when the frequency ratio is defined in terms of the grounded natural frequency that occurs when m_2 is infinitely large (the ideal rigid foundation); that is, $\omega_n = \sqrt{k/m_1}$.

We see from Eq. (4.7.13) that the behavior of this system is that of a

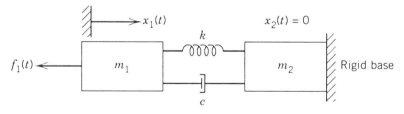

Fig. 4.7.3. Force transducer mounted on rigid foundation. (Courtesy of *Modal Analysis: The International Journal of Analytical and Experimental Modal Analysis*, Society for Experimental Mechanics.)

single DOF mechanical system and that it is identical in form to that of an accelerometer. When there is no mass attached to the seismic mass, we obtain the transducer's natural frequency as published by its manufacturer. When we attach the transducer to a large base and attach a structure to the other end of this mass, we increase mass m_1, and hence, lower the load cell's natural frequency accordingly. There is no loss of sensitivity in this case, only a loss in usable frequency range.

The fact that a force transducer in this environment behaves in a fashion similar to an accelerometer has given us a false sense of transducer behavior by leading us to expect that it will behave this way in all applications. Now, we turn our attention to the impulse hammer application, where the load cell has similar but somewhat different characteristics.

4.7.3 Force Transducer Attached to an Impulse Hammer

Figure 4.7.4 shows the equivalent dynamic model of a force transducer attached to an impulse hammer. In this case, we let base mass m_2 include the effective mass of the hammer while seismic mass m_1 includes the hammer's impact tip. For the hammer, we assume that the dominant input force comes from its inertia. Thus we can set force F_2 in Eq. (4.7.10) equal to zero. This gives us an output voltage expression of

$$E_f = \left[\frac{m_2 S_f}{(m_1 + m_2)}\right] H_f(\omega) F_1 = S_f^* H_f(\omega) F_1 \qquad (4.7.14)$$

where the effective voltage sensitivity S_f^* (volts/unit of force) is given by

$$S_f^* = \left[\frac{m_2}{m_1 + m_2}\right] S_f \qquad (4.7.15)$$

It is clear that hammer calibration can easily change with changing either mass m_1 or m_2, which is precisely what we do in most instances as we move from one test setup to another.

Masses m_1 and m_2 are changed to achieve different contact times and peak impulse forces, since impact time depends on interface spring constant, hammer mass, and test structure characteristics, while peak forces depend on the same variables as well as the initial impact velocity. We also see from Eqs. (4.7.5) that transducer natural frequency and damping are dependent on both tip (seismic) mass m_1 and hammer body mass m_2. These masses are used in Eqs. (4.7.5) to define the damping ratio ζ and frequency ratio r for use in Eq. (4.7.11), which describes the load cell's FRF in this case. The equations we developed here are identical to those

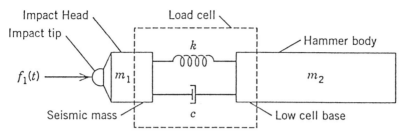

Fig. 4.7.4. Schematic showing masses m_1 and m_2 when attached to a hammer body. (Courtesy of *Modal Analysis: The International Journal of Analytical and Experimental Modal Analysis*, Society for Experimental Mechanics.)

obtained by Han and McConnell,[11] who showed that these equations describe impact hammer performance in a satisfactory manner over a broad range of mass ratios.

We note that different persons can have significantly different hand mass. This hand mass variation can cause measurement error when using lightweight hammers, since the assumption that F_2 is zero during impact is no longer true as some hand inertia is transferred to m_2. Any change in m_2 can change the effective voltage sensitivity S_f^*. Hence, it is recommended that the person using a lightweight impact hammer also be the person to use the hammer during calibration.

4.7.4 Force Transducer Used with Vibration Exciter and Structure

Let us consider the situation shown in Fig. 4.7.5a, where the force transducer is attached on one end to the structure under test and on the other end to the vibration exciter. The force transducer is attached to the structure through some kind of attachment device, which may be simply a small bolt or may be a much larger mounting block. Often we attach an accelerometer to the structure at this same location so we can measure the driving point FRF (such as receptance, mobility, or acceleration). In this model, we consider all masses (such as bolts, mounting blocks, and accelerometer mass) to be part of mass m_1. Similarly, m_2 represents all mass on the base side of the transducer. We find in this subsection that we are at liberty to select nearly any point below the transducer's base connection surface as this interface surface without altering our results. This is an important result, since it allows use to use any convenient

---

[11]S. Han and K. G. McConnell, "Effect of Mass on Force Transducer Sensitivity," *Experimental Techniques*, Vol. 10, No. 7, 1986, pp. 19–22.

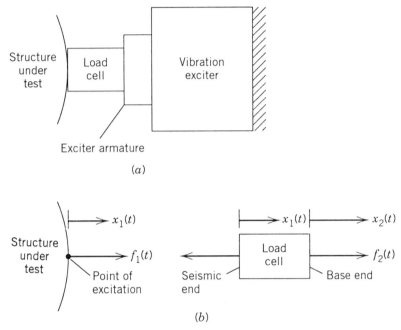

Fig. 4.7.5. A load cell used with a vibration exciter and structure under test. (*a*) Schematic of test setup. (*b*) FBD showing forces and motion for ideal case of no rotational motion. (Courtesy of *Modal Analysis: The International Journal of Analytical and Experimental Modal Analysis*, Society for Experimental Mechanics.)

system. Thus mass m_2 may or may not include part of the exciter's armature mass. We explore the vibration exciter's characteristics further in Chapter 6.

The force transmitted to the test structure at the transducer test structure interface is given by F_1, while the transducer's base force is given by F_2. We need to develop a relationship between these two forces for use in Eq. (4.7.10) in order to obtain the force transducer's output voltage. The structure's driving point receptance FRF $H_s(\omega)$ provides us with a useful relationship that brings the structure's response characteristics into the transducer's description. We know that the structure's amplitude of motion in Fig. 4.7.5b is given by

$$X_1 = H_s(\omega)F_1 \tag{4.7.16}$$

We also have another expression for amplitude X_1 in Eq. (4.7.6) so that equating Eqs. (4.7.6) and (4.7.16) gives us an expression for F_2 in terms of F_1 such that

$$F_2 = \left[\frac{\Delta(\omega)H_s(\omega) + S - m_2\omega^2}{S}\right]F_1 \qquad (4.7.17)$$

Equation (4.7.17) shows how the structure's driving point FRF and the transducer's $\Delta(\omega)$ terms, as well as transducer's dynamic stiffness S and base mass m_2, alter the input force F_1. Force F_2 often varies widely in order to produce the same input force F_1 with changing excitation frequency. Equation (4.7.17) clearly shows that the useful force available to excite the structure is dependent on the dynamic characteristics of the force transducer as well as the structure under test.

We can obtain the load cell's output voltage by inserting Eq. (4.7.17) into Eq. (4.7.10) and carefully manipulating the $H_f(\omega)$ terms in Eq. (4.7.10) to obtain

$$\begin{aligned} E_f &= S_f H'_f(\omega)[1 - m_1 H_s(\omega)\omega^2]F_1 \\ &= S_f H'_f(\omega)[1 + m_1 A_s(\omega)]F_1 \end{aligned} \qquad (4.7.18)$$

where $A_s(\omega)$ is the structure's driving point accelerance.
$S_f = S_q/C_f = S_z/(C_f k)$ is load cell voltage sensitivity (volts/unit of force).

The transducer FRF is given by

$$H'_f(\omega) = \underbrace{\left[\frac{jT_f\omega}{1 + jT_f\omega}\right]}_{\text{electrical}} \underbrace{\left[\frac{1}{1 + jc\omega/k}\right]}_{\text{mechanical}} \qquad (4.7.19)$$

Equation (4.7.19) contains a startling result. We see that the transducer has no mechanical resonance, only a small amount of signal attenuation and phase shift. This means that the $\Delta(\omega)$ term has dropped out of the measurement problem.

The mechanical term's significance can be seen when it is expressed as

$$1 + \frac{jc\omega}{k} = 1 + j2\zeta_{bt}\frac{\omega}{\omega_{bt}} = 1 + j\eta_{ft} \qquad (4.7.20)$$

where $\omega_{bt} = \sqrt{k/m_s}$ is a *bare* transducer's natural frequency.
$\zeta_{bt} = c/2\sqrt{km_s}$ is a *bare* transducer's damping ratio.
m_s is a *bare* transducer's seismic mass.
η_{ft} is the force transducer's structural damping.
The form of damping may be application dependent so that both types of terms are shown in Eq. (4.7.20). A load cell is said to be *bare* when its seismic end is not attached to any external structure or mass so that the

seismic mass is a free "bare" surface while the base is attached to a large mass. The load cell manufacturer quotes the *bare* transducer's natural frequency in the specifications. The phase shift associated with this term is usually less than one to three degrees for lightly damped transducers.

Clearly, Eq. (4.7.18) shows us that mass m_1 and the structure's driving point receptance $H_s(\omega)$ (or accelerance $A_s(\omega)$) cause a force measurement error. This error is seen to change from application to application due to the $H_s(\omega)$ (or $A_s(\omega)$) term, as well as to the change in the effective seismic mass m_1. We now look at this error for a simple single DOF structural system in order to obtain a feel for its relative importance.

We model a single DOF structural system with mass m_s, damping c_s, and stiffness k_s so that the second order differential equation of motion that describes its behavior is given by

$$m_s \ddot{x}_1 + c_s \dot{x}_1 + k_s x_1 = f_1(t) \tag{4.7.21}$$

The corresponding structural driving point FRF $H_s(\omega)$ is

$$H_s(\omega) = \frac{1}{k_s - m_s \omega^2 + jc_s \omega} \tag{4.7.22}$$

Now, let $\hat{E}_f$ be the true force voltage. Then, using Eq. (4.7.22) in Eq. (4.7.18), we obtain a voltage error ratio ER as

$$\frac{\hat{E}_f}{E_f} = ER = \left[\frac{k_s - m_s \omega^2 + jc_s \omega}{k_s - (m_1 + m_s)\omega^2 + jc_s \omega} \right]$$

$$= \left[\frac{1 - r^2 + j2\zeta_s r}{1 - (1 + M)r^2 + j2\zeta_s r} \right] \tag{4.7.23}$$

where M is the mass ratio of m_1/m_s.

ζ_s is the structural damping ratio.

$r = \omega/\omega_{ns}$ is the dimensionless frequency ratio.

Equation (4.7.23) is identical to one obtained by McConnell[12] when using a different method of analysis. Figure 4.7.6 is a plot of ER from McConnell's results for $M = 0.01$ and $\zeta_s = 0.01$ for frequencies in the vicinity of structural resonance when the transducer has 5 percent damping. From this plot, we see that force is first overestimated by 27 percent at point A and then underestimated by 22 percent at point B in the most critical

---

[12]K. G. McConnell, "Errors in Using Force Transducers," *Proceedings of the 8th International Modal Analysis Conference*, Vol. 2, Kissimmee, FL, 1990, pp. 884–890.

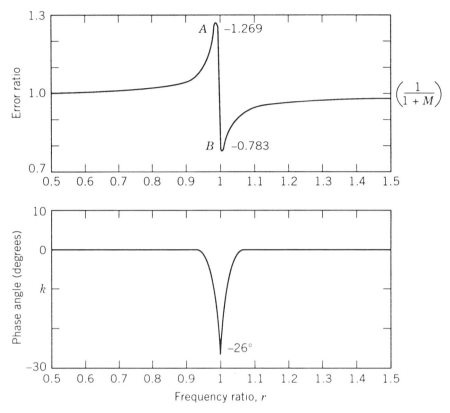

Fig. 5.7.6. Error ratio (magnitude and phase) as a function of frequency ratio r for mass ratio $M = 0.01$, structural damping ratio = 0.01, and transducer damping of 0.05. (Courtesy of *Modal Analysis: The International Journal of Analytical and Experimental Modal Analysis*, Society for Experimental Mechanics.)

frequency range, which is near the structure's resonance. These significant errors cause a shifting of the apparent structural resonance frequency as well as peak FRF values. We see that *ER* has significant phase shift in this frequency range as well. Thus the force transducer gives potentially serious errors in the frequency range where we need our data to be the best for most vibration analysis methods.

A further examination of Eq. (4.7.23) indicates that the measurement error asymptotically approaches a value of $1/(1 + M)$ for frequencies well above structural resonance. The maximum, minimum, and resonance values of *ER*, as well as phase at structural resonance, are given in Table 4.7.1 for several values of damping and mass ratios. From an examination of this table, it is obvious that mass m_1 (transducer seismic mass, attachment masses, and accelerometer mass) must be very small to avoid serious

TABLE 4.7.1. Maximum and Minimum Error Ratio Values for Different Mass and Structure Damping Ratios[a]

Mass Ratio M	Damping Ratio ζ	Maximum ER	Minimum ER	Resonance ER	Phase (degrees)
1	0.01	20.9	0.02	0.02	−174
1	0.05	7.07	0.10	0.01	−152
0.1	0.10	1.25	0.765	0.894	−26.9
0.1	0.05	1.56	0.610	0.707	−51.4
0.1	0.01	4.89	0.196	0.196	−135
0.01	0.05	1.05	0.949	0.995	−5.1
0.01	0.01	1.27	0.783	1.013	−26.0
0.001	0.05	1.005	0.995	1.001	0
0.001	0.01	1.025	0.975	1.001	−2.3

[a]The force transducer is assumed to have 5 percent damping.

force measurement errors when measuring forced response of low weight and lightly damped structures.

4.7.5 The Impedance Head

Figure 4.7.7 shows a different type of load cell, called an *impedance head*, where both force and acceleration are measured at the same point of the structure when attaching a single instrument. The name "impedance head" is not really correct, for these transducers are usually employed to measure either *driving point accelerance* $[a/F]$ or *driving point mobility* $[v/F]$, not *impedance* $[F/v]$.

As shown in Fig. 4.7.7, we see that the accelerometer's seismic mass and piezoelectric sensor are located at the top end, and that the load cell's piezoelectric sensor and seismic mass are located at the bottom end of the impedance head. Also, we see that the accelerometer measures the force transducer's base motion, not the structure's motion. It is not too difficult to envision an impedance head being used in an environment where the test system masses lower the force transducer's resonant frequency to be in the test frequency range so that force transducer base motion and structure motion are significantly different. Hence, the test data can be meaningless unless care is exercised.

In addition, impedance heads tend to be much larger than force transducers so that they have larger masses and mass moments of inertia. These larger inertia characteristics cause additional errors, particularly in situations where the structure has significant rotational motion. In these

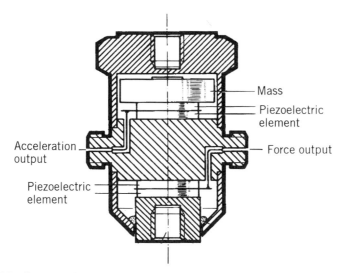

Fig. 4.7.7. Cross-section of an impedance head. (Courtesy of Bruel & Kjaer Instruments, Inc.).

cases, the load cell tends to pick up addition signals, due to bending sensitivity. Bending sensitivity is like cross-axis sensitivity for accelerometers. In short, it is the author's opinion that impedance heads can cause significant measurement errors, which could be avoided by using smaller load cells and accelerometers. In Section 4.8, we shall explore ways to remove these force transducer errors from our data by using either electronic or frequency domain methods.

4.8 CORRECTING FRF DATA FOR FORCE TRANSDUCER MASS LOADING

In Section 4.7 we found that the force transducer interacts with its environment. This interaction causes measurement errors and transducer characteristics that change. In this section, we look at the load cell in a different way and find that two of the three test environments reduce to a common error term and that this measurement error can be compensated for. This compensation can be done in either the frequency domain after the data is gathered and safely stored or in the time domain before storing the data by using either electronic signal addition or subtraction (depending on signal phasing). We explore both time and frequency domain compensation schemes in this section.

4.8.1 Consistent Force Transducer Model

A consistent force transducer model is developed here that can be used in all applications where the seismic mass is attached to the structure under test. This model is obtained by using the transducer's sensing end mass FBD, shown in Fig. 4.8.1a. The differential equation of motion for the seismic mass is the same as the first of the two equations in Eq. (4.7.1) and is written as

$$m_1\ddot{x}_1 + c(\dot{x}_1 - \dot{x}_2) + k(x_1 - x_2) = -f_1(t) \qquad (4.8.1)$$

Now, we recall that relative motion $z\ (= x_2 - x_1)$ is the quantity measured by the piezoelectric sensor. Then, we can express Eq. (4.8.1) as

$$c\dot{z} + kz = f_1(t) + m_1\ddot{x}_1 \qquad (4.8.2)$$

where relative motion z is related to the desired force $f_1(t)$ and the inertia force error term of $m_1\ddot{x}_1$. Recall that m_1 represents the combined masses on the transducer's sensing end, consisting of load cell seismic mass, attachment hardware mass, and driving point accelerometer mass. The accelerometer's mass is included even though the accelerometer is not located between the piezoelectric sensor and the structure. It may be located on the back side of the attachment point, as shown in Fig. 4.8.1b where m_1 is represented by all cross-hatched mass. The accelerometer's mass distorts the measured force and can be easily accounted for.

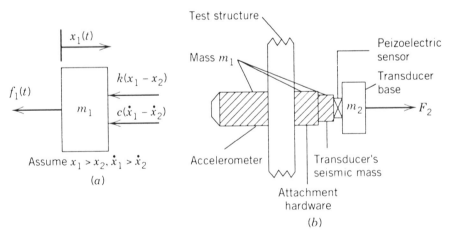

Fig. 4.8.1. (a) FBD of a force transducer's seismic mass m_1. (b) Schematic of load cell attached to structure under test showing all contributors to mass m_1.

CORRECTING FRF DATA FOR FORCE TRANSDUCER MASS LOADING

Now, if the motions and forces are sinusoidal with force $f_1(t)$ considered to be the reference phasor, and if the force transducer is attached to the structure so that $A_1 = A_s(\omega)$, then Eq. (4.7.2) becomes

$$SZ = F_1 + m_1 A_1 = (1 + m_1 A_s(\omega))F_1 \qquad (4.8.3)$$

where $S = k + jc\omega$ is the transducer's apparent stiffness (see Eq. (4.7.2)).
Z is the transducer's relative motion phasor.
F_1 is the magnitude of the excitation force and is the reference phasor in this analysis.
A_1 is the structure's acceleration response (magnitude and phase) phasor.
$A_s(\omega)$ is the structure's driving point acceleration.

We see that the error term in Eq. (4.8.3) is completely defined by the $m_1 A_s(\omega)$ term. This term is application dependent in the sense that m_1 may change from point to point in a given structure, and of course, usually changes with each structure tested, while $A_s(\omega)$ changes in magnitude and phase with frequency as well as from point to point in the same structure, not to mention from structure to structure.

If Eq. (4.8.3) is combined with Eqs. (4.7.8) and (4.7.9), we obtain the load cell's output voltage as

$$E_f = H'_f(\omega) S_f (1 + m_1 A_s(\omega)) F_1 \qquad (4.8.4)$$

where $H'_f(\omega)$ is the force transducer's FRF that is given by Eq. (4.7.19) and is written here for convenience:

$$H'_f(\omega) = \underbrace{\left[\frac{jT_f \omega}{1 + jT_f \omega}\right]}_{\text{electrical}} \underbrace{\left[\frac{1}{1 + jc\omega/k}\right]}_{\text{mechanical}} \qquad (4.8.5)$$

Equations (4.8.4) and (4.8.5) define the force transducer's performance and show that the error term (other than H'_f) is due to the combination of driving point acceleration $A_s(\omega)$ and mass m_1.

This theoretical model is most insightful for developing useful and practical correction schemes. It is also most surprising since the force transducer does not show any mechanical resonance characteristics in its output, while the transducer may contribute to test apparatus resonances. However, in measurement of the force transmitted, Eq. (4.8.4) is sufficient to model basic behavior provided there are no significant bending

226 TRANSDUCER MEASUREMENT CONSIDERATIONS

moments and/or shear forces being applied to the force transducer. Now, we look at ways to compensate for the $m_1 A_s(\omega)$ error term.

4.8.2 Correcting Driving Point Accelerance FRF in Frequency Domain

We recall that the driving point accelerance at the pth location is defined in terms of the complex ratio defined by

$$A_{pp}(\omega) = \frac{A_p}{F_p} \qquad (4.8.6)$$

where $A_{pp}(\omega)$ is the driving point accelerance at each frequency.
A_p is the driving point acceleration at each frequency.
F_p is the driving point force at each frequency.

Each of the quantities in Eq. (4.8.6) are complex (with real and imaginary parts or magnitude and phase) and the ratio must hold at every input-output frequency. Thus we can also think of A_p and F_p as representing frequency spectra of acceleration per hertz and force per hertz, respectively, when such a definition is desirable. We write $A_{pp}(\omega)$ as A_{pp} in the following paragraphs as a simpler notation.

A_{pp} in Eq. (4.8.6) represents what we want from our experimental measurements. Now, we look at what we really get when we put two voltage signals into a frequency analyzer. Then, we examine how we can correct the frequency analyzer's output to obtain what we want, namely, the accelerance as defined in Eq. (4.8.6).

In order to measure driving point accelerance, the frequency analyzer requires two input voltage signals. The force signal is defined by Eq. (4.8.4), while the acceleration signal is given by

$$E_{ap} = H_{ap}(\omega) \frac{S_{ap}}{g} A_p \qquad (4.8.7)$$

where A_p has units like m/s^2 or in./s^2 and g is the standard gravitational acceleration of 9.807 m/s^2 or 386 in./s^2, and S_{ap} is the voltage sensitivity in volts per g. Then, the voltage ratio becomes

$$\frac{E_{ap}}{E_{fp}} = \left(\frac{H_{ap}(\omega)}{H'_{fp}(\omega)}\right)\left(\frac{S_{ap}}{gS_{fp}}\right)\left(\frac{1}{1 + m_1 A_{pp}}\right)\left(\frac{A_p}{F_p}\right)$$

$$= (HI_{pp}(\omega))\left(\frac{S_{ap}}{gS_{fp}}\right)\left(\frac{1}{1 + m_1 A_{pp}}\right) A_{pp} \qquad (4.8.8)$$

where $HI_{pp}(\omega)$ is the *measurement systems FRF* given by

CORRECTING FRF DATA FOR FORCE TRANSDUCER MASS LOADING 227

$$HI_{pp}(\omega) = \frac{H_{ap}(\omega)}{H'_{fp}(\omega)} = \left(\frac{T_{ap}}{T_{fp}}\right)\left(\frac{1 + jT_{fp}\omega}{1 + jT_{ap}\omega}\right)\left(\frac{1 + jc\omega/k}{1 - r_a^2 + j2\zeta_a r_a}\right)_p \quad (4.8.9)$$

Here it must be emphasized that $H'_{fp}(\omega)$ and $H_{ap}(\omega)$ include the frequency analyzer's anti-aliasing filter characteristics as well as any instrumentation amplifier's attenuation and phase shifts. In other words, these instrument FRFs represent the overall measurement system characteristics.

We used Eq. (4.8.6) to define A_{pp} in the second line of Eq. (4.8.8). It is obvious that Eq. (4.8.8) contains a large number of terms and that what we want is the value of A_{pp}. The frequency analyzer often helps us by automatically scaling the output so that scale factors S_{ap} and S_{fp} are analyzer inputs as well. Then, the analyzer's display FRF, called the *measured FRF* and represented by $Hm_{pp}(\omega)$, can be expressed as

$$Hm_{pp}(\omega) = \left(\frac{gS_{fp}}{S_{ap}}\right)\left(\frac{E_{ap}}{E_{fp}}\right) = (HI_{pp}(\omega))\left(\frac{1}{1 + m_1 A_{pp}}\right) A_{pp} \quad (4.8.10)$$

Equation (4.8.10) shows us that the displayed FRF is not the desired driving point acceleration A_{pp}. There are two error terms, one due to instrumentation $HI_{pp}(\omega)$ and the other due to $m_1 A_{pp}$.

Fortunately, A_{pp} can be obtained directly from Eq. (4.8.10) as

$$A_{pp} = \frac{Hm_{pp}(\omega)}{HI_{pp}(\omega) - m_1 Hm_{pp}(\omega)} \quad (4.8.11)$$

It is clear from Eq. (4.8.11) that we need to know not only m_1 (the effective mass on the sensor's sensing end), but also the combined instrumentation system's FRF characteristics contained in $HI_{pp}(\omega)$. Equation (4.8.9) clearly shows that the load cell's time constant T_{fp} and the accelerometer's time constant T_{ap} play an important role at low frequencies and must be equal or very close. When these two time constants are equal, their effect drops out except that they lower the effective signal to noise ratio in each voltage signal at low frequencies. The mechanical terms in Eq. (4.8.9) show the importance of a reasonable match between the accelerometer and force transducer characteristics.

We would like to assume that $HI_{pp}(\omega) = 1$ so that Eq. (4.8.11) reduces to a simple estimate of the driving point acceleration given by

$$A_{pp}^e = \frac{Hm_{pp}(\omega)}{1 - m_1 Hm_{pp}(\omega)} \quad (4.8.12)$$

where the superscript e stands for estimate of $A_{pp}(\omega)$.

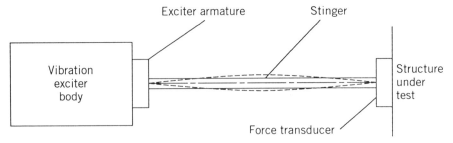

Fig. 4.8.2. Schematic showing flexible stinger between exciter armature and force transducer to reduce moments on the transducer.

Equations (4.8.11) and (4.8.12) do not take into account measurement errors that result from either bending moments or shearing forces acting on the load cell. Hence, it is important to use force transducers of low mass moments of inertia in order to reduce dynamically induced moments. In addition, the use of a stinger, as shown in Fig. 4.8.2, is helpful in reducing these bending moments. Otherwise, if the force transducer's base is attached directly to a vibration exciter's armature, we introduce the armature's entire mass and mass moment of inertia into the load cell's dynamics so that both dynamic shear and bending moments are easily created, loads that the force transducer may be sensitive to. Usually load cell sensitivity to shear forces and bending moments is quite small but it is also highly variable. Experience with one load cell showed that bending sensitivity was also highly nonlinear for larger moments. Consequently, shear and bending moments are undesirable and cannot be corrected for in so simple a manner as we have for inertia m_1. Now the question is: Can we do the same thing for transfer FRFs?

4.8.3 Correcting Transfer Accelerance FRFs in Frequency Domain

We recall that the transfer accelerance is defined in terms of

$$A_{qp}(\omega) = \frac{A_q}{F_p} \qquad (4.8.13)$$

We write $A_{pq}(\omega)$ as A_{pq} in the following paragraphs in order to simplify our notation.

As in the case for driving point accelerance, we need two voltage inputs for the frequency analyzer. One is the force voltage given by Eq. (4.8.4),

CORRECTING FRF DATA FOR FORCE TRANSDUCER MASS LOADING 229

while the acceleration voltage is given by Eq. (4.8.7) with subscript p changed to q. Then, the voltage ratio becomes

$$\frac{E_{aq}}{E_{fp}} = \left(\frac{H_{aq}(\omega)}{H'_{fp}(\omega)}\right)\left(\frac{S_{aq}}{gS_{fp}}\right)\left(\frac{1}{1+m_1A_{pp}}\right)\left(\frac{A_q}{F_p}\right) \quad (4.8.14)$$

$$= (HI_{qp}(\omega))\left(\frac{S_{aq}}{gS_{fp}}\right)\left(\frac{1}{1+m_1A_{pp}}\right)A_{qp}$$

from which we can write the analyzer's output FRF, $Hm_{qp}(\omega)$, as

$$Hm_{qp}(\omega) = \left(\frac{gS_{fp}}{S_{aq}}\right)\left(\frac{E_{aq}}{E_{fp}}\right) = (HI_{qp}(\omega))\left(\frac{1}{1+m_1A_{pp}}\right)A_{qp} \quad (4.8.15)$$

We can solve directly for the corrected transfer accelerance in terms of the measured FRF to obtain

$$A_{qp} = \frac{(1+m_1A_{pp})}{HI_{qp}(\omega)} Hm_{qp}(\omega) \quad (4.8.16)$$

where the instrumentation FRF is given by

$$HI_{qp}(\omega) = \frac{H_{aq}(\omega)}{H'_{fp}(\omega)} = \left(\frac{jT_{aq}\omega}{1+jT_{aq}\omega}\right)\left(\frac{1+jT_{fp}\omega}{jT_{fp}\omega}\right)\frac{(1+jc\omega/k)_p}{(1-r_a^2+j2\zeta_a r_a)_q} \quad (4.8.17)$$

Again we would like to think of our instrumentation as being ideal so that $HI_{pq}(\omega) = 1$. Then, we can simplify Eq. (4.8.16) to

$$A_{qp} = (1+m_1A_{pp})Hm_{qp}(\omega) \quad (4.8.18)$$

From both Eqs. (4.8.16) and (4.8.18), it is evident that we must first correct the driving point accelerance data to obtain our best estimate for $A_{pp}(\omega)$. Once we have done that, we are in position to correct all remaining acceleration FRFs.

This correction scheme is desirable since the four rules to implement it are clear. They are:

1. We need a good estimate for m_1.
2. We need to be sure that we have matching time constants in all measurement systems.

230 TRANSDUCER MEASUREMENT CONSIDERATIONS

3. We need to remain well below the accelerometer resonance frequency.
4. We need to have similar damping characteristics and phase shifts between the force transducer and the accelerometers employed.

Then, we can post process our measured frequency analyzer FRFs according to Eqs. (4.8.12) and (4.8.18) in a straightforward manner that requires little in the way of computation resources and needs to be done only once to obtain the corrected estimate of the structure's driving point and transfer FRFs. These corrections should be applied after the cross-axis sensitivity corrections of Section 4.6 have been applied to the data. The errors involved in our assumptions about $HI_{pp}(\omega)$ and $HI_{qp}(\omega)$ being unity need to be carefully examined.

The errors remaining are concerned with the effects of the addition of mass to the structure due to one or more accelerometers being mounted on the structure. This error can be estimated by attaching additional masses equal to that of the accelerometers to the structure and observing the effects on the measured natural frequencies. This needs to be done only during the initial tests in order to determine the structure's sensitivity to accelerometer mass. Recall that the accelerometer mass at drive point p is compensated for in the correction scheme outlined here if this accelerometer mass is included in mass m_1. Now we look at two electronic compensation schemes to correct for force measurement errors.

4.8.4 Electronic Compensation Using Seismic Acceleration

The idea of using electronic compensation to compensate for mass m_1 was suggested by Ewins.[13] We can write Eq. (4.8.4) (also combine Eq. (4.8.3) with Eqs. (4.7.8) and (4.7.9)) to obtain the force transducer voltage for location p as

$$E_{fp} = H'_{fp}(\omega)S_{fp}F_p + H'_{fp}(\omega)S_{fp}m_1A_p \qquad (4.8.19)$$

If we subtract from (or add to, depending on relative signal phasing) Eq. (4.8.19) a voltage proportional to the driving point acceleration from Eq. (4.8.7), we obtain

---

[13] D. J. Ewins, *Modal Testing: Theory and Practice*, Research Studies Press, Ltd., Letchworth, Hertfordshire, UK, 1986, pp. 142–146.

$$E = E_{fp} - GE_{ap}$$

$$= H'_{fp}(\omega)S_{fp}F_p + \left(H'_{fp}(\omega)S_{fp}m_1 - GH_{ap}(\omega)\frac{S_{ap}}{g}\right)A_p \quad (4.8.20)$$

$$\cong H'_{fp}(\omega)S_{fp}F_p$$

provided the amplifier gain G is adjusted so that

$$G = \left(\frac{H'_{fp}(\omega)}{H_{ap}(\omega)}\right)\left(\frac{S_{fp}m_1 g}{S_{ap}}\right) \quad (4.8.21)$$

Ideally, G should be a constant. However, we see that both instrument FRFs are involved. Thus in order to make this technique work, we need to have both time constants equal and the mechanical parts of these FRFs must be the same. These requirements are essentially the same four rules stated above for the FRF compensation scheme. Note that if the gain is properly adjusted through calibration and if we have satisfied the four transducer FRF rules, we can interpret the measured force as being correct. Then, the measured driving point accelerance is correct as far as this force transducer error is concerned. The transfer accelerances are also correct as measured except for the correction for accelerometer mass at location q. The other precautions, such as bending moments and shear forces, apply in this case as well.

It should be noted that some op-amps have small frequency shifts that occur between the input and output, shifts that cause no problems at low frequencies. At higher frequencies, however, these small phase shifts become larger, so that if one signal passes through an extra op-amp compared to the other signal, an additional phase shift occurs and Eq. (4.8.20) begins to break down due to this phase shift.

McConnell and Park[14] developed an electronic compensation scheme based on using a portion of the base acceleration signal that is subtracted from (or added to, depending on signal phasing) the force voltage. The only difference between McConnell and Park's scheme and Ewins' scheme is that force transducer resonance is not a factor in the Ewins' scheme.

---

[14] K. G. McConnell and Y. S. Park, "Electronic Compensation of a Force Transducer on an Accelerating Cylinder," *Experimental Techniques*, Vol. 21, No. 4, Apr. 1981, pp. 169–172.

4.8.5 Errors due to $HI_{pp}(\omega)$ Being Nonunity

While conducting experiments that used electronic compensation of forces according to the Ewins' technique discussed in Section 4.8.4 above, Hu[15] found large magnitude errors occurred only near resonance, the very place where good accuracy is required. This is not too surprising since we are dealing with a situation where the forces are small and the accelerations are large so that any compensation errors can become significant. Subsequently, McConnell and Hu[16] showed that these large errors are due to small positive phase shifts. Their explanation is outlined as follows.

If we substitute Eq. (4.8.10) into Eq. (4.8.12), we obtain an estimate for the driving point acceleration that includes the instrumentation FRF $HI_p(\omega)$ so that

$$A_{pp}^e = \frac{HI_p(\omega)A_{pp}(\omega)}{1 + m_1 A_{pp}(\omega)\{1 - HI_p(\omega)\}} \quad (4.8.22)$$

When $HI_p(\omega)$ = unity, Eq. (4.8.22) shows that the estimate is the driving point acceleration $A_{pp}(\omega)$. Now, assume that the driving point acceleration corresponds to that of a single DOF system with mass m_s, stiffness k_s, and damping c_s so that

$$A_{pp}(\omega) = \frac{-\omega^2}{k_s - m_s\omega^2 + jc_s\omega} \quad (4.8.23)$$

Next, we assume that the instrumentation FRF ratio has only phase shift errors so that

$$HI_p(\omega) = e^{-j\theta} = \cos(\theta) - j\sin(\theta) \quad (4.8.24)$$

and we define the error function as

$$\epsilon(\omega) = \left[\frac{A_{pp}^e(\omega) - A_{pp}(\omega)}{A_{pp}(\omega)}\right] \times 100 \quad (4.8.25)$$

---

[15] Ximing Hu, "Effects of Stinger Axial Dynamics and Mass Compensation Methods on Experimental Modal Analysis," Ph.D. Dissertation, Iowa State University, Ames, IA, Dec. 1991.

[16] K. G. McConnell and Ximing Hu, "Why Do Large FRF Errors Result from Small Relative Phase Shifts when Using Force Transducer Mass Compensation Methods?", *Proceedings of the 11th International Modal Analysis Conference*, Vol. II, Kissimmee, FL, Feb. 1993, pp. 845–859.

CORRECTING FRF DATA FOR FORCE TRANSDUCER MASS LOADING 233

Equation (4.8.25) can be written in terms of $HI_p(\omega)$ and $A_{pp}(\omega)$ by substituting Eq. (4.8.22) into Eq. (4.8.25) to obtain

$$\epsilon(\omega) = \left[\frac{\{HI_p(\omega) - 1\}\{1 + m_1 A_{pp}(\omega)\}}{1 + m_1 A_{pp}(\omega)\{1 - HI_p(\omega)\}}\right] \times 100 \qquad (4.8.26)$$

where $HI_p(\omega) = 1$ gives zero error. Substitution of Eqs. (4.8.23) and (4.8.24) into Eq. (4.8.26) gives the error function for this case as

$$\epsilon(r) = \left[\frac{-\{1 - \cos(\theta) + j\sin(\theta)\{1 - (1+M)r^2 + j2\zeta r\}}{\underbrace{1 - \{1 + M[1 - \cos(\theta)]\}r^2}_{\text{real}} + \underbrace{j2\zeta r\{1 - (Mr/2\zeta)\sin(\theta)\}}_{\text{imaginary}}}\right] \times 100$$

(4.8.27)

where $r = \omega/\omega_n$.
$\omega_n = \sqrt{k_s/m_s}$ is the structure's natural frequency.
$\zeta = c_s/2\sqrt{k_s m_s}$ is the damping ratio.
$M = m_1/m_s$ is the mass ratio.

An examination of Eq. (4.8.27) reveals that for positive phase angles, there is a critical phase shift for which both the real and the imaginary parts in the denominator go to zero at the same time. This critical angle (for small angle approximation of sine and cosine) and corresponding frequency ratio are given by

$$\theta_c \cong \left[\frac{2\zeta}{M\{1 - \zeta^2 M\}}\right] \text{ (radians)} \qquad (4.8.28)$$

and

$$r_c \cong \left[\frac{1}{1 - \zeta^2 M}\right] \qquad (4.8.29)$$

Equation (4.8.29) indicates that this super high resonant peak occurs at a frequency just slightly greater than unity since the quantity $1 - \zeta^2 M$ is very close to unity. For example, if the mass ratio is 0.10 and the damping ratio ζ is 1/2 percent, then the value of θ is 0.1 radians or 5.7 degrees and the value of r_c is 1.0000025. To give perspective to this value, consider that many multichannel frequency analyzers boast that their anti-aliasing filters are matched within ±5 degrees. We note that negative values of θ do not cause these undesirable results.

The upshot of all this is that the natural frequency estimate is essentially correct while the magnitudes become nonsense. So what is a practical scheme for using this technique? First, if the magnitudes remain within a factor of 20 percent, the natural frequency is correct and the magnitudes are mostly correct. Second, if the magnitude of the corrected curve increases dramatically compared to the original curve, then the natural frequency is correct but the magnitudes should be closer to those of the uncorrected data. One way to check if the analyzer anti-aliasing filters are to blame is to reverse the data channels and repeat the experiment. If the results repeat themselves, then the problem is a phase mismatch between the transducer channels themselves. McConnell and Abdelhamid[17] found a strain gage measurement system that had a linear phase shift with frequency by an amount of 0.026 degrees/Hz. The implication is clear: The correction errors will become worse at the higher natural frequencies. Hence, the low frequency modes may correct well while the higher modes will show some large errors.

If the relative phase shift is negative, the results from Eq. (4.8.27) indicate that maximum errors on the order of 10 percent occur for the values of M and ζ used in the above example. Also, note that the original paper showed that the errors were bounded by 1 to 10 percent for magnitude errors in $HI_p(\omega)$ of 10 percent. Hence, magnitude errors are not our major concern while positive relative phase shifts between the data channels on the order of 5 to 15 degrees are a concern.

A natural inclination is to use electronic compensation so the frequency domain errors noted above are not a concern, but recall that it was time electronic compensation that was being used when the error problem first came to our attention. Hence, both methods suffer from the same problem of positive relative phase shift. I believe that the frequency domain approach is superior, since the positive relative phase shift causes observable changes to take place while the time domain does not show any indication of trouble, particularly if we have no idea what good data looks like.

4.9 CALIBRATION

Transducer manufacturers provide calibration information concerning their instruments' voltage sensitivities. They usually employ secondary standards for calibration purposes that are traceable to the National Institute of Science and Technology, which uses absolute motion measuring techniques in calibrating the secondary standards. We can purchase calibration standards traceable to the National Institute of Science and Tech-

---

[17]K. G. McConnell and M. K. Abdelhamid, "On the Dynamic Calibration of Measurement Systems for Use in Modal Analysis," *Modal Analysis: International Journal of Analytical and Experimental Modal Analysis*, Vol. 2, No. 3, July 1987, pp. 121–127.

nology from most instrument manufacturers to use for our own calibration purposes.

Calibration is a process where we generate a known input to an instrument and then record its response so we can establish the instrument's input-output relationship in order to determine its voltage sensitivity and linearity over a range of frequencies. A transducer's voltage sensitivity is a very important quantity, which needs periodic checking in order to determine that the instrument is functioning properly and has not changed with use and/or misuse.

Static calibration of accelerometers can utilize the local value of gravity. This quick check method requires turning the instrument over from a $+1\,g$ (local) to $-1\,g$ (local) orientation. However, this gives a poor check on linearity and frequency response. Similarly, we can calibrate a force transducer using static loads such as weights. These weights are dependent on the local value of gravity since Newton's law of gravitational attraction relates weight W to the invariable mass quantity m by

$$W = mg \qquad (4.9.1)$$

In this case, the load cell's linearity to static loads can be established. Since the force transducer utilizes a piezoelectric sensor, these static loads are usually applied slowly, and then quickly removed, so that a near step input occurs.

In this section, we explore several commonly employed calibration techniques for accelerometers and force transducers, techniques that involve sinusoidal and transient calibrations.

4.9.1 Accelerometer Calibration—Sinusoidal Excitation

The sinusoidal motion generated by an electromagnetic vibration exciter provides a convenient input motion source over a broad range of frequencies and amplitudes. The accelerometer to be calibrated is mounted on top of a reference standard accelerometer, which is, in turn, mounted on the exciter's armature as shown in Fig. 4.9.1a. A detailed schematic of this mounting is shown in Fig. 4.9.1b, where it is seen that the reference accelerometer measures the *mounting surface motion*. The accelerometers are seen to be mounted *back to back*, so this calibration method is often referred to as *back to back calibration*. Of course, both accelerometers must be mounted to one another and the exciter using good mounting techniques (see Section 4.11), which involves using either greased surfaces or adhesives to ensure good mechanical bonding during calibration.

The reference accelerometer must be traceable to a national standard. Its output is often connected to a special charge amplifier that has special circuits to attenuate the accelerometer's resonance curve so that its sensi-

236 TRANSDUCER MEASUREMENT CONSIDERATIONS

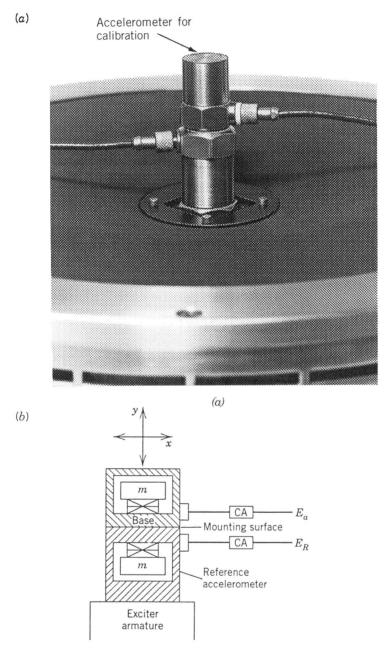

Fig. 4.9.1. (*a*) Photo showing accelerometer to be calibrated mounted on reference calibration accelerometer, which is, in turn, mounted on a vibration exciter's armature. (Courtesy Bruel & Kjaer Instruments, Inc.). (*b*) Cross-section of transducers shown in Fig. 4.9.1*a*.

tivity remains constant (usually 10.00 mV/g) over as wide a frequency range as possible. These attenuation filters can cause some phase shift problems, since we cannot filter a signal without some unwanted phase shifts occurring. Thus the reference accelerometer and charge amplifier come as a single unit. In addition, reference accelerometers are usually sensitive to accelerometer mass that is attached; consequently, their usable frequency range is limited. It is much easier to calibrate around 100 Hz than around 10,000 Hz. Thus we should only calibrate over the frequency range that is required for our needs in order to save time and money.

The output circuits are shown in Fig. 4.9.1b, where each transducer is connected to either a charge amplifier or a built-in voltage follower. The output voltages should be compared on an oscilloscope to ensure waveform purity as well as similarity. Otherwise, a simple RMS voltmeter can give very misleading results due to waveform distortions that contain other frequency components. The voltmeter employed to measure the output voltages should be used in the same range for both instruments in order to avoid voltmeter reading errors due to changing scales or ranges. Several techniques have been developed for reducing these measurement errors to achieve consistent calibration in the order of 0.1 percent while using FFT analyzers.[18]

The vibration exciter must have good y direction (see Fig. 4.9.1b) motion characteristics. Some exciter armatures tend to have a wobbly motion at certain frequencies, due to a combination of the armature suspension system and the location of the mounted mass center relative to the location of the armature's mass center. A wobbly motion causes excessive x direction motion (see Fig. 4.9.1b). Hence, it is good practice to use calibration equipment only for calibration purposes, in order to reduce wear and tear and to keep it in a state of excellent repair. The reference accelerometer should never be used in a test environment and should be recalibrated on a regular basis by an independent laboratory.

4.9.2 Accelerometer Calibration—Transient Excitation

The *gravimetric calibration*[19] method uses a "rigidly" mounted force transducer and an accelerometer mounted on a freely falling steel mass that is guided by a plastic tube, as shown in Fig. 4.9.2a. Each transducer is connected to either a charge amplifier or a built-in voltage follower, which

---

[18]T. R. Licht and H. Anderson, "Trends in Accelerometer Calibration," *Bruel & Kjaer Technical Review*, No. 2, 1987, pp. 23–42.

[19]R. W. Lally, "Gravimetric Calibration of Accelerometers," available from PCB Piezotronics, Inc., 3425 Walden Ave., Depew, NY 14043-2495.

238 TRANSDUCER MEASUREMENT CONSIDERATIONS

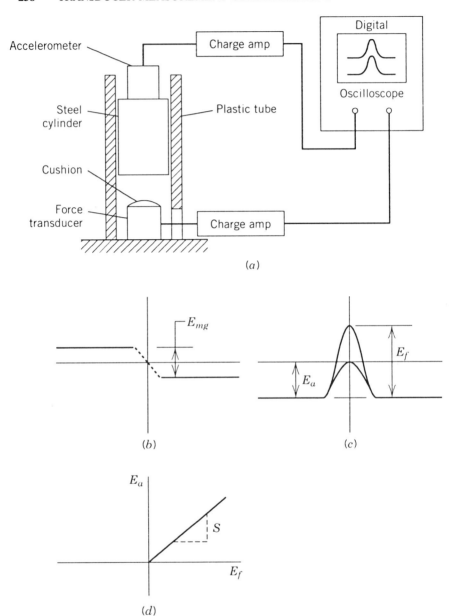

Fig. 4.9.2. Gravimetric calibration. (*a*) Schematic of test setup. (*b*) Second step showing static force voltage E_{mg}. (*c*) Third step impulse time histories E_f and E_a. (*d*) Third step impulse signals plotted against one another showing linearity and slope of E_f/E_a. (From J. W. Dally, W. F. Riley, and K. G. McConnell, *Instrumentation for Engineering Measurements*, 2nd ed., copyright © 1993 by John Wiley & Sons, New York. Reprinted by permission.)

is, in turn, connected to an oscilloscope (preferably digital), as shown in Fig. 4.9.2a.

This calibration method is based on Newton's second law of motion, which describes the impact force between the force transducer and the falling steel mass so that

$$F = ma = mg\left(\frac{a}{g}\right) \tag{4.9.2}$$

Three voltages are measured during three distinct steps in order to calibrate the accelerometer in a manner that is independent of force transducer sensitivity.

The first step consists of setting the steel mass with the accelerometer mounted upon the force transducer and allowing sufficient time for the force transducer's output voltage to drain off to zero due to its RC time constant. The second step consists of quickly removing the steel mass and measuring the force transducer's output voltage E_{mg}, as shown in Fig. 4.9.2b. The third step consists of dropping the steel mass onto the force transducer and simultaneously recording transient voltages of E_f for the force transducer and E_a for the accelerometer, as shown in Fig. 4.9.2c. We know that the output force and acceleration are related to their respective voltages divided by their respective voltage sensitivities. Thus we have

$$F_{mg} = mg = W = \left.\frac{E_{mg}}{S_f}\right\} \quad \text{second step} \tag{4.9.3}$$

and

$$\left.\begin{aligned} F &= \frac{E_f}{S_f} \\ \frac{a}{g} &= \frac{E_a}{S_a} \end{aligned}\right\} \quad \text{third step} \tag{4.9.4}$$

Now, if we insert Eqs. (4.9.3) and (4.9.4) into Eq. (4.9.2), we obtain

$$\frac{E_f}{S_f} = \frac{E_{mg}}{S_f}\frac{E_a}{S_a} \tag{4.9.5}$$

from which S_f cancels out so that the accelerometer's voltage sensitivity becomes

$$S_a = \left[\frac{E_a}{E_f}\right] E_{mg} \qquad (4.9.6)$$

It is evident from Eq. (4.9.6) that the accelerometer's voltage sensitivity is independent of the force transducer's sensitivity for a linear force transducer. We are at liberty to use the values of E_a and E_f at any common time during the impulse. However, we should use peak voltages since these voltages have the best signal to noise ratios. Since voltages E_a and E_f occur as a ratio, it is helpful to plot E_a versus E_f, as shown in Fig. 4.9.2d, and to extract the slope from this plot. This last step is conveniently done when using a digital oscilloscope that has straight line statistical analysis capabilities. Then, the slope represents the best average ratio of voltages. One can also see the linearity between signals so that significant nonlinearity or hysteresis is clearly evident from this plot.

The impulse time duration is controlled by the cushion material and falling mass. Since cushion materials are usually close to linear in behavior, the impact time history can be approximated by a half-sine function, as shown in Fig. 4.9.2c. The drop height and cushion material control the impulse force amplitudes. The major advantages of gravimetric calibration are its low cost and portability. It has two disadvantages. First, its results are dependent on the local value of gravity, since the reference voltage of E_{mg} is dependent on local value of weight or *mg*. Second, the calibration is limited in its frequency range and level of acceleration.

For those who need calibration to high frequencies and high g levels in the order of 10,000 to 200,000 g's, a commercially available system has been developed.[20] This technique employs impacting an aluminum rod with small projectiles to produce shock waves in the aluminum rod, shock waves that are measured with strain gages. The strain gage information is used to imply the acceleration levels that an accelerometer will experience when mounted on the end of the aluminum rod. Overall accuracies in the order of 6 percent can be obtained with this calibration method.

4.9.3 Force Transducer—Sinusoidal Excitation

The electromagnetic vibration exciter can be used to calibrate a force transducer over a wide range of frequencies and amplitudes by attaching

---

[20]R. D. Sill, "Shock Calibration of Accelerometers at Amplitudes to 100,000 g Using Compression Waves," Endevco Tech. Paper 283, available from Endevco, Rancho Viejo Rd., San Juan Capistrano, CA 92675.

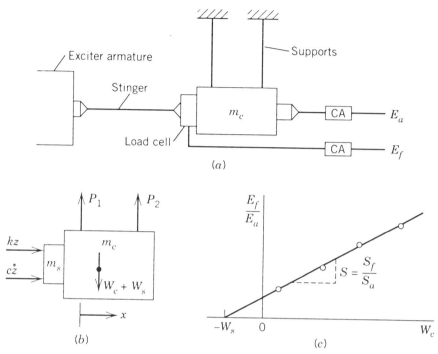

Fig. 4.9.3. Force transducer calibration. (*a*) Schematic of test setup. (*b*) FBD of calibration mass. (*c*) Plot of voltage ratio E_f/E_a versus calibration weight W_c. (From J. W. Dally, W. F. Riley, and K. G. McConnell, *Instrumentation for Engineering Measurements*, 2nd ed., copyright © 1993 by John Wiley & Sons, New York. Reprinted by permission.)

the load cell to the vibration exciter's armature through a stinger, as shown in Fig. 4.9.3*a*. The calibration mass m_c is suspended by support wires or cords, is attached to the load cell's seismic mass end, and is attached to an accelerometer on the right-hand side so that calibration mass m_c includes the accelerometer's mass.

The FBD in Fig. 4.9.3*b* shows that the piezoelectric sensor measures the total inertial load of the transducer's seismic mass m_s and calibration mass m_c so that the load cell's differential equation of motion is

$$c\dot{z} + kz = m_c\ddot{x} + m_s\ddot{x} \tag{4.9.7}$$

where x is the calibration inertia's motion. Thus we can write the force

242 TRANSDUCER MEASUREMENT CONSIDERATIONS

transducer's output voltage as

$$E_f = H'_f(\omega) S_f [m_c g + m_s g] \left[\frac{\ddot{x}}{g}\right] \quad (4.9.8)$$

while the accelerometer's output voltage is given by

$$E_a = H_a(\omega) S_a \left[\frac{\ddot{x}}{g}\right] \quad (4.9.9)$$

Now, if we look at the voltage ratio of E_f/E_a, we obtain

$$\frac{E_f}{E_a} = \frac{H'_f(\omega) S_f}{H_a(\omega) S_a} [m_c g + m_s g] = \frac{S_f}{S_a} [W_c + W_s] \quad (4.9.10)$$

where W_c is the calibration mass's weight (the value must not be dependent on the local value of g).
W_s is the seismic mass's weight.
The two transducer FRFs, $H'_f(\omega)$ and $H_a(\omega)$ in Eq. (4.9.10), must be nearly equal to unity for this scheme to work. This means that the low frequency time constants must be equal and we must limit the upper frequency used for calibration so that there are no magnitude and phase distortions due to ignoring the $H'_f(\omega)$ and $H_a(\omega)$ terms in Eq. (4.9.10).

If the calibration mass m_c is large (>100 times the seismic mass m_s), then we can solve directly for S_f from Eq. (4.9.10) to obtain

$$S_f = \left[\frac{E_f}{E_a}\right]\left[\frac{S_a}{W_c}\right] \quad (4.9.11)$$

Equation (4.9.11) indicates that we need either the slope of a plot of E_f versus E_a or the peak voltages (in order to have the best signal to noise ratio) to calculate the ratio E_f/E_a, the accelerometer's voltage sensitivity S_a, and calibration weight W_c. The calibration weight must be measured by using a balance type of scale that is independent of the local acceleration due to gravity. We see from Eq. (4.9.11) that this calibration is limited by the accuracy of the accelerometer's voltage sensitivity.

We can use Eq. (4.9.10) to obtain the load cell's effective seismic mass by plotting the voltage ratio as a function of the calibration weight W_c, as shown in Fig. 4.9.3c. The best straight line through these data points intersects the horizontal line at the seismic mass's weight W_s. We can also check the calibration given by Eq. (4.9.11) by noting that the straight line's slope S in Fig. 4.9.3c is S_f/S_a, so that the force transducer's voltage sensitivity is given by

$$S_f = SS_a \qquad (4.9.12)$$

By exercising reasonable care, accuracies better than ±1 percent can be obtained, along with reasonable values for the instrument's seismic mass.

4.9.4 Force Transducer—Transient Excitation

Impulse calibration is most often employed with load cells attached to hammers, as shown in Fig. 4.9.4. Again, calibration mass m_c is suspended as a parallel bar mechanism on two wires or cords and has an accelerometer attached to one end. The impulse force $f(t)$ acts on both the calibration mass and the impact hammer's tip. We saw in Section 4.7 that the output voltage for this impulse hammer is given by Eq. (4.7.14), which is written here for convenience as

$$E_f = \left[\frac{m_2 S_f}{(m_1 + m_2)}\right] H_f(\omega) F = S_f^* H_i(\omega) F \qquad (4.9.13)$$

where F is the input force and the transducer's effective voltage sensitivity S_f^* is given by Eq. (4.7.15) as

$$S_f^* = \left[\frac{m_2}{m_1 + m_2}\right] S_f \qquad (4.9.14)$$

and the transducer's FRF, $H_f(w)$, is given by Eq. (4.7.11) as

$$H_f(\omega) = \left[\frac{jT_f\omega}{1 + jT_f\omega}\right]\left[\frac{1}{1 - r^2 + j2\zeta r}\right] \qquad (4.9.15)$$

where T_f is the force transducer's RC time constant. The transducer's natural frequency and damping are defined in Eqs. (4.7.5).

We want to relate the load cell's output voltage to the accelerometer's output voltage. This is done using Newton's second law, as expressed by Eq. (4.9.2), and the acceleration voltage expression from Eq. (4.9.9) for the accelerometer's output voltage, along with Eq. (4.9.13), to obtain

$$E_f = \left[\frac{H_f(\omega)}{H_a(\omega)}\right]\left[\frac{S_f^*}{S_a}\right] m_c g E_a \qquad (4.9.16)$$

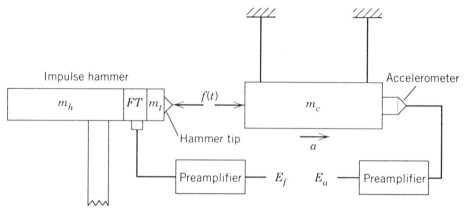

Fig. 4.9.4. Schematic of impulse force calibration of a load cell attached to an impulse hammer.

from which the force transducer's voltage sensitivity becomes

$$S_f^* = \left[\frac{H_a(\omega)}{H_f(\omega)}\right]\left[\frac{E_f}{E_a}\right]\left[\frac{S_a}{W_c}\right] \cong \left[\frac{E_f}{E_a}\right]\left[\frac{S_a}{W_c}\right] \quad (4.9.17)$$

In Eq. (4.9.17), we see that it is desirable to plot voltages E_f versus E_a as shown in Fig. 4.9.3d, so that we can use the slope as a statistical estimate. It also is clear from Eq. (4.9.17) that we need to have matching time constants in the instrument FRFs or to use frequencies above the higher of the low frequency cutoffs. Also, we should limit our input frequencies to be below 0.2 of the lowest transducer natural frequency. As we have shown in Chapter 2, the frequency content in the impulsive half-sine signal that is generated in this calibration has significant frequency components up to $1/T$ where T is the half-sine impulse duration. Thus by controlling the impulse duration, we can control the range of frequencies used for the calibration. This time duration is dependent on hammer mass, tip stiffness, and calibration mass.

The above technique is often used when working with frequency analyzers. It is convenient to set the accelerometer's voltage sensitivity in the analyzer, then strike the calibration mass five or more times to obtain an averaged acceleration FRF, and then adjust the force transducer sensitivity S_f^* in the analyzer to obtain an FRF readout value that is equal to $1/m_c$. This is a very good calibration technique for the analyzer's amplifier, anti aliasing filters (see Chapter 5), and analog to digital (A/D) characteristics are calibrated at the same time as the instrument systems. This is sometimes referred to as an *overall system calibration*.

Unfortunately, the above technique has been widely demonstrated using a hand held calibration mass of around 200 to 300 gm. When such a small mass is hand held, the hand constitutes a significant amount of unknown mass that varies with how hard the calibration mass is held, so that the calibration can easily be useless. In this case, the mass must be freely suspended on lightweight strings in order to eliminate the effects of hand mass on the calibration. It is clearly acceptable to hand hold a 400 pound or larger calibration mass if one is willing to do so.

A final pitfall has been observed with this calibration technique and was accidentally discovered by several students while taking a measurements course. They were asked to calibrate an impulse hammer using three different calibration masses with nominal weights of 50 lb, 5 lb, and 0.5 lb. They obtained consistent results from the 50 and 5 lb calibration weights, so that the accelerance FRF looked like those shown in Fig. 4.9.5a. When they used the 0.5 lb mass, however, they obtained an accelerance FRF that had a number of stair steps and proved to be inconsistent as shown in Fig. 4.9.5b. Why did this happen?

After some time, the students discovered that the voltage gains on the accelerometer and force transducer channels had never been changed when moving from the 50 lb weight to 0.5 lb weight. The students had adjusted the charge amp and analyzer gains for the 50 lb calibration weight so that the voltage signals were nearly equal, and they were using the top half of the A/D converter's range. By the time they got to the 0.5 lb calibration weight, these two voltages differed by a factor in the order of 100 to 1 so they were very unequal. Thus when one A/D converter was working at its top half (50 to 100 percent of its full scale), the other A/D converter was working at less than 1 percent of its full scale. Hence, the students were seeing the A/D converter's least significant digit, or several digits, of variation in the accelerance output. A further contribution to these stair step variations came from a poor signal to noise ratio in the smaller signal. Once the charge amplifier range and/or analyzer input amplifier gains were adjusted to improve the signal level, consistent force transducer calibration results were obtained for the small calibration weight as well.

The appearance of steps in this type of calibration is a clear indication that one of the A/D converters is working with signals that are too small. Thus either the analyzer's input voltage gains or the charge amplifier voltage gains need to be adjusted.

Finally, the curves in Figs. 4.9.5a and 4.9.5b show low frequency errors. These are due to a mismatch in the low frequency time constants between

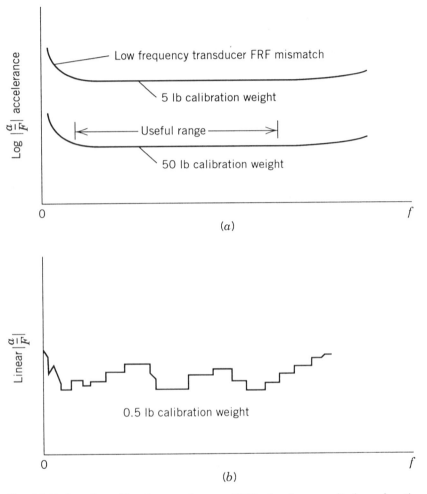

Fig. 4.9.5. Impulse calibration accelerance FRFs showing magnitude as function of frequency. (*a*) Two satisfactory calibrations. (*b*) Unsatisfactory calibration due to high noise and improper voltage range for the A/D converter.

the force transducer's FRF $H_f(\omega)$ and the accelerometer's FRF $H_a(\omega)$. These curves also show high frequency deviations from the central ideal response that is nearly constant over a broad range of frequencies. It is evident that such a calibration curve is very insightful, since we can see the frequency limits in both magnitude and phase when using these instruments.

4.9.5 Effects of Bending Moments on Measured Forces

McConnell and Varoto[21] reported an unusual measured response while trying to calibrate a force transducer for bending moment sensitivity using the physical setup shown in Fig. 4.9.6. The force transducer was mounted to a rigid base and had a $1 \times 1 \times 2.25$ in. steel bar attached. The impulse hammer IH was used to strike the bar at its center point 1 and at end points 2 and 3. The resulting input-output calibration FRFs are shown in

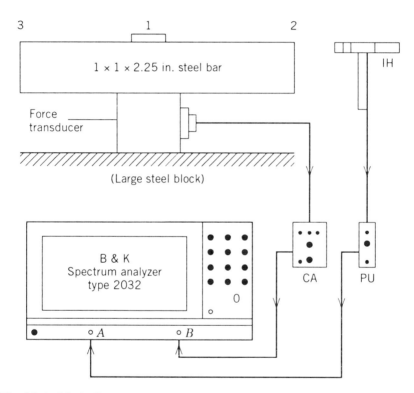

Fig. 4.9.6. Block diagram of the experimental setup. (Courtesy of the *Proceedings for the 11th International Modal Analysis Conference*, Society for Experimental Mechanics.)

---

[21]K. G. McConnell and P. S. Varoto, "A Model for Force Transducer Bending Moment Sensitivity and Response During Calibration," *Proceedings of the 11th International Modal Analysis Conference*, Vol. I, Kissimmee, FL, Feb. 1993, pp. 516–521.

248   TRANSDUCER MEASUREMENT CONSIDERATIONS

Fig. 4.9.7, where it is apparent that unusual behavior is occurring. It is anticipated that these curves should be horizontal straight lines that are greater and less than unity. The peak and valley for case 2 is quite different than that for case 3 in the 200 to 350 Hz frequency range. So what is happening?

The following is a brief summary of this paper. Figure 4.9.8 shows an uncoupled two DOF dynamic model that has differential equations of motion given by

$$m\ddot{y} + c_1\dot{y} + k_1 y = f(t) \tag{4.9.18}$$

$$I\ddot{\theta} + c_2\dot{\theta} + k_2\theta = lf(t) \tag{4.9.19}$$

where k_1 and c_1 are the translational motion spring and damping constants.

k_2 and c_2 are the rotational motion spring and damping constants.
$f(t)$ is the excitation force with magnitude f_0.
m is the bar mass.
I is the bar mass moment of inertia with respect to its mass center.

The bar is of length L. The coordinates y and θ describe the translational and rotational motions of the bar. The steady-state frequency domain solution of Eqs. (4.9.18) and (4.9.19) are complex phasors described by

$$A = \frac{f_0/k_1}{(1 - r_y^2) + j2\zeta_y r_y} \tag{4.9.20}$$

and

$$B = \frac{lf_0/k_2}{(1 - r_\theta^2) + j2\zeta_\theta r_\theta} \tag{4.9.21}$$

where dimensionless frequency ratios of

$$r_y = \frac{\omega}{\omega_y} \quad \text{and} \quad r_\theta = \frac{\omega}{\omega_\theta}$$

are used with

$$\omega_y = \left(\frac{k_1}{m}\right)^{1/2} = \text{translational natural frequency}$$

$$\omega_\theta = \left(\frac{k_2}{I_0}\right)^{1/2} = \text{rotational natural frequency}$$

CALIBRATION   249

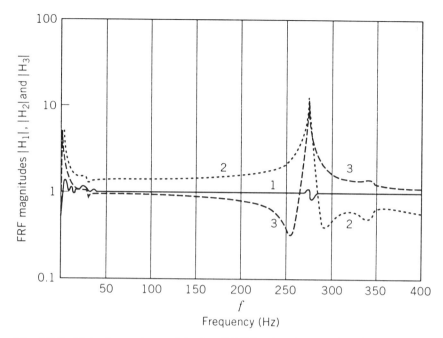

Fig. 4.9.7. Plot of experimental load cell FRFs when impacted at locations 1, 2, and 3 for the configuration shown in Fig. 4.9.6. (Courtesy of the *Proceedings for the 11th International Modal Analysis Conference*, Society for Experimental Mechanics.)

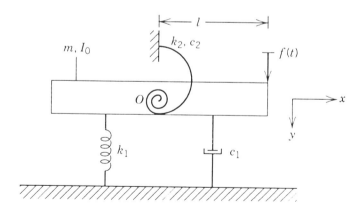

Fig. 4.9.8. A two DOF system model of the force transducer. (Courtesy of the *Proceedings for the 11th International Modal Analysis Conference*, Society for Experimental Mechanics.)

and where dimensionless damping ratios

$$\zeta_y = \frac{c_1}{2(mk_1)^{1/2}} = \text{translational damping ratio}$$

$$\zeta_\theta = \frac{c_2}{2(I_0 k_2)^{1/2}} = \text{rotational damping ratio}$$

are used for damping. It is assumed that the output voltage phasor is given by

$$E_f = S_y A + S_\theta B \qquad (4.9.22)$$

where S_y and S_θ are the transducer's voltage sensitivities to translational and rotational displacements. Now, substituting Eqs. (4.9.20) and (4.9.21) into Eq. (4.9.22) and using $S_F = S_y/k_1$ and $S_M = S_\theta/k_2$ as the voltage sensitivities to force and bending moments, we obtain

$$E_f = \left(\frac{(S_F + lS_M)(1 - \alpha r_\theta^2 + j\zeta_\theta \gamma \alpha r_\theta)}{[1 - (r_\theta/\beta)]^2 + j2\zeta_y(r_\theta/\beta)(1 - r_\theta^2 + j2\zeta_\theta r_\theta)} \right) f_0 \qquad (4.9.23)$$

where the dimensionless ratios of β, α, and γ are given by

$$\beta = \frac{\omega_y}{\omega_\theta} \qquad (4.9.24)$$

$$\alpha = \frac{\beta^2 S_F + lS_M}{\beta^2 (S_F + lS_M)} \qquad (4.9.25)$$

$$\gamma = \frac{\beta S_F + lS_M(\zeta_y/\zeta_\theta)}{\beta (S_F + lS_M)} \qquad (4.9.26)$$

Equation (4.9.23) has a number of important features. First, the apparent force sensitivity S_F at low frequencies is given by

$$S_{Fa} = S_F + lS_M \qquad (4.9.27)$$

This sensitivity depends on the sign of lS_M, so that a sensitivity higher than S_F occurs when lS_M is positive and a sensitivity lower than S_F occurs when lS_M is negative. This is the result that we would have intuitively anticipated.

Second, we see that there are two resonant conditions in the denominator, one corresponding to $r_\theta = 1$ for the torsional motion and one corresponding to $r_\theta/\beta = 1$ for the translational motion. Note that β is $\gg 1$ in this experiment.

Third, we see that a notch occurs when

$$\alpha r_\theta^2 = 1 \tag{4.9.28}$$

However, we see from Eq. (4.9.25) that the value of α is either greater than unity when lS_M is negative, or less than unity when lS_M is positive. Thus the notch can occur either before or after the torsional resonance.

Fourth, when $r_\theta \gg 1$, the apparent force sensitivity becomes

$$S_{Fa} \cong \alpha(S_F + lS_M) \cong S_F + \frac{lS_M}{\beta^2} \tag{4.9.29}$$

Equation (4.9.29) shows that the apparent force sensitivity approaches S_F when β is large compared to unity; the value is either slightly greater than S_F when lS_M is positive or slightly less than S_F when lS_M is negative.

A plot of Eq. (4.9.23) is shown in Fig. 4.9.9, where it is seen to have the same characteristics as the experimental data shown in Fig. 4.9.7. The parameter values were selected to be a reasonable match to the experimental values.

This example has shown the importance of understanding the effects of force transducer bending moment sensitivity, large mass moments of inertia, and low torsional natural frequencies. Obviously, the measurement errors were large enough in this instance to cause a ghost resonance to appear in a measured experimental FRF. The reader should be forewarned that some so-called *mechanical impedance heads* are designed with a large base mass in order to give the transducer *stability*. It now appears that this large base mass can cause torsional resonance to be reflected in the measured force. At the very least, this is an important possibility to be aware of.

4.10 ENVIRONMENTAL FACTORS

We purchase transducers that are designed to measure force or acceleration under a wide variety of conditions. There are a number of environmental factors that can alter or change a transducer's performance. We briefly mention some of these factors here.

4.10.1 Base Strain

The mounting surface of a structure can have significant bending strain, particularly when the transducer is mounted at a node point, so that there

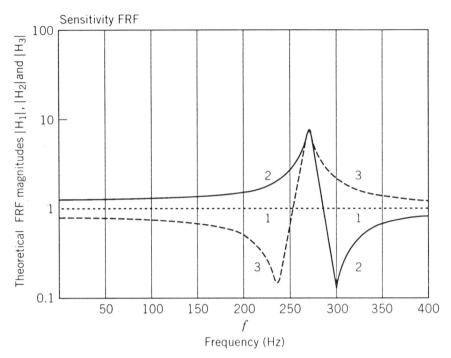

Fig. 4.9.9. Plot of theoretical force transducer FRFs when impacted at locations 1, 2, and 3 in Fig. 4.9.6. (Courtesy of the *Proceedings for the 11th International Modal Analysis Conference*, Society for Experimental Mechanics.)

is no motion to sense but a lot of bending strain or surface curvature. This surface curvature is transmitted to the piezoelectric sensing element through the transducer's base. Base strain can cause significant output signals when there is no motion.

Several designs have been developed to reduce or eliminate this sensitivity. Generally, compression designs have the most bending sensitivity, while shear designs usually have a minimum sensitivity to bending strains. Base strain sensitivity is like cross-axis sensitivity in that it is often orientation sensitive. The International Standards Organization (ISO) and American National Standards Institute (ANSI) standard test for base strain sensitivity at a level of $250\mu\epsilon$ at the base of a specially built cantilever beam. This test provides for a comparison basis between transducers, not an absolute sensitivity.

During a recent study of force transducers, Cappa and McConnell[22] discovered that force transducers can also be very base strain sensitive. It was found that this sensitivity varied from resonance to resonance since the amount of strain decreased with increasing natural frequencies and the same level of vibration. This sensitivity would never have been discovered if it were not for the unusual measurements that were being made. Thus while accelerometer base strain sensitivity is well documented, the force transducer's base strain sensitivity is not well documented but certainly exists.

4.10.2 Cable Noise

The cable that connects the transducer to its electronics can be a source of unwanted noise when using a charge amplifier type of interface electronics. There are three main noise sources: electromagnetic noise, ground loops, and triboelectric effects.

Electromagnetic effects are caused by power cables that carry large amounts of alternating electrical current. These cables are surrounded with large time varying magnetic fields that induce voltages in parallel running data cables. What could be more natural or tempting than to run a data cable parallel to power cables in the same cable tray? This is a serious mistake. It is recommended that the signal cables be run in a grounded steel conduit that is as far removed from the power cables as possible. Under no circumstances should excess cable be coiled neatly. We can inadvertently build ourselves a transformer with these neat coils of signal cable.

Ground loops are the natural result of having the accelerometer attached to a structure that is at a different electrical potential than the charge amplifier's ground. This potential difference causes a current to flow in the signal cable's shield and makes the transducer to appear to have a signal when none is present.

This problem is overcome by electrically isolating the transducer from the structure under test or by forcing both the structure under test and the charge amplifier to have the same ground point. Ground loops can be a most vexing problem. One rather strange ground loop source turned out to be a local radio station that electromagnetically excited the entire test structure. Once this source was recognized, a grounding stake was driven deeply into the earth to ground the entire structure and all electronic recording gear in order to adequately suppress this signal source.

Triboelectric effects are caused by damaged cables that generate an

---

[22]P. Cappa and K. G. McConnell, "Base Strain Effects on Force Measurements," *Proceedings, Spring Society for Experimental Mechanics*, Baltimore, MD, June 1994.

electrical charge when either the center core wire or the shield wire breaks its mechanical bond and "rubs" against the common electrical insulation material. The triboelectric effect often generates charge at the same frequency as the structure's vibration, since cable flexing occurs at the structure's test frequency.

A simple test method for detecting when cables are susceptible to triboelectric effects is to set the charge amplifier and recording gear to the voltage sensitivities that are to be used during the test. Then shake the transducer's cable and observe the recorded output signal. If there is a significant signal, then the cable should be given the *scissors award* whereby we convert it into two cables of equal length before discarding both. Otherwise, the discarded cable will come back into service at the most inappropriate time.

The triboelectric cable check should be made before and after each test. It is good practice to make sure that the cable is properly attached to the structure under test and that it is connected to a nonvibrating point at a point of minimum structural motion. We should do all that is possible to reduce cable flexing that leads to cable failure and triboelectric effects. There are no triboelectric effects for the built-in amplifier type of transducer system.

4.10.3 Humidity and Dirt

Generally, manufacturers seal the transducer at time of assembly so that dirt and humidity are not major problems for the transducer itself. Connections and connecting cables are another matter. Dirt and humidity on cable connections at the transducer end as well as the amplifier end have been known to cause significant measurement errors. More than once has the author run across connections that were filthy with oil and dirt. This oil and dirt (including ordinary greasy fingerprints) can cause a significant short to the circuit, drawing off important information from the charge generated. Hence, be sure to ensure that all connections are clean, clean, clean! Acetone cleaner is preferred over alcohol in most instances.

The connections can be protected from light humidity over relatively short time periods at modest temperatures by dipping the ends in silicon grease before assembly. For immersion applications, the assembled connector should be encased in acid free RTV silicone rubber and carefully cured according to instructions.

4.10.4 Mounting the Transducer

The method employed to mount the transducer to the structure's surface can be critical to achieving quality measurements. Transducers are often bolted to a structure's surface. Both the surface finish and its bending strain can influence how well the structure's motion is transferred to the transducer. It is recommended that the surface be polished to remove significant roughness and that a layer of high wax content grease be applied before the transducer is attached. The grease acts like an adhesive at higher frequencies where the transducer could resonate on the combination spring rate of the mounting bolt and the mounting surface. This apparent adhesive action is due to the grease's inability to flow at high frequencies. However, the use of grease to "bond" the transducer to the structure is limited by the grease's temperature characteristics.

Cementing of the transducer to the structure under test is often employed because we obtain a good bond while we also achieve electrical isolation if done with care. Epoxy cements are long lasting and are available for use in a wide range of temperatures. Cyanoacrylate cements are simple and fast to use with clean smooth surfaces but have a limited temperature range compared to the epoxy cements.

Double sided adhesive tapes provide a quick means of mounting lightweight accelerometers to structures. This mounting method is limited to low frequency applications and is recommended only in survey type tests where we are trying to find out what the test environment looks like. The hand held probe is not recommended except at low frequencies on large massive structures, since the added mass and damping of the hand will significantly affect the vibration response of a lightweight structure.

4.10.5 Nuclear Radiation

The use of accelerometers and load cells in the presence of nuclear radiation is so specialized that the transducer user should consult with the manufacturer before attempting to purchase and/or install a transducer. Usually instrument vendors have experience with your environment and know what transducer to recommend.

4.10.6 Temperature

Temperature causes a number of undesirable effects. First, most transducers are sensitive to rapid changes in temperature, and hence, generate false signals. Second, transducer voltage sensitivity usually changes with temperature. This change may be due to changes in charge sensitivity S_q as well as transducer capacitance. Third, there are temperature extremes beyond which damage may occur to the sensing element. Most piezoelectric sensor transducers can withstand temperatures up to 250°C. At higher temperatures, the piezoelectric material becomes depolarized and complete damage occurs once the Curie temperature is achieved.

4.10.7 Transducer Mass

The mass of an accelerometer can be large enough to cause a significant change in the structure's natural frequency. It is often difficult to estimate this sensitivity analytically. A simple field test for mass sensitivity consists of repeating a vibration test with a second mass equal to the accelerometer's mass attached as close as possible to the accelerometer. Then we note how much the resonant frequencies change. If they change more than a few percent, we may need to use a smaller and lighter weight accelerometer.

4.10.8 Transverse Sensitivity

We have paid significant attention to the cross-axis sensitivity of accelerometers and the effect of cross-axis sensitivity on the resulting FRFs. We have even developed a method to correct for accelerometer cross-axis sensitivity. Force transducers are sensitive to bending moments and shear forces. We have little experience in defining bending moment sensitivities, let alone prescribing methods of correction. Consequently, we must use all possible precautions to limit the significance of bending moments and shear forces in a given application through the use of stinger type of flexible structures to connect the force transducer to the exciter's armature.

4.11 SUMMARY

The objective of this chapter has been to review the pertinent characteristics of accelerometers and force transducers as commonly used in vibration measurements and analysis. The chapter began with the seismic mechanical model that shows accelerometers behave as a single DOF oscillator with two excitation forces, one due to gravity and one due to inertial excitation forces. The inertial excitation force is the one that causes the instrument to become sensitive to base acceleration, and hence, makes the instrument an accelerometer. The gravity force can cause a surprising low frequency measurement error when the accelerometer's sensing axis rotates while measuring a curvilinear acceleration motion.

The piezoelectric sensor is commonly used in both accelerometers and load cells, due to the sensor's small size, large stiffness, low damping, and high sensitivity. Thus the electrical characteristics of piezoelectric sensors and interface amplifiers have been examined in detail. The chapter began with a brief review of basic AC circuits and op-amps along with a description of the piezoelectric displacement charge sensitivity model. The charge amplifier's basic behavior is described in terms of two op-amps. The main advantages of charge amplifiers are that the time constant is controlled by its internal feedback resistance and capacitance, while the voltage sensitivity can be adjusted to obtain a standardized voltage sensitivity. The RC time constant leads to both magnitude and phase shift errors occurring at low frequency. The built-in voltage follower arrangement is simple to use but also has two time constants, one that is internal to the transducer's built-in voltage follower and the other external and affected by the user when the output is connected to one or more voltage recording devices. Thus it is the user's responsibility to assure that the output is not overloaded by connecting to a voltage recording device with a small input resistance, causing one of the time constants to be too small for the measurements to be made. A principle advantage of the built-in amplifier transducer is that the instrument is insensitive to triboelectric effects. Finally, this chapter has looked at the overall accelerometer FRF that consists of the low frequency piezoelectric RC time constant roll-off while the upper end is controlled by the mechanical system resonance. It is the user's responsibility to select an instrument system that is suitable for performing satisfactorily within these frequency limits.

It is often desirable to measure both linear and angular acceleration at a point in a structure. We have seen that, in principle, it requires two sensors placed a certain distance apart while mounted on a rigid member of considerable mass to achieve this measurement. A recently developed instrument that uses two piezobeams has been examined in detail; we have seen how a single transducer can be constructed to measure both linear and angular acceleration by mounting two piezobeams on a single center post. The outputs of these two piezobeams are combined in special ways to produce both linear and angular acceleration voltage signals from a single small lightweight transducer.

Next, we have looked at how the transducer system responds to transient inputs. First, we examined the phenomenon of mechanical "ringing" that occurs when a step like input is applied to the transducer. Then, the ramp-hold input was examined in detail to see how a transducer responds to rapidly changing ramp type inputs. It was seen that the ramp rise time must be at least five times the transducer's natural period. It was concluded that "transducer ringing" is a clear indication that the input is varying too rapidly for the transducer to respond accurately. It was also seen that the piezoelectric sensor affects long time measurements. This was shown by examining the response to a step input. The piezoelectric circuit's response exhibits an exponential decay that is controlled by the RC time constant. Rules for determining when the time constant is too short for a given transient were given for several typical transient signals. Particularly, the existence of an undershoot followed by an exponential decay was shown to be a clear indication of time constant problems.

Accelerometer cross-axis sensitivity can contaminate acceleration signals and the resulting FRF data. A theoretical model has been presented to explain the basic characteristics of cross-axis sensitivity. This model was expanded to include a tri-axial accelerometer, and a correction scheme was applied to experimentally measured data. It was seen for this particular case that cross-axis sensitivity contaminated the measured FRF to the point that a standard modal analysis package gave seven natural frequencies and distorted structural damping ratios when there should only be four. The corrected data gave four natural frequencies and damping values that correspond closely to those obtained under nearly ideal conditions that did not have cross-axis contamination of the data. It is clear from this experimental result that cross-axis sensitivity is a factor that we should be concerned with when dealing with unknown structures.

The force transducer can interact with the structure being measured and this interaction has been examined in detail. A general two DOF model was developed and applied to three common measurement environments that consist of a fixed base load cell, a force transducer mounted on an impulse hammer, and a load cell that is located between a structure and a vibration exciter. We found that the fixed base load cell behaves exactly like an accelerometer except that the seismic mass is dependent on connection masses employed. The resonant behavior in this case has misled us to believe that force transducer resonance is similar to that of an accelerometer in all situations. However, we saw for the impulse hammer case that the force transducer's sensitivity and natural frequency change with changes in either seismic or base masses. When the force transducer is connected between a structure and a vibration exciter, it exhibits a rather interesting behavior of having no natural frequency, and its output is dependent on the product of effective seismic mass and acceleration of the structure under test. This shows how the force transducer signals are altered by the very structural response that the force excites. This interaction is shown to lead to large errors around structural resonance, the very place we need the best readings in many vibration tests.

It is evident from Sections 4.7 and 4.8 that force transducers are dependent on their application environment. However, to understand what we have in terms of force transducer signals, we need to understand how the transducer is being used, and how we can compensate for and analyze the resulting signal. It is clear that we can use a very simple model, as given by Eq.(4.8.4), for all force transducer voltage output signals when the load cell's sensing end is attached to the structure under test. This signal contains an error that is due to and controlled by the structure under test through the $m_1 A_s(\omega)$ term. This term can be corrected for in either the frequency domain or the time domain. The methods for making this correction and the rules by which the corrections can be accomplished have been outlined. It should be evident that a good theoretical model is required to understand how to interpret the resulting experimental data and implement these data compensation schemes. We should note that small positive relative phase shifts between force and acceleration can cause these correction schemes to have large errors around resonances. However, the good news is that the frequency domain method at least indicates that the problem of small relative phase shifts is taking place, while the time domain method does not show this effect in an obvious manner.

Calibration of accelerometers and load cells is an important consideration. Both sinusoidal and transient methods have been presented for both instruments. Calibration is an extremely important measurement step that needs to be done to assure that the data is reasonably free of misinformation due to damaged or defective transducer systems. It is shown that force transducers perform in an unexpected manner when placed in an environment with large mass moments of inertia and bending moments.

Finally, the chapter ends with a review of many of the environmental factors that can cause serious measurement errors to occur. One of the most serious of these factors is the misuse of transducers by poorly trained personnel. This environmental factor can only be overcome through good personnel management and policies, as well as a continuously implemented high quality educational program.

REFERENCES

The following references are for general information.
1. Baek, T. H. and K. G. McConnell, "Thrust and Motion Measurement of a Rocket Propelled Bar," *Experimental Techniques*, Vol. 11, No. 12, 1987, pp. 24–27.
2. Broch, J. T., *Technical Vibration and Shock Measurements*, available from Bruel & Kjaer Instruments, Inc., Marlborough, MA, Oct. 1980.
3. Bruel & Kjaer Instruments, *Piezoelectric Accelerometer and Vibration Preamplifier Handbook*, Marlborough, MA, Mar. 1978.
4. Bruel & Kjaer Instruments, *Technical Review*, A quarterly publication, Marlborough, MA:
    (a) "Vibration Testing of Components," No. 2, 1958.
    (b) "Measurement and Description of Shock," No. 3, 1966.
    (c) "Mechanical Failure Forecast by Vibration Analysis," No. 3, 1966.
    (d) "Vibration Testing," No. 3, 1967.
    (e) "Shock and Vibration Isolation of a Punch Press," No. 1, 1971.
    (f) "Vibration Measurement by Laser Interferometer," No. 1, 1971.
    (g) "A Portable Calibrator for Accelerometers," No. 1, 1971.
    (h) "High Frequency Response of Force Transducers," No. 3, 1972.
    (i) "Measurement of Low Level Vibrations in Buildings," No. 3, 1972.

(j) "On the Measurement of Frequency Response Functions," No. 4, 1975.
(k) "Vibration Monitoring of Machines," No. 1, 1987.
(l) "Recent Developments in Accelerometer Design," No. 2, 1987.
(m) "Trends in Accelerometer Calibration," No. 2, 1987.

5. Change, N. D., *General Guide to ICP Instrumentation*, Available from PCB Piezotronics, Inc., Depew, NY.
6. Dally, J. W., W. F. Riley, K. G. McConnell, *Instrumentation for Engineering Neasurements*, 2nd ed., John Wiley & Sons, New York, 1993.
7. Doyle, James F. and James W. Phillips (Editors), *Manual on Experimental Stress Analysis*, 5th ed., Society for Experimental Mechanics, Bethel, CT, 1989.
8. Endevco Corporation, *Shock and Vibration Measurement Technology: An Applications-Oriented Short Course*, San Juan Capistrano, CA, 1987.
9. Herceg, Edward E., *Handbook of Measurement and Control*, available from Schaevitz Engineering, Pennsauken, NJ, 1976.
10. Kobayashi, Albert S. (Editor), *Handbook on Experimental Mechanics*, Prentice-Hall, Englewood Cliffs, NJ, 1987.
11. McConnell, K. G. and M. K. Abdelhamid, "On the Dynamic Calibration of Measurement Systems for Use in Modal Analysis," *International Journal of Analytical and Experimental Modal Analysis*, Vol. 2, No. 3, July 1987, pp. 121–127.
12. Magrab, Edward B. and Donald S. Blomquist, *The Measurement of Time-Varying Phenomena: Fundamentals and Applications*, Wiley-Interscience, a division of John Wiley & Sons, New York, 1971.
13. Pennington, D., *Piezoelectric Accelerometer Manual*, Endevco Corporation, San Juan Capistrano, CA, 1965.
14. Peterson, A. P. G. and E. E. Gross, Jr., *Handbook of Noise Measurement*, available from General Radio Company, Concord, MA, 1972.

5 The Digital Frequency Analyzer

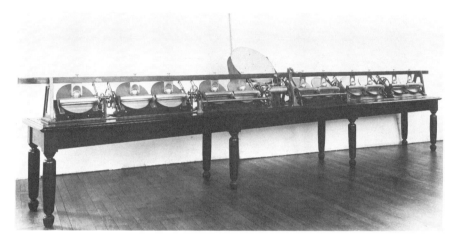

A mechanical Fourier frequency analyzer designed by Lord Kelvin (William Thomson) and his brother (Professor James Thomson) to analyze the frequency content of tides. This machine was built around 1876–1879 and is capable of developing the mean value and the first five frequency components; it is a marvel of brass gears, steel balls, and mechanical indicators. (Photo Courtesy of The National Museum of Science and Industry, London, U.K.).

5.1 INTRODUCTION

A key element in the development of the digital frequency analyzer was the invention of the fast Fourier transform (FFT) algorithm by Cooley and Tukey[1] in 1965. However, the rapid development of low cost high performance frequency analyzers since the late 1970's is due to the rapid development of high performance microprocessors and analog to digital (A/D) voltage converters. Digital frequency analyzers are replacing analog analyzers in most applications due to their high performance and versatility, the increased use of computer based data acquisition systems, and

---

[1] J. W. Cooley and J. W. Tukey, "An Algorithm for the Calculation of Complex Fourier Series," *Mathematics of Computation*, Vol. 19, 1965, pp. 297–301.

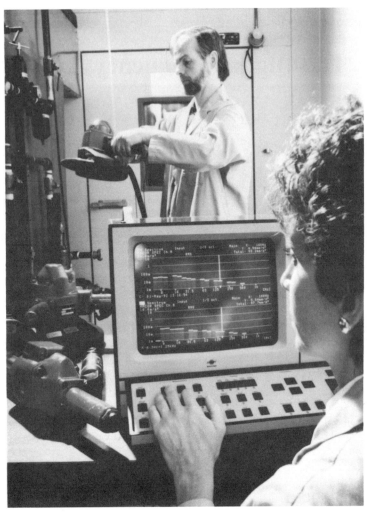

A modern frequency analyzer, capable of giving 800 frequency components in a few milliseconds, is being used to test industrial grinders to ISO-8662-1 and ISO-5349 standards for vibration transmitted to the worker. (Photo Courtesy of *Sound and Vibration*)>

significant cost reductions. Consequently, the analog analyzer is not considered in this book.

The modern digital frequency analyzer can process data in real time for an analysis bandpass from zero to X kHz where the upper limit frequency X is increasing each year as faster microprocessors are developed. These analyzers can present the frequency analysis in a number of useful displays with considerable data documentation. Unfortunately, many users have only a vague understanding of the processes involved in obtaining the frequency analysis, the meaning of the various displays that are readily available, and the manner in which the various quantities are interrelated when using this powerful instrument of vibration analysis.

We show that the entire process is based on the concepts of the Fourier series that we reviewed in Chapter 2. These concepts are used to explore the basic principles and operating characteristics of this instrument and to explain the meaning of such important concepts as *aliasing, window functions, digital filters, sample functions,* and *filter leakage*. We do not discuss the exact calculation process that is implied by the words *fast Fourier transform*. It is sufficient for our purposes of understanding to realize that this calculation process gives us the complex periodic Fourier series coefficients when the data is gathered according to certain rules. The resulting frequency spectrum is then displayed according to certain other rules, dependent on the type of frequency spectrum that we ask the machine to display. Our goal is to remove some of the mystery that surrounds digital frequency analyzers from a user's point of view for both single channel analyzers and dual channel analyzers. Multichannel analyzers are basically an extension of the dual channel analyzer.

5.2 BASIC PROCESSES OF A DIGITAL FREQUENCY ANALYZER

The digital frequency analyzer utilizes several basic processes. First, the analyzer samples a signal over time period T using an A/D converter that samples at a constant rate. Second, it is assumed that the captured signal is periodic with fundamental period T. Thus we calculate standard periodic Fourier series frequency components based on the assumption that only multiples of the fundamental frequency are present in the signal. Third, the analyzer manipulates these frequency components to produce different types of frequency analysis displays.

In this section, we examine the first two processes, that is, time sampling and the frequency component calculation assumption. The time sampling process leads us to the important concept of time domain multiplication of two signals is the same as frequency domain convolution of the Fourier transforms of these signals. This fundamental convolution process describes important analyzer data calculations that significantly affect the analyzer's performance and our understanding of why aliasing occurs, filters leak, and window functions generate digital filter characteristics. Important digital frequency analyzer characteristics are brought to light in this section.

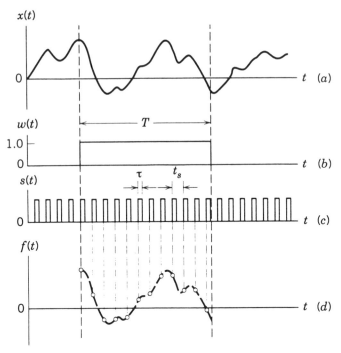

Fig. 5.2.1. The digital signal sampling process. (*a*) Voltage signal to be analyzed $x(t)$. (*b*) Window function $w(t)$. (*c*) Sample function $s(t)$. (*d*) Digitized version of analog signal to be analyzed $f(t)$. (From J. W. Dally, W. F. Riley, and K. G. McConnell, *Instrumentation for Engineering Measurements*, 2nd ed., copyright © 1993 by John Wiley & Sons, New York. Reprinted by permission.)

5.2.1 The Time Sampling Process

The FFT calculates the complex Fourier series frequency components by using N digital data points that are assumed to represent a time history that is periodic in time period T. The FFT calculation process requires that N be a power of 2, with 2^{10} (= 1024) and 2^{11} (= 2048) being commonly used values. Now we need to look at the process of obtaining these N digital samples that are to represent the original analog signal.

The digital signal sampling process is illustrated in Fig. 5.2.1, where the original analog signal is represented by $x(t)$. This original signal is sampled with an A/D converter at equal time increments over a specific time period. The resulting series of discrete points, called $f(t)$ in Fig. 5.2.1*d*, are used to represent the original signal $x(t)$. It is evident that the digital points may not give an adequate representation of the original signal if the sample rate is too slow. However, if the sample rate is fast enough, the digital representation can be very good.

BASIC PROCESSES OF A DIGITAL FREQUENCY ANALYZER

The sample process is defined in terms of a *window function* $w(t)$ that controls the sample length T and the *sample function* $s(t)$ that controls the time when a sample of $x(t)$ is taken. The window function shown here consists of a rectangular function that is zero everywhere before time is zero and after time period T and that is unity over the region 0 to T. This function selects the portion of $x(t)$ that is to be analyzed and gives each point equal weighting (or importance). We find that $w(t)$ also controls the digital analyzer's filter characteristics, which can be changed by using different window functions that emphasize some data points more than others.

The sample function consists of a number of rectangular pulses with unity height, with time duration τ, and with a spacing of t_s seconds. The rectangular pulse duration τ is small compared to repeat period t_s since only one point of $x(t)$ is selected for conversion at a given instant. The corresponding fundamental sample frequency is

$$f_s = \frac{1}{t_s} \tag{5.2.1}$$

We need a relationship between the highest frequency to be analyzed and the A/D converter's sample frequency so we can judge when we are sampling fast enough for a given problem.

The time when the pth sample is taken is given by

$$t_p = pt_s \tag{5.2.2}$$

where p ranges from 0 to $(N-1)$, that is, the first data point is assumed to occur at time $t = 0$ and the last data point is collected at time $t = (N-1)t_s$. It appears that there are $(N-1)$ data points, but recall that the first one corresponds to $p = 0$ and not to $p = 1$. The window function's duration T is related to the sample time[2] by

$$T = (N)t_s \tag{5.2.3}$$

and it controls the analyzer's fundamental frequency, since $f_0 = 1/T$.

The output digitized function $f(t)$ is related to the window function $w(t)$, the sample function $s(t)$, and the original function $x(t)$ by multiplication so that

$$f(t) = w(t)s(t)x(t) \tag{5.2.4}$$

---

[2]Often $(N-1)t_s$ is used in error. Equation (5.2.3) is correct.

is a sequence of discrete numbers that are equally spaced in time, as shown in Fig. 5.2.1d. Equation (5.2.4) raises a couple of questions. How is a frequency analysis of the $f(t)$ discrete numbers related to the frequency components of $x(t)$? What effect do the window and sample functions have?

5.2.2 Time Domain Multiplication and Frequency Domain Convolution

We want to relate the frequency components of $f(t)$ to those of $x(t)$ but we see from Eq. (5.2.4) that if we take the Fourier transform of $f(t)$ we are also taking the Fourier transform of a product of several time functions. In order to explore these ideas, let us consider the product of two time functions such that

$$y_1(t) = y_2(t)y_3(t) \tag{5.2.5}$$

where each time function $y_i(t)$ has its own Fourier spectrum $Y_i(\omega)$ as defined by Eqs. (2.4.4) and (2.4.5) so that

$$Y_i(\omega) = \int_{-\infty}^{\infty} y_i(t) e^{-j\omega t} dt \tag{5.2.6}$$

and

$$y_i(t) = \frac{1}{2\pi} \int_{-\infty}^{\infty} Y_i(\omega) e^{j\omega t} d\omega \tag{5.2.7}$$

are the fundamental Fourier transform pair. Now, if we substitute Eq. (5.2.5) into Eq. (5.2.6), we obtain

$$Y_1(\omega) = \int_{-\infty}^{\infty} y_1(t) e^{-j\omega t} dt = \int_{-\infty}^{\infty} y_2(t)y_3(t) e^{-j\omega t} dt \tag{5.2.8}$$

Now, substituting appropriate Fourier transform expression forms for $y_2(t)$ and $y_3(t)$ from Eq. (5.2.7) into Eq. (5.2.8) and carefully manipulating and interpreting of the terms, we obtain

$$Y_1(\omega) = \frac{1}{2\pi} \int_{-\infty}^{\infty} Y_2(\nu)Y_3(\omega - \nu) d\nu \tag{5.2.9}$$

where ν is a dummy frequency variable of integration. Equation (5.2.9)

BASIC PROCESSES OF A DIGITAL FREQUENCY ANALYZER 267

is called *frequency domain convolution* and is the frequency domain equivalent to time domain convolution where we found that multiplication in the frequency domain gives time domain convolution (see Section 3.4, particularly Eq. (3.4.9)). The meaning of Eq. (5.2.9) is that for each frequency ω, integration of $Y_2(\nu)Y_3(\omega - \nu)$ over the entire range of ν gives a unique value for $Y_1(\omega)$. Equation (5.2.9) is fundamental to understanding the consequences of time multiplication that occurs in the time sampling process.

5.2.3 Sample Function Multiplication Gives Aliasing

At this point we want to examine the sample function's effects on the resulting frequency analysis. To this end we write Eq. (5.2.4) as

$$f(t) = s(t)x(t) \tag{5.2.10}$$

so that we are ignoring the window function's role at the moment. Equation (5.2.10) has the same form as Eq. (5.2.5) where $y_1(t) = f(t)$, $y_2(t) = s(t)$, and $y_3(t) = x(t)$. In order to utilize the equivalency of Eqs. (5.2.9) and (5.2.10), we need to determine an expression for the frequency spectrum for the sample function, that is, $S(\omega)$.

Recall that in Section 2.4 we obtained a Fourier series expression for the frequency spectrum of a function identical to the sample function with a periodic frequency f_s (or ω_s) so that

$$S(p\omega_s) = \frac{2}{\beta}\left[\frac{\sin(p\pi/\beta)}{(p\pi/\beta)}\right] = \frac{2}{\beta}\text{sinc}\left(\frac{p\pi}{\beta}\right) \tag{5.2.11}$$

where $\beta = t_s/\tau$ is very large in this application. Equation (5.2.11) indicates that this frequency spectrum has components that are spaced ω_s apart and vary according to the sinc function. Now, we know that the sample function $s(t)$ includes all time, and hence, all of the pulses. In order to include all pulses, we rewrite Eq. (5.2.11) as

$$S(\omega) = \int_{-\infty}^{\infty} s(t)e^{-j\omega t}\,dt = \lim_{n \to \infty} 2\tau n \,\text{sinc}\left(\frac{p\omega_s \tau}{2}\right) = \delta(\omega - p\omega_s) \tag{5.2.12}$$

where n is the number of rectangular sample pulses.
  p is the frequency component of interest.

THE DIGITAL FREQUENCY ANALYZER

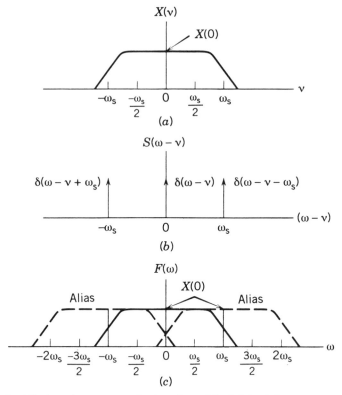

Fig. 5.2.2. Aliasing due to sample function. (*a*) Original frequency spectrum $X(\nu)$. (*b*) Sample function frequency spectrum $S(\omega - \nu)$. (*c*) Resulting frequency spectrum $F(\omega)$ showing spectrum overlap regions due to $\pm\omega_s$ sample function frequency components. (From J. W. Dally, W. F. Riley, and K. G. McConnell, *Instrumentation for Engineering Measurements*, 2nd ed., copyright © 1993 by John Wiley & Sons, New York. Reprinted by permission.)

Equation (5.2.12) shows us that the sample function frequency spectrum consists of Dirac delta functions (with unity area but no width) that are spaced ω_s apart, as shown in Fig. 5.2.2b. The Dirac delta functions in Eq. (5.2.12) are to be convolved with the input frequency spectrum of $X(\nu)$.

We would like to apply the frequency domain convolution concept contained in Eq. (5.2.9) to the situation shown in Fig. 5.2.2 where the original frequency spectrum of $X(\nu)$ is shown in Fig. 5.2.2a and the sample function frequency spectrum, consisting of only three out of all possible Dirac delta functions, is shown in Fig. 5.2.2b. First, let us look at the center Dirac delta function (the one corresponding to $\omega - \nu = 0$ or to p in Eq. (5.2.12) being 0). Then Eq. (5.2.9) becomes

$$F(\omega) = \frac{1}{2\pi}\int_{-\infty}^{\infty} \delta(\omega - \nu)X(\nu)\,d\nu = \frac{X(\omega)}{2\pi}\int_{\omega-\nu<0}^{\omega-\nu>0} \delta(\omega - \nu)\,d\nu = \frac{X(\omega)}{2\pi}$$
(5.2.13)

since the Dirac delta function is zero everywhere except where its argument $(\omega - \nu)$ is zero and it has unity area when integrated over zero from $\omega - \nu < 0$ to $\omega - \nu > 0$. Equation (5.2.13) shows that the center portion of the output frequency spectrum is the same as the input spectrum except for the scale factor of 2π, which can easily be removed. Thus we see that the solid center curve in Fig. 5.2.2c comes from the convolution of $S(\omega - \nu)$ and $X(\nu)$ for those values of ν that make the quantity $(\omega - \nu)$ zero.

We can repeat the process of Eq. (5.2.13) when we use the values of ν that satisfy the relationship of $\omega - \nu = \omega_s$. Then, $\delta(\omega - \nu)$ in Eq. (5.2.13) is replaced by $\delta(\omega - \nu - \omega_s)$, which corresponds to the Dirac delta function that is centered at $(+\omega_s)$ since p in Eq. (5.2.12) is 1. In this case, we generate the dashed curve that represents $X(\omega - \omega_s)$, which is centered on $(+\omega_s)$ as shown in Fig. 5.2.2c. We call this curve an *alias* since it is a false result. Similarly, if we use the values of ν that satisfy $\omega - \nu = -\omega_s$, we are dealing with the Dirac delta function that is centered at $(-\omega_s)$. Then, Eq. (5.2.13) generates the alias curves that are centered at $(-\omega_s)$.

We see in Fig. 5.2.2c that $F(\omega)$ contains the original frequency spectrum $X(\nu)$ not only once, centered at $\omega - \nu = 0$, but many times, centered at $\pm\omega_s$, $\pm 2\omega_s$, and so on, since $S(\omega)$ contains all such Dirac delta functions. This means that the curves in Fig. 5.2.2c must be added together to obtain the resultant output frequency spectrum. This addition presents no problem if the input spectrum $X(\nu)$ is limited to having frequency components that are less than half of the sample frequency, that is $\omega_{max} < \omega_s/2$, so that there is no spectrum overlap. We call $\omega_s/2$ the *Nyquist frequency*. We obtain undesirable results if the original signal contains frequencies that exceed the Nyquist frequency. In this case, the alias frequency components add to the actual frequency components to produce a false (incorrect) frequency spectrum.

It would be delightful if we could counteract the aliasing problem digitally after capturing the data. Unfortunately this is not possible, so we are forced to control aliasing by employing a sharp roll-off *anti-aliasing* electronic filter, as shown in Fig. 5.2.3. This electronic filter is applied to the electronic signal before it is sampled so that the high frequency components above the Nyquist frequency are significantly removed from the measurement, as shown. The ideal is to be down about 80 dB (for 16 bit A/D converters) at the break frequency ω_b as shown. This means that the filter is down about 40 dB at the Nyquist frequency. If we use a filter with a falloff rate of 120 dB/octave, the break frequency is about $0.4\omega_s$,

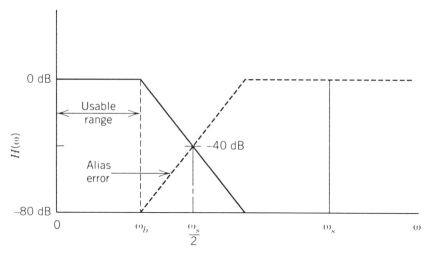

Fig. 5.2.3. Ideal electronic anti-aliasing filter characteristics showing removal of bias errors and usable frequency range.

so that we can use up to 40 percent of the sample frequency. Thus a 1 kHz sample rate allows us to analyze signals up to 400 Hz. Since we use different sample rates, we must have tunable electronic filters that change with the sample rate, a feature that dramatically increases their cost.

The use of anti-aliasing filters sounds very good and presents a simple solution. However, an electronic filter that rolls off at 120 dB/octave is a complicated electronic device that involves a number of poles and zeros in its design in order to keep the response relatively flat up to the break frequency. The curve shown in Fig. 5.2.3 is an idealization. The real filter oscillates about the desired straight line and has significant phase shifts as well between input and output. Thus what we have is a situation where we are analyzing a signal $y(t)$ that may be significantly different from input $x(t)$, since the output frequency spectrum from the anti-aliasing filter is given by

$$Y(\omega) = H(\omega)X(\omega) \qquad (5.2.14)$$

where $X(\omega)$ is the input frequency spectrum and $H(\omega)$ is the anti-aliasing filter FRF characteristics. Hence, these filters can have significant effect on the signal that is processed. It is good practice to analyze a signal with several different sample rates to establish just how much the anti-aliasing filter may be distorting the results in a given situation.

The above discussion involves considerable mathematical sophistication

BASIC PROCESSES OF A DIGITAL FREQUENCY ANALYZER 271

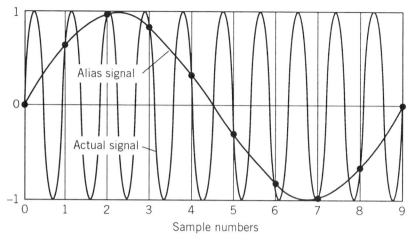

Fig. 5.2.4. A physical explanation for aliasing where the sample rate is too small for signal. Only 10 data points are taken where at least 50 are required to begin to describe the actual signal.

at times. A simple physical explanation for the aliasing phenomenon is shown in Fig. 5.2.4. In this case, the unfiltered signal is a sinusoid with a frequency that is about 10 times the sample frequency so we are sampling much too slowly. However, if we look at the data points sampled, they plot as part of a perfect sinusoid that appears to exist while in reality it does not. The frequency analyzer process looks at these data points and incorrectly calculates that the frequency is one-tenth the actual frequency and proudly reports this incorrect fact in its frequency spectrum display. This is why we must either use an anti-aliasing filter or we must carefully evaluate the signal to be certain that we are sampling fast enough to avoid aliasing problems.

5.2.4 The Window Function Creates the Digital Filter Characteristics

For our discussion here, we assume that we have valid data without any aliasing problems, so that we can write Eq. (5.2.4) as

$$f(t) = w(t)x(t) \qquad (5.2.15)$$

We investigate the meaning of the window function $w(t)$ in Eq. (5.2.15) in light of the time domain multiplication and frequency domain convolution results of Eq. (5.2.9). In view of how Eq. (5.2.5) creates Eq. (5.2.9), we

272   THE DIGITAL FREQUENCY ANALYZER

see from Eq. (5.2.15) that Eq. (5.2.9) can be written as

$$F(\omega) = \frac{1}{2\pi} \int_{-\infty}^{\infty} W(\omega - \nu) X(\nu) \, d\nu \qquad (5.2.16)$$

from which we see that the window function's frequency spectrum $W(\omega - \nu)$ is convolved with the input frequency spectrum $X(\nu)$.

In order to see how Eq. (5.2.16) works, consider the situation shown in Fig. 5.2.5 where $W(\omega - \nu)$ has unity magnitude over a frequency bandwidth from $-\pi B$ to πB for a total width of $2\pi B$ rad/s or B Hz and has zero magnitude over all other frequencies. Now let ω have a specific value of ω_1 and let us look at what happens to points A and C in Fig. 5.2.5b. Point A corresponds to $\omega_1 - \nu_A = -\pi B$ in Fig. 5.2.5b, so that in Fig. 5.2.5a we have $\nu_A = \omega_1 + \pi B$ and point A becomes A' as shown. Similarly, point C corresponds to $\omega_1 - \nu_B = \pi B$ in Fig. 5.2.5b, so that in Fig. 5.2.5a point C becomes point C', which is located at $\nu_B = \omega_1 - \pi B$, as shown. Now if $\omega = -\omega_1$, we find that points A and C in Fig. 5.2.5b become points A'' and C'' centered around $-\omega_1$, as shown in Fig. 5.2.5a.

Application of Eq. (5.2.16) to the situation shown in Fig. 5.2.5 for $W(\omega - \nu)$ being an ideal rectangular filter of width $2\pi B$ gives

$$F(\omega_1) = \frac{1}{2\pi} \int_{\omega_1 - \pi B}^{\omega_1 + \pi B} X(\nu) \, d\nu = \overline{X(\omega_1)} B \qquad (5.2.17)$$

where $F(\omega_1)$ represents the cross-hatched area in Fig. 5.2.5a centered at $\nu = \omega_1$ divided by 2π. This area can be represented by the average value of $\overline{X(\omega_1)}$ times the filter bandwidth B in Hz. We can plot the value of $F(\omega_1)$ as a single point D in Fig. 5.2.5c. Similarly, point E results when $\omega = -\omega_1$.

Clearly we see that $W(\omega - \nu)$ acts like a filter in examining the frequency content of $X(\nu)$ for changing values of ω. This means that ω plays the role of *filter center frequency* and that $F(\omega)$ is proportional to the frequency averaged value of $X(\nu)$ at each value of ω. Equation (5.2.17) clearly shows us that the value of $F(\omega)$ becomes closer and closer to the value of $X(\nu)$ as B becomes smaller and smaller and that the two are identical when B = unity. We also see that both $F(\omega)$ and $X(\nu)$ have the same units, since $W(\omega - \nu) \, d\nu/2\pi$ has no units, due to the fact that the units of $W(\omega - \nu)$ are those of 1/Hz and cancel in the integration process.

The results of Eq. (5.2.16) suggest that it is possible to multiply a time history by an appropriate window function in the time domain and obtain an output in the frequency domain that is automatically filtered by a particular digital filter. This observation illustrates the role that window

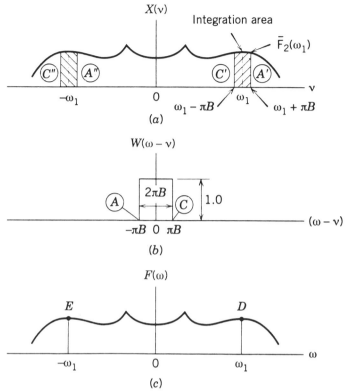

Fig. 5.2.5. Illustration of window function controlling the digital filter characteristics. (*a*) Original frequency spectrum $X(\nu)$. (*b*) Ideal filter characteristics $W(\omega - \nu)$. (*c*) Output frequency spectrum. (From J. W. Dally, W. F. Riley, and K. G. McConnell, *Instrumentation for Engineering Measurements*, 2nd ed., copyright © 1993 by John Wiley & Sons, New York. Reprinted by permission.)

functions play in digital frequency analysis and why we employ a number of different window functions that are most useful for different types of signals that we are interested in.

5.2.5 Filter Leakage

The digital frequency analyzer is based on the assumption that the signal is periodic with fundamental period T regardless of the facts involved. This means that we are trying to describe a signal in terms of periodic Fourier series frequency components that are multiples of fundamental frequency

$$\omega_0 = \frac{2\pi}{T} = 2\pi f_0 = \left(\frac{2\pi}{N}\right) f_s \qquad (5.2.18)$$

This works well when the signal is an integer multiple of ω_0 (or f_0). However, the signal is usually of a frequency that is not an integer multiple of f_0. This means that there are discontinuities in magnitude and slope at the ends of the sample that is analyzed. This mismatch in frequency between signal and analyzer frequency components creates magnitude and slope discontinuities in the sample function $f(t)$, which cause *filter leakage* to occur.

We can illustrate this mismatching of signal ends by considering the situation shown in Fig. 5.2.6, where two analysis periods, each of length T, are shown. In the upper plot, sinusoidal signal A fits precisely within period T so that it has a continuous magnitude and slope at the beginning and the end. Because signal A fits perfectly within period T, the repeat cell is identical to that in the basic cell. However, the lower plot shows a signal discontinuity between the magnitudes and slopes at the beginning and the end of the basic sample cell. The analyzer assumes that the repeat cell is exactly like the basic cell so that the analyzer thinks that it is analyzing the periodic signal shown in the lower plot instead of a sinusoid that is periodic with a slightly different frequency. The discontinuity in magnitude and slope will cause additional frequency components to be calculated in order to account for the signal discontinuity that occurs at its ends.

We can explore the effects of the periodic assumption illustrated in Fig. 5.2.6 further by considering a sinusoid that is expressed by

$$x(t) = \sin(N_c \omega_0 t) \qquad (5.2.19)$$

where N_c is the number of cycles within the analysis window of duration T.
ω_0 is the analyzer's fundamental frequency.

Equation (5.2.19) is clearly a periodic function. The magnitude of the pth Fourier series coefficient for this signal is given by

$$X_p = \sqrt{2\left(\frac{\text{sinc}^2[(p + N_c)\pi]}{(1 - N_c/p)} + \frac{\text{sinc}^2[(p - N_c)\pi]}{(1 + N_c/p)}\right)} \qquad (5.2.20)$$

where p can have both positive and negative integer values. When N_c is an integer, only $p = N_c$ and $p = -N_c$ give values of unity while all other

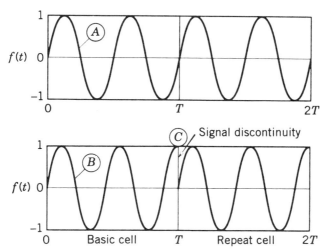

Fig. 5.2.6. Plot of two sinusoidal time histories of slightly different frequencies where curve A fits perfectly within analysis window of period T, while curve B has a magnitude and slope discontnuity at its ends. (From J. W. Dally, W. F. Riley, and K. G. McConnell, *Instrumentation for Engineering Measurements*, 2nd ed., copyright © 1993 by John Wiley & Sons, New York. Reprinted by permission.)

values of X_p are zero. Also, we see that Eq. (5.2.20) contains the sinc function, which is characteristic of a rectangular window.

In order to understand the sinc function's significance, we need to obtain the Fourier transform of the rectangular window function that has unity height and duration T. This transform gives us

$$W(\omega) = \text{sinc}\left(\frac{\omega T}{2}\right) \quad (5.2.21)$$

where the digital filter characteristic is a continuous function of the argument $(\omega T/2)$, an argument that becomes $[(\omega - \nu)T/2]$ when applied as a digital filter. We illustrate the significance of Eqs. (5.2.20) and (5.2.21) for different values of N_c.

The frequency components from Eq. (5.2.20) are plotted in Fig. 5.2.7 for $N_c = 10$, 10.25, and 10.5, while the corresponding digital filter functions as described by Eq. (5.2.21) are plotted in Fig. 5.2.8. The first 20 frequency components are shown in each case. When N_c is an integer and is equal to 10, we obtain a single nonzero frequency component at $p = 10$, as shown in Fig. 5.2.7a. In this case, the function being analyzed fits perfectly within the rectangular window function of period T and there is

276 THE DIGITAL FREQUENCY ANALYZER

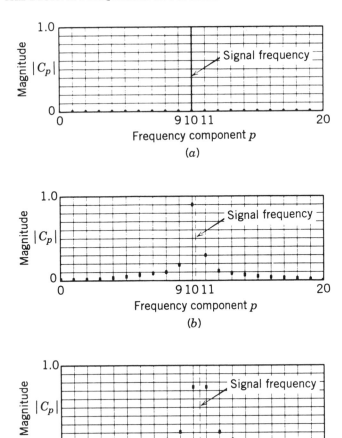

Fig. 5.2.7. Analyzer frequency components of a sinusoidal signal with N_c cycles in the analysis window of period T. (a) $N_c = 10$. (b) $N_c = 10.25$. (c) $N_c = 10.50$. Note filter leakage in plots *b* and *c*. (From J. W. Dally, W. F. Riley, and K. G. McConnell, *Instrumentation for Engineering Measurements*, 2nd ed., copyright © 1993 by John Wiley & Sons, New York. Reprinted by permission.)

no filter leakage. We also see in Fig. 5.2.8*a* that the signal frequency lies exactly at the center of the digital filter function as required by Eq. (5.2.16). This digital filter function is zero where all other analyzer frequencies intersect with the digital filter function so that only when $p = N_c$ does a frequency component result. This is an ideal result from the frequency analyzer.

When N_c is a noninteger of 10.25, significant mismatching in magnitude

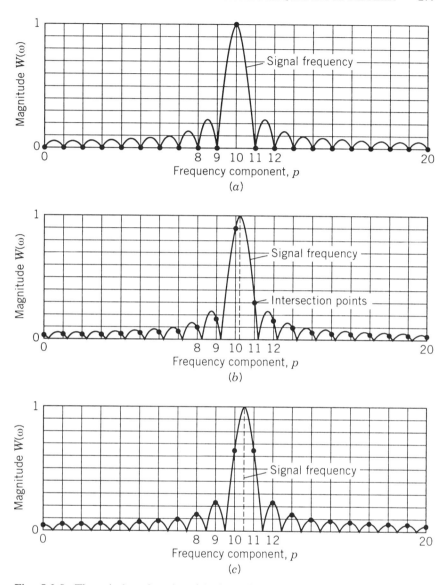

Fig. 5.2.8. The window function (sinc($\omega T/2$)) is shifted for values of $N_c = 10$, 10.25, and 10.5, according to Eq. (5.2.16). The frequency components in Fig. 5.2.7 correspond to intersections of component lines with the window function. (a) $N_c = 10$. (b) $N_c = 10.25$. (c) $N_c = 10.5$.

and slope occurs at the signal's ends within the rectangular window function. In this case, we obtain the results shown in Fig. 5.2.7b, where there are many new frequency components, and yet the signal's frequency has been changed only 2.5 percent. The actual frequency ω lies between the analyzer's tenth ($10\omega_0$) and eleventh ($11\omega_0$) frequency components. We find that the same amplitudes occur at the intersection of the sinc function and the frequency analyzer's frequency components, as shown in Fig. 5.2.8b when the sinc function is shifted to be a maximum over the actual frequency, as required by Eq. (5.2.16).

A similar result occurs when $N_c = 10.5$, as shown in Figs. 5.2.7c and 5.2.8c. In this case, there are two equally large frequency components at the center, instead of one large one and smaller adjacent ones. When this happens, the actual frequency is midway between the adjacent frequency components, namely, the 10th and 11th components.

It is clear that there is more to interpreting a frequency spectrum than looking for the highest peak value and declaring that is the magnitude and the frequency of the signal being analyzed. We see that changing frequency ω by 5 percent causes a completely different set of information to appear; compare Figs. 5.7.7a and 5.7.7c. The peak components have a magnitude of 0.637 for $N_c = 10.5$, compared to unity for $N_c = 10$, an error of nearly 46 percent. What magnitude do we report when a slight change in frequency can cause such large variations in available numbers? Which frequency do we report? The tenth or the eleventh component? Interpretation of these results should make the reader aware that we need to understand the fundamental processes utilized in order to correctly interpret what the information means.

Obviously, changing the window function changes the digital filter characteristic, which in turn changes the output frequency spectrum. We delay looking at this question of different window functions until Section 5.4, where we will look at the practical application of different window functions.

5.3 DIGITAL ANALYZER OPERATING PRINCIPLES

In the previous section, we saw that the sampling process generated a signal that is the product of the original function, the window function, and the sample function. This multiplication in the time domain is equivalent to convolution in the frequency domain. This basic process leads to aliasing errors and the concept of filter leakage, since we are trying to express all frequencies in terms of only a few Fourier series frequency components. We saw that the aliasing errors can be controlled in principle by using electronic anti-aliasing filters on the signal before it is sampled in the analyzer. It must be realized that anti-aliasing filters can significantly alter the signal being analyzed. An important consequence of the convolution integral is the fact that the window function controls the digital filter characteristics. The implications of using four commonly employed window functions for different signals will be addressed in Section 5.4.

In this section, we first look at a typical operating block diagram for a digital frequency analyzer and then look at how the internal calculation results are manipulated to provide different kinds of analysis and displays.

5.3.1 Operating Block Diagram

One possible block diagram of a single channel 1024 data point digital frequency analyzer is shown in Fig. 5.3.1. The voltage input signal is represented by $x(t)$. This signal is applied to the input attenuator and tunable anti-aliasing filter section. The attenuator is used to adjust the overall signal level to be compatible with the A/D converter's full scale voltage range. It is desirable to operate the A/D converter as near to full scale as practical, dependent on the signal type. For example, when dealing with most random signals, the full scale should be at least four to five times the RMS value in order to measure the highest peak values. However, most analyzers have an indication when the signal is over ranging. Unfortunately, they give no indication when they are being operated in an under range condition, often leading to poor signal resolution. Setting the proper attenuation levels is an important first step in any signal analysis so that the A/D converter is operating in an optimum manner to utilize the instrument's full dynamic range.

The output from the attenuator anti-aliasing filter block is the voltage $y(t)$. Recall that $y(t)$ is different from $x(t)$, due to the anti-aliasing filter (see Eq. (5.2.14)) and attenuator gains. The $y(t)$ signal is often mixed with a low level random noise in order to reduce or hide the A/D converter's operating noise. Next, the A/D converter samples the signal to produce a series of numbers that are acquired at the converter's sample frequency f_s. These numbers are stored in the 1024 word data memory and represent signal $x(t)$ as a set of numbers that we can call $y(t)$.

When the 1024 word data memory is full, the numbers are multiplied by the window function and are stored in the analysis memory as the signal we have called $f(t)$. The digital spectrum calculator processes the signal that is stored in the analysis memory, using the periodic FFT calculation scheme to generate the periodic Fourier series frequency components, and then stores the calculated spectrum in the spectrum average memory. The display and interface section process the spectrum average memory data for use by the analyzer's cathode ray tube (CRT) display, the digital plotter, and the host computer.

280 THE DIGITAL FREQUENCY ANALYZER

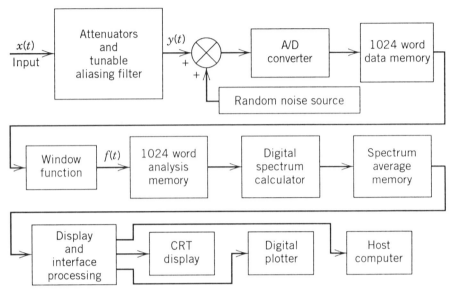

Fig. 5.3.1. Block diagram showing a digital frequency analyzer's basic processes. (From J. W. Dally, W. F. Riley, and K. G. McConnell, *Instrumentation for Engineering Measurements*, 2nd ed., copyright © 1993 by John Wiley & Sons, New York. Reprinted by permission.)

5.3.2 Internal Calculation Relationships

Today we can purchase inexpensive A/D plug-in boards for personal computers (PCs) that often use multiplexed A/D converters. These converters often cause time and phase shifts between channels, due to input signal multiplexing. In addition, they often come with software that contains unknown FFT calculation schemes that use unknown scaling laws while calculating the frequency spectra. The system's performance characteristics need to be carefully checked. In this section, we present a set of equations that can be used to verify how a frequency analyzer works.

The FFT algorithm calculates periodic Fourier series frequency components that are equivalent to Eq. (2.3.3) when t is zero so that

$$X_p = \frac{1}{T} \int_0^T x(t)\, e^{-jp\omega_0 t}\, dt \qquad (5.3.1)$$

applies to the data within the data window. We can create a digital Fourier transform (DFT) version of Eq. (5.3.1) by observing that the nth data point $f(t) = f(n)$ occurs at time $t_n = nt_S = nT/(N)$, that integration on t

can be replaced by summation on n, that $\omega_0 T = 2\pi$, and that dt is the same as t_s. Substitution of all of these changes into Eq. (5.3.1) gives

$$X_p = \frac{1}{N} \sum_{n=0}^{N-1} f(n) \, e^{-j(2\pi/N)pn} = \frac{1}{N} \sum_{n=0}^{N-1} w(n) y(n) \, e^{-j(2\pi/N)pn} \quad (5.3.2)$$

where we have used our previous notation so that $w(n)$ is window function $w(t)$ and $y(n)$ is the electronically filtered signal $y(t)$. We have shown in Section 2.3 that X_p occurs in complex conjugate pairs (see Eq. (2.3.4)) so that we have

$$\begin{aligned} X_p &= a_p + jb_p \\ X_{-p} &= a_p - jb_p = X_p^* \end{aligned} \quad (5.3.3)$$

The magnitude and phase of the pth frequency component are calculated from (see Eq. (2.3.5))

$$\begin{aligned} |X_p| &= \sqrt{a_p^2 + b_p^2} = \sqrt{X_p X_p^*} \quad \text{magnitude} \\ \tan \phi_p &= \frac{b_p}{a_p} \quad \text{phase} \end{aligned} \quad (5.3.4)$$

The frequency analyzer displays only one-sided frequency spectra so that p ranges from 0 to N_d where N_d is the highest frequency component that is displayed. The magnitude and phase of these $N_d + 1$ frequency components need to be calculated and displayed. To do this, we recall from Eq. (2.2.8) that the pth harmonic's magnitude is related to the magnitude of X_p by

$$B_p = 2|X_p| = 2\sqrt{X_p X_p^*} \quad (5.3.5)$$

This peak magnitude is related to the root mean square amplitude by

$$B_{\text{RMS}_p} = 0.707 B_p = \sqrt{2}|X_p| = \sqrt{2 X_p X_p^*} \quad (5.3.6)$$

We can use Eqs. (5.3.5) and (5.3.6) to calculate either the peak or RMS frequency spectra magnitudes for p ranging from 1 to N_d. The phase angle can be calculated from Eq. (5.3.4) for each of these frequency spectra displays. The mean value has a special relationship since the mean is the same for both peak and RMS frequency spectra displays. This relationship is

$$B_0 = X_0 \quad (5.3.7)$$

We see from the above equations that either we need to store the frequency components a_p and b_p, so we can calculate magnitude and phase, or we need to store the magnitude $|X_p|$ and the phase ϕ_p. With this information at our command, we can calculate any frequency spectra information that we desire.

The mean square of the signal is calculated from Parseval's formula as given by either Eqs. (2.3.6) or (2.3.7). However, we need to modify Parseval's formula for use in a digital analyzer to be

$$A_{RMS}^2 = X_0^2 + 2 \sum_{p=1}^{N_d} |X_p^2| = B_0^2 + \sum_{p=1}^{N_d} \frac{B_p^2}{2} = B_0^2 + \sum_{p=1}^{N_d} B_{RMS_p}^2 \qquad (5.3.8)$$

Note that the summation covers only the frequencies analyzed and not all frequencies, as required by Parseval's original formula. However, Eq. (5.3.8) is extremely useful in manipulating frequency spectral information.

The frequency components are also called *spectral lines* and *frequency lines*. These spectral lines are Δf Hz apart in the frequency spectra display. The value of Δf is given by

$$\Delta f = \frac{\omega_0}{2\pi} = \frac{1}{T} \qquad (5.3.9)$$

so that the pth spectral frequency is

$$f_p = p \, \Delta f \qquad (5.3.10)$$

for $p = 0, 1, 2, \ldots, N_d$. We should not confuse frequency component *spacing* with *effective analyzer bandwidth*. The effective analyzer bandwidth will be considered in Section 5.4.

We have looked at the simple formulas that can be used to calculate the various frequency components as manipulated in the frequency analyzer. It should be emphasized that the frequency components calculated are precisely those that will recreate the N data points that make up the discrete N point function $f(t)$. There is no information contained in the analyzer beyond this minimum input information. Now we look at how we can scale these calculated results and produce frequency spectra displays that are suitable for periodic, transient, and random time histories.

5.3.3 Display Scaling

Our goal in this section is to present the concepts on which we can take the same frequency analysis information as described by Eqs. (5.3.3) and (5.3.4) and create suitable frequency spectra for use with periodic, transient, and random time histories. We are also interested in how we should interpret the frequency spectra units in terms of the signal's units.

Periodic Time Histories The periodic time history is ideally suited for display purposes, since its frequency spectrum is made up from Fourier series frequency components. The only problem we have to deal with is filter leakage since the signal's frequency components may lie between those available from the frequency analyzer. In other words, the signal's fundamental period T_0 is different from the analyzer's fundamental period T.

The usual frequency display is either one of peak frequency components versus frequency or one of RMS frequency components versus frequency. Equation (5.3.5) can be used to calculate the peak component display, while Eq. (5.3.6) can be used to calculate the RMS component display. We must keep in mind that Eq. (5.3.7) must be used to calculate the mean value in either case. The signal's overall mean square or RMS value is calculated by using the modified Parseval formula given by Eq. (5.3.8).

The units of the display are those of $x(t)$. Thus if $x(t)$ represents acceleration in g's, then the units displayed should be g's peak or RMS, dependent on whether the frequency spectrum is peak or RMS. Many spectrum analyzers allow us to input the transducer's voltage sensitivity (in terms of mV/g or mV/kN, etc.) so the units are automatically displayed when the peak or RMS output format is selected.

Transient Time Histories The transient time history requires the use of Fourier transforms that generate continuous frequency spectra. It was shown in Section 2.4 that the Fourier series frequency components can be used to estimate the Fourier transform's spectral density under certain rules. First, the transient signal should occupy 10 percent or less of the sample period T. Second, the spectral density components are related to the Fourier Series frequency components by

$$F(p\,\Delta f) = F(\omega) = X_\mathrm{p}T = |X_p|T\,e^{j\phi_p} = F_p\,e^{j\phi_p} \qquad (5.3.11)$$

where the magnitude of the pth frequency component becomes

$$F_p = |X_p|T \qquad (5.3.12)$$

The phase angle ϕ_p is the same as that given by Eq. (5.3.4). The frequency spectra made up of the magnitudes F_p will have the same values as the continuous frequency spectra at each frequency line.

The display will have units of $x(t)$/Hz in this case. Thus if $x(t)$ represents force in newtons, then the display units are those of newtons/Hz. Again, many analyzers will display these units automatically once the voltage

284 THE DIGITAL FREQUENCY ANALYZER

sensitivity is entered and the transient spectral density display is selected. The interpretation of this display is that of a spectral density that corresponds to transient Fourier transform analysis.

The transient signal's mean square is given by Eq. (2.4.10). This equation can be written in terms of the analyzer's parameters as

$$\text{mean square} = X_0^2 T + 2 \sum_{p=1}^{N_d} F_p^2 \Delta f = X_0^2 T + 2 \sum_{p=1}^{N_d} |X_p|^2 T \quad (5.3.13)$$

since $\Delta f = 1/T$.

Random Time Histories The random time history is characterized by the dual sided auto-spectral density (ASD) $S(\omega)$, which, unfortunately, is also called power spectral density (PSD). $S(\omega)$ is defined over all frequencies from $-\infty$ to $+\infty$. We find it more convenient to use the single sided ASD $G(\omega)$, which is defined over all positive frequencies of $0 < \omega < \infty$ by

$$G(\omega) = \begin{cases} S(0) & \omega = 0 \\ 2S(\omega) & \omega > 0 \end{cases} \quad (5.3.14)$$

where $G(\omega)$ represents the mean square/Hz. Recall from Section 2.8 that Parseval's formula for the mean square and the ASD are related. Thus we can estimate the value of $G(\omega)$ by using the square of the RMS frequency components B_{RMS_p} multiplied by time T such that

$$\left. \begin{array}{l} G(p\omega_0) = G(p\,\Delta f) = (B_{\text{RMS}_p})^2 T \\ = 2|X_p|^2 T = 2 X_p X_p^* T \end{array} \right\} \quad \text{for } p = 1, 2, \ldots, N_d \quad (5.3.15)$$

and

$$G(0) = X_0^2 T \quad \text{for } p = 0$$

Equation (5.3.15) shows us that all phase information is lost in the ASD relationship since $X_p X_p^*$ contains only magnitude information.

Recall from Section 2.8 that the mean square, the auto-correlation function, and the ASD are related by

$$R_{xx}(0) = \frac{1}{2\pi} \int_0^\infty G(\omega)\, d\omega \quad (5.3.16)$$

This continuous integral can be written in terms of the summation paameters from Eq. (5.3.15) to give

$$R_{xx}(0) = B_0^2 + \sum_{p=1}^{N_d} (B_{\text{RMS}_p})^2 = X_0^2 + \sum_{p=1}^{N_d} 2|X_p|^2 \qquad (5.3.17)$$

Equation (5.3.17) is identical to Parseval's formula. This result indicates that we can calculate the mean square using Parseval's formula for discrete frequency components that come out of the frequency analyzer's Fourier series calculation scheme. The only requirement is that the signal must be analyzed many times in order to satisfy the random signal's long averaging times. Additional corrections must be applied to the values given by Eqs. (5.3.15) and (5.3.17) that are dependent on the type of window function employed. These additional corrections will be discussed in detail in Section 5.4 for four commonly employed window functions. It is important to recognize how these corrections are to be applied to periodic, transient, and random signals.

The units employed in an ASD frequency spectrum are those of $x(t)^2$/Hz. Thus if $x(t)$ has acceleration units of g's, then the ASD display has units of g^2/Hz.

The equations of this section can be used to check out the calculations employed in a single channel frequency analyzer. In some instruments, the factors employed by the system's designers may be slightly different from those employed here. The user is well advised to know just how his/her analyzer works; this can be determined by using several different types of signals and different methods of measurement and analysis in order to understand the meaning of a given frequency spectrum as displayed by the analyzer. Such knowledge can prevent embarrassing moments — not to mention dangerous interpretations — from occurring.

5.4 FACTORS IN THE APPLICATION OF A SINGLE CHANNEL ANALYZER

We need to interpret frequency spectra. The selection of an appropriate window function for a given signal is an important consideration in any measurement interpretation. Consequently, we need to understand the characteristics of window functions, their corresponding digital filters, and how these filters work with different signal types.

In this section, we look at a number of factors that affect the use of a single channel analyzer. First, we consider four parameters that are often used to characterize filters. Second, we examine the characteristics of four commonly employed digital filters and see how they can be applied to sinusoidal, transient, and random signals. Third, we examine the idea of spectral line uncertainty when dealing with random measurements, and finally, we make some general recommendations on which window function to use with a given type of signal.

5.4.1 Filter Performance Characteristics

The Dirac delta function is an ideal filter since it examines one frequency component at a time. However, this filter is impossible to implement in a practical way. Our next best alternative for an ideal filter is the rectangular filter shown in Fig. 5.4.1a. This filter has unity height over a frequency bandwidth of B Hz that ranges from the lower frequency f_1 to its upper frequency f_u and is zero at all other frequencies. Then, we see that all frequencies within bandwidth B are transmitted, and those outside bandwidth B are eliminated. Figure 5.4.1b shows a practical filter that can be constructed and is described by four characteristics of *center frequency*, *bandwidth*, *ripple*, and *selectivity*. Now, let us look at each of these characteristics.

Center Frequency Filters are often classified as *constant bandwidth* and *constant percentage bandwidth*. The constant bandwidth center frequency is given by the arithmetic mean of

$$f_0 = \frac{f_1 + f_u}{2} \tag{5.4.1}$$

This center frequency is expressed in terms of multiples of the line spacing Δf so that the pth center frequency is given by

$$f_0 = p\,\Delta f \tag{5.4.2}$$

The constant percentage bandwidth center frequency is given by

$$f_0 = \sqrt{f_1 f_u} \tag{5.4.3}$$

and is not generally used in vibration work. The constant percentage bandwidth filter is often used in acoustics.

Bandwidth The filter's bandwidth is a measure of the filter's resolution or ability to separate frequency components, particularly when dealing with random signals. Both the *effective noise bandwidth* and *3 dB bandwidth* definitions are commonly used.

The *effective noise bandwidth* B comes directly from the convolution

FACTORS IN THE APPLICATION OF A SINGLE CHANNEL ANALYZER

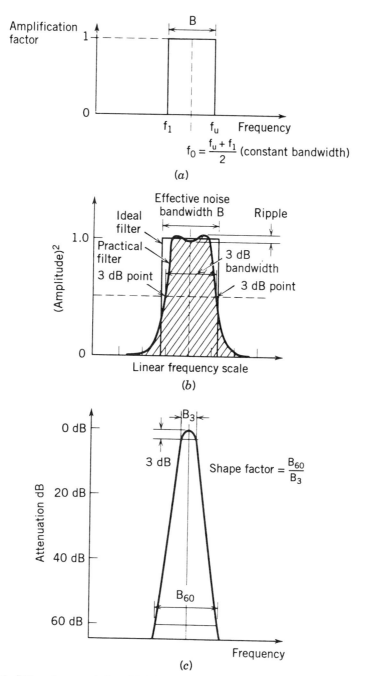

Fig. 5.4.1. Filter characteristics. (*a*) Ideal. (*b*) Practical. (*c*) Shape factor. (Courtesy of Bruel & Kjaer Instruments, Inc.)

theorem described by Eq. (5.2.9). In this case, the ASD curve is assumed to be constant with a unity value. Then if the filter characteristic is squared and integrated according to Eq. (5.2.9), the resulting cross-hatched area in Fig. 5.4.1b can be replaced by an ideal filter with bandwidth B and unity height as shown.

The *3 dB bandwidth* is simply the frequency difference in hertz between the points where the filter amplitude is down 3 dB (0.707 of its peak value) or where the square of the filter is 0.5 of its peak value. In many filters, the difference between the 3 dB bandwidth and the effective noise bandwidth B is small, with the effective noise bandwidth always being the larger of the two.

Ripple The ripple represents the maximum amount of uncertainty that exists across the top of the filter within one bandwidth, as shown in Fig. 5.4.1b. This quantity can be easily expressed in terms of percent since the ideal filter amplitude is unity.

Selectivity A filter's selectivity is a measure of its ability to differentiate between two closely spaced frequency components that vary widely in amplitude. The *shape factor* is one parameter that is used to describe a filter's selectivity, as shown in Fig. 5.4.1c. The shape factor is described in terms of the ratio of the 60 dB down bandwidth (B_{60}) compared to the 3 dB bandwidth (B_3) so that

$$\text{shape factor} = \frac{B_{60}}{B_3} \quad (5.4.4)$$

The shape factor covers a dynamic range of 1000 to 1 and is greater than unity with values close to unity being desirable.

5.4.2 Four Commonly Employed Window Functions

Most digital frequency analyzers employ two or more different window functions. Four commonly used functions are called *rectangular, Hanning, Kaiser–Bessel,* and *flat top*; they can be defined in terms of a general window function given by

$$\left. \begin{aligned} w(t) &= a_0 - a_1 \cos(\omega_0 t) + a_2 \cos(2\omega_0 t) \\ &\quad - a_3 \cos(3\omega_0 t) + a_4 \cos(4\omega_0 t) \end{aligned} \right\} \quad \text{for } 0 < t < T \quad (5.4.5)$$

$$w(t) = 0 \quad \text{elsewhere}$$

FACTORS IN THE APPLICATION OF A SINGLE CHANNEL ANALYZER 289

TABLE 5.4.1. Window Function Coefficients

Function	Coefficients				
	a_0	a_1	a_2	a_3	a_4
Rectangular	1	—	—	—	—
Hanning	1	1	—	—	—
Kaiser–Bessel	1	1.298	0.244	0.003	—
Flat top	1	1.933	1.286	0.388	0.032

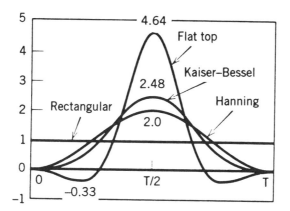

Fig. 5.4.2. Time domain descriptions of rectangular, Hanning, Kaiser–Bessel, and flat top window functions. (Courtesy of Bruel & Kjaer Instruments, Inc.)

where ω_0 is the analyzer's fundamental frequency. The a_i coefficients are chosen such that the area under each window is the same. These coefficients are given in Table 5.4.1, while a plot of each function is shown in Fig. 5.4.2. It is evident in Fig. 5.4.2 that the rectangular window treats all time data equally, while the other window functions start and end at zero in order to reduce filter leakage due to magnitude and/or slope discontinuities at the window ends. These window functions place different amounts of emphasis on the middle data points.

The four filter's performance factors of bandwidth, ripple, and selectivity are given in Table 5.4.2, while the actual digital filter characteristics are shown in Fig. 5.4.3. The bandwidth performance factors in Table 5.4.2 are given in terms of frequency component line spacing of Δf. It is evident that the noise bandwidth and 3 dB bandwidths are quite close, with the 3 dB bandwidth is always smaller than the noise bandwidth. The filter ripple is measured at $\pm \Delta f/2$ for each digital filter, as shown in Fig. 5.4.3, and is seen to decrease rapidly with increased window function complexity. Obviously, the Hanning window is far superior to the rectangular window,

TABLE 5.4.2. Bandwidth, Ripple, and Selectivity Characteristics of Four Common Window Functions

Window Type	Bandwidth		Ripple (dB)	Highest Sidelobe (dB)	Sidelobe Falloff (dB/decade)	B_{60}	Shape Factor
	Noise	3 dB					
Rectangular	$1.00\,\Delta f$	$0.89\,\Delta f$	3.92	−13.3	20	$665\,\Delta f$	750
Hanning	$1.50\,\Delta f$	$1.44\,\Delta f$	1.42	−31.5	60	$13.3\,\Delta f$	9.2
Kaiser–Bessel	$1.80\,\Delta f$	$1.71\,\Delta f$	1.02	−66.6	20	$6.1\,\Delta f$	3.6
Flat top	$3.77\,\Delta f$	$3.72\,\Delta f$	0.01	−93.6	0	$9.1\,\Delta f$	2.5

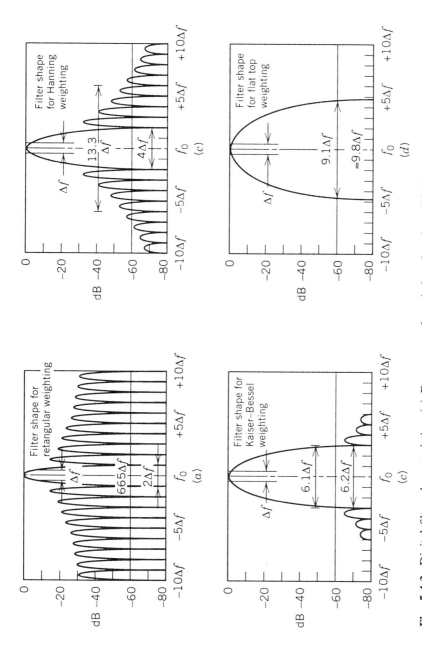

Fig. 5.4.3. Digital filter characteristics. (*a*) For rectangular window function. (*b*) For Hanning window function. (*c*) For Kaiser–Bessel window function. (*d*) For flat top window function. (Courtesy of Bruel & Kjaer Instruments, Inc.)

while the low ripple of the flat top window is clear evidence for its name. The last four columns in Table 5.4.2 deal with selectivity, which has more meaning when we examine a plot of the different digital filters.

The digital characteristics of each of these filters are shown in Fig. 5.4.3 for a $\pm 10 \, \Delta f$ range of frequencies centered at frequency f_0 and covering an 80 dB dynamic range. The meaning of the *highest sidelobe* and *sidelobe fall-off rate per decade* in Table 5.4.2 becomes clear as a measure of filter selectivity for comparison purposes. The meaning of 60 dB bandwidth (B_{60}) is also clearly evident when comparing the different filter characteristics, while the shape factor values are easily seen to fall into place. The first filter zero for the center lobe is shown at the bottom of each center lobe, and is seen to increase as the center lobe becomes wider with a corresponding decrease in side lobes. A comparison of the filters shows us clearly that the attempt to reduce leakage leads to reduction of the side lobes at the expense of the width of the filter's center lobe, a fact that is reflected in the increase in noise bandwidth in Table 5.4.2. Thus the analyzer designer is limited in just what can be done to reduce leakage and still maintain other desirable characteristics. We see that these limitations cause us to use different filters for different purposes.

5.4.3 Window Comparison for Use with Sinusoidal Signals

A significant use of digital filters is to analyze periodic signals. In this subsection, we look at the analysis of a single sinusoid and also the analysis of two closely spaced sinusoids that have widely different amplitudes. It is hoped that this experience will give the reader a feel for what can be achieved using digital analysis techniques.

Single Sinusoid An 800 line digital frequency analyzer, which can implement the four window functions, is used to analyze a single sinusoid when set up in the 3.2 kHz baseband mode so that $\Delta f = 4$ Hz. The results of this analysis are shown in Fig. 5.4.4. Two sinusoidal frequencies are analyzed, one at a time. The first sinusoid is at 1600 Hz and corresponds to a "best case" situation. The second sinusoid is at 1602 Hz and represents the "worst case" situation that corresponds to maximum filter leakage. The best case is represented as the 0 dB line so that ripple is evident at the peak frequency components. Also, note that each spectrum is displayed with a 60 dB (1000 to 1) amplitude range.

The rectangular window function has a single line spectrum in the best case situation shown in Fig. 5.4.4a. In the worst case situation, the rectangular window has a peak spectral value that is down 3.9 dB (a value of 0.637 compared to unity), due to filter ripple, and has leakage over a

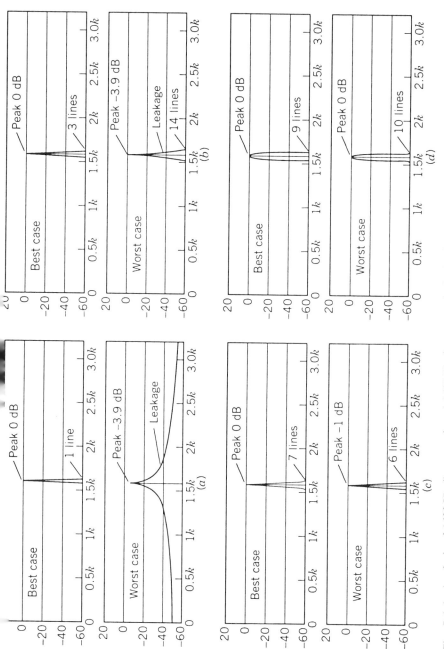

Fig. 5.4.4. Analysis of 1600 Hz (best case) and 1602 Hz (worst case) single sinusoid with analyzer set on 3200 Hz baseband range ($\Delta f = 4$ Hz). (a) Rectangular window function. (b) Hanning window function. (c) Kaiser–Bessel window function. (d) Flat top window function. (Courtesy Bruel & Kjaer Instruments, Inc.)

wide frequency range. The most significant leakage components are those within the −40 dB region.

The Hanning window function has a peak spectral value of 0 dB in the best case (Fig. 5.4.4b) but three lines are displayed compared to the single line for the rectangular window. The three lines come from the center lobe shown in Fig. 5.4.3b. The worst case has a peak spectral value that is 1.4 dB down, due to filter ripple (a value of 0.851 compared to unity and a value of 0.637 for the rectangular window). The worst case shows that leakage involves only 14 lines compared to the rectangular window. It is seen that this is a considerable improvement in accuracy and leakage over the rectangular window.

The Kaiser–Bessel window function response has a best case value of 0 dB (Fig. 5.4.4c) and involves seven lines compared to three for the Hanning window. This shows the "center lobe width effect" on the frequency spectrum's appearance. The worst case response is down 1 dB due to filter ripple (a value of 0.891 compared to unity, 0.851 for the Hanning window, and 0.637 for the rectangular window). The worst case involves only 6 frequency lines as shown, compared to the Hanning window's 14 lines.

The flat top window function analysis has a best case value of 0 dB (Fig. 5.4.4d) and a worst case analysis of 0 dB. The number of spectral lines involved changed from nine to ten lines where all lines lie within the center lobe. It appears that the flat top analysis gives the most consistent estimate of the signal's magnitude but has poor frequency resolution due to the flat top; see Fig. 5.4.3.

Analysis of Two Closely Spaced Sinusoids An effective way to demonstrate the performance of the four window functions to distinguish between sinusoidal frequency components is to analyze a periodic signal composed of two sinusoids that are close together in frequency and significantly different in amplitudes under worst case (maximum filter leakage) conditions. A worst case occurs when an 800 line frequency analyzer is operating in the 3.2 kHz baseband condition so that $\Delta f = 4$ Hz and the two signals have frequencies of 1626 Hz (midway between 1624 and 1628 Hz analyzer frequency lines) and 1650 Hz (midway between 1648 and 1652 Hz analyzer frequency lines). The 1626 Hz signal's amplitude is considered to be the reference quantity with a 0 dB value, while the 1650 Hz signal is 40 dB smaller (a factor of 1/100) than the 1626 Hz signal's amplitude. The frequency analysis of this periodic signal by the four different window functions is shown in Fig. 5.4.5, where the indicated values are measured at the 1624 Hz and 1648 Hz spectral lines.

The rectangular window function analysis is shown in Fig. 5.4.5a. Since this is a worst case analysis, the 1626 Hz spectral component has a peak value that is down 3.9 dB at 1626 Hz, while the 1650 Hz signal component

FACTORS IN THE APPLICATION OF A SINGLE CHANNEL ANALYZER   295

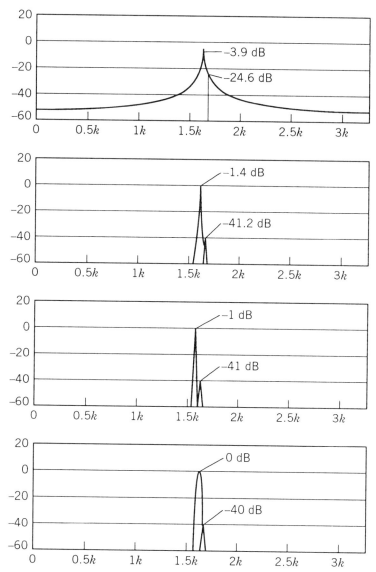

Fig. 5.4.5. Separating 1626 Hz (0 dB) signal and 1650 Hz (−40 dB) signal under "worst case" (maximum leakage), conditions. (*a*) Using rectangular window function. (*b*) Using Hanning window function. (*c*) Using Kaiser–Bessel window function. (*d*) Using flat top window function. Values shown correspond to 1624 and 1648 Hz readings. (Courtesy Bruel & Kjaer Instruments, Inc.)

is completely covered by filter leakage that has a value of −24.6 dB instead of nearly −40 dB. Note that the two equal frequency components at the peak indicate that the actual frequency is midway between the two spectral components. This case clearly shows how filter leakage can completely mask a small signal at a frequency close to a large signal. Only the peak frequency is clearly defined with a 3.9 dB error. The higher frequency component needs to have an amplitude that is closer to one tenth of the larger amplitude (down less than 20 dB) to be seen in this analysis.

The Hanning window function analysis is shown in Fig. 5.4.5b. There are two clear indications of frequency components, one around 1624 and 1628 Hz frequency lines and the other around 1648 and 1652 Hz frequency lines. Note how these peaks have a square type flat top, an indication that the actual frequency component lies halfway between the two frequency lines. Both signals should be attenuated 1.4 dB due to ripple. However, the lower peak is not attenuated the full 1.4 dB due to leakage from the larger component. We would conclude that the ratio of peaks is 97.7 to 1 instead of 100 to 1 and that our results are far better than for the rectangular window. This analysis resolves both frequencies with equal accuracy.

The Kaiser–Bessel window function frequency analysis is shown in Fig. 5.4.5c. In this case, the peak values are both down exactly 1 dB from what they should read and the ratio of peaks calculates out to be 100 to 1, as it should. This indicates that the filter leakage is so small that it has no effect on the reading of the smaller amplitude. The amplitude error is due to the signals being worst case, as indicated by two equal readings. This signal analysis is superior to the Hanning window function. This window function shows the signal's frequency components with equal magnitude accuracy and the same frequency resolution.

The flat top window function frequency analysis is shown in Fig. 5.4.5d where the true signal amplitudes are indicated. The frequency resolution is not nearly as good as for either the Hanning or the Kaiser–Bessel window functions since the flat top window includes too many frequency lines to give sharp frequency resolution; see Fig. 5.4.3.

The results of this frequency analysis show that the frequency components must be spaced at least five to six lines apart (24 Hz in this case is exactly six lines), and the amplitude ratio must be less than 100 to 1 if the Hanning and Kaiser–Bessel window functions are to resolve the frequencies and the flat top window function is to give accurate amplitudes. As the two frequency amplitudes become closer together, the frequencies can be closer together, but never less than four lines apart. If the frequencies are closer than four lines, then a zoom analysis is required, as discussed in Section 5.6. Note that the flat top window function will have difficulty with amplitudes if the lines are closer than four to five lines apart; see Fig. 5.4.3.

5.4.4 Spectral Line Uncertainty

We are concerned with the amount of variation that occurs in our signal analysis, particularly when dealing with random signals. Consider the ASD curve shown in Fig. 5.4.6a. The actual ASD curve $G(f)$ is shown along with frequency component estimate $\bar{G}(\nu)$. Unfortunately, this estimate has a *normalized error* ϵ_r that depends on a number of measurement factors, so the measured ASD $\bar{G}(\nu)$ is bounded by

$$G(\nu)(1 - \epsilon_r) < \bar{G}(\nu) < G(\nu)(1 + \epsilon_r) \tag{5.4.6}$$

We need to determine what factors influence the normalized error ϵ_r since we anticipate that different types of errors will occur in relatively flat regions compared with peak and valley regions.

We use the filter scheme shown in Fig. 5.4.6b to illustrate how we can estimate ϵ_r for a given reading. It is assumed that the signal $x(t)$ is a stationary ergodic random signal. Signal $x(t)$ passes through an electronic filter that has effective bandwidth B_e and center frequency ν and generates an output voltage signal $x(t, \nu, B_e)$. This voltage signal is then squared, then integrated, and finally time averaged over time period T_r so that the output has a time averaged value of $\bar{A}^2(\nu, B_e)$. This output value is divided by the effective bandwidth B_e to generate the estimate of the frequency spectrum. We can describe this process mathematically by

$$\bar{G}(\nu) = \frac{1}{B_e T_r} \int_0^{T_r} x^2(t, \nu, B_e)\, dt = \frac{\bar{A}^2(\nu, B_e)}{B_e} \tag{5.4.7}$$

where

$$\bar{A}^2(\nu, B_e) = \frac{1}{T_r} \int_0^{T_r} x^2(t, \nu, B_e)\, dt \tag{5.4.8}$$

is the mean square estimate passing through the filter and $\bar{G}(\nu)$ is the analog estimate of $G(\nu)$. We anticipate that if we let time $T_r \to \infty$ and B_e approach near 0, then the estimate becomes $G(\nu)$ itself so that

$$G(\nu) = \lim_{\substack{T_r \to \infty \\ B_e \to 0}} \frac{1}{B_e T_r} \int_0^{T_r} x^2(t, \nu, B_e)\, dt \tag{5.4.9}$$

Equation (5.4.9) implies that long averaging times and very small filter bandwidths are required to achieve an accurate estimate of $G(f)$.

The variance of $\bar{A}^2(\nu, B_e)$ is a measure of the uncertainty of the output

298 THE DIGITAL FREQUENCY ANALYZER

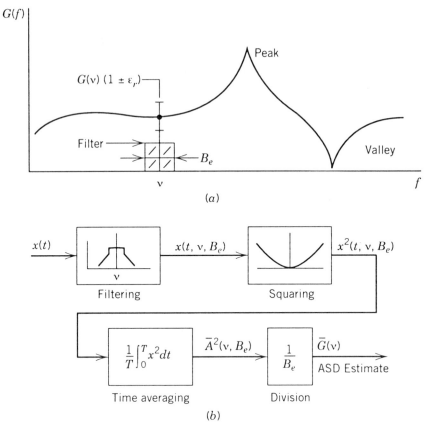

Fig. 5.4.6. (a) ASD curve showing filter position and measurement uncertainty at frequency ν. (b) Basic frequency analysis process used to estimate spectral density at center frequency ν.

spectral density. Bendat and Piersol[3] showed that the variance of $\bar{A}^2(\nu, B_e)$ can be written and approximated by

$$\text{Var}[B_e \bar{G}^2(\nu)] = B_e^2 \, \text{Var}[\bar{G}^2(\nu)] \cong \frac{B_e^2 G^2(\nu)}{B_e T_r} \quad (5.4.10)$$

since B_e is a constant. Dividing the right-hand side by $G^2(\nu)$ gives a normalized error that is defined by

---

[3] J. S. Bendat and A. G. Piersol, *Random Data: Analysis and Measurement Procedures*, 2nd ed., John Wiley & Sons, New York, 1986.

FACTORS IN THE APPLICATION OF A SINGLE CHANNEL ANALYZER   299

$$\epsilon_r[\bar{G}(f)] = \left[\frac{\text{Var}[\bar{G}^2(\nu)]}{G^2(\nu)}\right]^{1/2} \cong \frac{1}{\sqrt{B_e T_r}} \quad (5.4.11)$$

Equation (5.4.11) is valid for all frequencies ν and applies to the ASD values. Bendat and Piersol also show that the normalized error for the RMS frequency spectrum is one half the ASD error, or in other words,

$$\epsilon_r[\bar{A}(\nu, B_e)] \cong \frac{1}{2\sqrt{B_e T_r}} \quad (5.4.12)$$

where $\bar{A}(\nu, B_e)$ represents the measured RMS spectral density. This means that the RMS spectral variance should be about half as large as the ASD variance. In both cases, the product of $B_e T_r$ is the key parameter to reducing spectral line measurement uncertainty.

Bias Errors in the Estimate When a filter is used to measure the ASD of a signal that contains a sharp peak or valley (as often occurs in vibrations), a bias error results, as shown in Fig. 5.4.7. The problem occurs when the effective filter bandwidth is too large relative to the peak (or valley), so that the peaks are underestimated and the valleys are overestimated, as shown. Consequently, we see that the area under the filter is presented as the average height $\bar{G}(f_0)$, rather than as the peak value $G(f_0)$, so that a bias of $\{b\bar{G}(f_0)\}$ occurs. In other words, area A_1 (above the filter rectangle) is equal to the sum of the two areas A_2 and A_3 (that are not included under the curve $G(f)$) so that the rectangular filter area is the same as the area under the $G(f)$ curve over bandwidth B_e. Then we can

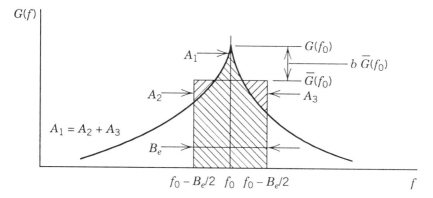

Fig. 5.4.7. Resonant ASD curve and filter interaction showing source of peak (and valley) bias error.

300   THE DIGITAL FREQUENCY ANALYZER

express the peak value as

$$G(f_0) = \bar{G}(f_0) + b\bar{G}(f_0) \tag{5.4.13}$$

Bendat and Piersol (see footnote 2) have indicated that using a Taylor series expansion of $G(f)$ around its peak value allows the bias error to be estimated from

$$b\bar{G}(f_0) \cong \frac{B_e^2}{24} G''(f) \tag{5.4.14}$$

where $G''(f)$ is the second derivative of the ASD curve. This approximate formula overestimates the error so it is conservative when the derivatives are large.

We can apply Eq. (5.4.14) to a highly resonant mechanical system where its 3 dB bandwidth is related to the mechanical damping ζ and natural frequency f_n by

$$B_n = 2\zeta f_n \tag{5.4.15}$$

Then the *normalized bias error* at resonance becomes

$$\epsilon_b[\bar{G}(f_n)] = \frac{b\bar{G}(f_n)}{G(f_n)} = -\frac{1}{3}\left(\frac{B_e}{B_n}\right)^2 \tag{5.4.16}$$

An analysis of Eq. (5.4.16) shows that if $B_n = 3B_e$, then a 3.7 percent bias error results, while $B_n = 4B_e$ gives approximately a 2 percent bias error. This severe requirement on filter bandwidth B_e is a further reason for developing an interest in *zoom* frequency analysis, described in Section 5.6.

Digital Analyzer Error Estimates   We have shown the errors as seen from an analog filter viewpoint. When dealing with digital instruments, these concepts must be translated into the number of window samples required to achieve the same normalized error. Bendat and Piersol (see footnote 2) have presented a detailed statistical analysis that utilizes chi-square statistical concepts. Equivalent results can be obtained directly from Eqs. (5.4.11) and (5.4.12) without resorting to chi-square statistical concepts. We have shown conceptually in Eq. (5.4.9) that a long time record is required in order to obtain a good estimate of the spectral density at a given frequency. Suppose that we analyze a random signal that consists of N_r statistically independent analyzer windows of length T so that the entire record length becomes

FACTORS IN THE APPLICATION OF A SINGLE CHANNEL ANALYZER 301

$$T_r = N_r T \qquad (5.4.17)$$

The analyzer has an effective bandwidth B_e, as given in Table 5.4.2 (the noise bandwidth column for four common window functions). Here, we see that the effective bandwidth is Δf so that B_e is greater than or equal to Δf, that is

$$B_e \geq \Delta f = \frac{1}{T} \qquad (5.4.18)$$

Then if we substitute Eqs. (5.4.17) and (5.4.18) into Eq. (5.4.11), we find the normalized error is controlled by

$$B_e T_r = N_r B_e T \geq N_r \qquad (5.4.19)$$

so that the effective $B_e T_r$ product is always greater than or equal to N_r. In addition, the maximum error is given by Eq. (5.4.11), as opposed to Eq. (5.4.12). Thus as a practical matter, we can estimate our ASD errors for digital analysis to be bounded by

$$\epsilon \leq \frac{1}{\sqrt{N_r}} \qquad (5.4.20)$$

It is clear from Eq. (5.4.20) that approximately 100 samples must be obtained to have a normalized error less than 10 percent.

Some analyzers use the *number of degrees of freedom n* to set the number of window samples to use in calculating an average frequency spectra. The number of degrees of freedom definition comes from the chi-square statistical analysis and is given in this case by

$$n = 2N_r \qquad (5.4.21)$$

Users must be aware which definition is being used with their instrument.

Linear averaging is used in calculating each spectral amplitude. If the standard formula for calcuiating the average is used, then an error results if the process stops before the average is complete, since the standard formula divides by N_r. This problem is overcome when the pth frequency component for the rth sample $X_p(r)$ is averaged by using the following formula:

$$|\overline{X_p(k)}|^2 = \frac{1}{k}|X_p(k)|^2 + \frac{k-1}{k}|\overline{X_p(k-1)}|^2 \qquad \text{for } 1 \leq k \leq N_r$$

$$(5.4.22)$$

where the bar indicates the average value of all previous magnitudes. The advantage of Eq. (5.4.22) is that the averaging process can be stopped and started at any value of k and the correct average value up to that value of k is obtained.

5.4.5 Recommended Window Usage

We have examined four different window functions in this section, functions that are readily available in many frequency analyses. We look at the usefulness of these window functions for each type of signal.

Periodic Signals

The Rectangular Window This window function is poorly suited for general use with periodic signals. There are, however, special periodic signals where the rectangular window is excellent. These are pseudorandom signals (which are composed of the same frequency components as the analyzer) and order tracking experiments (where the sample rate is a multiple of the fundamental event frequency such as 128 times shaft speed). In both of these cases there is no filter leakage.

The Hanning Window This window function is recommended for general periodic analysis.

The Kaiser–Bessel Window This window function is recommended for periodic signals that require frequency selectivity.

The Flat Top Window This window function is excellent at determining amplitudes but is poor in frequency resolution. It is fully effective at measuring spectral amplitudes when the frequency components are at least five to six line spacings apart.

Transient Signals

The Rectangular Window This window function is required in transient measurements, except in special cases where an exponential window function is used. The exponential window function case is discussed in Chapter 7 for impulse testing applications.

The Hanning Window This window function is used in special transient applications where randomly spaced repetitive impulses are used and the vibration decays quickly within the window period T so that many (usually five or more) impulses occur within the window period. Otherwise, a Hanning window should not be used in transient measurements without careful consideration.

The Kaiser–Bessel and the Flat Top Windows These window functions are inappropriate to use with transient signals and are not recommended.

Random Signals

The Rectangular Window This window function is poorly suited for use with random signals. Only with pseudorandom signals is its use recommended, since this is a special case with zero filter leakage when properly implemented.

The Hanning Window This window function is recommended as a general purpose window for use with random signal analysis.

The Kaiser–Bessel Window This window function gives good frequency resolution but is not recommended for general purpose random signal analysis since, in some analyzers, it slows down the analyzer's processing speed compared to using the Hanning window.

The Flat Top Window This window function is not recommended for use with random signals.

This section has attempted to introduce the many factors that enter into frequency analysis in order to obtain satisfactory results. It is recommended that any user carefully learn the features of his or her instrument by analyzing known signals. It is through careful experimentation with known signals that real expertise begins to develop. Sometimes it is not obvious which type of filter to use with a given signal, for the signal may be a mixture of several basic types such as periodic and random. Then more than one type of analysis may be necessary in order to extract and understand the signal's important characteristics.

5.5 OVERLAPPING SIGNAL ANALYSIS TO REDUCE ANALYSIS TIME

In Section 5.4, we saw that random signals require a large number of individual frequency spectra that must be averaged together in order to reduce spectral uncertainty. It was also seen that the Hanning, Kaiser–Bessel, and flat top window functions essentially ignore a portion of the signal by tapering to zero at their ends. The time required by the analyzer to process one window of data into an averaged frequency spectrum is called the *process calculation time* T_c. So the question becomes: Can data windows be overlapped so that we can obtain the averaged frequency spectrum faster? If so, what are the implications of using overlapped signal analysis with random signals? This section addresses the basic issues of overlapping, of effective filter ripple, of effective bandpass averaging time

304 THE DIGITAL FREQUENCY ANALYZER

product BT and effective spectral averaging, and of real time processing limits. All of this analysis will be done with the window function in the time domain.

Overlapping can be used with all window functions. However, it turns out that the Hanning window is ideally suited for this application. Consequently, the Hanning window function is emphasized in this section. The other window functions are mentioned only for comparison purposes.

5.5.1 Overlapping and Ripple

The idea of window overlapping is illustrated in Fig. 5.5.1a, where three time shifted window functions are shown relative to the reference window function. The reference window function extends from $t = 0$ to $t = T$. Now we want to shift the window function a portion of the window period T and describe this shift in terms of *shift factor r*. Then, the first time shifted window function starts at $t = rT$ and ends at $(1 + r)T$. The second time shifted window function starts at $2rT$ and ends at $(1 + 2r)T$. The third time shifted window function starts at $3rT$, and so on. The window shift factor is r, and the amount of window overlap OL is related by

$$OL = (1 - r) \qquad (5.5.1)$$

where OL is often expressed as a percentage.

Now we are interested in the net effect of using overlapped window functions. The signal to be analyzed is random, which means that we are interested in how the square of the signal behaves. This means that behavior of the square of the overlapping window functions for any shift value r is of interest. Gade and Herlufsen[4] have shown that the effective square of the window function $w_e^2(t)$ is the average of the linear sum of the window functions at a given time t so that

$$w_e^2(t) = \frac{1}{N} \sum_{i=1}^{N} w^2(t - irT) \qquad (5.5.2)$$

where r is the shift factor.
T is the window length.
i is the window number in the sum.
N is the number of windows involved at time t.

Figure 5.5.1b shows the results of Eq. (5.5.2) when applied to the Hanning window functions of Fig. 5.5.1a when $r = 1/4$. Note how the effective

---
[4]S. Gade and H. Herlufsen, "Use of Weighting Functions in DFT/FFT Analysis (Part II)," *Technical Review*, Nos. 3 and 4, 1987, available from Bruel & Kjaer Instruments.

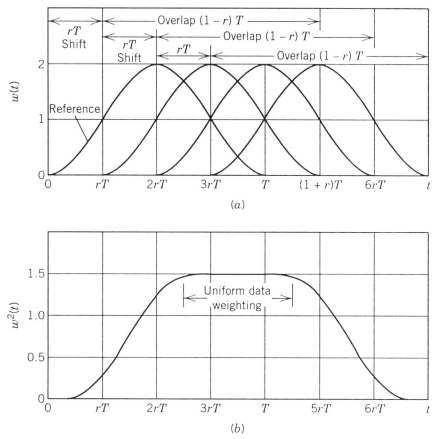

Fig. 5.5.1. Hanning window function $w(t)$. (*a*) Definition of window shift and overlap ($r = \tfrac{1}{4}$). (*b*) Plot of average value of sum of $w^2(t)$ from Fig. 5.5.1*a* showing uniform data weighting with tapered ends for Hanning window.

window function becomes a constant value of 1.5 for values of $t > \cong 0.7T$. This curve begins to return to zero around $1.1T$ if more window functions are not added. Thus we anticipate that all time data in the central portion are treated equally with only the beginning and ending portions being discounted.

The effective window function varies with overlap. Figure 5.5.2 shows the plot of overlapped window functions $w(t)$ and effective square of the window function $w_e^2(t)$ as calculated from Eq. (5.5.2) for overlap of $OL = 0$, 50, 66.7, and 75 percent. The 0 percent overlap case is shown in Fig. 5.5.2*a* where the corresponding $w_e^2(t)$ curve is seen to have a wavy shape that varies from zero to four. Thus some data is effectively ignored while

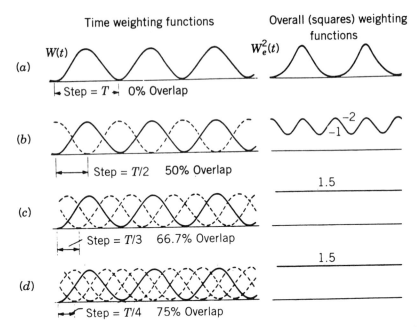

Fig. 5.5.2. Hanning window function and average squared window function. (a) For 0 percent overlap. (b) For 50 percent overlap. (c) For 66.7 percent overlap. (d) For 75 percent overlap. (Courtesy Bruel & Kjaer Instruments, Inc.)

other data is given great emphasis. Gade and Herlufsen[3] show that the effective time for the Hanning window is $0.375T$ when there is no overlap.

When $r = 0.5$ (50 percent overlap), Eq. (5.5.2) reduces to

$$w_e^2(t) = 1 + \cos^2\left(\frac{2\pi}{T}t\right) \qquad (5.5.3)$$

for the Hanning window function, which is plotted in Fig. 5.5.2b. The squared equivalent window (called power weighting) in Eq. (5.5.3) varies from 1 to 2 so that this window has a 3 dB ripple. A remarkable result is that the ASD effective filters are flat when $r = \frac{1}{3}$ and $\frac{1}{4}$. The ripple dB are plotted as a function of shift factor r in Fig. 5.5.3, where it is clearly seen that the ripple is negligible for r less than $\frac{1}{3}$ and is always zero when r is a fraction given by

$$r = \frac{1}{n} \qquad (5.5.4)$$

for $n \geq 3$. Thus it is common to use either $\frac{2}{3}$ or $\frac{3}{4}$ overlap.

The reason that the ripple behaves as shown in Fig. 5.5.3 is that phase shift takes place between spectral components that are measured by different time shifted windows. It is not difficult to show (using Fourier

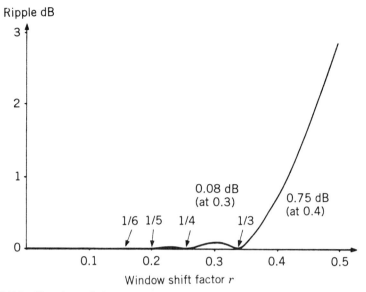

Fig. 5.5.3. Hanning window squared and averaged showing ripple as a function of window shift factor r. (Courtesy Bruel & Kjaer Instruments, Inc.)

series definitions) that all frequency spectrum components have a basic phase shift due to window time shift of rT that is given by

$$\theta = r(2\pi) \qquad (5.5.5)$$

In addition, the squared equivalent Hanning window function becomes

$$w_e^2(t) = 1.5 - 2\cos(\Omega t) + 0.5\cos(2\Omega t) \qquad (5.5.6)$$

from which we see that this function is made up of three terms, a constant of 1.5, a frequency at Ω ($= 2\pi/T$), and a frequency at 2Ω. Now when Eq. (5.5.6) is convolved with the random signal's frequency spectrum, we want only the 1.5 constant value to be effective. This means that each spectral component must cancel when convolved with each window function frequency of Ω and 2Ω; otherwise these window function frequency components (Ω and 2Ω) will generate unwanted information in the resulting frequency analysis.

Consider the case when $r = 0.5$ so that only two window functions overlap, meaning that a given frequency component with amplitude A is analyzed twice, once by window 1 (called amplitude A_1) and once by window 2 (called amplitude A_2). Then, the basic phase shift from Eq. (5.5.5) gives $\theta = \pi$ for frequency Ω, while $\phi = 2\theta = 2\pi$ for frequency

308   THE DIGITAL FREQUENCY ANALYZER

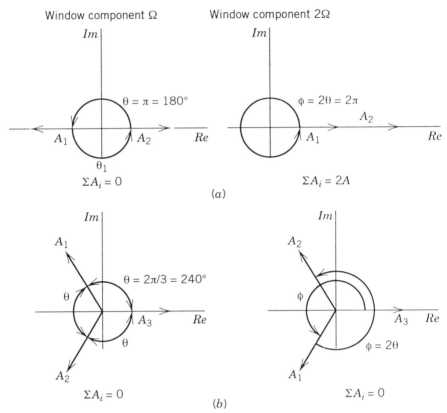

Fig. 5.5.4. Addition of the same frequency component from overlap analysis due to window function frequency components of Ω and 2Ω. (*a*) $r = \frac{1}{2}$. (*b*) $r = \frac{1}{3}$.

2Ω. These conditions are plotted in Fig. 5.5.4*a* for window frequency components Ω and 2Ω. In the Ω case, the two measured components are 180 degrees apart so they add to zero. In the 2Ω case, the two measured components add. Thus we see that the two components cancel in one case and add in the other to give a ripple of 3 dB.

Similarly, when $r = 1/3$, there are three frequency components that add at each window frequency, one from each window analysis. The corresponding basic angle from Eq. (5.5.5) is $\theta = 2\pi/3$ (120 degrees) in this case so that $\phi = 4\pi/3$ (240 degrees). Then the three frequency components add to zero for both the Ω and 2Ω window function frequencies, as shown in Fig. 5.5.4*b*. The same happens for $r = \frac{1}{4}, \frac{1}{5}, \frac{1}{6}$, and so on. Thus the effective filter is flat (with a maximum ripple less than 0.1 dB, see Fig. 5.5.3) over all the entire time domain if $r \leq \frac{1}{3}$. This is a very desirable result.

TABLE 5.5.1. Four Window Function Correlation Coefficients Versus Window Overlap

Window Type	Overlap (percent)			
	100[a]	75	50	25
Rectangular	1.00	0.7500	0.500	0.2500
Hanning	1.00	0.6592	0.1667	0.0075
Kaiser–Bessel	1.00	0.5389	0.0735	0.0014
Flat top	1.00	0.0455	0.0153	0.0005

[a]Same as 0 percent overlap.

5.5.2 Effective Bandwidth Time Product and Measurement Uncertainty

We have seen in Section 5.4.4 that the ASD relative standard deviation is related to the effective bandwidth time product of $B_e T_r = n_d$. The effective BT product for a single analysis of a given record becomes

$$BT_e = \frac{B_e T_r}{n_d} \cong 1 \quad (5.5.7)$$

which shows that a single window analysis of a random signal has a very high uncertainty where the uncertainty is as large as the measured value itself. Now we need to establish how overlapping affects the values of BT_e. Note that BT_e is the effective BT product on a per record basis.

The effective BT value is related to the correlation between the overlapped windows. This correlation between two overlapping window functions can be written as

$$c(r) = \frac{\int_0^T w(t) w\{t + (1-r)T\} \, dt}{\int_0^T w^2(t) \, dt} \quad (5.5.8)$$

The four window function correlation values are shown in Table 5.5.1 for overlaps of 100, 75, 50, and 25 percent. It is clear from Table 5.5.1 that the rectangular and Hanning window functions have significantly better correlation for a given amount of overlap. Harris[5] and Welch[6] have shown that the *effective time BT* product for a given record can be estimated

---

[5]F. J. Harris, "On the Use of Windows for Harmonic Analysis with the Discrete Fourier Transform," *Proceedings of the IEEE*, Vol. 66, No. 1, Jan. 1978, pp. 51–83.
[6]P. D. Welch, "The use of Fast Fourier Transform for the Estimation of Power Spectra," *IEEE Transactions on Audio Electro Acoustics*, Vol. AU-15, June 1967, pp. 70–73.

from

$$BT_e(50 \text{ percent}) \cong \frac{1}{1 + 2c^2(50 \text{ percent})} \qquad (5.5.9)$$

for 50 percent overlap and

$$BT_e(75 \text{ percent}) \cong \frac{1}{1 + 2c^2(75 \text{ percent}) + 2c^2(25 \text{ percent})} \qquad (5.5.10)$$

for 75 percent overlap when the number of records analyzed $N_r > 10$. Measured effective BT values are given by Gade and Herlufsen (see footnote 3), where 100 estimates of an auto spectrum using 100 averages and 40 frequency lines were used in calculating BT_e. These measured values, and those from Eqs. (5.5.9) and (5.5.10), are given in Table 5.5.2, where it is seen that they agree within a few percent for each window function.

The results of Table 5.5.2 for the Hanning window function show that 75 percent overlap gives four averages during one window period T, compared to one average without overlapping. Clearly, we see that the effective overlap analysis is at least twice as fast, since four window records are analyzed with $BT_e \cong 0.52$ to give an effective BT value of 2.08 ($BT = 4 \times 0.52 \cong 2.08$) in the same time that one analysis without overlap would take. Another way of looking at this result is as follows: If we used $N_r = 100$ averages and 0 percent overlap, we would use $100T$ seconds to do the analysis. Now, if we use 75 percent overlap and $N (= 200)$ averages, we have effective N_r averages of 104 (since $N_r = N \times BT_e = 200 \times 0.52 \cong 104$), but these averages take only $50T$ seconds ($NrT = 200 \times 0.25 \times T$ since $r = 0.25$) to do the analysis. Thus in this case, we are being twice as effective, a significant improvement in time performance where the same uncertainty results in half the time.

5.5.3 Real Time Analysis

Real time analysis occurs when the data is processed fast enough so that the data is sampled continuously at equal time increments. This means that the calculation time T_c must be less than window time T when there is no overlap. A convenient parameter to describe real time processing capabilities is the *real time bandwidth*, since the baseband (frequency range of analysis from 0 to f_{max}) is used to set the analyzer's operating condition. When the baseband is less than the real time bandwidth, the analyzer is capable of operating in real time. Table 5.5.3 shows the real time bandwidths for the Bruel & Kjaer Model 2032 dual channel frequency

TABLE 5.5.2. Effective BT_e per Record for Random Signal Analysis

Window Type	Overlap (percent)					
	0		50		75	
	Theoretical	Measured	Theoretical	Measured	Theoretical	Measured
Rectangular	1.00	0.954	0.660	0.674	0.363	0.368
Hanning	1.00	0.995	0.947	0.940	0.520	0.535
Kaiser–Bessel	1.00	1.009	0.989	0.996	0.628	0.628
Flat top	1.00	0.990	1.00	1.00	0.995	0.994

^{a}To obtain effective number of averages, multiply number of averages by this number to obtain n_d so that $n_d = BT_eN$.

TABLE 5.5.3. Real Time Bandwidths[a] for Bruel & Kjaer Model 2032 When Operating in Single and Dual Channel Modes

Window Type	Bandwidth (kHz)	
	Single Channel	Dual Channel
Rectangular	17.7	5.8
Hanning	16.6	5.4
Kaiser–Bessel	10	3.8
Flat top	10	3.8

[a]For auto spectra of random signals with zero window overlap.

analyzer for both single and dual channel operation when doing ASD random signal analysis with no overlap for each window function. It is clear from this table that the rectangular and Hanning windows are superior in speed. The basebands available in the 2032 start at 25.6 kHz and decrease by factors of two in a sequence of 12.8 kHz, 6.4 kHz, 3.2 kHz, 1.6 kHz, and so on. Thus it is evident that 12.8 kHz is the largest baseband that is also a real time bandwidth for the Hanning window function when operating in the single channel mode, while 3.2 kHz is the largest real time baseband for dual channel operation. Now let us look at what happens when we use overlap operations.

When there is window function overlap, the calculation time is limited by the shift time of rT since all processing must be complete before the next window of data is ready for processing. This means that we must satisfy the relationship of

$$T_c \leq rT \qquad (5.5.11)$$

A practical amount of overlap is $\frac{2}{3}$ or $\frac{3}{4}$. The calculation time for the 800 line Bruel & Kjaer Model 2032 dual channel analyzer when operating in the single channel mode and using the Hanning window is approximately 50 ms. This means that $\Delta f \cong 20$ Hz so that the *real time bandwidth* becomes 16.6 kHz (800 times Δf), a value in agreement with Table 5.5.3. When $\frac{2}{3}$ overlap is used, $r = \frac{1}{3}$ and the overlap real time bandwidth becomes 16.6/3, or 5.5 kHz. However, when $\frac{3}{4}$ overlap is used, $r = \frac{1}{4}$ and the overlap real time bandwidth becomes $16.6/4 \cong 4.15$ kHz. Since the baseband selection choices are either 6.4 or 3.2 kHz for this instrument in this range of frequencies, it is obvious that either $\frac{2}{3}$ or $\frac{3}{4}$ overlap can operate in real time when the 3.2 kHz baseband or less is selected and cannot operate in real time if the 6.4 kHz baseband or higher is selected.

When the Bruel & Kjaer Model 2032 operates in the dual channel mode, the calculation time T_c increases by a factor of nearly three as seen in Table 5.5.3. This increase in processing time is due to the fact that two channels of data must be analyzed as well as cross-channel information processed. The values in Table 5.5.3 are presented as typical values that are found in manuals. We must know how to interpret and use this information. The dual channel frequency analyzer and its operating characteristics will be discussed in detail in Sections 5.8 and 5.9.

5.6 ZOOM ANALYSIS

We have shown that frequency spectrum resolution is controlled by line spacing Δf. We have seen a couple of instances in the previous sections where we needed to increase the analyzer's frequency resolution. Common instances requiring zoom analysis are: (1) signals that contain multiples of a relatively low frequency such as occur in ball bearings, (2) signals that contain two or more closely spaced resonance peaks, (3) signals that contain a large number of harmonics of one or more fundamental frequencies, as occur in gear boxes, and so on. The question is: How can frequency resolution be increased while maintaining the same baseband frequency range of 0 to $f_{\max}$? We know that using a slower sample frequency will increase the window period and decrease Δf, but this approach also reduces the baseband frequency range, a solution that is often unacceptable.

One way to increase the frequency resolution while maintaining baseband frequency range is to increase the data window length by an integer factor of N while using the same sample frequency f_s. Then the new spectral line spacing Δf_n becomes

$$\Delta f_n = \frac{f_s}{N_r} = \frac{f_s}{NN_0} = \frac{\Delta f}{N} \quad (5.6.1)$$

where N_0 is the number of data points used in the baseband frequency analysis (usually 1024 or 2048), N is the integer resolution expansion factor that must be a power of 2, and N_r is the total number of data points in the record. Equation (5.6.1) shows that the resolution is increased by a factor of N so that we can zoom in on a given range of frequencies in order to obtain a more detailed description in the local frequency domain, as shown in Fig. 5.6.1.

The zooming process in Fig. 5.6.1 is similar to using a zoom lens in photography, where a small portion of the picture is enlarged to obtain finer detail. Thus the name of *zoom frequency analysis*, since we are looking in greater detail at a small portion of the baseband frequency

314 THE DIGITAL FREQUENCY ANALYZER

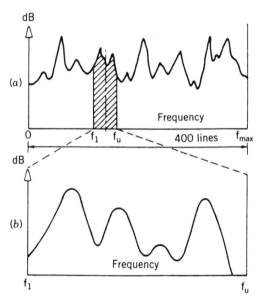

Fig. 5.6.1. Illustration of zoom expansion. (*a*) Baseband analysis 0 to f_{max}. (*b*) N times zoom expansion of cross-hatched region in Fig. 5.6.1*a*, from f_l to f_u, using the same number of spectral lines in the display. (Courtesy of Bruel & Kjaer Instruments, Inc.)

spectrum. In photography, the zooming process reduces the amount of light available, so that more time is required to expose the film. Consequently, a *smeared* photograph can easily occur if either the object or the camera moves during the longer exposure time. A similar phenomenon occurs during zoom frequency analysis, since we need to record a time history that is N times longer. Consequently, the signal may change slightly during the long sampling time and cause a *smeared frequency spectrum*.

Two basic methods have evolved to accomplish a practical implementation of zoom FFT processing. The first method uses the *heterodyning* or frequency shift concept, a process that is based on the fact that multiplication in the time domain causes convolution in the frequency domain. The second method involves digitally recording a signal that is N times longer than the standard data window of N_0 (usually 1024 or 2048) data points.

This longer signal requires N times more data memory as well as an N times larger frequency spectrum memory. The time required to do an FFT calculation is of the order of $\{N_r \log(N_r)\}$. If the zoom is by a factor of 10, the calculation time is about 12.8 times longer when $N_0 = 4096$ data points and about 13.3 times longer when $N_0 = 512$ data points.

We briefly explore the heterodyning method first, and then explore the long record method in greater detail, since it has other useful applications

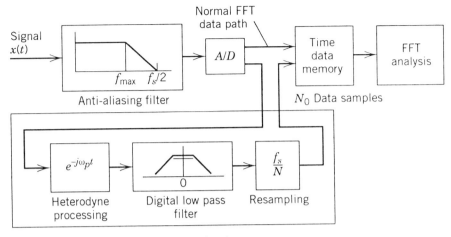

Fig. 5.6.2. Block diagram of heterodyne (frequency shift) zoom processing.

besides zoom analysis. These applications will be developed further in Section 5.7. Finally, we look at an application of zoom analysis.

5.6.1 Zoom FFT Analysis Using the Heterodyning Method

This is the most commonly implemented FFT zoom method, due to its lower initial cost. The process consists of using the anti-aliasing filter and A/D converter in the standard manner, as shown in Fig. 5.6.2, where the sample frequency f_s remains the same so that the baseband frequency range is unaltered and we can zoom in on any part of that frequency range. Now, instead of placing the digital data directly into the time data memory, the sampled data must first pass through a zoom processor before storage. On entry to the zoom processor, the data is multiplied by heterodyning time function of

$$f(t) = e^{-j\omega_p t} \qquad (5.6.2)$$

where $\omega_p = 2\pi f_p$ is the base frequency component about which the zoom is to be expanded.

The time multiplication and frequency convolution theorem can be used to show that the frequency spectrum's frequency scale is shifted f_p Hz to the left so that the apparent zero frequency is at f_p, as shown in Fig. 5.6.3. The cross-hatched baseband region in Fig. 5.6.3 is to be expanded by the zoom process. This is done by using digital low pass filtration and filtered signal resampling. The portion of the baseband

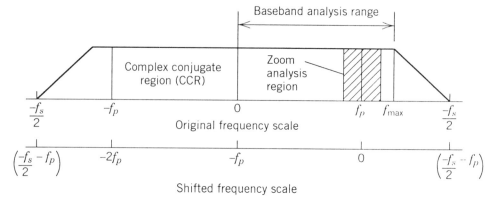

Fig. 5.6.3. Baseband frequency spectrum showing original and shifted frequency scales, and zoom analysis region.

frequency spectrum that is to be zoomed in on is limited by a digital low pass filter that removes all frequencies outside of the region of interest. The digitally filtered time data is resampled once out of every N data points, and these resampled data are put into the time data memory for use in FFT analysis. Hence, the data record covers a long time record of length $T_r = NT$ but only N_0 data points are being analyzed where these N_0 data points represent that portion of the time history with frequency components within the zoom window. The FFT analysis of the N_0 data points from a total NN_0 data points gives the zoom analysis with the same number of spectral lines (usually 400 Δf), but these lines are spaced Δf_n apart, as required by Eq. (5.6.1). This zoom spectrum is centered on f_p.

The positive features of heterodyne zoom FFT analysis are large zoom factors and real time analysis with minimum equipment expense. The negative features are that the original time function is not stored (and hence, is lost for further data analysis without reprocessing the analog signal), and that process errors accumulate at center frequency f_p.[7] The loss of signal data during processing can be a serious problem when dealing with transient signals or with low frequency signals that require long sampling times. In these cases, the analog data must be stored on magnetic tape for reanalysis, a process that introduces additional equipment cost and a source of noise problems. In addition, if there are several regions that require zoom analysis, the sampling process must be done for each region of interest, a time consuming process that adds indirectly to the cost of this less expensive type of frequency analyzer.

The accumulation of errors at center frequency f_p can cause significant

---

[7]Private conversations with D. Robb, Imperial College, London, U.K., 1991.

measurement errors, so it is good practice to set f_p away from the region of frequencies that are of most interest. This takes discipline on the user's part, for it is natural to select the frequency range of interest to be in the zoom display's center when setting up a zoom analysis. This setup process involves either selecting f_p and the zoom range or selecting the lower and upper frequencies in the zoom window. In either case, be aware of this center frequency error and keep the most interesting frequency range away from f_p in order to minimize this inherent zoom analyzer error, which occurs only when this method of zoom analysis is employed.

5.6.2 Long Time Record Zoom FFT Analysis

The long time record zoom analysis requires considerable digital memory to store the time data and the accumulated high resolution frequency spectrum, which has line spacing Δf_n as given by Eq. (5.6.1). A long time record consisting of N window periods T_0 that is sampled with sample frequency f_s so as to maintain the baseband's maximum frequency (usually $f_{max} = f_s/2.56$) is shown in Fig. 5.6.4. We see from this figure that the time record's length is given by

$$T_r = N_r \Delta t = NN_0 \Delta t \tag{5.6.3}$$

as we string together one standard window length after another. The problem is how to process this large time record without having to do an FFT analysis on all N_r data points at one time. We now look into the mathematical logic by which we can process this large amount of data N times with standard sets of digital data of length N_0 in order to speed up the processing time, obtain the same detailed frequency spectrum that we would associate with this long time history, and use less hardware memory.

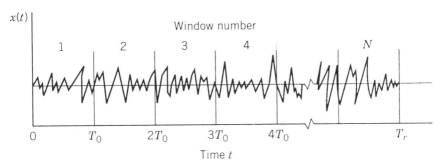

Fig. 5.6.4. Time history showing how N window samples are defined in long record storage.

318   THE DIGITAL FREQUENCY ANALYZER

The pth Fourier series frequency component for this large data window is given by

$$X_p = \frac{1}{T_r} \int_0^{T_r} x(t) \, e^{-jp\omega_o t} \, dt \qquad (5.6.4)$$

when $x(t)$ is a continuous signal. However, since we have the digital version of $x(t)$, we can write the digital equivalent of Eq. (5.6.4) as

$$X_p = \frac{1}{N_r} \sum_{k=0}^{N_r} x(t_k) \, e^{-jpk(2\pi/N_r)} \qquad (5.6.5)$$

where p ranges from 0 to the Nyquist value of $N_r/2$. These frequency components are spaced Δf_n Hz apart, and there are N times as many components as we would obtain by a standard analysis of N_0 data points, namely, $N_0/2$. Now, we can write index k in terms of two other indices, q and m, such that

$$k = Nq + m \qquad (5.6.6)$$

where N is the number of standard window samples collected, m is the sample offset that varies from 0 to $(N-1)$, and q is the new summation index that varies from 0 to $(N_0 - 1)$. Equation (5.6.6) shows that k takes on every Nth value as q varies from 0 to $(N_0 - 1)$ for each offset value m. However, all N_r data points are represented by Eq. (5.6.6) in N separate offset collections as m varies from 0 to $(N-1)$. Thus, N_r data points can be divided up into N sums of N_0 data points that are spaced N points apart with a summation on m from 0 to $(N-1)$.

Now, inserting Eqs. (5.6.3) and (5.6.6) in Eq. (5.6.5) and using the consequences of Eq. (5.6.6) in rearranging the summation process, we can express Eq. (5.6.5) as

$$X_p = \frac{1}{N} \sum_{m=0}^{N-1} e^{-jpm(2\pi/N_r)} \left[\frac{1}{N_0} \sum_{q=0}^{N_0-1} x(Nq + m) \, e^{-jpq(2\pi/N_0)} \right] \qquad (5.6.7)$$

The terms inside the brackets represent the standard FFT analysis of N_0 data points. The term before the left-hand bracket represents a phase shift correction, since the data used in the Fourier analysis is offset by m. Consequently, if we let

$$X_{pm} = \frac{1}{N_0} \sum_{q=0}^{N_0-1} x(Nq + m) \, e^{-jpq(2\pi/N_0)} \qquad (5.6.8)$$

be the frequency spectrum that results from one standard FFT analysis of record length N_0 and offset m, then we can write Eq. (5.6.7) as

$$X_p = \frac{1}{N} \sum_{m=0}^{N-1} e^{-jpm(2\pi/N_r)} X_{pm} \qquad (5.6.9)$$

Finally, we see from Eq. (5.6.8) and Fig. 5.6.5 that the frequency components repeat themselves every N_0 components, so that the p subscript on X_{pm} in Eqs. (5.6.8) and (5.6.9) can be written as

$$p' = p - nN_0 \qquad (5.6.10)$$

where $0 < n < (N/2 - 1)$ so that p' always ranges from 0 to $(N_0 - 1)$. This means that we need to store only one N_0 set of regular and complex conjugate region (CCR in Figs. 5.6.3 and 5.6.5) frequency spectra data and then recycle these frequency components according to Eq. (5.6.10). In addition, let us define the phase shift function by

$$PS(N_r, p, m) = e^{-jpm(2\pi/N_r)} \qquad (5.6.11)$$

Then, Equation (5.6.9) becomes

$$X_p = \frac{1}{N} \sum_{m=0}^{N-1} PS(N_r, p, m) X_{(p-nN_0)m} \qquad (5.6.12)$$

Equation (5.6.12) shows that a complete FFT of all of the data occurs when N independent FFTs are obtained and are averaged according to Eq. (5.6.12). The frequency components given by Eq. (5.6.12) are the same as those obtained from Eq. (5.6.5) (by direct Fourier analysis of the entire time record). The time required to do this process is approximately N times that for a single sample of N_0 points. When $N = 10$, the time is approximately 11 times, compared to 13.3 times to do the FFT on the entire record once, a significant time savings. It is also found that this method saves on the amount of memory required to achieve the zoom analysis.

We can now summarize the meaning of the above equations in terms of the process outlined in Fig. 5.6.6. First, we begin with an analog signal that passes through a standard anti-aliasing filter and is sampled at the standard sample rate f_s by the A/D converter. Second, we sample N window's worth of data for a total of N_r data points that are stored permanently in memory. Third, we selectively sample this N_r data record and take N_0 data points that are offset by m (where m varies from 0 to $N - 1$ as we repeat this partial sampling process) to obtain the mth data

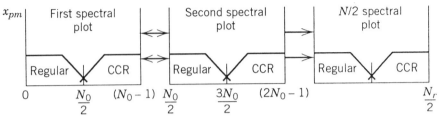

Fig. 5.6.5. The frequency components that repeat regular and complex conjugate parts, which are combined to give the $N_r/2$ frequency components calculated from Eq. (5.6.12).

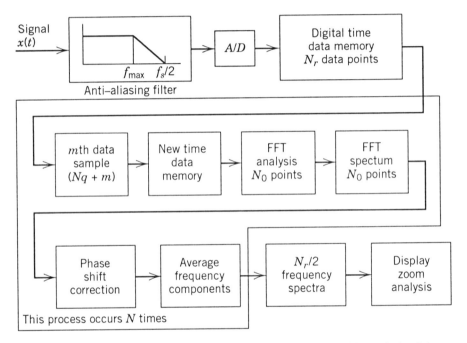

Fig. 5.6.6. Block diagram showing long time record zoom FFT analysis giving $N_r/2$ frequency components Δf_n Hz apart. The display is usually 400 or 800 lines.

set and temporarily store this data in the new time data memory. Fourth, we calculate the mth FFT and temporarily store this N_0 spectral information. Fifth, this spectral information is phase shifted and the $N_r/2$ potential frequency components are averaged according to Eq. (5.6.12). Sixth, this process of sampling the permanently stored time data, computing FFTs, and averaging the phase shifted frequency spectrum is repeated N times. As a practical matter, the value of N is usually 10 or less, a major limitation for this method.

5.6.3 Zoom Analysis with and Without Sample Tracking

We have mentioned the effects of slight frequency variations within the data window and how these variations can cause frequency spectral smearing while doing zoom analysis. We examine a typical rotating machine type of zoom analysis where slight rotational speed changes cause spectral smearing, see its effect on the resulting spectra, and then see how this smearing can be avoided.

A typical experimental setup is shown in Fig. 5.6.7, where a rotating machine is being monitored. The accelerometer measures the machine's vibration level at one point while a sensor (usually magnetic) is used to monitor shaft rotational speed.

Two basic methods are used to generate the required external sample rates. One method measures a single pulse every revolution, and then uses a tracking frequency multiplier to achieve a sufficiently fast sample rate. The other method uses a high quality gear that is attached to the shaft. The sensor pulses due to gear tooth passage can be used as an externally generated pulse train. A serious problem can arise when using this gear arrangement, due to the shaft's torsional vibrations. This vibration can cause additional modulation of the external sample rate that is not related to the average shaft speed. Hence, the more expensive tracking frequency multiplier is recommended unless the user knows that torsional vibrations are not taking place in the shaft where the gear is attached. An additional disadvantage of the gear method occurs if the sample frequency must be changed significantly, since the gear must be changed in this case.

The question of a sufficiently fast sample rate can present an interesting problem. Usually, we should do standard baseband measurements in order to determine the range of frequencies of interest. Then the minimum

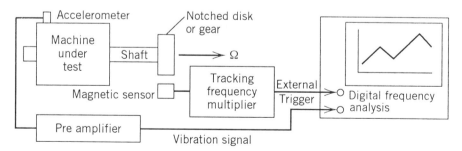

Fig. 5.6.7. Typical experimental setup for measuring vibration of rotating machines using magnetic sensor to generate external trigger signal to frequency analyzer.

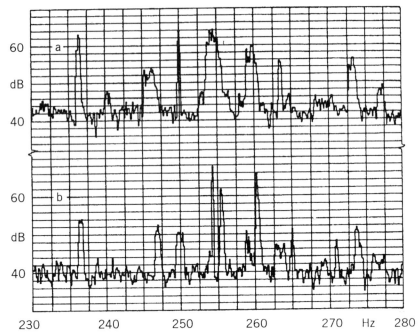

Fig. 5.6.8. Zoom frequency analysis of small electric motor showing spectral smearing. (*a*) Internal sample frequency showing smearing. (*b*) External sample frequency without smearing due to small changes in rotational speed with load. (Courtesy Bruel & Kjaer Instruments, Inc.).

external sample rate should be at least three times the maximum frequency of interest, since $f_s = 2.56 f_{max}$, so that the frequency of interest is included. Remember that the frequency analysis is in terms of this basic external rate so that order analysis results, namely, the frequencies are multiples of the shaft's rotational rate.

Two zoom frequency spectra from the frequency analysis of an accelerometer mounted on a small electric motor[8] are shown in Fig. 5.6.8, where the frequency range is 230 to 280 Hz. The top frequency spectrum was obtained using the analyzer's internal sample rate, while the lower frequency spectrum was obtained by using an externally generated sample rate. There are some interesting differences in the results. First, the 250 Hz frequency component corresponds to the fifth harmonic of the 50 Hz line frequency so that the top spectrum shows a sharp peak while the bottom

---

[8]N. Thrane, "Zoom-FFT," *Technical Review*, No. 2, Available from Bruel & Kjaer Instruments, 1980.

spectrum shows a smeared peak. Second, peaks around 235, 245, 260, and 273 Hz in the lower spectrum are smeared in the upper spectrum. Third, the peak around 263 Hz in the upper spectrum is nearly nonexistent in the lower spectrum. Finally, the upper spectrum peak around 253.5 Hz becomes two peaks at 253.6 Hz and 254.2 Hz in the lower spectrum. It is clear from examining these two zoom spectra that the interpretation of results can be quite misleading and that proper synchronization with slight speed variations can be a significant consideration in making zoom measurements on rotating machines. We shall explore additional measurement considerations and spectral smearing in the next section.

5.7 SCAN ANALYSIS, SCAN AVERAGING, AND MORE ON SPECTRAL SMEARING

Once a time history is safely stored in digital memory, there are additional ways this data can be analyzed. In this section we explore the implications of obtaining long time records that can be analyzed at our leisure by using scan analysis and scan averaging. We also explore spectral smearing due to nonstationary signals further. The scan analysis and scan averaging are independent of any zoom analysis that may be employed, but all results are affected by spectral smearing, a characteristic that may be present in the analysis of all real signals.

The long time record frequency analyzer has several important features and advantages. For example, we can do a scan analysis where we quickly look at the baseband frequency content over the length of the record. This analysis may show regions where a zoom analysis is called for. Then the long time record type of zoom analysis can be performed. Once this zoom analysis is done, we can easily display and explore any portion of the high resolution frequency spectra from 0 to f_{max}. There is no need to do any more analysis, since the entire zoom spectra are available in the high resolution spectral memory. Clearly, the data is recorded in real time while the analysis is not done in real time. Moreover, many more types of analysis can be performed on the same original time history data without rerunning the experiment. This can be a real advantage when the experiment is either a single transient or extremely expensive to operate.

For discussion purposes in this section, we assume that the analyzer has characteristics similar to the Bruel & Kjaer Model 2033 single channel frequency analyzer. The standard window size $N_0 = 1024$ data points, the analyzer displays 400 spectral lines, and the window expansion factor $N = 10$. It takes the analyzer about 110 ms to do one standard FFT analysis from start to completion of averaging.

5.7.1 Scan Analysis

The scan analysis process is shown in Fig. 5.7.1. This process starts with recording a long term data signal that contains $10N_0$ (10 lk windows) data points. This signal may be stationary or nonstationary periodic, a transient, a stationary random, or a nonstationary random signal. We can select any portion of this time record for analysis by applying either a rectangular or Hanning window function to the part selected, as shown. This sample portion contains N_0 data points on which the instrument performs a single FFT frequency analysis; it then displays a corresponding frequency spectrum, as indicated in Fig. 5.7.1. Under certain scan averaging conditions, the Hanning window function can be used for transient signals as well as periodic and random signals.

The particular portion selected for analysis can be obtained either manually or automatically. When in the automatic scanning mode, either the rectangular or the Hanning window is stepped along according to Table 5.7.1. *Shift factor r* (from Section 5.5) and *step size N_s* (number of data points by which the window function is shifted for each analysis) are related to one another by

$$N_s = N_0 r = 1024r \qquad (5.7.1)$$

where N_0 is the number of data points in a standard FFT window (1024 in this example). The number of spectral averages that fit within a scan

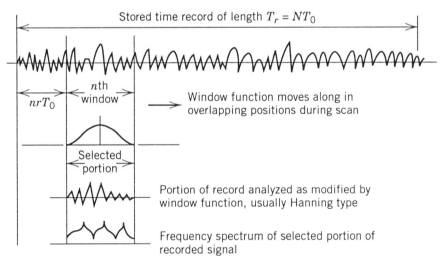

Fig. 5.7.1. Scan analysis procedure illustration showing standard window function being moved along and selecting different portions of signal for analysis.

TABLE 5.7.1. Scan Analysis and Hanning Window Scan Averaging Parameters for Use with Bruel & Kjaer Model 2033 Analyzer

	Scan Analysis			Scan Average	
Shift Factor r	Step Size (Data Points)	Number of Spectra per Scan	Scan Time (seconds)	BT_e Product	Effective Record Length (windows)
1	1024	10	1	10	—
$\frac{1}{2}$	512	19	2	18.5	—
$\frac{1}{4}^a$	256	37	4	20	$9\frac{1}{4}$
$\frac{1}{8}$	128	73	8	20	$9\frac{1}{8}$
$\frac{1}{16}$	64	145	15	20	$9\frac{1}{16}$
$\frac{1}{32}$	32	289	30	20	$9\frac{1}{32}$
$\frac{1}{64}$	16	577	60	20	$9\frac{1}{64}$
$\frac{1}{128}$	8	1153	120	20	$9\frac{1}{128}$

Source: This table adapted from N. Thrane, "Zoom-FFT," *Technical Review*, Available from Bruel & Kjaer Instruments, No. 2, 1980.
^{a}There is no advantage in using shift factors less than $\frac{1}{4}$.

of a record of length T_r for a given shift factor is obtained from the fact that $T_r = (N_a - 1)rT_0 = NT_0$, so that

$$N_a = \frac{N-1}{r} + 1 \qquad (5.7.2)$$

where N_a is the number of spectral averages per scan.
 N is the number of windows in record time T_r.

Equations (5.7.1) and (5.7.2) can be used to generate the data in the first three columns in Table 5.7.1. The first column shows the shift factor that goes from 1 to a value of $\frac{1}{128}$ by factors of two. Then, the second column is calculated from Eq. (5.7.1), while the third column gives the number of spectra per scan across the time data, as calculated from Eq. (5.7.2). The fourth column shows the amount of time required to complete a scan with a given step size. The scan time is the number of spectra per scan times the approximately 0.11 seconds of processing time for one FFT analysis, and is seen to range from approximately 1 second to 120 seconds.

The automatic scan is particularly useful in analysis of transient signals, such as a rapid machine run up test. As the analysis window proceeds through the data, rapid changes in frequency spectra become readily visible that might otherwise be lost in any averaging scheme. An additional benefit from a scan is that the *scan average* is automatically generated during the scan analysis. The scan average is a linear average of all

326   THE DIGITAL FREQUENCY ANALYZER

frequency spectra generated, is useful for transient, periodic, and random signals, and is helpful in defining regions for zoom analysis of the same time data. This zoom analysis would consist of 4000 spectral lines for the specifications noted above, namely $N = 10$, and the basic window has 400 spectral lines. Now we need to look more carefully at what a scan average means.

5.7.2 Scan Averaging and Resulting Frequency Spectra

In Section 5.5, we found that we can analyze random signals in a satisfactory manner when the Hanning window is overlapped with a shift factor $r \leq \frac{1}{3}$. Figure 5.7.2 is a plot of the Hanning window function squared for shift factors of 1, $\frac{1}{2}$, and $\frac{1}{4}$. We see from Table 5.7.1 that the shift factor decreases by factors of two, so that $r = \frac{1}{4}$ is the first shift factor that meets the flat window weighting requirement, namely, that all data be given equal weight over the flat central portion of the time record, as shown in Fig. 5.7.2. The square of the Hanning window function varies from 0 to 4 in each curve shown.

In Section 5.5 we found that ASD uncertainty requires the BT_e product to be greater than 70. For the case of 25 percent shift factor (75 percent overlap), the effective time constant per analysis from Table 5.5.2 is 0.52, so that analysis of 37 records gives a BT_e value of about 20, a value that is on the border of being marginal for acceptable random signal analysis. A BT_e of 20 translates into an uncertainty of ±15 percent for 95 percent confidence limits. This value of BT_e remains the same for all other shift factor values, as shown in Table 5.7.1. Consequently, we are able to

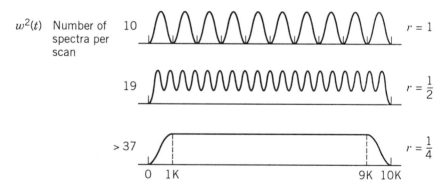

Fig. 5.7.2. Plot of square of Hanning window function for $r = 1, \frac{1}{2}$, and $\frac{1}{4}$, giving spectra per scan of 10, 19, and 37. Note the minimum magnitude is 0 while the maximum is 4. (Courtesy Bruel & Kjaer Instruments, Inc.)

obtain good estimates on periodic signals but only borderline acceptable estimates on random signals using this technique. Thus the periodic signals can be evaluated for further zoom analysis if required.

It is also evident from Table 5.7.1 that there is little to be gained by using shift values less than $\frac{1}{4}$, which requires approximately 4 seconds for analysis, a time that is independent of bandwidth employed. This analysis time is controlled by the number of spectra per scan and the FFT processing time, two factors that have nothing to do with the time required to acquire the data. The BT_e value remains constant and we see that the effective record length changes only slightly with significantly more averages and analysis time.

A significant use of scan average with this type of analyzer is the analysis of transient signals that fit within the long input signal memory. We have learned previously that transient analysis requires a rectangular type of window function. We see in Fig. 5.7.2 that the use of a Hanning window function and the 37 window scan average gives us a nearly rectangular window function over the central 80 percent portion of the window function. This window function is flat with a $\{1 - \cos(\omega t)\}^2$ taper on each end. The end taper reduces the effective analysis time T_{eff} from $10T_0$ to around $9.25T_0$ for the 37 window scan average. This effective time is obtained by finding the area under the analysis curve and equating that area to T_{eff} times central portion window height.

The rules for doing a transient analysis in this manner are few. First, the transient signal must occupy the central portion of the effective window function shown in Fig. 5.7.2 for a 37 window scan average. Second, there must be at least 28 window scans in the average so that shift factor $r \leq \frac{1}{3}$. The 37 scan average ($r = \frac{1}{4}$) satisfies this requirement. Third, frequency spectra scaling is correctly evaluated since subtle scaling changes often occur as we change from one type of analysis to another. These scaling issues depend on the particular instrument employed, for the instrument designer may or may not take scaling into account.

5.7.3 Scan Average Analysis of a Transient Signal

Let us see how the scan analysis works on a nearly ideal transient signal, as shown in Fig. 5.7.3a.[9] This signal consists of a "chirp" type sound that has 3 cycles of a 980 Hz sinusoid that is nearly centered in the 2 second time window. The analysis baseband is 0 to 2 kHz so that $\Delta f = 5$ Hz and a single data window period of $T_0 = 0.2$ second.

---

[9]This example is directly from N. Thrane, "Zoom-FFT," *Technical Review*, No. 2, Available from Bruel & Kjaer Instruments, 1980.

328 THE DIGITAL FREQUENCY ANALYZER

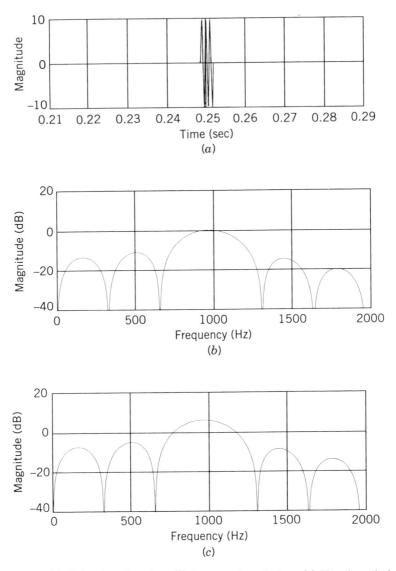

Fig. 5.7.3. (*a*) Chirp time function. (*b*) Rectangular window. (*c*) Hanning window. (Courtesy of Bruel & Kjaer Instruments, Inc.)

We can think of the chirp signal as being composed of a 980 Hz sine wave over the entire window multiplied by a rectangular weighting function that has a unity value over a 3.06 ms time period out of the entire 2 second window period T_r. Now we found in Chapter 2 that the frequency

SCAN ANALYSIS, SCAN AVERAGING, AND SPECTRAL SMEARING 329

components of a unit magnitude rectangular pulse with duration of T and with a repeat period (which is the same as the window period) T_r so that $T_r = \beta T$, are given by Eq. (2.4.3) as

$$X_p = \frac{1}{\beta} \frac{\sin(p\pi/\beta)}{(p\pi/\beta)} = \frac{1}{\beta} \text{sinc}\left(\frac{\omega T}{2}\right) \qquad (5.7.3)$$

In this example, $T = 0.00306$ second so that $\beta = 654$. The sinc function in Eq. (5.7.3) will be zero whenever the argument is a multiple of π, an event that occurs about every 327 Hz. The convolution theorem suggests that the resulting frequency analysis of this signal (composed of the product of a 980 Hz sinusoid and a rectangular weighting function) should be the sinc function centered at 980 Hz. There should be zeros at 327 Hz intervals away from the 980 Hz center frequency, such as 0, 327, 653, 1307, 1634, and 1961 Hz. The baseband scan average analysis in Figs. 5.7.3b and 5.7.3c support this observation.

Now, let us look more closely at the resulting scan average analysis in Figs. 5.7.3b and 5.7.3c. First, the signal is nearly centered in the time window as required. Second, the scan average uses 37 spectra in the scan average as required in Table 5.7.1. The rectangular window function is used to obtain the ASD spectra in Fig. 5.7.3b, while the Hanning window function is used to obtain the ASD spectra in Fig. 5.7.3c. Since the rectangular window function has a uniform window weighting of unity over the entire window, the analyzer is adjusted so the reading is 0 dB. Then, when the Hanning window scan average analysis is used, the reading at 980 Hz is seen to be 6 dB higher. This 6 dB comes about from the squared window function shown in Fig. 5.7.2, which has a value of 4 over the central portion, compared to unity for the rectangular window. Then, $10 \log(4) = 6$ dB accounts for the increased reading value. It is clear that we should carefully read the analyzer's manual in order to understand if any corrections are required.

5.7.4 More on Spectral Smearing

The following real life experimental result illustrates the type of factors that enter into spectral smearing, factors that may be hidden at first glance. A spring in a particular application was experiencing a high failure rate. The spring in Fig. 5.7.4a was exposed to a displacement time history similar to that in Fig. 5.7.4b, where the spring was compressed rapidly, held for nearly $T_0/4$ seconds, and then quickly released. This loading cycle is repeated every T_0 seconds. The spring's natural frequency is about 250 Hz as installed with $\delta(t) = 0$. A strain gage was attached near the

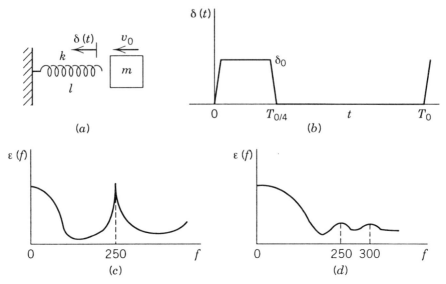

Fig. 5.7.4. A real system that exhibited smeared frequency spectrum, no resonance peak, and floating window function. (*a*) Spring. (*b*) Ideal input motion. (*c*) Anticipated strain frequency spectrum. (*d*) Actual measured strain frequency spectrum.

spring's midpoint (close to $0.6l$) and a strain frequency spectrum similar to Fig. 5.7.4c was expected. The strain frequency spectrum turned out to be more like that in Fig. 5.7.4d. In place of a sharp resonance peak near 250 Hz, two small lumps were present, one close to 250 Hz and the other close to 300 Hz. None of these measurements would explain the spring's high failure rates. So what is wrong with the measurements, and maybe, with our understanding of these results?

First, it was found that in one operating mode, mass m in Fig. 5.7.4a had a velocity in the order of 1.8 meters/second and severely impacted the spring's end, contrary to the normal cam motion as described by $\delta(t)$. This impact loading certainly excited the spring into axial resonance. Second, the strain gage location is close to a strain node point for the first mode of vibration, so that any resonant readings that occurred were severely attenuated. This strain gage location is adequate for measuring the nearly static strain induced by displacement $\delta(t)$, consistent with the measured frequency spectrum. Third, further analysis and experiment showed that when the spring was fully compressed by an amount of δ_0, its first natural frequency changed from around 250 Hz to around 300 Hz. Hence, it is concluded that we had a situation where the signal's frequency is close to 300 Hz for about one-quarter of repeat period T_0 and is close to 250 Hz for three-quarters of repeat period T_0.

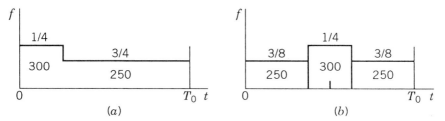

Fig. 5.7.5. Definition of signal frequency distribution in time domain. (*a*) Case 1. (*b*) Case 2.

In order to understand what may be happening in the frequency analysis that would contribute to smearing out any signs of a resonant peak and give us two relatively flat lumps, two theoretical signal cases were simulated. In case 1, a signal is generated where the first quarter period is at 300 Hz and the remaining three quarter periods are at 250 Hz, as shown in Fig. 5.7.5a. In case 2, the same signal is time shifted so that the 300 Hz portion is at the midpart of the data window, as shown in Fig. 5.7.5b. Now we analyze both signals, using first a rectangular window function and then a Hanning window function.

The rectangular window function analysis results are shown in Fig. 5.7.6a for both the original and time shifted signals. It is not surprising that the analysis results are the same, for the signal data remains unchanged when time shifted; that is, we are analyzing the same data so that the frequency component magnitudes remain the same.

The Hanning window function analysis results are shown in Fig. 5.7.6b, where there is a dramatic difference between frequency spectra, depending on where the 300 Hz frequency signal is relative to the center of the Hanning window function. In case 1, the 300 Hz components are nearly absent since the Hanning window function essentially removes the 300 Hz signal as it occurs at the beginning of the window; see Fig. 5.7.5a. The 250 Hz component has a large amplitude (close to 9), along with some apparent filter leakage.

In case 2, the 300 Hz frequency portion is in the center of the Hanning window function so that this frequency becomes more evident in the frequency spectrum and the 250 Hz component is significantly reduced. It should not be too hard to envision a situation where the signal being analyzed continually shifts within the data window as many averages are being taken so that the 300 Hz portion of the signal moves about relative to window's center. If the 300 Hz frequency portion of the signal had a somewhat higher value, then the analysis could generate nearly equal lumps, one centered around 250 Hz and the other around 300 Hz. Consequently, there would be no definite resonance peak to measure in this

332 THE DIGITAL FREQUENCY ANALYZER

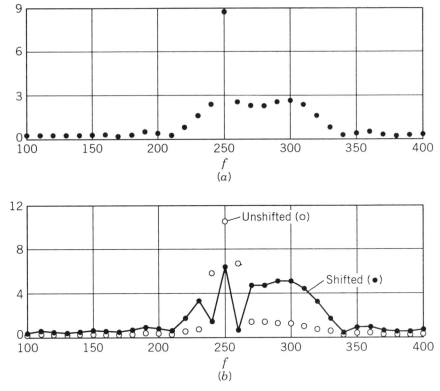

Fig. 5.7.6. Frequency analysis of a signal composed of a 300 Hz sine wave for one-fourth of the analysis window and a 250 Hz sine wave for three-fourths of the analysis window with the 300 Hz portion first (unshifted case) and in the center portion (shifted case). (*a*) Rectangular window function. (*b*) Hanning window function.

case, only two bumps that are near the two dominant frequencies that come and go in time. Also, in the real case, the frequencies did not change instantly as in this theoretical mode, but they changed in a rapid and continuous manner.

In summary, a number of factors contributed to the unexpected strain frequency spectrum. First, the strain gage location was a poor choice for measuring the first natural frequency resonant response, since it was mounted too close to the strain node point. Consequently, the resonance signal was small and nearly hidden in signal noise. Second, the spring's geometry and mass distribution changed sufficiently to cause the natural frequency to change nearly 20 percent, from approximately 250 Hz to approximately 300 Hz. Third, the frequency analysis used a Hanning window function so that the 300 Hz short time sinusoid portion of the record floated in and out of the center of the window function since no synchronization was attempted in the experiment.

Once these factors were understood, a better test could be conducted by using a better strain gage location, and by using a frequency analysis based on the rectangular window function so that the position of a given frequency component has little or no effect on the resulting frequency analysis. More window leakage is the price we have to pay in order to eliminate the floating signal in the window function phenomenon.

5.8 THE DUAL CHANNEL ANALYZER

The single channel frequency analysis discussion was directed at understanding the fundamental processes and issues involved in interpreting vibration frequency spectra. On the other hand, we need to measure a structure's frequency response function (FRF). The frequency response function relates the structure's response to a given excitation and is the fundamental relationship for describing a linear system. The simplest FRF requires that we use a dual channel frequency analyzer, one channel to measure the input or excitation and one channel to measure the output or response. It is obvious that we can have multiple inputs and multiple outputs, but this multichannel situation is a duplication of dual channel FRF processing. We concentrate on the dual channel instrument as a basis of understanding the basic processes involved. Bendat and Piersol[10] discuss multichannel processing in considerable detail.

Recall from Section 3.4 that linear system input and output are related as shown in Fig. 5.8.1. In this case, the input time history is $x(t)$, the characteristic impulse response function is $h(t)$, and the output time history is $y(t)$. These three quantities are related in the time domain by the convolution (also called Duhamel or Faltung) integral, which is given by

$$y(t) = \int_{-\infty}^{t} x(\tau) h(t - \tau) \, d\tau \qquad (5.8.1)$$

Then, if we use $\mathscr{F}[x(t)]$ to represent the Fourier transform (or Fourier

---

[10]J. S. Bendat and A. G. Piersol, *Random Data: Analysis and Measurement Procedures*, 2nd ed., John Wiley & Sons, New York, 1986.

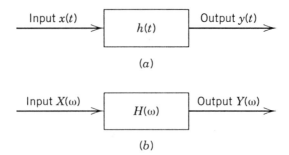

Fig. 5.8.1. Linear system input-output models. (*a*) Convolution time domain. (*b*) Multiplication in frequency domain.

series depending on signal type) of $x(t)$, we can write

$$X(\omega) = \mathscr{F}[x(t)]$$
$$H(\omega) = \mathscr{F}[h(t)] \qquad (5.8.2)$$
$$Y(\omega) = \mathscr{F}[y(t)]$$

The inverse Fourier transform of $X(\omega)$ can be written as $\mathscr{F}^{-1}[X(\omega)]$ so that Eqs. (5.8.2) become

$$x(t) = \mathscr{F}^{-1}[X(\omega)]$$
$$h(t) = \mathscr{F}^{-1}[H(\omega)] \qquad (5.8.3)$$
$$y(t) = \mathscr{F}^{-1}[Y(\omega)]$$

One of the main concepts that came out of Section 3.4 is the frequency domain input output FRF relationship, which is given by

$$Y(\omega) = H(\omega)X(\omega) \qquad (5.8.4)$$

Equation (5.8.4) is the basis for obtaining a structure's FRF, implying that we can estimate the FRF by simply measuring the input and output frequency spectra.

Equations (5.8.1) through (5.8.4) are limited by four basic assumptions. First, the system is linear so that doubling or halving the input doubles or halves the output at any given frequency. Second, the system is stable so that

$$\int_{-\infty}^{+\infty} |h(\tau)| \, d\tau < \infty$$

that is, the integral is bounded. Third, the system is physically possible so that it does not respond before the excitation is applied, namely, $h(\tau) = 0$ for $\tau < 0$. Fourth, the system is time invariant so that $h(\tau)$ and $H(\omega)$ do not change with time, that is, the system's mass, damping, and stiffness remain constant.

In this section, we look at how we can implement Eq. (5.8.4) in a practical way as well as have confidence in the quality of the measured frequency spectra and corresponding estimated FRF by utilizing the coherence function. We consider both ideal and actual estimates used in signal processing in this section. Sections 5.9 will discuss the effects on the results of noise at the input, output, and both input and output signals.

5.8.1 Ideal Input-Output Relationships

The FRF in Eq. (5.8.4) can be estimated directly for sinusoidal and transient signals, but not for random signals. Thus we need a calculation methodology that is appropriate for all types of signals. It turns out that three different methods can be employed. We develop each method in order.

FRF Estimates The first estimate of $H(\omega)$ is called $H_1(\omega)$ and is obtained by multiplying Eq. (5.8.4) by $X^*(\omega)$ to obtain[11]

$$X^*(\omega)Y(\omega) = H(\omega)X^*(\omega)X(\omega) \qquad (5.8.5)$$

[11]Before proceeding, we need to be clear about the meaning of $X(\omega)$ and $Y(\omega)$ in Eq. (5.8.5). Recall that we examined in Section 5.3.2 how frequency spectra are calculated in a frequency analyzer. For example, $X(\omega)$ can represent the Fourier series frequency spectrum that corresponds to a periodic function. In that case, $X(\omega)$ has input variable's units such as g's of acceleration or force in pounds or newtons.

If the input is a transient type, then $X(\omega)$ is the Fourier transform type of frequency spectrum with terms that are calculated by multiplying the analyzer's discrete frequency components by the window period T so that it has units either of g's/Hz or newtons/Hz. When we multiply $X(\omega)$ times its complex conjugate, we obtain an autospectrum which can have units of $(g's/Hz)^2$ or $(newtons/Hz)^2$. This set of units is often expressed in terms of $(g's$ or $Newtons)^2$-s/Hz and is called *energy density*. This means that energy density is the Fourier transform frequency spectra squared.

Finally, if we divide the energy density by the window period T or multiply it by the line spacing Δf, we obtain the random ASD auto-spectrum or cross-spectrum, depending on which side of Eq.

(5.8.5) we are working with. Hence, the important point is that both sides of Eq. (5.8.5) must be treated the same. If they are, then we can use this equation as our basis of calculation and must bear in mind that we can have auto- and cross-spectral densities that are interpreted in terms of the signal types being measured. This means that both analyzer channels must process the quantities in an identical manner. Hopefully the system designer made sure that we the user cannot use the energy density definition for one channel and the ASD definition for the other channel. If that happens, a constant error is introduced into our FRF measurement.

Assuming that we respect the condition to process both channels of data in the same manner, then Eq. (5.8.5) becomes

$$S_{xy}(\omega) = H(\omega) S_{xx}(\omega) \tag{5.8.6}$$

where $S_{xy}(\omega)$ is the dual sided ($\pm\omega$'s) cross spectrum between input and output.

$S_{xx}(\omega)$ is the dual sided auto-spectrum of the input signal.

Generally, only the single sided frequency spectra ($+\omega$'s) of $G_{xy}(\omega)$ and $G_{xx}(\omega)$ are used. Thus we obtain from Eq. (5.8.6) that

or

$$H(\omega) = \frac{S_{xy}(\omega)}{S_{xx}(\omega)} \equiv H_1(\omega) \quad \text{for } -\infty < \omega < +\infty$$

$$H(\omega) = \frac{G_{xy}(\omega)}{G_{xx}(\omega)} \equiv H_1(\omega) \quad \text{for } 0 < \omega < +\infty \tag{5.8.7}$$

The second estimate of $H(\omega)$ is called $H_2(\omega)$ and is obtained by multiplying each side of Eq. (5.8.4) by $Y^*(\omega)$ to obtain

$$Y^*(\omega) Y(\omega) = H(\omega) Y^*(\omega) X(\omega)$$
$$S_{yy}(\omega) = H(\omega) S_{yx}(\omega) \tag{5.8.8}$$

Equation (5.8.7) shows that $H(\omega)$ can be estimated from the dual sided auto-spectrum $S_{yy}(\omega)$ (or single sided auto-spectrum $G_{yy}(\omega)$) and the dual sided cross-spectrum $S_{yx}(\omega)$ (or single sided cross-spectrum $G_{yx}(\omega)$). This calculation method gives

$$H(\omega) = \frac{S_{yy}(\omega)}{S_{yx}(\omega)} \equiv H_2(\omega) \quad \text{for } -\infty < \omega < +\infty$$

or (5.8.9)

$$H(\omega) = \frac{G_{yy}(\omega)}{G_{yx}(\omega)} \equiv H_2(\omega) \quad \text{for } 0 < \omega < +\infty$$

The values obtained by $H_1(\omega)$ and $H_2(\omega)$ are often different, depending on the measurement situation, particularly signal noise as discussed in Section 5.9. Theoretically, the phases in both estimates of $H(\omega)$ are the same since $G_{yx}(\omega)$ and $G_{xy}(\omega)$ are complex conjugates of one another, that is, $G_{yx}(\omega) = G^*_{xy}(\omega)$ or $G_{xy}(\omega) = G^*_{yx}(\omega)$.

A third calculation method is called $H_a(\omega)$ and is obtained by using only the magnitudes of both sides of Eq. (5.8.4) to obtain

$$Y^*(\omega)Y(\omega) = H^*(\omega)H(\omega)X^*(\omega)X(\omega)$$

or

$$S_{yy}(\omega) = |H(\omega)|^2 S_{xx}(\omega) \tag{5.8.10}$$

Equation (5.8.10) involves only the auto-spectra $S_{xx}(\omega)$ and $S_{yy}(\omega)$ so that there is no phase information whatsoever in this estimate of $H(\omega)$; only magnitudes are determined. Thus from Eq. (5.8.10), we obtain

$$|H(\omega)|^2 = \frac{S_{yy}(\omega)}{S_{xx}(\omega)} \equiv |H_a(\omega)|^2 \quad \text{for } -\infty < \omega < +\infty$$

or (5.8.11)

$$|H(\omega)|^2 = \frac{G_{yy}(\omega)}{G_{xx}(\omega)} \equiv |H_a(\omega)|^2 \quad \text{for } 0 < \omega < +\infty$$

as the estimate for the FRF's magnitude.

Coherence We need to relate how well the output of the structure under test is related to the input to the structure. We turn to the concepts developed in Section 2.5 on correlation coefficient for guidance. Recall that the normalized correlation coefficient ρ_{xy} is given by Eq. (2.5.7) as

$$\rho_{xy} = \frac{\sigma_{xy}}{\sigma_x \sigma_y} \tag{5.8.12}$$

where σ_{xy} is the cross-correlation coefficient.
σ_x is the input signal's variance.
σ_y is the output signal's variance.
The correlation coefficient describes how well the value of y is related to

the value of x. When ρ_{xy} is ± 1, the correlation is unity. In the case of the above equations, we see that we are dealing with mean square and cross-spectral quantities. Thus we use the *coherence function* $\gamma^2(\omega)$ at each frequency ω as a measure of how well the output is linearly related to the input. The coherence function corresponds to the square of the correlation coefficient and is given by

$$\gamma^2(\omega) = \frac{|G_{xy}(\omega)|^2}{G_{xx}(\omega)G_{yy}(\omega)} = \frac{|S_{xy}(\omega)|^2}{S_{xx}(\omega)S_{yy}(\omega)} \qquad (5.8.13)$$

This coherence function is a measure of the linearity of the output to the input at every frequency ω and is rated on a 0 to 1 scale. Note that the coherence function is based on statistical averages in the quantities $G_{xy}(\omega)$, $G_{xx}(\omega)$, and $G_{yy}(\omega)$. If we apply Eq. (5.8.13) to a single measurement, the coherence is unity, even in the presence of noise, since there is no information available to indicate that the output is not due to the input. It is only after several measurements are averaged that the coherence concept can detect the lack of relationship.

Inserting Eqs. (5.8.6) and (5.8.8) into Eq. (5.8.13) shows that coherence can be expressed by

$$\gamma^2(\omega) = H_1(\omega)/H_2(\omega) \qquad (5.8.14)$$

since $G_{xy}(\omega)$ and $G_{yx}(\omega)$ are complex conjugates. The implication of Eq. (5.8.14) is that $H_1(\omega) \leq H_2(\omega)$ since coherence is always ≤ 1. Thus we would anticipate that $H_1(\omega)$ tends to underestimate the actual FRF and that $H_2(\omega)$ tends to overestimate the actual FRF when coherence is less than unity. We explore the effect of uncorrelated noise on estimates $H_1(\omega)$ and $H_2(\omega)$ in Section 5.9.

5.8.2 Actual Input-Output Estimates for a Digital Analyzer

The above equations outline the kinds of calculation processes that we would like to do at each frequency ω. Unfortunately, all we have available for our calculations are the auto- and cross-spectra as estimated by the dual channel frequency analysis that is done at discrete frequencies of $\omega_p = p(2\pi \Delta f)$ where Δf is the analyzer's frequency line spacing. In the following equations, a hat ($\hat{}$); is placed over each measured estimate of a given quantity in order to distinguish the measured from the theoretical quantities.

The auto spectra $G_{xx}(\omega)$ and $G_{yy}(\omega)$ are estimated from the mathematical expectation[12] of the individually measured frequency spectra so that

$$G_{xx}(\omega) = E[\hat{X}^*(\omega)\hat{X}(\omega)] = \lim_{n_d \to \infty} \frac{1}{n_d} \sum_{p=1}^{n_d} \hat{X}_p^*(\omega)\hat{X}_p(\omega) \quad (5.8.15)$$

and

$$G_{yy}(\omega) = E[\hat{Y}^*(\omega)\hat{Y}(\omega)] = \lim_{n_d \to \infty} \frac{1}{n_d} \sum_{p=1}^{n_d} \hat{Y}_p^*(\omega)\hat{Y}_p(\omega) \quad (5.8.16)$$

for each frequency ω_p where n_d is the number of window data blocks analyzed. Similarly, the cross-spectral densities are estimated from

$$G_{xy}(\omega) = E[\hat{X}^*(\omega)\hat{Y}(\omega)] = \lim_{n_d \to \infty} \frac{1}{n_d} \sum_{p=1}^{n_d} \hat{X}_p^*(\omega)\hat{Y}_p(\omega) \quad (5.8.17)$$

and

$$G_{yx}(\omega) = E[\hat{Y}^*(\omega)\hat{X}(\omega)] = \lim_{n_d \to \infty} \frac{1}{n_d} \sum_{p=1}^{n_d} \hat{Y}_p^*(\omega)\hat{X}_p(\omega) \quad (5.8.18)$$

We know that we cannot take an infinite number of window data blocks. Thus we need to try and grasp the type of averaging that is implied by Eqs. (5.8.15) through (5.8.18). If we look at Eqs. (5.8.15) and (5.8.16), we see that the product of either $\hat{X}_p^*(\omega)\hat{X}_p(\omega)$ or $\hat{Y}_p^*(\omega)\hat{Y}_p(\omega)$ is always real and positive. This means that the different components add up only along the real axis of a real and imaginary axis plot, as shown in Fig. 5.8.2. In this case, the average value given by Eqs. (5.8.15) and (5.8.16) is simply the real axis length divided by the number of vectors that are added, n_d. In this case, we see that the addition of a new vector has only a small effect on the resulting average value and uncertainty of the estimation drops rather quickly with the number of averages, n_d.

---

[12]See J. S. Bendat and A. G. Piersal, *Random Data: Analysis and Measurement Procedures*, 2nd ed., John Wiley & Sons, New York, 1986.

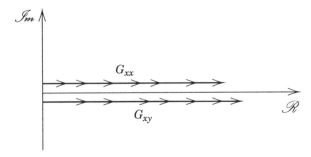

Fig. 5.8.2. Addition of auto-spectra vectors at a given frequency during averaging process showing only real values. (Courtesy of Bruel & Kjaer Instruments, Inc.)

5.8.3 Auto-Spectra and Cross-Spectra Averaging

The cross-spectra averaging offers more interesting uncertainty possibilities. Let us rewrite Eq. (5.8.17) as

$$G_{xy}(\omega) \cong \frac{1}{n_d} \sum_{p=1}^{n_d} \hat{X}_p^*(\omega)\hat{Y}_p(\omega) = \frac{1}{n_d} \sum_{p=1}^{n_d} |\hat{G}_{xy}(\omega)|_p \, e^{j\Delta\phi(\omega)_p} \quad (5.8.19)$$

where the relative phase shift is given by

$$\Delta\phi(\omega)_p = [\phi_y(\omega) - \phi_x(\omega)]_p \quad (5.8.20)$$

and the cross-spectral magnitude is given by

$$|\hat{G}_{xy}(\omega)|_p = |\hat{X}_p^*(\omega)||\hat{Y}_p(\omega)| \quad (5.8.21)$$

Equation (5.8.19) shows us that each cross-spectrum sample has magnitude given by Eq. (5.8.21) and a phase angle given by Eq. (5.8.20), and that the estimate is the average of the sum of these cross-spectral estimate vectors. These cross-spectral vectors can be plotted on real and imaginary axes, as shown in Fig. 5.8.3. In the ideal case, the phase angle $\Delta\phi$ is always the same and the magnitudes of $\hat{G}_{xy}$ are the same for each data window so we obtain the picture shown in Fig. 5.8.3a, a very neat and tidy result. When the phase and magnitudes have some uncertainty, then we obtain a plot similar to Fig. 5.8.3b. In this case, the average is obtained by taking the magnitude and phase corresponding to point A, which results when n_d quantities are added vectorially, giving a solid straight line as an estimate. Dividing the length by n_d gives the average magnitude while the

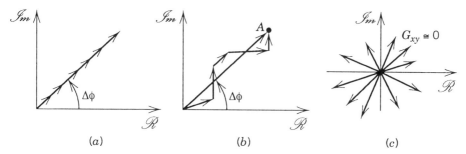

Fig. 5.8.3. Diagram showing how cross-correlation vectors add in magnitude and phase for various signal pairs. (*a*) Ideal addition (no uncertainty). (*b*) Common situation (modest uncertainty). (*c*) Near zero correlation. (Courtesy of Bruel & Kjaer Instruments, Inc.)

angle $\Delta\phi$ is obtained from the real and imaginary coordinates that apply to point A.

When the phase becomes uncorrelated, we see that the magnitudes plot in nearly every direction with the sum adding up to near zero magnitude, as shown in Fig. 5.8.3c. In this case the output is uncorrelated to the input and the coherence goes toward zero. Recall that the auto-spectra always add to positive values, while the magnitude of the cross-spectra can go to near zero, as shown. Thus for the case in Fig. 5.8.3c, there is high uncertainty in the corresponding FRF's validity.

5.8.4 Some Reasons Coherence Is Less Than Unity

There are four common reasons that cause the coherence to be less than unity. These reasons are: (1) non linear structural response, (2) digital filter leakage (called resolution bias error), (3) time delays between signals, and (4) uncorrelated noise in the measurements of $x(t)$ and $y(t)$. We take a look at the implications of the first three reasons in this section and delay consideration of the fourth reason to Section 5.9.

Nonlinear Structures Nonlinear structures cause low coherence in two ways. First, the phase angle between input and output is dependent on the amplitude of response, so that in random and transient measurements, this phase angle changes from data window to data window. Consequently, we have a situation where the phase angle changes from one cross-correlation vector to another. This type of phase angle uncertainty is shown in Fig. 5.8.3b.

Second, when nonlinear systems are excited at one frequency, they usually generate responses at frequencies that are multiples of the apparent structural resonance frequency (though sometimes they generate subharmonic frequencies as well). This causes these higher (multiple) frequencies to have response spectra that do not correlate with the input signal's spectra. Consequently, the coherence will be less than unity at these higher frequencies.

Resolution Bias Error The resolution bias error has been discussed in detail in Section 5.4.4 for random signals. In Fig. 5.4.7, it was seen that the area under the FRF curve that is contained within the filter bandwidth is interpreted in terms of the filter bandwidth. Consequently, the FRF's peak value is underestimated because the digital filter is too wide for the resonant peak under measurement. We can explain this leakage by another argument in the time domain.

Suppose, for example, that the excited vibration does not decay to zero within the data window. Then there is filter leakage in the frequency analysis due to chopping off the response signal due to data window length T. If the frequency analyzer's effective baseband is decreased so that there is more time to capture the response in the data window, then there is less measurement bias and the coherence improves. The data window length is increased by either lowering the baseband frequency or using zoom analysis.

This increase of effective analysis time is shown in Fig. 5.8.4 for a random excitation experiment. In Fig. 5.8.4a, the baseband is 12.8 kHz ($\Delta f = 16$ Hz) and the 1040 Hz resonance shows a reading of 17 dB while the coherence is 0.759. In Fig. 5.8.4b, the analysis has been changed to an 800 Hz bandwidth analysis ($\Delta f = 1$ Hz) through use of the zooming process. In this case, the peak is seen to occur at 1033 Hz and to have a peak value of 19.1 dB while the coherence has increased to 0.999. Recall that zooming increases the time record analyzed so that the response has more time to decay, and thus it reduces the filter leakage over the much longer data window.

We should point out that *coherence cannot detect leakage if the time functions are deterministic* and repeat themselves in each measurement window. In this case, the input and output frequency spectra, as well as the cross-spectra, are the same for each sample except for the uncorrelated noise in the measurements. This statement presumes that there is no uncorrelated noise in the measurement signal.

Time Delays in Signals There are vibration measurements where time delays occur in either the system under test or in the measurement system employed. There are instances where a linear phase lag with frequency

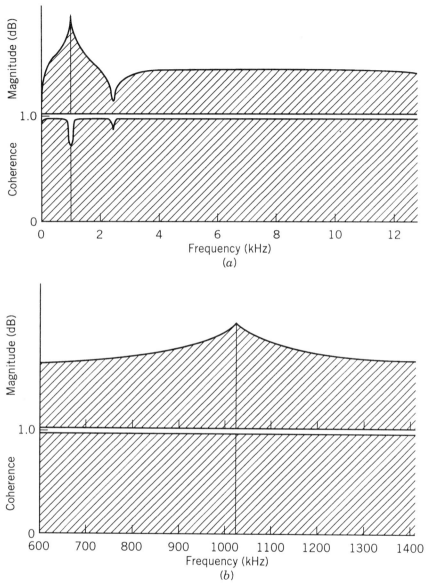

Fig. 5.8.4. Effect of record length per data window on FRF magnitude and corresponding coherence. (*a*) Baseband 0–12.8 kHz ($\Delta f = 16$ Hz). (*b*) Zoom 608 Hz to 1408 ($\Delta f = 1.0$ Hz) when analyzing a random excitation of a liner system, $n = 100$ samples. (Courtesy Bruel & Kjaer Instruments, Inc.)

that occurred in the measurement amplifiers[13] had nothing to do with the structure under test. For such situations, the coherence is estimated to be biased by

$$\gamma^2(\omega) \cong (1 - \tau/T)^2 \qquad (5.8.22)$$

where τ is the amount of time shift and T is the window length. Thus it is necessary to have time shift capability in a dual channel analyzer so that this phenomenon can be accounted for. In the case cited here, a time shift of 80 μs corrected the situation.

5.8.5 Operating Block Diagram

In order to measure, frequency analyze the signals, and calculate the required quantities, the dual channel analyzer is usually organized as shown in Fig. 5.8.5. The two input channels each contain matched anti-aliasing filters and synchronized A/D converters that are used to record the time histories. These time histories are multiplied by appropriate window functions and the corresponding FFT calculations are performed

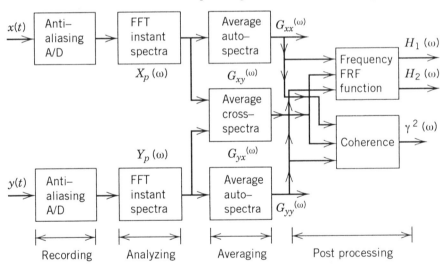

Fig. 5.8.5. Block diagram of dual channel frequency analyzer set up to generate FRFs.

---

[13]K. G. McConnell and M. K. Abdelhamid, "On the Dynamic Calibration of Measurement Systems for Use in Modal Analysis," *International Journal of Analytical and Experimental Modal Analysis*, Vol. 2, No. 3; 1987, pp. 121–127.

to generate the instant frequency spectra for each channel. These instant frequency spectra are then used to create the averaged auto-spectra and cross-spectra as shown. These averaged auto- and cross-spectra are then used to generate the FRF and the coherence function. Once this information is available, auto-correlation functions, impulse response functions, and cross-correlation functions are nearly instantly available. Based on this flow diagram, it is evident why we often speak of recording, analysis, averaging, and post processing types of analysis steps or processes.

It is worth noting that the calculation scheme presented here is universal and can be applied to periodic, transient, and random signals.

5.9 THE EFFECTS OF SIGNAL NOISE ON FRF MEASUREMENTS

In Section 5.8, the type of calculation methods that can be applied to any type of signal for use in digital analyzers when measuring a FRF for a linear system have been developed. We found that there were three different ways to estimate the FRF; two of them included magnitude and phase results while the third gave only magnitudes. In addition, it was mentioned that four factors affected the measurement result. Three of these factors were discussed in Section 5.8.

In this section, we address the fourth factor, that is, the effect of uncorrelated signal noise, when the noise is in the input signal, the output signal, or both signals simultaneously. In addition, we look at the situation where one input is measured while another input occurs that is not measured. The unmeasured input may be either correlated or uncorrelated with the measured input. The reader who is interested in pursuing this topic further for multiple inputs and outputs should carefully study Bendat and Piersol.[14]

5.9.1 Noise in the Input Signal

Let us consider the measurement situation shown in Fig. 5.9.1. In this case, the linear system is denoted by $h(t)$ and $H(\omega)$, the actual input signal is $f(t)$ with frequency spectrum of $F(\omega)$, the input noise is $n(t)$ with frequency spectrum of $N(\omega)$, the measured input signal is $x(t)$ with frequency spectrum $X(\omega)$, and the output signal is $y(t)$ with frequency

---

[14]J. S. Bendat and A. G. Piersol, *Random Data: Analysis and Measurement Procedures*, 2nd ed., John Wiley & Sons, New York, 1986.

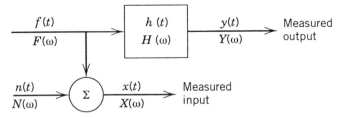

Fig. 5.9.1. Linear model with noise $n(t)$ in the input signal.

spectrum $Y(\omega)$. Output $y(t)$ is linearly related to the actual input $f(t)$. Noise $n(t)$, actual input $f(t)$, and output $y(t)$ are *uncorrelated* so that the corresponding cross-spectral densities are zero, that is

$$G_{nf}(\omega) = G_{ny}(\omega) = 0 \qquad (5.9.1)$$

Noise $n(t)$ is usually due to instrumentation transducers, cables, amplifiers, A/D converters, and so on.

We can develop the required relationships to estimate the actual FRF $H(\omega)$ by starting with the pth estimate. Then, we have

$$\hat{Y}_p(\omega) = H(\omega)\hat{F}_p(\omega) \qquad (5.9.2)$$

and

$$\hat{X}_p(\omega) = \hat{F}_p(\omega) + \hat{N}_p(\omega) \qquad (5.9.3)$$

The averaged auto-spectral estimates (summing on p from 0 to n_d, see Eqs. (5.8.15) and (5.8.16)) then become

$$\hat{G}_{xx}(\omega) = \hat{G}_{ff}(\omega) + \hat{G}_{nn}(\omega) \qquad (5.9.4)$$
$$\hat{G}_{yy}(\omega) = |H(\omega)|^2\hat{G}_{ff}(\omega) = |H(\omega)|^2(\hat{G}_{xx}(\omega) - \hat{G}_{nn}(\omega)) \qquad (5.9.5)$$

when Eq. (5.9.4) is inserted into Eq. (5.9.5). The cross-spectral estimate (see Eq. (5.8.17)) becomes

$$\hat{G}_{xy}(\omega) = \hat{X}^*(\omega)\hat{Y}(\omega) = \hat{G}_{fy}(\omega) = H(\omega)\hat{G}_{ff}(\omega) \qquad (5.9.6)$$

since $n(t)$ is uncorrelated with $f(t)$ and $y(t)$. Then, if we use the definition of $H_1(\omega)$ given by Eq. (5.8.6), we obtain an estimate for $H(\omega)$ that is given by

$$\hat{H}_1(\omega) = \frac{\hat{G}_{xy}(\omega)}{\hat{G}_{xx}(\omega)} = \frac{H(\omega)\hat{G}_{ff}(\omega)}{\hat{G}_{ff}(\omega) + \hat{G}_{nn}(\omega)} \qquad (5.9.7)$$

$$= \frac{H(\omega)}{(1 + [\hat{G}_{nn}(\omega)/\hat{G}_{ff}(\omega)])}$$

Equation (5.9.7) indicates that $H_1(\omega)$ is underestimated when there is uncorrelated noise, since the denominator term is always greater than or equal to unity. Recall that the auto-spectrum has magnitude only; hence the phase is measured correctly since it comes from the cross-spectrum, which is error free in this case. The smaller the uncorrelated noise on the input signal, the closer $H(\omega)$ is estimated by the measured frequency spectrum.

The estimate $\hat{H}_2(\omega)$ is given by Eq. (5.8.8) as

$$\hat{H}_2(\omega) = \frac{\hat{G}_{yy}(\omega)}{\hat{G}_{yx}(\omega)} = \frac{|H(\omega)|^2 \hat{G}_{ff}(\omega)}{H^*(\omega)\hat{G}_{ff}(\omega)} = H(\omega) \qquad (5.9.8)$$

since $|H(\omega)|^2 = H^*(\omega)H(\omega)$. Equation (5.9.8) shows that $H_2(\omega)$ is insensitive to input signal noise when this noise is uncorrelated with the actual input or output time histories, a result that is important to note as demonstrated by Mitchell.[15]

The third method of estimating the FRF magnitude is given by Eq. (5.8.11) as

$$|H_a(\omega)|^2 = \frac{\hat{G}_{yy}(\omega)}{\hat{G}_{xx}(\omega)} = \frac{|H(\omega)|^2 \hat{G}_{ff}(\omega)}{\hat{G}_{nn}(\omega) + \hat{G}_{ff}(\omega)}$$

from which we obtain

$$|H_a(\omega)| = |H(\omega)| \frac{1}{(1 + [\hat{G}_{nn}(\omega)/\hat{G}_{ff}(\omega)]^{0.5}} \qquad (5.9.9)$$

Equation (5.9.9) shows that the FRF is underestimated by the square root of 1 plus the auto spectral ratio of noise to signal.

The corresponding value of the coherence is given by Eq. (5.8.14), which, when the values of $H_1(\omega)$ and $H_2(\omega)$ are inserted from the above

---

[15]L. D. Mitchell, "Improved Method for the FFT Calculation of the Frequency Response Function," *Journal of Mechanical Design*, Vol. 104, Apr. 1982.

equations, gives

$$\gamma^2(\omega) = \frac{H_1(\omega)}{H_2(\omega)} = \frac{1}{(1 + [\hat{G}_{nn}(\omega)/\hat{G}_{ff}(\omega)])} \quad (5.9.10)$$

It is obvious that the coherence function is sensitive to the input signal's noise relative to the actual signal at each frequency ω. The decrease in coherence is dependent on the noise to signal ratio. Thus we desire to have a high signal to noise ratio so that the coherence will be close to unity. The input signal to noise ratio, $IS/N(\omega)$, is related to the measured coherence by rearranging Eq. (5.9.10) to obtain

$$IS/N(\omega) = \frac{\hat{G}_{ff}(\omega)}{\hat{G}_{nn}(\omega)} = \frac{\gamma^2(\omega)}{1 - \gamma^2(\omega)} \quad (5.9.11)$$

It is clear from Eq. (5.9.11) that good input signal to noise ratio results when the coherence is close to unity. This signal noise ratio is the error term in each of the above equations except $H_2(\omega)$, which measures the FRF correctly regardless of the input signal's noise as long as this noise is uncorrelated and the output signal is noise free.

5.9.2 Noise in the Output Signal

The measurement situation for noise in the output signal is shown in Fig. 5.9.2. In this case, the input signal is $x(t)$ and $X(\omega)$, the linear system is represented by $h(t)$ and $H(\omega)$, the output signal is $o(t)$ and $O(\omega)$, the noise is $m(t)$ and $M(\omega)$, and the measured output signal is $y(t)$ and $Y(\omega)$. The noise $m(t)$, the system's output $o(t)$, and the input $x(t)$ are uncorrelated so that

$$\hat{G}_{mo}(\omega) = \hat{G}_{mx}(\omega) = 0 \quad (5.9.12)$$

while the input $x(t)$ and the output $o(t)$ are correlated. The pth estimated output frequency spectrum is related to the noise and input by

$$\hat{Y}_p(\omega) = \hat{M}_p(\omega) + \hat{O}_p(\omega) = \hat{M}_p(\omega) + H(\omega)\hat{X}_p(\omega) \quad (5.9.13)$$

Then, if we average over a number of window samples, we obtain the estimate of the output auto-spectrum as

$$\hat{G}_{yy}(\omega) = \hat{G}_{mm}(\omega) + \hat{G}_{oo}(\omega) = \hat{G}_{mm}(\omega) + |H(\omega)|^2 \hat{G}_{xx}(\omega) \quad (5.9.14)$$

THE EFFECTS OF SIGNAL NOISE ON FRF MEASUREMENTS

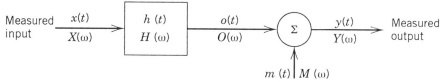

Fig. 5.9.2. Linear model with noise $m(t)$ in the output signal.

and the cross-spectrum as

$$\hat{G}_{xy}(\omega) = \hat{G}_{xo}(\omega) + \hat{G}_{xm}(\omega) = H(\omega)\hat{G}_{xx}(\omega) \quad (5.9.15)$$

due to the uncorrelated cross-spectrum given by Eq. (5.9.12). We see that Eq. (5.9.15) is the definition for $H_1(\omega)$ so that

$$H_1(\omega) = \frac{\hat{G}_{xy}(\omega)}{\hat{G}_{xx}(\omega)} = H(\omega) \quad (5.9.16)$$

This is an ideal result since the uncorrelated noise has no effect on the measured FRF estimate in this case.

The value of $H(\omega)$ as estimate from $H_2(\omega)$ becomes

$$\hat{H}_2(\omega) = \frac{\hat{G}_{yy}(\omega)}{\hat{G}_{yx}(\omega)} = \frac{|H(\omega)|^2 \hat{G}_{xx}(\omega) + \hat{G}_{mm}(\omega)}{H^*(\omega)\hat{G}_{xx}(\omega)}$$

which can be reduced to

$$\hat{H}_2(\omega) = H(\omega)\left(1 + \frac{\hat{G}_{mm}(\omega)}{\hat{G}_{oo}(\omega)}\right) = H(\omega)\left(1 + \frac{\hat{G}_{mm}(\omega)}{|H(\omega)|^2 \hat{G}_{xx}(\omega)}\right) \quad (5.9.17)$$

Equation (5.9.17) shows that $H_2(\omega)$ is higher than the actual FRF value $H(\omega)$ by an amount that reflects the noise to signal ratio in the structure's output signal. We note that the phase determined from the $\hat{H}_2(\omega)$ estimate is correct even though its magnitude is in error. We also note that the $|\hat{H}(\omega)|^2 G_{xx}(\omega)$ term can cause serious errors when $H(\omega)$ is a deep valley that is too near to a high peak, due to window filter leakage from the high peak. The obviousness of this statement is clearly reflected by comparing Figs. 5.4.3a and 5.4.3b for a peak and valley frequency difference within the limits of 10 Δf.[16]

The magnitude of the frequency spectra given by the ratio of autospectra gives

$$|\hat{H}_a(\omega)|^2 = \frac{\hat{G}_{yy}(\omega)}{\hat{G}_{xx}(\omega)} = \frac{|H(\omega)|^2 \hat{G}_{xx}(\omega) + \hat{G}_{mm}(\omega)}{\hat{G}_{xx}(\omega)}$$

which reduces to

$$|\hat{H}_a(\omega)| = |H(\omega)| \left(1 + \frac{\hat{G}_{mm}(\omega)}{\hat{G}_{oo}(\omega)}\right)^{0.5} \quad (5.9.18)$$

It is clear from Eq. (5.9.18) that the FRF magnitude is overestimated by the square root of 1 plus the noise to signal ratio, an error that is smaller than that for $\hat{H}_2(\omega)$ in Eq. (5.9.17).

Similarly, if we solve for the coherence using Eqs. (5.9.16) and (5.9.17), we obtain

$$\gamma^2(\omega) = \frac{H_1(\omega)}{H_2(\omega)} = \frac{1}{(1 + [\hat{G}_{nn}(\omega)/\hat{G}_{oo}(\omega)])} \quad (5.9.19)$$

We can use Eq. (5.9.19) to solve for the output signal to noise ratio, $OS/N(\omega)$, in terms of the coherence function. The result is

$$OS/N(\omega) = \frac{\hat{G}_{oo}(\omega)}{\hat{G}_{mm}(\omega)} = \frac{\gamma^2(\omega)}{1 - \gamma^2(\omega)} \quad (5.9.20)$$

which is identical in form to that obtained in Eq. (5.9.11) for input signal to noise ratio when the noise is on the input signal side of the measurement system. Note that Eqs. (5.9.11) and (5.9.20) are restricted to the nearly ideal case of either input or output uncorrelated noise, respectively.

5.9.3 Noise in the Input and Output Signals

The case where there is uncorrelated noise in both the input and output signals is shown in Fig. 5.9.3. In this case, the variables are defined the same as in Figs. 5.9.1 and 5.9.2, but we note that $n(t)$, $m(t)$, and $f(t)$ are

---

[16] See a paper by K. G. McConnell and P. S. Varoto, "The Effects of Windowing on FRF Estimations for Closely Spaced Peaks and Valleys", *Proceedings 13th International Modal Analysis Conference*, Nashville, TN, February 1995, Vol. I, pp. 769–775, for more details.

THE EFFECTS OF SIGNAL NOISE ON FRF MEASUREMENTS   351

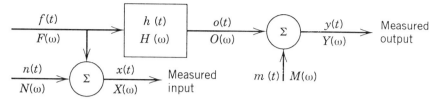

Fig. 5.9.3. Linear model with noise $n(t)$ in the input signal and noise $m(t)$ in the output signal.

uncorrelated while $n(t)$, $m(t)$, and $o(t)$ are uncorrelated. Then it follows that

$$G_{nm}(\omega) = G_{nf}(\omega) = G_{mf}(\omega) = G_{no}(\omega) = G_{mo}(\omega) = 0 \quad (5.9.21)$$

hold true for all equations that apply to this case.

The averaged estimates can be written as

$$\hat{G}_{xx}(\omega) = \hat{G}_{ff}(\omega) + \hat{G}_{nn}(\omega) \quad (5.9.22)$$

$$\hat{G}_{yy}(\omega) = \hat{G}_{mm}(\omega) + \hat{G}_{oo}(\omega) = \hat{G}_{mm}(\omega) + |H(\omega)|^2 \hat{G}_{ff}(\omega) \quad (5.9.23)$$

$$\hat{G}_{xy}(\omega) = \hat{G}_{fo}(\omega) = H(\omega)\hat{G}_{ff}(\omega) \quad (5.9.24)$$

Then, using Eqs. (5.9.22) through (5.9.24), we can determine

$$\hat{H}_1(\omega) = \frac{\hat{G}_{xy}(\omega)}{\hat{G}_{xx}(\omega)} = H(\omega) \frac{1}{(1 + [\hat{G}_{nn}(\omega)/\hat{G}_{ff}(\omega)])} \quad (5.9.25)$$

$$\hat{H}_2(\omega) = \frac{\hat{G}_{yy}(\omega)}{\hat{G}_{xy}^*(\omega)} = H(\omega)\left(1 + \frac{\hat{G}_{mm}(\omega)}{\hat{G}_{oo}(\omega)}\right)$$

$$= H(\omega)\left(1 + \frac{\hat{G}_{mm}(\omega)}{|H(\omega)|^2 \hat{G}_{ff}(\omega)}\right) \quad (5.9.26)$$

$$|H_a(\omega)|^2 = \frac{\hat{G}_{yy}(\omega)}{\hat{G}_{xx}(\omega)} = |H(\omega)|^2 \frac{(1 + [\hat{G}_{mm}(\omega)/\hat{G}_{oo}(\omega)])}{(1 + [\hat{G}_{nn}(\omega)/\hat{G}_{ff}(\omega)])} \quad (5.9.27)$$

The corresponding coherence function becomes

$$\gamma^2(\omega) = \frac{H_1(\omega)}{H_2(\omega)} = \frac{1}{(1 + [\hat{G}_{nn}(\omega)/\hat{G}_{oo}(\omega)])(1 + [\hat{G}_{nn}(\omega)/\hat{G}_{ff}(\omega)])} \quad (5.9.28)$$

In this case, we see from Eq. (5.9.28) that we cannot solve for either the input or output signal to noise ratio, since they both contribute to the decrease in coherence at any given frequency.

Equations (5.9.25), (5.9.26), and (5.9.27) show that $H_1(\omega)$ underestimates $H(\omega)$, $H_2(\omega)$ overestimates $H(\omega)$, and $H_a(\omega)$ lies somewhere between $H_1(\omega)$ and $H_2(\omega)$. Thus we have that

$$\hat{H}_1(\omega) < H(\omega) < \hat{H}_2(\omega) \qquad (5.9.29)$$

so that the actual FRF is bounded by these two estimates. In the absence of noise, these three estimates give the system's FRF. We find that $H_2(\omega)$ is better for estimating peaks than $H_1(\omega)$ since the input signal to noise ratio tends to deteriorate at resonance. This deterioration of input signal is caused by a complicated interaction that occurs between the structure under test and the vibration exciter so that the force available to excite the system at resonance "drops out." A thorough discussion of force dropout will be presented in Chapters 6 and 7.

Similarly, the coherence tends to decrease at an antiresonant condition where the output signal is small so that the signal to noise ratio decreases. In this case, $H_1(\omega)$ gives a better estimate of the measured FRF. Consequently, we need to examine the measured FRF with caution in order to determine the cause for a drop in coherence near resonance and antiresonance, as shown in Fig. 5.8.4, in order to obtain the best estimate of the structure's FRF. Remember that we are dealing with random signals so that the signal at any given frequency is much smaller than the overall RMS voltage. This means that the signal may be too close to the noise floor when the drop in input force or output motion occurs, leading to low signal to noise ratios at certain frequencies. We have seen that these frequencies often occur at the structure's resonance or antiresonance condition. The most important case for us is the resonant condition. However, from Eq. (5.9.26) we see that a deep valley next to a peak can appear as a peak in the $H_2(\omega)$ estimate. Thus we need to examine both $H_1(\omega)$ and $H_2(\omega)$ in order to detect such problems.

5.9.4 More Than One External Input

There are instances when we are exciting a structure that has other excitation inputs. For example, this type of situation can occur when exciting a structure that is tied to a foundation that is excited by passing traffic. In this case, we would expect the second input to be uncorrelated to input excitation $x(t)$ if the vibration exciter is not directly attached to the foundation. On the other hand, if the exciter is attached to the foundation, then the two inputs may be correlated. We explore the simplest of these situations in terms of the model shown in Fig. 5.9.4. In this situation, $x(t)$

THE EFFECTS OF SIGNAL NOISE ON FRF MEASUREMENTS

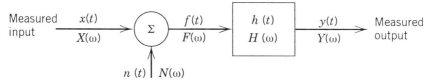

Fig. 5.9.4. Linear model with unknown input signal $n(t)$.

is the measured input while $n(t)$ is the unmeasured external input, which may or may not be correlated to the measured input. We explore both the case when $n(t)$ and $x(t)$ are uncorrelated and the case when they are correlated.

Uncorrelated Inputs In the uncorrelated case, $n(t)$ and $x(t)$ have a zero cross-spectral density given by

$$\hat{G}_{nx}(t) = 0 \tag{5.9.30}$$

while the pth output amplitude spectral density is estimated by

$$\hat{Y}_p(\omega) = H(\omega)\hat{F}_p(\omega) = H(\omega)(\hat{X}_p(\omega) + \hat{N}_p(\omega)) \tag{5.9.31}$$

Then, the corresponding auto- and cross-spectra become

$$\hat{G}_{yy}(\omega) = H^*(\omega)H(\omega)(\hat{G}_{xx}(\omega) + \hat{G}_{nn}(\omega)) \tag{5.9.32}$$

and

$$\hat{G}_{xy}(\omega) = H(\omega)\hat{G}_{xx}(\omega) \tag{5.9.33}$$

It is clear that Eq. (5.9.33) gives $H_1(\omega)$ immediately as

$$\hat{H}_1(\omega) = \frac{\hat{G}_{xy}(\omega)}{\hat{G}_{xx}(\omega)} = H(\omega) \tag{5.9.34}$$

while $H_2(\omega)$ becomes

$$\hat{H}_2(\omega) = \frac{\hat{G}_{yy}(\omega)}{\hat{G}_{yx}(\omega)} = H(\omega)\left(1 + \frac{\hat{G}_{nn}(\omega)}{\hat{G}_{xx}(\omega)}\right) \tag{5.9.35}$$

354 THE DIGITAL FREQUENCY ANALYZER

and $H_a(\omega)$ gives

$$|\hat{H}_a(\omega)|^2 = |H(\omega)|^2 \left(1 + \frac{\hat{G}_{nn}(\omega)}{\hat{G}_{xx}(\omega)}\right) \quad (5.9.36)$$

It is evident from Eqs. (5.9.34) and (5.9.35) that $H_2(\omega)$ overestimates $H(\omega)$ by an amount that is dependent on the magnitude of the noise to signal ratio, that $H_1(\omega)$ has no bias error and is equal to $H(\omega)$, and that both estimates of $H_1(\omega)$ and $H_2(\omega)$ have the correct phase angle. Estimate $H_a(\omega)$ lies between estimates $H_1(\omega)$ and $H_2(\omega)$ as before.

Correlated Inputs In the correlated inputs case, $\hat{G}_{nx}(\omega)$ is nonzero. Then the above equations become

$$\hat{Y}_p(\omega) = H(\omega)\hat{F}_p(\omega) = H(\omega)(\hat{X}_p(\omega) + \hat{N}_p(\omega)) \quad (5.9.37)$$

$$\hat{G}_{yy}(\omega) = H^*(\omega)H(\omega)(\hat{G}_{xx}(\omega) + \hat{G}_{nn}(\omega) + \hat{G}_{xn}(\omega) + \hat{G}_{nx}(\omega)) \quad (5.9.38)$$

$$\hat{G}_{xy}(\omega) = H(\omega)(\hat{G}_{xx}(\omega) + \hat{G}_{xn}(\omega)) \quad (3.9.39)$$

so that the FRF estimates become

$$\hat{H}_1(\omega) = \frac{\hat{G}_{xy}(\omega)}{\hat{G}_{xx}(\omega)} = H(\omega)\left(1 + \frac{\hat{G}_{xn}(\omega)}{\hat{G}_{xx}(\omega)}\right) \quad (5.9.40)$$

$$\hat{H}_2(\omega) = \frac{\hat{G}_{yy}(\omega)}{\hat{G}_{yx}(\omega)} = H(\omega)\left(1 + \frac{\hat{G}_{nn}(\omega) + \hat{G}_{xn}(\omega)}{\hat{G}_{xx}(\omega) + \hat{G}_{nx}(\omega)}\right) \quad (5.9.41)$$

$$|\hat{H}_a(\omega)|^2 = |H(\omega)|^2 \left(1 + \frac{\hat{G}_{nn}(\omega) + \hat{G}_{xn}(\omega)}{\hat{G}_{xx}(\omega) + \hat{G}_{nx}(\omega)}\right) \quad (5.9.42)$$

Equations (5.9.40) through (5.9.42) show that each estimate has a built-in bias error due to correlation between inputs. This is why we need to isolate the vibration exciter's base from the foundation to which we attach the test structure. Otherwise, correlated inputs occur through the common foundation.

This correlated input situation is illustrated in Fig. 5.9.5a, where a simple mass on a spring structure is attached by two springs to a foundation, to which, in turn, is attached a vibration exciter. The input force is measure by the force transducer FT, while the structure's response is measured by the accelerometer. However, there are no rigid foundations in real life, so that the reaction force between the exciter and the foundation is transmitted to the structure's springs through foundation motion. This motion causes a force that correlates with the measured force that is

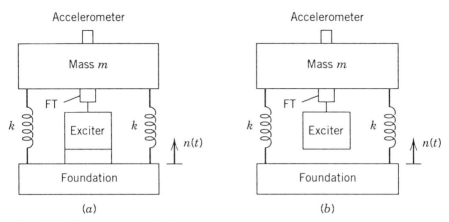

Fig. 5.9.5. Test systems. (*a*) Correlated vibration exciter attached to foundation. (*b*) Uncorrelated vibration exciter isolated from foundation.

applied to the mass. Consequently, the exciter must be supported on soft isolation springs, Fig. 5.9.5*b*, in order to break this path for transfer of excitation energy to the structure. Otherwise, there is no way to achieve the desired test aims without incurring considerable measurement errors due to correlated noise.

There are a large number of equations in this section. These equations are summarized in Table 5.9.1. Recall that the input and output signal to noise ratio can be calculated from the coherence only when noise occurs on either the input signal alone or the output signal alone. When there is noise on both input and output signals at the same time, then we cannot calculate the signal to noise ratio, as is clearly evident from the coherence equation in Table 5.9.1. Finally, we should be aware that low coherence indicates that we have a testing problem. High coherence, on the other hand, does not guarantee that there are no testing problems, only that we have not detected any problems.

5.10 SUMMARY

This chapter has described the processes employed in a digital frequency analyzer, beginning with data sampling where a continuous analog signal is converted into a finite number of discrete data points in the data window. The sample function selects the points gathered by the A/D converter, while the window function controls the data window's size. It has been shown that multiplication of two time functions generates frequency domain convolution. Aliasing occurs when discrete points are used to represent an analog signal, a natural consequence of the frequency

TABLE 5.9.1. Summary of FRF Estimates for Random Signals

Quantity	Input and Output[a]		External Input[b]	
	Uncorrelated Noise		Uncorrelated	Correlated
$H_1(\omega)$	$\dfrac{H(\omega)}{(1 + [\hat{G}_{nn}(\omega)/\hat{G}_{ff}(\omega)])}$		$H(\omega)$	$H(\omega)\left(1 + \dfrac{\hat{G}_{xn}(\omega)}{\hat{G}_{xx}(\omega)}\right)$
$H_2(\omega)$[c]	$H(\omega)\left(1 + \dfrac{\hat{G}_{mm}(\omega)}{\hat{G}_{oo}(\omega)}\right)$		$H(\omega)\left(1 + \dfrac{\hat{G}_{nn}(\omega)}{\hat{G}_{xx}(\omega)}\right)$	$H(\omega)\left(1 + \dfrac{\hat{G}_{nn}(\omega) + \hat{G}_{xn}(\omega)}{\hat{G}_{xx}(\omega) + \hat{G}_{nx}(\omega)}\right)$
$\gamma^2(\omega)$	$\dfrac{1}{(1 + [\hat{G}_{mm}(\omega)/\hat{G}_{oo}(\omega)])(1 + [\hat{G}_{nn}(\omega)/\hat{G}_{ff}(\omega)])}$		Single coherence does not apply to this case of excitation[d]	

[a]$n(t)$ is the uncorrelated input noise, $m(t)$ is the uncorrelated output noise, $x(t)$ is the measured input signal, $y(t)$ is the measured output signal, $f(t)$ is the actual input to system under test, $o(t)$ is the actual output of the system under test. For input signal noise only, set $G_{mm}(\omega) = 0$. For output signal noise only, set $G_{nn}(\omega) = 0$.
[b]The external input $n(t)$ is not measured while input $x(t)$ and output $y(t)$ are measured.
[c]Note that $G_{oo}(\omega)$ and $G_{xx}(\omega)$ contain the $|H(\omega)|^2$ type terms.
[d]Multiple coherence functions are needed in this case, as outlined by Bendat and Piersol [1].

domain convolution theorem or, in other words, of the signal being sampled too slowly. An anti-aliasing filter is employed to remove aliasing by removing the high frequency components before sampling. Digital filters have filter leakage, a phenomenon that is associated with trying to represent a frequency component that does not exist in the analyzer by a number of adjacent frequency components instead. The amount of filter leakage is dependent on the digital filter's characteristics.

In Section 5.3 we have considered operating principles and calculation procedures that are employed to analyze the data and to create displays of various kinds of frequency spectra. We have found Parseval's formula to be useful in manipulating the discrete Fourier frequency components that are used to represent periodic frequency spectra, transient frequency spectra, and random ASD frequency spectra. The major difference between displays is a matter of either multiplying the discrete frequency components by the window period T or dividing them by the spectral line spacing Δf.

Section 5.4 has been concerned with factors that influence our application of a single channel analyzer. The first item we have considered is how to characterize and compare different window functions and corresponding digital filters. Factors such as center frequency, bandwidth, ripple, and selectivity have been used in comparing four commonly employed window functions. These are rectangular, Hanning, Kasier–Bessel, and flat top. Next we have applied these four filters to a single sinusoidal signal in order to understand how they analyzed or interpreted this most fundamental of signals. Then we have looked at the resulting analysis of two sinusoids that are closely spaced in frequency and have amplitudes that varied by 100 to 1, a spread of 40 dB, when using the four window functions. This exercise allowed us to understand the frequency resolution of periodic signals by the various digital filters. We have examined the basis of spectral line uncertainty that arises when we attempt to analyze a random signal. In general, it is found that the BT product (filter bandwidth B times window length T) is equal to the number of data windows or records that are sampled and averaged to obtain the ASD. The spectral line error is bounded by one over the square root of the number of data window samples averaged. Section 5.4 ends with a recommendation as to which window function to use with periodic, transient, and random signals.

In Section 5.5 we have looked at the use of window overlapping in order to speed up data processing. It has been found that overlapping can cause significant filter ripple to occur. However, when the shift factor is one-third or less, so that the signal is overlapped by two-thirds or more, the ripple drops off rapidly and is either close to zero or zero. We have found that a 75 percent overlap speeds up the frequency analysis by a factor of two for the same spectral line uncertainty. Finally, the idea of real time analysis has been explored.

358 THE DIGITAL FREQUENCY ANALYZER

Section 5.6 has addressed the issue of zoom analysis, and two different methods, the heterodyning method and the long record method, have been presented. In both methods, the same sample frequency is employed in order to maintain the analyzer's baseband width. The heterodyning method multiplies the sampled signal by a sinusoidal window function, which shifts the frequency domain axis by the window function's frequency. Then the over sampled digital data is passed through a low pass digital filter and every Nth data point is kept for frequency analysis purposes. The result is to zoom in on a particular region of the frequency spectra. The length of data sampled is increased by a factor of N. One shortcoming of this zoom method is that the original data is lost during data processing so that we need to repeat the long sampling time for each zoom analysis. In the long record method, all of the data is recorded at one time, and then analyzed using offset sample sets, so that a large detailed frequency spectrum is quickly generated. The major advantages of this method are that the original sampled data is retained for further analysis if so desired, and that we can look at any portion of the detailed frequency spectrum for zoom analysis purposes. One of the problems with zoom analysis is frequency spectral smearing due to small changes in the signal's frequency during the long time that the data is sampled. This smearing can be overcome in some instances by using external triggering so that the sample rate changes with a machine's operating speed.

Section 5.7 has been concerned with scan analysis, scan averaging, and more spectral smearing. Once a long time history is recorded in digital memory, this data can be analyzed in a number of different ways. Scan analysis can provide a detailed picture of how the frequency spectrum changes with position along the time record. Then scan averaging can be used to analyze either transient or random signals while using a Hanning window function. Finally, spectral smearing from actual field data is discussed to show how a rapid change from one natural frequency to a second natural frequency on a periodic basis effectively smears out and suppresses both resonant peaks, a classic case where resonance peaks are not evident in the frequency analysis when the springs are actually resonating.

Section 5.8 has introduced us to the idea of a dual channel frequency analyzer so that we can measure a system's FRF. The ideal input-output relationships for linear systems have been examined in order to develop a calculation scheme that can be applied to periodic, transient, and random signals. The coherence is introduced as a measure of the correlation of the output signal with the input signal. Then we have looked at how the analyzer actually estimates the auto-spectra and cross-spectra, as well as at how averaging processes are utilized. Then we have examined three reasons for coherence being less than unity: nonlinear structures, digital filter leakage that leads to resolution bias error, and time delays between signals. This section has ended with a typical operating block diagram that shows the major steps as the dual channels of data are processed.

Finally, Section 5.9 has addressed the issue of random signal noise sources and how these noise sources affect the different FRF estimates. First, we have looked at the effects of uncorrelated noise in the input signal on FRF estimates. Second, we have looked at the effects of uncorrelated noise in the output signal on FRF estimates. Third, we have looked at the effects of uncorrelated noise in the input and output signals on the FRF estimates. Finally, we have explored the effects of an external system input that is not measured. In one case, this external input is uncorrelated with the measured input. In the other case, it is correlated with the measured input. It is seen that totally different results are obtained, so it is clear why it is important to isolate a vibration exciter from the foundation that supports the structure under test. The analysis shows that $H_1(\omega)$ is the lower bound estimate, while $H_2(\omega)$ is the upper bound estimate of the system's actual FRF $H(\omega)$. The higher the coherence, the closer together these two estimates are. However, we must remember that while low coherence is an indication of measurement problems and high coherence is usually an indication of good quality measurements, there are situations where coherent noise in both measurements will give high coherence and poor measurements. This can be the case when the vibration exciter and the structure are attached to a common foundation.

It should be obvious that a great deal of knowledge is required to understand the processes and errors that occur in obtaining reliable frequency spectra, interpreting frequency spectra, and obtaining FRF estimates. The reader is encouraged to study several of the references in order to gain deeper understanding of how to properly use such a powerful tool as the multichannel digital frequency analyzer.

REFERENCES

The following references are for general information.
1. Bendat, J. S. and A. G. Piersol, *Random Data: Analysis and Measurement Procedures*, 2nd ed., John Wiley & Sons, New York, 1986.
2. Broch, J. T., *Mechanical Vibration and Shock Measurements*, available from Bruel & Kjaer Instruments, Inc., Marlborough, MA, Oct. 1980.
3. Broch, J. T., *Non-Linear Systems and Random Vibration*, available from Bruel & Kjaer Instruments, Inc., Marlborough, MA, Jan. 1972.
4. Bruel & Kjaer Instruments, *Technical Review*, a quarterly publication available from Bruel & Kjaer Instruments, Inc., Marlborough, MA.

(a) "Laboratory Tests of the Dynamic Performance of a Turbocharger Rotor-Bearing System," No. 4, 1973.
(b) "On the Measurement of Frequency Response Functions," No. 4, 1975.
(c) "An Objective Comparison of Analog and Digital Methods of Real-Time Frequency Analysis," No. 1, 1977.
(d) "Digital Filters and FFT Technique," No. 1, 1978.
(e) "Measurement of Effective Bandwidth of Filters," No. 2, 1978.
(f) "Discrete Fourier Transforms and FFT Analyzers," No. 1, 1979.
(g) "Zoom-FFT," No. 2, 1980.
(h) "Cepstrum Analysis", No. 3, 1981.
(i) "System Analysis and Time Delay Spectrometry," Part I, No. 1; Part II, No. 2; 1983.
(j) "Dual Channel FFT Analysis," Part I, No. 1; Part II, No. 2; 1984.
(k) "Hilbert Transform," No. 3, 1984.
(l) "Vibration Monitoring of Machines," No. 1, 1987.
(m) "Signals and Units," No. 3, 1987.
(n) "Use of Weighting Functions in DFT/FFT Analysis," Part I, No. 3; Part II, No. 4; 1987.

5. Cooley, J. W. and J. W. Tukey, "An Algorithm for the Calculation of Complex Fourier Series," *Mathematics of Computation*, Vol. 19, 1965, pp. 297–301.
6. Crandall, S. H. and W. D. Mark, *Random Vibration in Mechanical Systems*, Academic Press, New York, 1963.
7. Ewins, D. J., *Modal Testing: Theory and Practice*, Research Studies Press Ltd., Letchworth, Hertfordshire, UK (Also available from Bruel & Kjaer Instruments, Inc., Marlborough, MA).
8. Harris, F. J., "On the Use of Windows for Harmonic Analysis with the Discrete Fourier Transform," *Proceedings of the IEEE*, Vol. 66, No. 1, Jan. 1978, pp. 51–83.
9. Kobayaski, Albert S. (Editor), *Handbook on Experimental Mechanics*, Prentice-Hall, Englewood Cliffs, NJ, 1987.
10. McConnell, K. G. and M. K. Abdelhamid, "On the Dynamic Calibration of Measurement Systems for Use in Modal Analysis," *International Journal of Analytical and Experimental Modal Analysis*, Vol. 2, No. 3, 1987, pp. 121–127.
11. Mitchell, L. D., "Improved Methods for the FFT Calculation of the Frequency Response Function," *Journal of Mechanical Design*, Vol. 104, Apr. 1982.
12. Newland, D. E., *An Introduction to Random Vibrations and Spectral Analysis*, Longman Group Limited, London, 1975.
13. Pandit, S. M., *Modal and Spectrum Analysis: Data Dependent Systems in State Space*, John Wiley & Sons, New York, 1991.
14. Papoulis, A., *Signal Analysis*, McGraw-Hill, New York, 1977.
15. Peterson, A. P. G. and E. E. Gross, Jr., *Handbook of Noise Measurement*, available from General Radio Company, Concord, MA, 1972.
16. Randall, B., *Application of B & K Equipment to Frequency Analysis*, 2nd ed.,

Available from Bruel & Kjaer Instruments, Inc., Marlborough, MA, Sept. 1977.
17. Rhodes, J. D., *Theory of Electrical Filters*, John Wiley & Sons, New York, 1976.
18. Welch, P. D., "The Use of Fast Fourier Transform for the Estimation of Power Spectra," *IEEE Transactions on Audio Electro Acoustics*, Vol. AU-15, June 1967, pp. 70–73.

6 Vibration Exciters

6.1 INTRODUCTION

Many different types of vibration exciter schemes are employed in vibration testing. It is a rather well kept secret that these exciters signifi-

A test engineer is positioning accelerometers on a launch vehicle that is attached to a vibration exciter in preparation for a system level vibration test using a vibration control and analysis system. (Photo Courtesy of *Sound and Vibration*).

364 VIBRATION EXCITERS

cantly interact with the structure under test, and hence, can significantly influence test results. This is particularly true in the region around structural resonance, the very region where test results are usually most critical.

Vibration exciters can take any number of forms: a static load can be applied to a structure and then suddenly released; a machine can have its own internal excitation sources of rotating unbalances or multi piston pumps; a vehicle carrying a test item can be rammed into the proverbial "brick wall;" a carefully controlled experiment can be carried out in a highly controlled laboratory environment; and so on. In this chapter we explore the more common excitation techniques that are employed and try to describe their characteristics. It is found that there is a significant interaction between the structure under test and the exciter, an interaction that has profound testing consequences that are often either overlooked or not understood by the user.

6.1.1 Static Excitation Schemes

One common method of excitation is to apply a static load to a structure, and then suddenly release the load, as shown in Fig. 6.1.1. This excitation technique appears to be very attractive, since we can apply the load slowly with a convenient loading mechanism over the time period from 0 to T_0, let any residual vibration decay away over the time period T_0 to T_1, and then release the load quickly through simple releasing mechanisms, causing a nearly step loading to occur at time T_1, as shown. Such a loading has the capability of exciting a wide range of frequencies. What is often missed in this approach is that the same loading limitations are imposed on static loading schemes as occur for any dynamic point loading scheme.

We can illustrate this limitation by considering the modal model developed in Chapter 3 for continuous systems. Recall that the pth modal

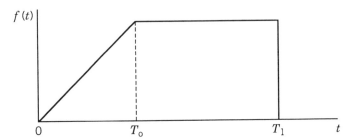

Fig. 6.1.1. A slowly applied near static load that is suddenly released to excite a structure.

INTRODUCTION

response is given by Eq. (3.7.19) as

$$m_p \ddot{q}_p + C_p \dot{q}_p + k_p q_p = Q_p f(t) \tag{6.1.1}$$

where q_p is the pth generalized coordinate.
m_p is the pth modal mass.
C_p is the pth modal damping.
k_p is the pth modal stiffness.
Q_p is the pth generalized excitation force.
$f(t)$ is the time dependent variation of the external loading.

The corresponding generalized excitation force, Q_p in Eq. (6.1.1), is given by Eq. (3.7.20):

$$Q_p = \int^l P(x) U_p(x) \, dx \tag{6.1.2}$$

where $U_p(x)$ is the pth mode shape.
$P(x)$ is the spatial distribution of the excitation force.

The corresponding particular solution to Eq. (6.1.1) is given by Eq. (3.7.21) as

$$u(x,t) = \sum_{p=1}^{N} U_p(x) H_p(\omega) Q_p e^{j\omega t} \tag{6.1.3}$$

where $H_p(\omega)$ is the pth modal FRF.

Equation (6.1.3) describes the structure's response as it is slowly loaded over time period 0 to T_1 in Fig. 6.1.1; it also describes the deflections as the dynamic response decays away, and it describes the structure's static deflection that remains after reaching static equilibrium. These static deflections constitute the initial conditions when the step input unloading is applied at time T_1.

The important idea here is to examine the consequences of applying a point static load. When a point load is applied at location x_0, this load can be represented by

$$P(x,t) = P_0 \, \delta(x - x_0) f(t) \tag{6.1.4}$$

where P_0 is the force's magnitude and $\delta(x - x_0)$ is the Dirac delta function. It is clear from substituting Eq. (6.1.4) into Eq. (6.1.2) that the pth generalized excitation force becomes

$$Q_p = P_0 U_p(x_0) \tag{6.1.5}$$

The significance of Eq. (6.1.5) is that those modes that have node points at $x = x_0$ (i.e., $U_p(x_0) = 0$) do not participate in the resulting vibration when load P_0 is suddenly released *since only those modes that responded in generating the static deflections are available when the structure is released.* Consequently, the modes corresponding to node points at $x = x_0$ are not present in the structure's response when the load is released!

A simple experiment with a guitar can effectively illustrate this phenomenon. The string is plucked at the midpoint for one test and, say, at the quarter point for another test. A frequency analysis of the resulting sound will show the absence of every second harmonic when the midpoint is deflected and released. Every fourth harmonic is absent when the quarter point is deflected and released. The same result is obtained if the straight line initial static deflection of the guitar string is analyzed in terms of the sinusoidal mode shapes. Those natural frequencies with node points at the point of load application will be absent from the list of modes that are excited.

Thus we should not be surprised when the static load sudden release excitation technique does not excite all natural frequencies that should occur in a given range of frequencies! Only in cantilever beam type structures will such an excitation technique excite all modes when static loads are applied and released at the beam's free end, a location that has no node points. Hence, even static load sudden release excitation schemes are subject to the same limits that are imposed by the structure's mode shapes on dynamic excitation forces.

6.1.2 Dynamic Impulse Loading Schemes

There are many ways to generate dynamic loads that utilize a part of the structure. For example, the vibrations of large ships are excited by the so called "drop anchor" method, where the anchor is released and allowed to free fall, and then is suddenly stopped. This impulse excitation method excites a number of the ship's flexural and torsional modes of vibration and has proven to be a simple and effective excitation means for these large structures.

Another common method of impact loading uses a second structure to impact with the structure under test. A simple model of such a test is shown in Fig. 6.1.2 and can be thought of in terms of two single degree of freedom (DOF) systems. System 1 consists of mass m_1, damping c_1, and stiffness k_1 and represents the structure under test. System 2 consists of mass m_2, damping c_2, and stiffness k_2 and represents the impacting structure. Each system has its own uncoupled natural frequency that is given by

INTRODUCTION 367

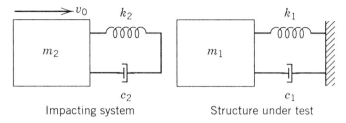

Fig. 6.1.2. Definition of impacting structure (system 2) and structure under test (system 1) for conducting impulse tests.

$$\omega_{11} = \sqrt{\frac{k_1}{m_1}} \qquad (6.1.6)$$

and

$$\omega_{22} = \sqrt{\frac{k_2}{m_2}} \qquad (6.1.7)$$

and it is assumed that $\omega_{22} \gg \omega_{11}$. System 2 is assumed to have an initial velocity of v_0 before it impacts system 1, which is initially at rest. Spring k_2 and damper c_2 constitute the interface between the two systems during impact, and we assume that spring k_2 and damper c_2 can only support compressive loads. Furthermore, if we assume that m_1 is very large compared to m_2, then we can show that the impact force is given by

$$F(t) = \begin{cases} \dfrac{m_2 v_0 \omega_n}{\sqrt{1-\zeta^2}} e^{-\zeta \omega_n t} \sin(\omega_d t + \varphi) & \text{for } 0 < t < T/2 \\ 0 & \text{for } t > T/2 \end{cases} \qquad (6.1.8)$$

where $\omega_n = \omega_{22}$.

$$\zeta = \frac{c_2}{2\sqrt{k_2 m_2}}$$

$$\omega_d = \omega_n \sqrt{1-\zeta^2}.$$

Also,

$$\tan \varphi = \frac{2\zeta\sqrt{1-\zeta^2}}{1-2\zeta^2} \qquad (6.1.9)$$

368 VIBRATION EXCITERS

For very light damping, Eq. (6.1.8) reduces to be nearly the same as the well known undamped half sine impulse that is given by

$$F(t) = \begin{cases} m_2 v_0 \omega_n \sin(\omega_n t) & \text{for } 0 < t < T/2 \\ 0 & \text{for } t > T/2 \end{cases} \quad (6.1.10)$$

Equation (6.1.8) shows that the peak force F_0 is controlled by the initial momentum of $m_2 v_0$ and impacting structure's natural frequency ω_n. This relationship works well as long as there is a large inertia mismatch between the impacting structure and the impacted structure and $\omega_{11} \ll \omega_{22}$, so that impact time impact duration is short compared to the structure's natural period, usually less than $0.1T$.

A rather interesting phenomenon occurs when masses m_1 and m_2 are nearly equal, for we have observed a double strike (double impact) phenomenon taking place where structure 1 appears to stop vibrating after half of a cycle. Why?

Consider a traditional linear impulse momentum analysis where we ignore the effects of k_1 and c_1. Let V_1 be the velocity of mass m_1 after the impact, V_2 be the velocity of the impacting structure after impact, and e be the traditional coefficient of restitution. Also assume that V_1 and V_2 are in the same direction as v_0. Then we can write the velocities of masses m_1 and m_2 after impact as

$$V_1 = \left(\frac{m_2(1+e)}{m_1 + m_2}\right) v_0 = \left(\frac{M(1+e)}{M+1}\right) v_0 \quad (6.1.11)$$

and

$$V_2 = \left(\frac{m_2}{m_1 + m_2} - \frac{m_1 e}{m_1 + m_2}\right) v_0 = \left(\frac{M-e}{1+M}\right) v_0 \quad (6.1.12)$$

where the system mass ratio M is given by

$$M = m_2/m_1 \quad (6.1.13)$$

When mass ratio M is small, we clearly see from Eqs. (6.1.11) and (6.1.12) that structure 1 has only a small percentage of the impact velocity v_0 transmitted to it, while structure 2 is seen to rebound with a velocity that is essentially $-ev_0$. However, we see that the described phenomenon of double strikes will occur whenever $V_2 \geq 0$, since structure 2 remains in the vicinity of structure 1 so that structure 2 is impacted a second time by the rebounding structure 1. In fact, when mass ratio M is equal to the coefficient of restitution, Eq. (6.1.12) indicates that the impacting structure is at rest and will remain so until struck by system 1 during its return stroke one half natural period later.

Thus double strikes are an indication of an effective mass ratio that is close to the coefficient of restitution e. We need to make M smaller by reducing the impacting mass m_2. If $M <$ about 1/5 (note the old five and ten rule quoted earlier), then the impacting structure (system 2) will depart with a velocity around 50 to 60 percent of its incoming velocity v_0 and should be clear of the structure under test (system 1) by the time structure 1 has passed though about 70 percent of its natural period. Consequently, reducing the impacting structure's mass will solve the double strike problem. This is a marvelous experiment to spring on novices shortly after introducing them to impulse testing, for they are completely puzzled by what is occurring. Once they have learned the concept, they should be fully aware that this phenomenon can occur with any type of structure if the correct conditions exist, and hopefully, they will recognize the problem when it occurs and properly correct the test mass ratio. Then, double strikes are gone.

The idea of shock excitation uses the same arrangement as shown in Fig. 6.1.2. In this case, the test item is mounted on mass m_2 (which should be at least ten times greater than the test item mass) and the resulting half sine acceleration is used as the input loading. If we divide Eq. (6.1.10) by m_2, we obtain the acceleration of m_2 as

$$a_2(t) = \begin{cases} v_0 \omega_n \sin(\omega_n t) & \text{for } 0 < t < T/2 \\ 0 & \text{for } t > T/2 \end{cases} \quad (6.1.14)$$

where the peak input acceleration is estimated by

$$a_0 = v_0 \omega_n \quad (6.1.15)$$

From Eq. (6.1.15), it is evident that it is easier to increase initial velocity v_0 in order to increase the magnitude of the maximum acceleration than it is to increase ω_n. The range of frequencies excited by this excitation technique is controlled by the impact frequency ω_n. A frequency analysis of a half sine time history shows that the first zero occurs in the frequency spectrum at a frequency of $1.5/T$ (or $1.5f$ since $f = 1/T$) where T is one-half of the natural period of the impact frequency, ω_n.

It should be evident that many different impact excitation schemes can be cooked up using this type of approach, leading to vertical drop, linear rail, and pendulous devices. The half sine impact occurs naturally with these linear systems. As the initial velocity increases, the half sine may become clipped, due to nonlinear spring behavior. However, this clipping only changes the input frequency spectrum, not the overall test concept. In fact, near triangular impact shapes have been developed using specially shaped lead impact surfaces. The reader who has such impact requirements should consult impact machine manufacturers.

6.1.3 Controlled Dynamic Loading Schemes

A number of different controlled dynamic loading schemes are employed in vibration testing. A typical block diagram is shown in Fig. 6.1.3, where there is a controller, power or energy source, an exciter device, and instruments that measure either input force and/or input motion and/or output motion. The input signals may be either periodic (most often sinusoidal), transient, or random. The controller can be either the analog type or a digital type. The controller sends out a signal to the energy source, which in turn drives the vibration exciter. Usually either the input force $P(t)$ or the input acceleration $a(t)$ is measured, amplified by a preamplifier and fed back to the controller, so that the proper control signal is provided to drive the energy source. There are situations where the structure's response is also fed back in order to provide limits on the output responses.

In the sections that follow, the emphasis will be primarily on the exciter's characteristics, and on how the exciter and the structure under test interact, a topic that has not received as much attention as it should, for significantly different behavior is found to occur in the most critical region, namely, around the structure's resonances. The control aspects will be discussed in more detail at the end of this chapter.

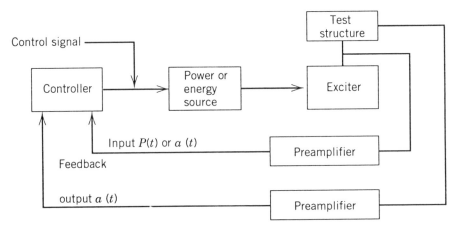

Fig. 6.1.3. Block diagram of controlled dynamic test system.

6.2 MECHANICAL VIBRATION EXCITERS

Mechanical vibration exciters are often divided into two different types, called *direct-drive* and *rotating unbalance*. The direct-drive type can be used as either a *reaction* exciter or a *vibration shake table*. The rotating unbalance exciter can be used only as a reaction type exciter. A reaction exciter develops its excitation force through an inertial loading that is caused by accelerating a reaction mass.

The direct-drive mechanical exciter consists of a table that is guided to have rectilinear motion and is driven by either a crank slider mechanism, a Scotch yoke mechanism, or a cam type mechanism that operates between points A and C, as shown in Fig. 6.2.1a. This mechanism moves the table relative to the base with nearly sinusoidal motion in the case of the crank slider driver, sinusoidal motion in the case of a Scotch yoke mechanism, and, in the case of a cam mechanism, variable types of motion, dependent on the cam design employed. The direct-drive exciter can be used in two ways, either as a reaction type when the exciter's base is attached to the structure under test, or as a shake table when the test item is attached to the exciter's table.

The rotating unbalance vibration exciter is a reaction type and is shown schematically in Fig. 6.2.1b, where two unbalance masses rotate in opposite directions in order to generate a dynamic reaction force that acts only in the y direction. Both the direct-drive and rotating unbalance mechanical exciter share a common set of basic concepts and characteristics. The concepts and characteristics of these two basic exciter types are explored in this section.

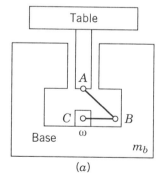
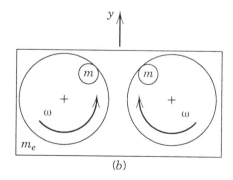

Fig. 6.2.1. Schematic of two basic mechanical vibration exciters. (*a*) Direct-drive. (*b*) Rotating unbalance.

6.2.1 The Direct-Drive Exciter Model

The direct-drive mechanical vibration exciter can be modeled as shown in Fig. 6.2.2. We see shortly that this model can represent two basic test situations, depending on whether the exciter's base is attached to the test structure or the test structure is attached to the exciter's table while its base is attached to a support foundation. System 1 consists of m_1, c_1, and k_1. Mass m_1 represents the exciter's base mass plus either a foundation mass or a test structure's mass. Similarly, c_1 and k_1 represent either a foundation or a test structure damping and stiffness, respectively. The table is modeled as system 2 and consists of table mass m_2, damping c_2 due to relative motion between table and the machine's base, and spring k_2, which represents the flexibility of the connecting drive links. The relative mechanical motion that is generated by the exciter mechanism is represented by $y(t)$, while $x_1(t)$ and $x_2(t)$ represent the base and table motions, respectively.

The equations of motion for this system can be written down immediately (see Chapter 3) as

$$m_1\ddot{x}_1 + C_{11}\dot{x}_1 + C_{12}\dot{x}_2 + K_{11}x_1 + K_{12}x_2 = -k_2 y(t)$$
$$m_2\ddot{x}_2 + C_{21}\dot{x}_1 + C_{22}\dot{x}_2 + K_{21}x_1 + K_{22}x_2 = k_2 y(t)$$
(6.2.1)

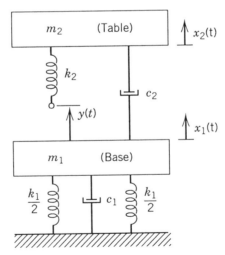

Fig. 6.2.2. Model of direct-drive mechanical vibration exciter mounted on a foundation or structure under test.

where

$$[C] = \begin{bmatrix} c_1 + c_2 & -c_2 \\ -c_2 & c_2 \end{bmatrix} \quad \text{and} \quad [K] = \begin{bmatrix} k_1 + k_2 & -k_2 \\ -k_2 & k_2 \end{bmatrix} \quad (6.2.2)$$

represent the damping and stiffness matrices.

The steady-state response of this exciter is obtained by assuming that the relative input mechanical motion between points A and C in Fig. 6.2.1a is given by

$$y(t) = y_0 e^{j\omega t} \quad (6.2.3)$$

The corresponding base and table motions are assumed to be given by

$$\begin{aligned} x_1(t) &= A_1 e^{j\omega t} \\ x_2(t) &= A_2 e^{j\omega t} \end{aligned} \quad (6.2.4)$$

where A_1 and A_2 are response phasors with magnitude and phase. Insertion of Eqs. (6.2.3) and (6.2.4) into Eq. (6.2.1) converts Eq. (6.2.1) into an algebraic equation from which we can solve for both A_1 and A_2.

However, before solving for A_1 and A_2, it is convenient to work with dimensionless terms. We start by defining the system's natural frequencies in terms of uncoupled natural frequencies, namely, for the exciter's base (system 1), we have

$$\omega_{11} = \sqrt{k_1/m_1} \quad (6.2.5)$$

and for the exciter's table (system 2), we have

$$\omega_{22} = \sqrt{k_2/m_2} \quad (6.2.6)$$

Also, let us define the mass ratio M as

$$M = m_2/m_1 \quad (6.2.7)$$

and the natural frequency ratio β as

$$\beta = \omega_{22}/\omega_{11} \quad (6.2.8)$$

Then the determinant of the coefficients in Eq. (6.2.1) when Eqs. (6.2.3) and (6.2.4) are substituted becomes

$$\Delta(\omega) = [K_{11} - m_1\omega^2 + jC_{11}\omega][K_{22} - m_2\omega^2 + jC_{22}\omega] - [K_{12} + jC_{12}\omega]^2$$
$$= \frac{k_1 k_2}{\beta^2} \Delta(r) \qquad (6.2.9)$$

where $\Delta(r)$ is the dimensionless expression of $\Delta(\omega)$ given by

$$\Delta(r) = [(1 + \beta^2 M) - r^2 + j(2\zeta_1 + 2\zeta_2 \beta M)r][\beta^2 - r^2 + j2\zeta_2 \beta r]$$
$$- [\beta^2 M][\beta + j2\zeta_2 r]^2 \qquad (6.2.10)$$

where r is the dimensionless frequency ratio given by

$$r = \frac{\omega}{\omega_{11}} \qquad (6.2.11)$$

The natural frequencies correspond to the real part of either Eqs. (6.2.9) or (6.2.10) being zero, so we have

$$r_1^2 = \left(\frac{\omega_1}{\omega_{11}}\right)^2 = \tfrac{1}{2}[1 + (1 + M)\beta^2 - \sqrt{(1 + (1 + M)\beta^2)^2 - 4}] \qquad (6.2.12)$$

as the first natural frequency ω_1 and

$$r_2^2 = \left(\frac{\omega_2}{\omega_{11}}\right)^2 = \tfrac{1}{2}[1 + (1 + M)\beta^2 + \sqrt{(1 + (1 + M)\beta^2)^2 - 4}] \qquad (6.2.13)$$

as the second natural frequency ω_2. It should be pointed out that the quantity within the square root is always a real positive number so that the natural frequencies are also real positive numbers.

In a typical application, β should be large compared to unity and M should be small compared to unity. Under these conditions, we can approximate Eq. (6.2.12) as

$$r_1 \cong \frac{1}{\sqrt{1 + M}} \qquad (6.2.14)$$

so that the fundamental natural frequency is less than ω_{11}. This natural frequency corresponds to masses m_1 and m_2 moving together in phase like a single rigid body with spring k_1 as the only actively deformed spring. Thus the first mode shape corresponds to rigid body motion of the two masses. The second natural frequency given by Eq. (6.2.13) is approximated as

$$r_2 \cong \beta\sqrt{1 + M} \tag{6.2.15}$$

in which case the two masses m_1 and m_2 move out of phase with one another in such a way so that $x_1 \cong - Mx_2$, spring k_2 is the active spring, and spring k_1 has no significant effect on the responses. It is evident from Eq. (6.2.15) that ω_2 is greater than ω_{22}. Thus the fundamental natural frequency is less than ω_{11} while the second natural frequency is greater than ω_{22}.

Now we can write down the solutions for the base motion as

$$A_1 = \left[\frac{k_2}{\Delta(\omega)}\right] m_2\omega^2 y_0 = \left[\frac{\beta^2 M r^2}{\Delta(r)}\right] y_0 \tag{6.2.16}$$

and for the table motion as

$$A_2 = \left[\frac{k_1 - m_1\omega^2\omega + jc_1\omega}{\Delta(\omega)}\right] k_2 y_0 = \left[\frac{1 - r^2 + j2\zeta_1 r}{\Delta(r)} \beta^2\right] y_0 \tag{6.2.17}$$

It is evident from Eq. (6.2.16) that the base motion is excited by a reaction force that corresponds to the exciter table's inertia force given by

$$\text{reaction force} = m_2(\omega^2 y_0) = m_2 \text{ (acceleration)} \tag{6.2.18}$$

The table's motion in Eq. (6.2.17) is related to y_0 but is modified by the bracketed quantity that depends on frequency ratio r.

Typical base and table motions are shown in Figs. 6.2.3a and 6.2.3b, respectively. The dimensionless quantities in Eqs. (6.2.16) and (6.2.17) are plotted for $\beta = 40$, $M = 0.5$, $\zeta_1 = 0.01$, and $\zeta_2 = 0.01$. A larger mass ratio M is used for illustration purposes so that the response's salient features show more clearly. Usually $M \ll 0.5$.

Now let us examine the base response shown in Fig. 6.2.3a. We see that the base motion increases along a line with a slope of 12 dB/octave below the fundamental natural frequency at r_1, passes through a resonance around frequency r_1, becomes nearly constant between r_1 and r_2, passes through a second resonance around frequency r_2, and finally drops off at 12 dB/octave for frequencies above r_2. The resonances occur at $r_1 \cong 0.82$ and $r_2 \cong 0.49$, as predicted by Eqs. (6.2.14) and (6.2.15), respectively.

The table response in Fig. 6.2.3b starts at unity, passes through resonance at r_1, passes through a sharp dip at $r = 1$, becomes nearly constant in the region between r_1 and r_2, passes through resonance at r_2, and drops

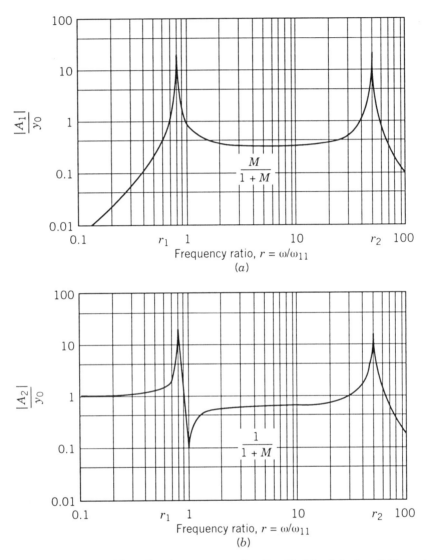

Fig. 6.2.3. Direct-drive vibration response for $\beta = 40$, $M = 0.5$, $\zeta_1 = 0.01$, and $\zeta_2 = 0.01$. (*a*) Exciter base motion. (*b*) Exciter table motion.

off at 12 dB/octave above r_2. The significant dip at $r = 1$ is caused by the numerator in Eq. (6.2.17) becoming small when $r \cong 1$. At this frequency, the base (system 1) is acting like a *dynamic absorber* to the exciter table (system 2). In fact, the base motion has near unity amplitude that is nearly 180 degrees out of phase to the table motion. Now we look at using this system as a vibration exciter.

6.2.2 The Direct-Drive Vibration Exciter Table

There are two basic ways that the direct-drive vibration exciter system can be installed in order to excite a test item that is attached to the exciter table. The first method is to use a lightly damped foundation (called a lightly damped foundation case), while the second method is to use a heavily damped foundation (called the heavily damped foundation case). We briefly explore both cases when only an *inertial load* is attached to the exciter table.

Bare Table with Lightly Damped Foundation This case is illustrated in Fig. 6.2.3. Figure 6.2.3a shows two resonant peaks with a flat region between r_1 and r_2 where the magnitude of A can be estimated from

$$A_1 = \frac{M}{1 + M} y_0 \qquad (6.2.19)$$

or about 0.33 for the plot shown. Figure 6.2.3b shows that the table's motion is relatively constant in the frequency region from $r \cong 2$ to $r \cong 30$. The approximate magnitude of A_2 in this mid frequency region is given by

$$A_2 = \left[\frac{1}{1 + M}\right] y_0 \qquad (6.2.20)$$

or about 0.67 for the graph shown. It is obvious that this input motion is also affected by the dynamic characteristics of the structure attached to the exciter table. Hence, this is the best input response we can hope for. Consequently, the exciter's useful range is limited to being above the foundation's natural frequency r_1 and below the exciter's internal natural frequency r_2.

Note that the addition of any mass to the exciter's table will increase m_2, and hence, reduce the useful frequency range by lowering ω_2. In addition, adding a test structure of mass m_s and stiffness k_s will cause an additional "dynamic absorber" action to show up at the test structure's natural frequency $\omega_s = \sqrt{k_s/m_s}$ in the table's input motion.[1] Consequently, mechanical vibration test machines do not have constant motion amplitude over the frequency range of interest.

Bare Table with a Highly Damped Foundation The use of a highly damped foundation can significantly increase the effective range of the exciter, as

---

[1] C. M. Harris and C. S. Crede (Editors), *Shock and Vibration Handbook*, Vol. 2, McGraw-Hill, New York, Chapter 25 by K. Unholtz, "Vibration Testing Machines," pp. 25-7–25-9.

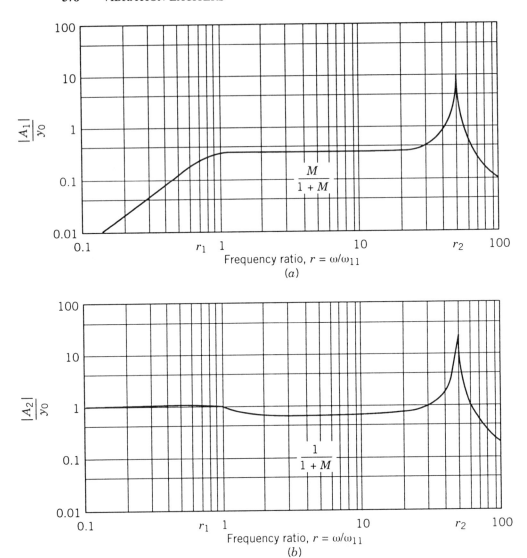

Fig. 6.2.4. Direct-drive response for $\beta = 40.9$, $M = 0.5$, $\zeta_1 = 0.7$, and $\zeta_2 = 0.01$. (a) Exciter base motion. (b) Table motion.

shown in Fig. 6.2.4, where ζ_1 is increased to 0.7 while the mass ratio remains at 0.5. The mass ratio is left large in order to show the effectiveness of foundation damping. In this case, the base motion increases at a rate of 12 dB/octave below r_1, flattens out at a nearly constant value given by Eq. (6.2.19) in the interval between r_1 and just below r_2, passes through resonance around r_2, and then falls off at 12 dB/octave above r_2. The table motion is seen to start out at amplitude y_0 below r_1 and fall to an amplitude given by Eq. (6.2.20) in the region between r_1 and r_2, pass through resonance around r_2, and then drop out at 12 dB/octave. It is clear that the usable frequency range is increased by increasing the foundation damping. It is also seen from Eqs. (6.2.19) and (6.2.20) that it is a good idea to make the foundation mass large compared to the table mass so that M is small. Then $A_1 \cong M y_0$ and $A_2 \cong y_0$.

The major complaint about this second mounting method is the amount of vibration that is transmitted to the surrounding structures. Consequently, we should install such exciters on vibration isolators with light damping while using as small a value of M as practical. Then, with the lower value of ω_1 controlled by m_1 and k_1 and the value of ω_2 controlled by the exciter design, we obtain the largest possible usable range for a given exciter.

6.2.3 The Direct-Drive Reaction Type Vibration Exciter

The direct-drive reaction type of application follows directly from Fig. 6.2.3. In this case, the exciter's base is attached to the structure under test. This means that mass m_1 is the sum of both the structure's mass m_s and the exciter base's mass m_b. Then, Fig. 6.2.3a shows how the structure reacts to the inertial excitation force of $m_2 \omega^2 y_0$ from low frequencies to significantly above r_1. This is precisely the type of response we would expect from such an excitation force. Unfortunately, there can be a significant error in the structure's measured natural frequency since we measure ω_1 as given by

$$\omega_1 = \sqrt{\frac{k_1}{m_s + (m_b + m_2)}} \qquad (6.2.21)$$

as estimated from Eq. (6.2.14). It is clear that $(m_b + m_s)$ represents the exciter's mass that must be compared to the structure's mass m_s as a source of error in the measured natural frequency. This same frequency error limitation applies to rotating unbalance vibration exciters as well.

6.2.4 The Rotating Unbalance Exciter

The rotating unbalance exciter is described in most vibration textbooks. For our purposes, we can represent this exciter as two counterrotating masses m with eccentricity e, as shown in Fig. 6.2.5. In this case, the total mass m_t is given by

VIBRATION EXCITERS

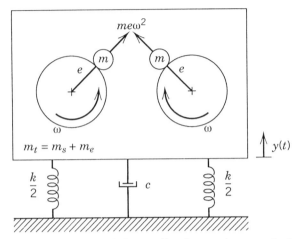

Fig. 6.2.5. Model of rotating unbalance vibration exciter attached to structure under test.

$$m_t = m_s + m_e \tag{6.2.22}$$

where m_s is the test structure's mass and m_e is the exciter's mass that includes the two rotating unbalance masses m. For this system, it is well known that the differential equation of motion is given by

$$m_t \ddot{y} + c\dot{y} + ky = 2me\omega^2 \sin(\omega t) \tag{6.2.23}$$

It is clear that we have two inertia forces of $me\omega^2$ adding together to excite the test structure, as shown in Fig. 6.2.5. The horizontal components of the two rotating inertia forces are seen to cancel one another. Thus we see why it is advantageous to use two counterrotating unbalances to generate the excitation force.

The steady-state amplitude for Eq. (6.2.23) is given by

$$A = \frac{2me\omega^2}{k - m_t\omega^2 + jc\omega} = \left[\frac{2me}{m_t}\right] \frac{r^2}{1 - r^2 + j2\zeta r} \tag{6.2.24}$$

which is plotted in Fig. 6.2.6 for $\zeta = 0.05$ and $2m/m_t = 1$. It is clear from this plot that the amplitude of response increases at a rate of 12 dB/octave below the natural frequency, peaks around the natural frequency, and levels out at an amplitude given by

$$A = \left[\frac{2me}{m_t}\right] \tag{6.2.25}$$

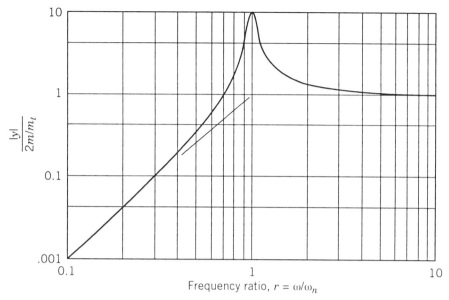

Fig. 6.2.6. Magnitude of structure's response when excited by rotating unbalance exciter. $\zeta = 0.05$, $m = 0.5$, $m_t = 1$, $e = 1$, so that $2m/m_t = 1$. True natural frequency is actually higher than that measured.

Equation (6.2.25) indicates that the final vibration state is one where the excitation inertia forces are counterbalanced by the inertia force due to the total mass. It is also clear that the resonance will be underestimated, due to the mass of the exciter being included in the structure's total apparent mass. When dealing with such a test system, we can estimate the effects of the exciter's mass on the test results as follows. Since the exciter's mass affects each mode shape in a complex structure differently, we can attach a mass equal to the exciter's mass to the structure at the same location, repeat the experiment, and note how much the natural frequencies have changed. These changes are a crude estimate of the exciter's mass on the measured natural frequency.

6.2.5 Driving Torque Considerations

It is one thing to assume that the input frequency ω is constant and it is quite another to drive a mechanical vibration exciter with a constant input frequency. For the two mass rotating exciter, the torque required to maintain a constant rotating speed can be estimated by

$$T \cong -\left[\frac{2m}{m_t}\left\{\frac{r^2}{1-r^2+j2\zeta r}\right\}+1\right]me^2\omega^2\sin(2\omega t) \qquad (6.2.26)$$

While Eq. (6.2.26) is approximate, it can give us some insights. First, the torque varies as a sinusoid at twice the excitation frequency. Second, the low speed torque magnitude is approximately $me^2\omega^2$. Third, as the excitation frequency approaches the structure's natural frequency, r approaches unity and the resonant term dominates the bracketed terms. When this happens, a very complicated interaction begins to occur between the excitation frequency ω, the structure's response, and the exciter's drive motor torque capacity.

Often we are unable to actually measure a structure's peak response at resonance. This can happen if the exciter motor is open loop so that it has no precise speed control other than the amount of power supplied. Such an exciter will suddenly accelerate through the structure's resonant frequency as this frequency is approached from below. If we reduce power to lower the excitation frequency and sneak up on resonance from above, we find that there is a limit to the magnitude of motion that we can excite. Consequently, we are left with a hole in the response plot right around resonance. The amount of frequency variation due to this torque variation can be controlled by using very large inertia wheels in the exciter's design as well as by using drive motors that have significant speed control. These comments about torque apply to the direct-drive vibration exciters as well.

Mechanical vibration exciters have limited application for general purpose testing due to limited frequency range, structural rigidity, large exciter mass, bearing wear (leading to banging and impact loads), speed control around resonance, and so on. However, there are many low frequency applications where they are ideally suited, particularly when the test frequency and input levels are nearly constant. One of their most significant competitors for low frequency testing is the electrohydraulic vibration exciter that is discussed in Section 6.3.

6.3 ELECTROHYDRAULIC EXCITERS

The electrohydraulic vibration exciter is generally used for low frequency excitation environments that require large amounts of force and relatively low velocities. The frequency range varies from near 0 Hz on the low end to 40 to 100 Hz on the high end. The exact range depends on a number of factors such as pump and servovalve flow rate capacity. We look at these factors in more detail in this section.

The electrohydraulic vibration exciter was originally developed as a hydraulic materials testing machine. Fatigue testing applications required that these machines operate at higher frequencies in order to do the tests more rapidly. Other applications include wave generators for conducting fluid structure interaction studies, earthquake simulators, and so on. As this technology developed, the upper usable frequency range increased until the electrohydraulic vibration exciter became a reality. In these applications, the electrohydraulic cylinder becomes the generator of motion $y(t)$ in Figs. 6.2.1 and 6.2.2. Thus all of the dynamic considerations of Section 6.2 apply to these devices, which require a reaction mass and a moving table mass.

6.3.1 Electrohydraulic System Components

A typical closed loop hydraulic vibration exciter block diagram is shown in Fig. 6.3.1. The major components are the system controller, which can be either analog or digitally based, the valve driver, the servovalve, the hydraulic pump (power supply), the hydraulic actuator, the moving table, and the instrumentation. The error signal is used to drive the valve driver, which is an electrohydraulic device that actuates the larger servovalve, which in turn controls the amount of oil that is forced into the hydraulic actuator. The hydraulic actuator drives the table, which in turn excites the test item. The input force and/or motion is used as a feedback to establish that the desired motion is taking place.

The hydraulic power supply usually consists of a high performance hydraulic pump with appropriate pressure relief valves, oil coolers, and so on. The flow rate performance of such pump systems in conjunction with an appropriate servovalve is a function of frequency, as shown in

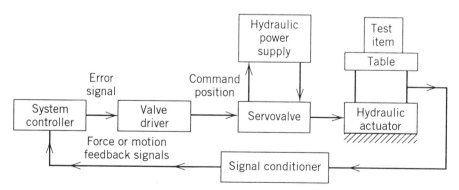

Fig. 6.3.1. Typical closed loop control system with major components.

384 VIBRATION EXCITERS

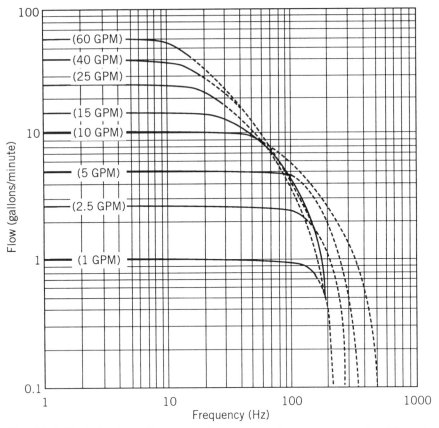

Fig. 6.3.2. Typical volume flow versus frequency performance curves for different servovalves and actuators. (Adapted from MTS Maintenance Manual. Courtesy MTS Systems Corp.)

Fig. 6.3.2. The dashed area of the curves at the higher frequencies reflect the effects of the servovalve, hydraulic actuator, and test system dynamics.

The hydraulic connections of a four-way servovalve and the actuator are shown in Fig. 6.3.3, where servovalve position α controls the direction of oil flow from the pump's high pressure supply side line P to the actuator piston and from the piston low pressure side to the pump's reservoir line R. When α is positive, the spool is moved to the right of the neutral position so that the high pressure oil is admitted to the right-hand side of the actuator piston and the low pressure oil is drained to the reservoir on the left-hand side of the piston. This causes the piston to move to the left with a positive velocity v or to develop a positive force F on the table of test item.

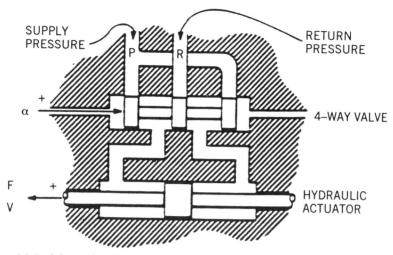

Fig. 6.3.3. Schematic of servovalve and actuator to describe system variables. (Courtesy MTS Systems Corp.)

The spool position α is scaled so that it has a maximum value of $\pm$ unity. The scaled force $\bar{F}$ varies between -1 and 1 and is defined by

$$\bar{F} = \frac{F}{F_{max}} \qquad (6.3.1)$$

where F_{max} is the maximum actuator force that corresponds to the full effective supply pressure drop occurring across the actuator. Similarly, the scaled velocity $\bar{v}$ varies between -1 and 1 and is defined by

$$\bar{v} = \frac{v}{v_{max}} \qquad (6.3.2)$$

where v_{max} is the maximum velocity with full effective supply pressure drop across the servovalve. The relationship between the scaled force, scaled velocity, and spool position is[2]

$$\bar{F} = \frac{|\bar{v}|}{\bar{v}} \left[1 - \frac{\bar{v}^2}{\alpha^2} \right] \qquad (6.3.3)$$

---

[2] See A. J. Clark, 'Sinusoidal and Random Motion Analysis of Mass Loaded Actuators and Valves," *Proceedings of the National Conference on Fluid Power*, Volume XXXVII, 39th annual meeting, Los Angeles, CA, 1983.

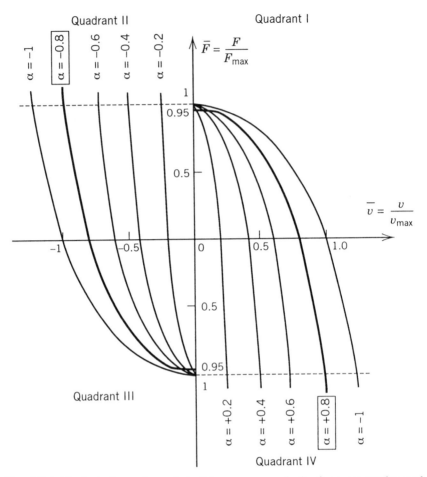

Fig. 6.3.4. Typical phase plane plot of force versus velocity for a servovalve and actuator for constant control positions α. (Courtesy MTS Systems Corp.)

The relationship described by Eq. (6.3.3) is shown in Fig. 6.3.4, where $\bar{F}$ is plotted against $\bar{v}$ for constant values of α. The system is usually designed to operate so that $-0.8 < \alpha < 0.8$ in order to provide some reserve capacity for control purposes. It is obvious from Fig. 6.3.4 and Eq. (6.3.3) that the relationship among these variables is nonlinear. In quadrants II and IV the velocity is greater than unity, a result due to the servovalve acting as an energy dissipating device. Note that the odd symmetry of the curves can cause a change in slope as the velocity passes through zero. Thus it is desired to flatten out the force curve to be near 0.95 of the maximum value. It is obvious that a carefully designed controller is required for this application.

TABLE 6.3.1. Hypothetical System Parameters

Feature		Parameters
Size of table platform		6.0×6.0 m
Specimen mass		50,000 kg
Table mass		40,000 kg
Actuator moving mass		5,000 kg
Test specimen height		6.0 m
Test specimen overturning moment		1000 kN-m
Table motion limits:		
Displacement (meters)	Longitudinal, lateral	±0.2
	Vertical	±0.1
Velocity (meters/second)	Longitudinal, lateral	±1.0
	Vertical	±0.5
Acceleration (g)	Longitudinal, lateral	±1.0
	Vertical	±0.5
Frequency of operation		0.4–40 Hz
Actuator force (per unit)		±250 kN
Foundation mass		4,000,000 kg
Foundation dimensions		$16 \times 16 \times 7$ m

6.3.2 Application to Earthquake Simulator

The parameters and a typical configuration for a hypothetical earthquake motion shaking table, as described by Clark,[3] are given in Table 6.3.1 and shown in Fig. 6.3.5, respectively. This is a six degree-of-freedom system that has three linear and three angular motions. The large test table is 6 meters square with a mass of 40,000 kg and is connected to the 4,000,000 kg reaction mass by 12 (250 kN) hydraulic actuators. The reaction mass is isolated from the earth by an independent air spring suspension system that has a natural frequency of 0.8 Hz longitudinally and 1 Hz for rotational motion with damping ratios of 16 and 20 percent, respectively. Table 6.3.2 shows the natural frequencies and the controlling phenomena for each of these frequencies.

In order to demonstrate the type of interaction that can occur, the longitudinal table acceleration per unit of error voltage (see Fig. 6.3.1) FRF is plotted in Fig. 6.3.6 (in dB) for this system over a 0.2 to 200 Hz bandwidth. These plots represent the characteristics of the "plant to be controlled." The unloaded response is shown in Fig. 6.3.6a, where a resonance is observed around 18 Hz. The near linear increase of 20 dB

---

[3]A. J. Clark, "Dynamic Characteristics of Large Multiple Degree of "Freedom" Shaking Tables," *Earthquake Engineering, Tenth World Conference*, Balkema, Rotterdam, Holland, 1992, pp. 2823–2828.

388 VIBRATION EXCITERS

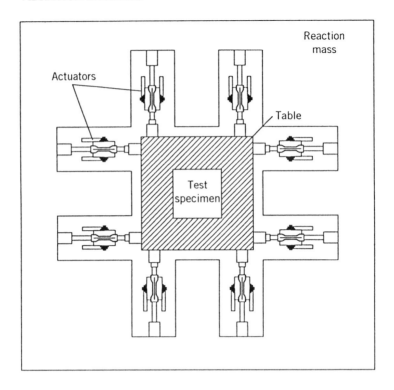

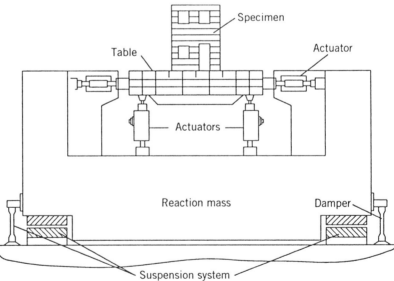

Fig. 6.3.5. Mechanical configuration of a typical earthquake simulator showing 12 actuators mounted with full spherical swivel bearings at each end. (Courtesy MTS Systems Corp.)

TABLE 6.3.2. Natural Frequencies and Damping Ratios

Controlling Phenomenon	Natural Frequency (Hz)		Damping (Percent)	
	Long	Pitch	Long	Pitch
Air spring suspension	0.80	1.0	16	20
Reaction mass flexibility	60	70	1.0	1.5
Load train compliance	40	60	5.0	5.0
Servovalve response	100	100	70.0	70.0
Oil column compliance	20	40	10.0	15.0
Actuation lateral bending	45	50	5.0	5.0
Table internal flexibility	110	90	1.0	1.0
Specimen flexibility	10	80	1.0	1.0

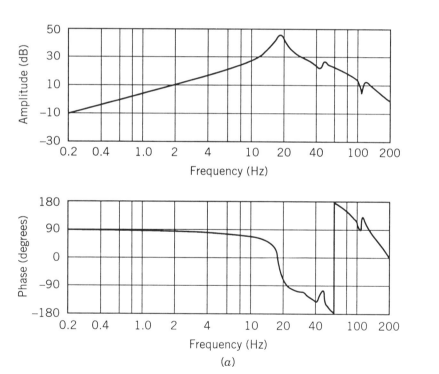

Fig. 6.3.6. Typical exciter table response FRF for longitudinal motion. (*a*) Unloaded. (*b*) Loaded with 5000 kg rigid mass. (Courtesy MTS Systems Corp.)

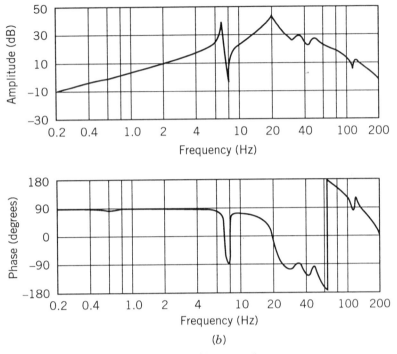

Fig. 6.3.6. (*Continued*).

per decade shows how the servovalve input voltage and table inertia interact. There is a small resonance bump around 45 Hz that corresponds to lateral bending of the actuator.

When the 50,000 kg test item that has a horizontal natural frequency of 10 Hz and a vertical natural frequency of 80 Hz is installed on the exciter's table, we see in Fig. 6.3.6b that the test item's natural frequency has dropped to about 7 Hz from 10 Hz. Also, we see that an 8 Hz dynamic absorber valley following the resonant peak. We also see that the oil column compliance resonance is clearly present at 20 Hz. The falloff after 20 Hz is nearly the same in each case. However, a close inspection shows that other resonant frequencies are present in this response as well.

A comparison of Figs. 6.3.6a and 6.3.6b clearly shows how the test item's dynamic characteristics change the "plant to be controlled." The upshot of this comparison shows the challenges that face the control system designer. What signals should be used for feedback? How should these signals be modified in order to maintain system control regardless of the dynamic characteristics of the test structure? Fortunately, as users we do not have to concern ourselves with these issues.

6.4 THE MODELING OF AN ELECTROMAGNETIC VIBRATION EXCITER SYSTEM

The electrodynamic vibration exciter is extremely popular for use in vibration testing, due to the wide range of force vectors and frequencies that we can obtain. The sizes available range widely; there are very small exciters that are capable of developing less than 1 lb peak sinusoidal forces from near 0 to over 10 kHz, and there are large systems that develop 40,000 lb of sinusoidal force over a more limited frequency range from near 0 to approximately 3 kHz.

The general test arrangement of an electrodynamic vibration exciter is shown in Fig. 6.4.1. The *controller* generates an output voltage $V(t)$ that is the input voltage to the power amplifier. Voltage $V(t)$ depends on the feedback information of either acceleration and/or force. Different controller schemes will be discussed later in this chapter.

The power amplifier generates an output voltage $E(t)$ and corresponding current $I(t)$ that drives the vibration exciter. The power amplifier has two modes of operation called *voltage mode*, where $E(t)$ is proportional to the input voltage $V(t)$, and *current mode*, where $I(t)$ is proportional to

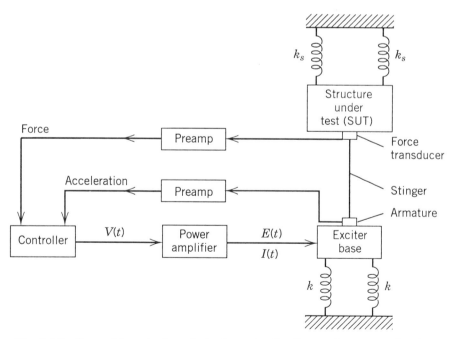

Fig. 6.4.1. General layout of an electrodynamic vibration exciter system showing controller, feedback of force and/or acceleration, power amplifier, stinger, and exciter and structure support systems.

the input voltage $V(t)$. The electrodynamic vibration exciter converts the electrical power into heat and mechanical power, which is then transmitted to the structure under test (SUT).

The *armature* is the exciter's moving element, and it is connected to the SUT by a *stinger*. The stinger is a connecting link that is axially stiff but flexurally weak in order to reduce the transmission of bending moments to the armature on one end and the force transducer and SUT on the other end. Often the armature has an accelerometer built into it, so this acceleration may be used for feedback purposes. Similarly, a force transducer and/or accelerometer may be attached to the SUT at its excitation point for feedback purposes as well.

The SUT may be either *grounded*, in which case the support springs k_s are ideally rigid, or *ungrounded*, in which case the support springs k_s are very soft compared to the structure's stiffness. In the ungrounded case, the structure appears to be free in space and has low frequency rigid body vibration modes that are dependent on the SUT's mass and rigid body mass moments of inertia as well as support springs k_s. It is important that the highest of these low frequency rigid body vibration modes be less than one-fifth the lowest natural frequency of the free structure. Otherwise, these support springs will significantly influence the structure's first and possibly the second of the freely supported natural frequencies. Similarly, the vibration exciter may be grounded or ungrounded by its support springs k.

The typical electrodynamic exciter is constructed like the one shown schematically in Fig. 6.4.2. The exciter consists of two main parts, a heavy exciter base containing magnetic fields and a moving element or *armature*.

The armature consists of three parts called *table*, *spider*, and *coil*. The table is the armature's end that is attached to the SUT. The spider is the structure that connects the armature's table and coil together; it may consist of numerous columns or of a thin cylindrical shell. The coil consists of wire wrapped around a lightweight nonmagnetic core in the form of an aluminum or magnesium thin shell. In some cases a solid aluminum conductor in the form of a cylindrical shell is used for the coil in place of the heavier copper wire. In any event, it is important to produce a strong magnetic field in the gap between the concentric cylindrical poles in which the coil moves freely. The armature is supported on flexures (springs k_f) that are stiff to any lateral and rotational motion but relatively soft for axial motion, as shown. The flexures are attached to the *table guide ring*, which is attached to the exciter's base.

The exciter's base is usually massive and has a magnetic field flux path as shown. This magnetic flux is due to a permanent magnet in smaller exciters (up to about 100 lb capacity) and a *field coil* (as shown) for the large exciter units. A strong stray magnetic field may exist around the table top due to the gap. This field is reduced by using a *DeGaussing coil*

AN ELECTROMAGNETIC VIBRATION EXCITER SYSTEM

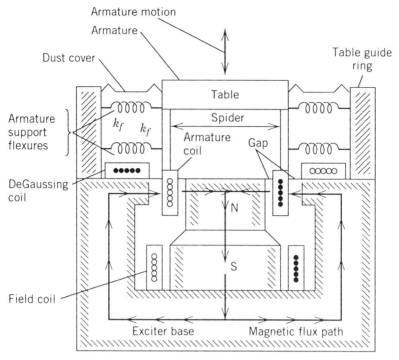

Fig. 6.4.2. Cross section of an electrodynamic exciter showing base, magnetic flux path, armature table and coil, and gap.

that produces an equal and opposite magnetic field. The armature assembly is protected from foreign material getting into the magnetic gap by a flexible *dust cover*. Cooling air is often blown over the armature and field coils to remove heat. Also note that some armatures form an air spring between the table top and the magnet's cylinder, with a small gap around the coil for air leakage when the spider is a closed thin shell.

In this section, we develop the various parameters that control the exciter's behavior. Since there are many parameters, we explore their characteristics in isolated combinations and use these results to reduce the model's complexity in order to understand the types of effects each parameter has. The interaction of the exciter with the structure under test will be discussed in detail in Chapter 7.

6.4.1 Exciter Support Dynamics

In this subsection, we are interested in exploring the dynamics of the vibration exciter when mounted in its environment. The low frequency dynamic model can be represented by the two DOF system shown in Fig.

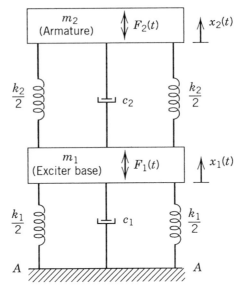

Fig. 6.4.3. Low frequency dynamic model of an electrodynamic exciter consisting of base mass and armature mounted on foundation A–A.

6.4.3 where $F_1(t)$ and $F_2(t)$ represent the electrodynamic force that acts on both the armature and the exciter base at the same time so that they are equal and opposite. Hence,

$$F_1(t) = -F_2(t) = -F(t) \tag{6.4.1}$$

Spring k_1 and viscous damper c_1 represent the exciter's foundation. Spring k_2 and damper c_2 represent the armature's support flexures (k_f in Fig. 6.4.2). The electromagnetic interaction is not considered in this analysis since we are concerned only with how the basic mechanical system behaves in response to a sinusoidal force acting between the armature and exciter base.

The differential equations of motion that govern this system's behavior are given by Eqs. (6.2.1) so that we have

$$\begin{aligned} m_1\ddot{x}_1 + C_{11}\dot{x}_1 + C_{12}\dot{x}_2 + K_{11}x_1 + K_{12}x_2 &= -F(t) \\ m_2\ddot{x}_2 + C_{21}\dot{x}_1 + C_{22}\dot{x}_2 + K_{21}x_1 + K_{22}x_2 &= F(t) \end{aligned} \tag{6.4.2}$$

where

$$[C] = \begin{bmatrix} c_1 + c_2 & -c_2 \\ -c_2 & c_2 \end{bmatrix} \quad \text{and} \quad [K] = \begin{bmatrix} k_1 + k_2 & -k_2 \\ -k_2 & k_2 \end{bmatrix} \quad (6.4.3)$$

represent the damping and stiffness matrices. If we assume that $F(t)$ is a sinusoidal phasor with frequency ω and that $x_1(t)$ and $x_2(t)$ are sinusoidal phasors of the same frequency, then Eqs. (6.4.2) have steady-state phasor solutions given by

$$X_1 = \left[\frac{m_2}{\Delta(\omega)}\right]\omega^2 F_0 = \left[\frac{r^2}{\Delta(r)}\right][\beta^2 M]\frac{F_0}{k_2} \quad (6.4.4)$$

for the foundation motion $x_1(t)$ and

$$X_2 = \left[\frac{k_1 - m_1\omega^2 + jc_1\omega}{\Delta(\omega)}\right]F_0 = \left[\frac{1 - r^2 + j2\zeta_1 r}{\Delta(r)}\right][\beta^2]\frac{F_0}{k_2} \quad (6.4.5)$$

for the armature's motion $x_2(t)$. Variables ω_{11}, ω_{22}, M, β, $\Delta(\omega)$, $\Delta(r)$, and r are defined in Eqs. (6.2.5) through (6.2.11), namely,

$$\omega_{11} = \sqrt{\frac{k_1}{m_1}} \quad \text{and} \quad \omega_{22} = \sqrt{\frac{k_2}{m_2}} \quad (6.4.6)$$

for the uncoupled natural frequencies,

$$M = \frac{m_2}{m_1} \quad \text{and} \quad \beta = \frac{\omega_{22}}{\omega_{11}} \quad (6.4.7)$$

for the mass ratio and frequency ratio; and

$$\Delta(\omega) = [K_{11} - m_1\omega^2 + jC_{11}\omega][K_{22} - m_2\omega^2 + jC_{22}\omega] - [K_{12} + jC_{12}\omega]^2$$

$$= \frac{k_1 k_2}{\beta^2}\Delta(r) \quad (6.4.8)$$

for the characteristic frequency equation, where $\Delta(r)$ is expressed as

$$\Delta(r) = [(1 + \beta^2 M) - r^2 + j(2\zeta_1 + 2\zeta_2 \beta M)r][\beta^2 - r^2 + j2\zeta_2 \beta r]$$
$$- [\beta^2 M][\beta + j2\zeta_2 r]^2 \quad (6.4.9)$$

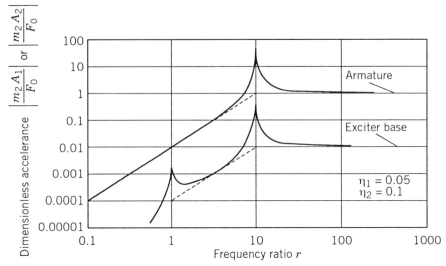

Fig. 6.4.4. Magnitude of armature and base acceleration as a function of dimensionless frequency ratio r for $M = 0.01$ and $\beta = 10$ when exciter is mounted on a soft base.

and r is the dimensionless frequency ratio given by

$$r = \frac{\omega}{\omega_{11}} \tag{6.4.10}$$

First, we need to establish typical values for M and β. An analysis of one manufacturer's product line shows M to range from 0.5 percent to 1.2 percent with a mean value of 0.8 percent for exciters that range from 2 lb force vector to over 1000 lb force vector. Thus for convenience, we use a 1 percent mass ratio for calculation purposes. This analysis also shows that typical armature natural frequencies ω_{22} range from 20 to 90 Hz.

We need to see how a typical system acts dynamically when we use two different installation concepts. The foundation natural frequencies can be either low around 2 Hz or much higher around 200 Hz. The 2 Hz foundation frequency corresponds to an isolated vibration exciter, while a 200 Hz foundation natural frequency corresponds to an attempt to *rigidly mount* the exciter's base. Thus we have $M = 0.01$ and $\beta = 10$ for the isolated vibration exciter and $M = 0.01$ and $\beta = 0.10$ for the "rigidly mounted" vibration exciter. The different response curves for the armature and exciter base are shown in Figs. 6.4.4 and 6.4.5.

Figure 6.4.4 is a plot of a dimensionless acceleration frequency response function (FRF) for the exciter's base that is obtained from Eq. (6.4.4) as

$$\frac{m_2 A_1}{F_0} = -\left[\frac{\omega^2}{\Delta(\omega)}\right] = -\left[\frac{r^4}{\Delta(r)}\right] M \quad (6.4.11)$$

where A_1 is the base acceleration while the armature's dimensionless accelerance FRF is obtained from Eq. (6.4.5) as

$$\frac{m_2 A_2}{F_0} = -\left[\frac{k_1 - m_1\omega^2 + jc_1\omega}{\Delta(\omega)}\right] m_2 \omega^2 = \left[\frac{1 - r^2 + j2\zeta_1 r}{\Delta(r)}\right] r^2 \quad (6.4.12)$$

where A_2 is the armature's acceleration. The damping term $(2\zeta_1 r)$ is replaced by loss factor η_1 and $(2\zeta_2 r)$ is replaced by loss factor η_2, since experimental evidence by Tomlinson[4] on a number of exciters shows the armature's damping is nearly constant with frequency. The air spring between the table and the magnet with the small air gap at the coil (see Fig. 6.4.2) may contribute significantly to this structural damping. Such arrangements are known to produce both a significant spring rate and hysteresis losses when the air is forced in and out through a small hole. For calculation purposes, η_1 and η_2 are assumed to be 0.05 and 0.1, respectively. Frequency ratio r is given by Eq. (6.4.10) and is based on the exciter's base natural frequency ω_{11}, which depends primarily on k_1 and m_1.

The armature's acceleration is seen to increase at the rate of 40 dB/decade up to the armature's flexure natural frequency ω_{22} ($r = 10$), where armature resonance occurs, and then to flatten out to a value of unity for $r \gg 10$. Based on the armature's loss factor, this acceleration curve should peak around 10 but it is closer to 70. Thus the more lightly damped exciter base motion influences the measured armature acceleration, a result that can be overlooked when trying to estimate armature damping from the peak of the armature's acceleration curve compared to the horizontal line. Also, note that the armature does not show a resonance around $r = 1$. However, a close examination shows a slight discontinuity in the acceleration curve that corresponds to the exciter's base passing through its first resonance at $r = 1$.

The base acceleration is seen to increase at a rate of 80 dB/decade below $r = 1$, pass through a base resonance at $r = 1$, then increase at a rate of 40 dB/decade between $r = 1$ and $r = 10$, pass through the armature's flexure resonance, and finally approach a value of 0.01 for increasing frequencies above $r = 10$. We see that the exciter base motion is 0.01

---

[4]Private conversation with Professor G. R. Tomlinson from University of Manchester, Manchester, U.K., July 1991.

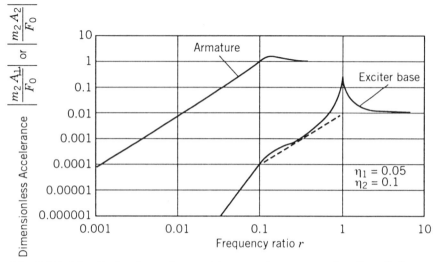

Fig. 6.4.5. Magnitude of armature and base accelerance as a function of dimensionless frequency ratio r for $M = 0.01$ and $\beta = 0.1$ when exciter is mounted on a stiff base.

compared to the armature's motion of unity. This is precisely the response that we would anticipate for two masses connected by a spring and experiencing equal and opposite excitation forces when in free space. In other words, the dynamic behavior above $r = 10$ is the same as that obtained for a system in free space without any foundation spring k_1.

The case shown in Fig. 6.4.5 corresponds to $\beta = 0.1$. In this case, the armature's flexure resonance occurs at $r = 0.1$ while the support system resonance occurs at $r = 1$. Again the armature's acceleration increases at approximately 40 dB/octave, passes through a slight resonance, and then takes on a value of unity. The exciter base follows a similar response, as it did previously, with no apparent resonance corresponding to the armature's resonance, where only a break in slope occurs. In this case, the armature's acceleration shows much higher damping than in the previous case at the armature's flexure resonance frequency. Again, we can be deceived about the armature's damping, but this is the best mounting for measuring it without resorting to measuring the relative motion. Also, we see that both accelerations become the same as in Fig. 6.4.4 at high frequencies, namely, 0.01 for the base and 1 for the armature.

The only disadvantage of the case shown in Fig. 6.4.5 is that more vibration energy is transmitted to the supporting foundation, since the support spring k_1 is much stiffer. The preferred installation is to have soft springs k_1 in order to minimize the forces transmitted to the surrounding environment. In this case, we can treat the exciter's base as having acceler-

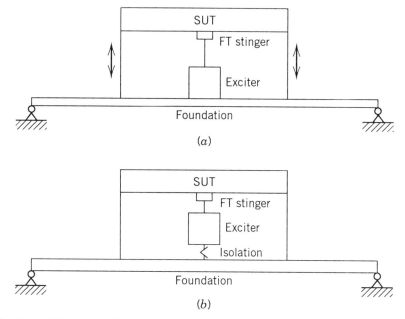

Fig. 6.4.6. Two ways of connecting the SUT and the foundation. (*a*) Coupling of exciter and structure through common foundation motion. (*b*) Isolation of exciter from structure's foundation.

ation that is equal to M times the armature's acceleration. *Thus we shall assume in our further models that the exciter's base is fixed.*

A further reason for using isolated vibration exciters is shown in Fig. 6.4.6. Here we model the foundation as a simply supported beam or floor panel. In Fig. 6.4.6a both the exciter and the SUT are attached to the same foundation so that any foundation motion is transmitted directly to the SUT. Thus the force transducer is measuring only part of the force transmitted to the SUT. In Fig. 6.4.6b, the isolated exciter base reduces dramatically the energy that is transmitted to the foundation. Thus it is good testing practice to use softly suspended exciters in order to break the foundation energy path.

6.4.2 Armature Dynamics

The armature is modeled as shown in Fig. 6.4.7. In this model, m_1 ($=m_t$) is the table mass and k_1 is the flexure's stiffness. k_2 is the spider's stiffness and m_2 ($=m_c$) is the coil's mass. Then we see that the armature mass m_a

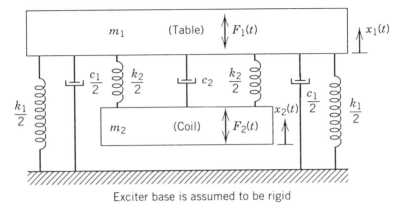

Fig. 6.4.7. Dynamic model of an electrodynamic exciter's armature structure composed of table and coil masses, flexure stiffness k_1 and spider stiffness k_2. Damping is shown as viscous but is structural in reality.

is given by

$$m_a = m_1 + m_2 = m_t + m_c \tag{6.4.13}$$

The exciter's base is assumed to be rigidly mounted to the earth, consistent with the results obtained above in Section 6.4.1. Then the equations of motion for the armature are given by

$$m_1\ddot{x}_1 + C_{11}\dot{x}_1 + C_{12}\dot{x}_2 + K_{11}x_1 + K_{12}x_2 = F_1(t)$$
$$m_2\ddot{x}_2 + C_{21}\dot{x}_1 + C_{22}\dot{x}_2 + K_{21}x_1 + K_{22}x_2 = F_2(t) \tag{6.4.14}$$

where $[c]$ and $[k]$ are described by Eq. (6.4.3). The steady-state acceleration response phasors to Eq. (6.4.14) are given by

$$A_1 = \frac{A_{11}(\omega)}{m_a} F_1 + \frac{A_{12}(\omega)}{m_a} F_2 \quad \text{(table)} \tag{6.4.15}$$

for the table and

$$A_2 = \frac{A_{21}(\omega)}{m_a} F_1 + \frac{A_{22}(\omega)}{m_a} F_2 \quad \text{(coil)} \tag{6.4.16}$$

where A_1 and A_2 are the acceleration phasors.

$A_{pq}(\omega)$ are the armature's dimensionless acceleration response phasors at p due a force at q,

F_1 and F_2 are the forces applied to the table and coil, respectively. The dimensionless accelerance's are given by

$$A_{11}(\omega) = \frac{m_a A_1}{F_1} = \frac{-r^2(1+M)(\beta^2 - r^2 + j\beta\eta_2)}{\Delta(r)} \quad (6.4.17)$$

for the table's acceleration due to force F_1,

$$A_{12}(\omega) = A_{21}(\omega) = \frac{m_a A_1}{F_2} = \frac{-r^2(1+M)(\beta^2 + j\beta\eta_2)}{\Delta(r)} \quad (6.4.18)$$

for either the coil's or the table's acceleration due to either F_1 or F_2, and

$$A_{22}(\omega) = \frac{m_a A_2}{F_2} = \frac{-r^2(1+M)[(1+M\beta^2) - r^2 + j(\eta_1 + \beta M \eta_2)}{M\Delta(r)}$$

$$(6.4.19)$$

for the coil's acceleration due to force F_2. The dimensionless variables used in these accelerance equations are defined in Eqs. (6.4.6) through (6.4.10).

We can gain insight into the armature's behavior if we plot these three dimensionless accelerance functions. The armature's support system is assumed to have $\eta_1 = 0.05$ as before, the armature's damping is assumed to be light so that $\eta_2 = 0.001$, and the mass ratio M is usually around 5 to 20 percent of the armature's mass. The frequency ratio ranges from around 100 to over 200. For ease of computation and plotting, we assume $M = 0.1$ and $\beta = 100$. The resulting curves are shown in Fig. 6.4.8.

All three dimensionless accelerance FRFs, $A_{11}(\omega)$, $A_{12}(\omega)$, and $A_{22}(\omega)$, are shown in Fig. 6.4.8. These three curves are identical as r ranges from 0.1 to 10. In this range, the curves grow at 40 dB/decade, pass through armature suspension resonance at $r_1 \cong 1/\sqrt{1+M} = 0.95$ (this resonance is controlled by k_1 and armature mass m_a), and then become constant at 0 dB (unity magnitude). Ideally, we would like the response to continue at 0 dB for all higher frequencies, but unfortunately, these curves diverge considerably from unity in the range of $10 < r < 1000$.

First, consider the coil's dimensionless accelerance $A_{22}(\omega)$, which predicts the coil's motion due to the force acting on the coil. This accelerance is seen to dip significantly at $r_a \cong 31.6$, as predicted by Eq. (6.4.19). Then it resonates at $r_2 \cong \beta\sqrt{1+M} = 105$, and finally reaches a plateau given by $(1+M)/M \cong 11$ (20.8 dB). The dip corresponds to the table mass m_t being a *dynamic absorber* to the coil. Thus, the coil responds over a wide dynamic range of approximately 100 dB (1×10^5) and has considerable acceleration after coil resonance r_2.

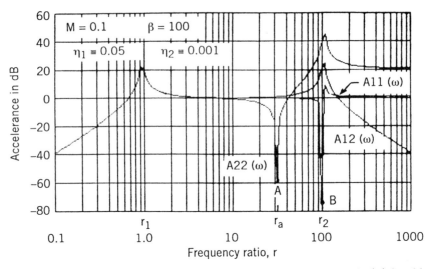

Fig. 6.4.8. Dimensionless accelerance plots for exciter armature: $A_{11}(\omega)$ is table response to excitation force F_1; $A_{12}(\omega)$ is table response due to coil excitation; $A_{22}(\omega)$ is coil response due to coil force. Note dynamic absorber action in $A_{11}(\omega)$ and $A_{22}(\omega)$ at points B and A, respectively.

Second, consider the table's dimensionless accelerance $A_{11}(\omega)$, which predicts the table's motion due to external forces that act upon the table's surface. This accelerance dips significantly at $r_b = \beta = 100$, as required by Eq. (6.4.17). This dip is followed by resonance at $r_2 \cong 105$, and then takes on a constant value of $(1 + M)$ for $r \gg r_2$. The reason that $A_{11}(\omega)$ takes on a value larger than unity and $A_{22}(\omega)$ takes on a value larger than 10 in this case is the fact that the total armature mass m_a is used in normalizing the results instead of table mass m_t or the coil mass m_c, respectively.

Third, consider the table's dimensionless transfer accelerance $A_{12}(\omega)$, which represents the table's acceleration due to force on the coil. We see that this response continues through a resonance at r_2 and then drops off at 40 dB/decade. It is clear from this curve that it is impossible to effectively drive the table from the coil end for frequencies above $r = 300$, at which point the accelerance is down 20 dB. This shows that the practical upper limit for effective use of an exciter is controlled by this second resonance at r_2. This resonance is due to the light coil mass resonating against the larger table mass through the armature spring k_2, namely, $r_2 \cong \beta\sqrt{1 + M} = \omega_{22}\sqrt{1 + M/\omega_{11}}$. Consequently, this second resonant frequency is greater than ω_{22} for any mass that we might attach to the table, since the bare table will have the largest possible mass ratio M.

The consequence of this analysis is that there is a region starting at 0 frequency up to just below r_2 where the *bare table motion behaves as* though the armature is a rigid body mounted on springs so that between r_1 and r_2 the bare table acceleration is given by

$$a_{\text{bare}} = \frac{F_0}{m_a} \qquad (6.4.20)$$

A quick glance at $A_{12}(\omega)$ shows us the bare table characteristics as we should be able to measure them with an accelerometer. This simple curve is often misunderstood in that often it is thought that the armature behaves this way when the exciter is attached to a structure. We shall see that this is not the case. However, we can make a reasonable assumption based on accelerance $A_{12}(\omega)$, namely, *the armature can be considered to be a rigid body of mass m_a mounted on springs so long as we use only frequencies that are less than one-half of the bare table coil resonance r_2.*

6.4.3 Electromechanical Coupling Relationships

The electrical voltage and current and the mechanical force and motion interact at the coil-magnetic field interface. This interaction is controlled by the Ampère and Lenz laws. There are linear and nonlinear versions of these results, depending on the exciter's design.

Ampère's Law relates coil force F_c to conductor current I. This force is developed by the interaction between the magnetic field of strength B that exists across the gap and the magnetic field due to current flowing in the coil's conductors. The resulting force is

$$F_c = (Bln)I = K_f I \qquad (6.4.21)$$

where l is the length of one coil, n is the number of coils, and K_f is the force-current constant that is equal to Bln. Simple magnetic theory shows that the field strength B is related to the flux linkage ψ by

$$B = \frac{d\psi}{dx} \qquad (6.4.22)$$

Tomlinson[5] has shown that the effective field strength B is nonlinear with

---

[5]G. R. Tomlinson, "Force Distortion in Resonance Testing of Structures with Electro-Dynamic Vibration Exciters," *Journal of Sound and Vibration*, Vol. 63, No. 3, 1979, pp. 337–350.

coil position x such that

$$B = \frac{d\psi}{dx} = B_0\left[1 - \left\{\frac{x + x_0}{x_{\max}}\right\}^2\right] \quad (6.4.23)$$

where B_0 is the largest effective field strength.

x_0 is an offset position about which the coil moves with amplitude x.

$x_{\max}$ is the maximum permissible table motion.

Combining Eqs. (6.4.21) and (6.4.23) shows that the force current relationship is non-linear such that

$$F_c = (B_0 ln)\left[1 - \left\{\frac{x + x_0}{x_{\max}}\right\}^2\right]I = K_f\left[1 - \left\{\frac{x + x_0}{x_{\max}}\right\}^2\right]I \quad (6.4.24)$$

When the motion of x is small compared to $x_{\max}$, Eq. (6.4.24) appears linear for all practical purposes. However, we find that this is not the case when we test low frequency structures at resonance, a case where x can be on the order of $x_{\max}$.

Lenz's law relates the voltage induced in a conductor (called the *back emf*—electromotive force) when it moves in a magnetic field with strength B with a velocity of $\dot{x}$ so that

$$E_{\text{bemf}} = (Bln)\dot{x} = K_v \dot{x} \quad (6.4.25)$$

where l is the coil's length per turn.

n is the number of turns.

K_v is the back emf voltage constant.

The nonlinear behavior is obtained when the nonlinear characteristics of Eq. (6.4.23) are substituted into Eq. (6.4.25) to obtain

$$E_{\text{bemf}} = (B_0 ln)\left[1 - \left\{\frac{x + x_0}{x_{\max}}\right\}^2\right]\dot{x} = K_v\left[1 - \left\{\frac{x + x_0}{x_{\max}}\right\}^2\right]\dot{x} \quad (6.4.26)$$

Thus a nonlinear back emf voltage is developed for large amplitude armature motions like those that occur when testing low frequency structures.

Before leaving this topic, we need to note that some authors make a big point about K_f being equal to K_v. This is true when SI units are employed but untrue if pounds are used for force and inches/second are used for velocity. In the first case, the ratio of K_f/K_v gives (N-m/s)/watts or watts/watts, which is unity, while in the second case the K_f/K_v ratio gives (in.-lb/s)/(watts), which is not unity. Thus we keep the subscripts

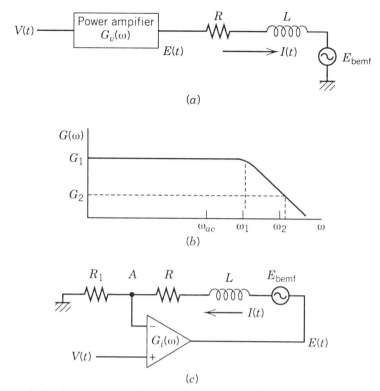

Fig. 6.4.9. Basic schematic circuits of power amplifier connected to vibration exciter, showing voltage and current mode of operation. (*a*) Amplifier in voltage mode of operation. (*b*) Amplifier FRF showing gain effects on usable frequencies. (*c*) Amplifier in current mode of operation.

on K_f and K_v throughout this work so there is no mistake which constant is being employed in a given equation, and the reader is free to use any convenient system of units.

6.4.4 Power Amplifier Characteristics

Conceptually, we use two modes of amplifier behavior called *voltage mode* and *current mode*. The basic circuits for each mode of operation are shown in Fig. 6.4.9, where the coil's electrical characteristics are included as well. These characteristics have been found to consist of coil resistance and possibly some amplifier output resistance, dependent on design, which are lumped into resistance R, in place coil inductance L, and back emf E_{bemf}.

Voltage Mode of Operation The voltage mode circuit is shown in Fig.

6.4.9a, where the power amplifier is represented by a gain $G_v(\omega)$ like the one shown in Fig. 6.4.9b. The amplifier's output voltage $E(t)$ is related to its input voltage $V(t)$ in the frequency domain by the gain FRF so that

$$E(\omega) = G_v(\omega)V(\omega) \qquad (6.4.27)$$

The amount of gain employed can change $G_v(\omega)$'s characteristics as shown. For example, if the gain is G_1, then the curve is flat until break frequency ω_1 where the curve begins to fall off at 20 dB/decade, which is typical for many amplifiers.[6] If the gain is lowered to G_2, the break frequency increases to ω_2. The implication of the 20 dB/decade slope is that lowering the gain by a factor of 10 increases the break frequency by a factor of 10. What we need to avoid is bringing the break frequency too close to the armature's coil resonance ω_{ac}, since $\omega_1 \gg \omega_{ac}$ allows us to treat the amplifier as a gain constant with no significant phase shifts. In a well designed system, the designer has taken these factors into account.

The differential equation relating the coil drive voltage $E(t)$ to the current is obtained by summing the voltage drops. This gives

$$RI + L\dot{I} + E_{\text{bemf}} = E(t) \qquad (6.4.28)$$

which on substitution of Eq. (6.4.25) becomes the fundamental voltage, current, and velocity relationship of

$$RI + L\dot{I} + K_v\dot{x} = E(t) \qquad (6.4.29)$$

which shows how the electrical current and voltage are interacting with mechanical velocity.

Current Mode of Operation The current mode of operation is shown schematically in Fig. 6.4.9c, where the coil characteristics are connected across the power amplifier's feedback when configured for constant current operation. In this arrangement, the amplifier's output voltage $E(t)$ is adjusted to maintain current I through resistor R_1 so that the voltage at point A is the same as the input voltage $V(t)$. In this way, the current is related to the input voltage in the frequency domain by

$$I(\omega) = G_i(\omega)V(\omega) \qquad (6.4.30)$$

independent of frequency, provided $G_i(\omega)$ has a frequency characteristic

---

[6] R. J. Smith, *Electronics: Circuits and Devices*, 3rd ed., John Wiley & Sons, New York, 1987.

similar to that shown in Fig. 6.4.9b. As we shall see, the test structure's motion covers a very wide dynamic range, which limits the applicability of Eq. (6.4.30) since the maximum output voltage is limited to E_{max}. Nonlinear effects, as described by Eqs. (6.4.24) and (6.4.26), will further erode the validity of these simple linear models.

In Section 6.5, we shall integrate all of these equations and explore the bare table response of an exciter. We shall consider the exciter SUT interaction in Chapter 7.

6.5 AN EXCITER SYSTEM'S BARE TABLE CHARACTERISTICS

In this section, we explore test system behavior that combines both mechanical and electrical characteristics when the power amplifier is used in either the current mode or the voltage mode of operation. The vibration table is bare or unconnected to any SUT for these considerations so that the vibration exciter's characteristics are evident. We shall consider the exciter's interaction with a grounded single DOF structure in Section 6.6.

6.5.1 The Electrodynamic Model

In Section 6.4, we developed a low frequency single DOF dynamic model of the armature that is valid up to about 0.5 of coil resonance. This model is shown in Fig. 6.5.1a. The differential equation of motion for the mechanical system is given by

$$m_a \ddot{x} + c_a \dot{x} + k_a x = F_c(t) = K_f I(t) \qquad (6.5.1)$$

where x is the armature motion.
  m_a is the armature mass.
  c_a is the armature damping.
  k_a is the armature stiffness.
  $F_c(t)$ is the coil excitation force that is dependent on coil current $I(t)$ (see Eq. (6.4.21)).
  K_f is the linear electromagnetic force current constant.

The input voltage $E(t)$ is related to the coil current $I(t)$, coil resistance R, coil inductance L, and back emf E_{bemf}, as shown in Fig. 6.5.1b, by Eq. (6.4.29) as

$$RI + L\dot{I} + K_v \dot{x} = E(t) \qquad (6.5.2)$$

where the back emf, as given by Eq. (6.4.25), is used. Now, we are

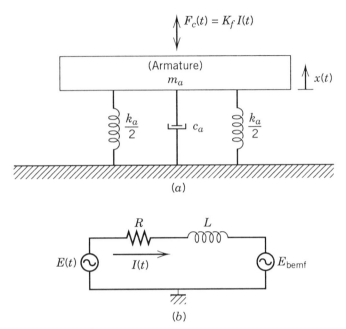

Fig. 6.5.1. The mechanical and electrical components of an exciter system. (*a*) Single DOF model of exciter's armature. (*b*) The coil's electrical circuit.

interested in how these equations interact when we use the power amplifier in either the current or voltage mode of operation.

6.5.2 Current Mode Power Amplifier

The current mode of operation allows us to assume that $I(t)$ is the reference phasor with magnitude I_0. Then Eqs. (6.5.1) and (6.5.2) can be transformed into algebraic frequency domain equations given by

$$(k_a - m_a\omega^2 + jc_a\omega)X = K_f I_0 \tag{6.5.3}$$

and

$$(R + jL\omega)I_0 + jK_v\omega X = E_0 \tag{6.5.4}$$

where X and E_0 are the unknown armature motion and required excitation voltage phasors, respectively. The solution of Eq. (6.5.3) gives the table motion, while the solution to Eq. (6.5.4) gives the voltage required to drive the coil with current. It is clear from Eq. (6.5.4) that the excitation voltage depends on armature response X.

The dimensionless accelerance is of interest since accelerance is a measure of the maximum acceleration level that the exciter can produce. The dimensionless accelerance is obtained from Eq. (6.5.3) as

$$A(\omega) = \frac{m_a(-\omega^2 X)}{K_f I_0} = \frac{-r^2}{1 - r^2 + j2\zeta_a r} = \frac{-r^2}{1 - r^2 + j\eta_a} \quad (6.5.5)$$

where r is the dimensionless frequency ratio based on armature natural frequency of $\sqrt{k_a/m_a}$.
ζ_a is the armature's dimensionless damping ratio.
η_a is the armature's loss factor.

The effects of using either viscous or structural damping in Eq. (6.5.5) are shown as follows. The excitation voltage needed to produce a constant magnitude current input is obtained from Eq. (6.5.4) in terms of a dimensionless voltage ratio of

$$\begin{aligned} E(\omega) &= \frac{E_0}{RI_0} = 1 + j\left[\beta_1 + \frac{2\zeta_e}{1 - r^2 + j\eta_a}\right]r \\ &= 1 + j\left[\beta_1 + \frac{2\zeta_e}{1 - r^2 + j2\zeta_a r}\right]r \end{aligned} \quad (6.5.6)$$

dependent on using either a structural or viscous damping model. β_1 is the dimensionless frequency ratio given by

$$\beta_\ell = \omega_a/\omega_\ell \quad (6.5.7)$$

where ω_ℓ is the break frequency determined by R and L in Eq. (6.5.4), so that

$$\omega_\ell = R/L \quad (6.5.8)$$

The *electro-magnetic damping ratio* ζ_e is given by

$$\zeta_e = \frac{C_m}{2\sqrt{k_a m_a}} = \frac{C_m}{2\omega_a m_a} \quad (6.5.9)$$

where C_m is the *electromagnetic damping* given by

$$C_m = \frac{K_v K_f}{R} \quad (6.5.10)$$

This electrodynamic damping term results from the back emf current being dissipated by the coil circuit's effective resistance. In the current amplifier case, the amplifier presents an infinite impedance to the back emf voltage so no energy can be dissipated. This infinite impedance occurs in the form of increasing excitation voltage in order to maintain the required current regardless of back emf generated. However, in the voltage mode, the back emf voltage generates a current that is dissipated by the circuit's resistance. This dissipation shows up as linear viscous damping, as will be demonstrated in the next subsection. Thus the m subscript indicates that this is a *magnetic* damping term. One should note that C_m can be highly nonlinear for large armature motions, due to magnetic field nonlinearities as discussed in Section 6.4; see Eqs. (6.4.24) and (6.4.26). These nonlinearities may contribute to the apparently consistent structural damping value that was measured by Tomlinson.

Equation (6.5.6) shows that armature resonance requires a large input voltage to maintain the required current input to the armature. This ratio is significantly influenced by the ratio of $C_m/c_a = (2\zeta_e)/(2\zeta_a)$. We consider the constant voltage mode of operation before we make a graphical comparison of system performance characteristics.

6.5.3 Voltage Mode Power Amplifier

In this subsection, we assume that the reference phasor in Eqs. (6.5.1) and (6.5.2) is the voltage $E(t)$ with magnitude of E_0. Then the frequency domain expressions of Eqs. (6.5.1) and (6.5.2) are the same as given by Eqs. (6.5.3) and (6.5.4). In this case, we need to solve for I_0 from Eq. (6.5.4) for insertion into Eq. (6.5.3). The resulting table motion equation can be written in two forms, dependent on how armature damping is modeled. First, assuming a viscous damping model, we obtain

$$A_v(\omega) = \frac{m_a(\omega^2 X)}{K_f(E_0/R)} = \frac{-r^2}{1 - (1 + M_L)r^2 + j[2(\zeta_a + \zeta_e) + \beta_1(1 - r^2)]r} \quad (6.5.11)$$

where ζ_a and ζ_e are the armature and electromagnetic viscous damping ratios.

M_L is the dimensionless mass ratio given by

$$M_L = m_\ell/m_a \quad (6.5.12)$$

m_ℓ is the *inductive mass* term due to coupling between electrical resistance and inductance with viscous mechanical damping given by

$$m_\ell = Lc_a/R \tag{6.5.13}$$

Equation (6.5.13) clearly shows us that increasing armature damping can increase the inductive mass, a term that can influence the armature's behavior.

If the armature damping is structural, then the dimensionless accelerance becomes

$$A_s(\omega) = \frac{m_a(-\omega^2 X)}{K_f(E_0/R)} = \frac{-r^2}{1 - (1 + M_L)r^2 + j[\eta_a + 2\zeta_e r + \beta_\ell(1 - r^2)r]} \tag{6.5.14}$$

where η_a is the armature's structural damping.

It should be pointed out that Eqs. (6.5.11) and (6.5.14) are approximate models, since at armature resonance the dominant damping is controlled by C_m in Eq. (6.5.10). This linear damping term is really nonlinear in that both K_f and K_v are amplitude dependent. Hence, the calculated responses are approximate at best, which points up how we can be misled in attempting to measure armature damping. Now we compare these responses.

6.5.4. Comparison of Bare Table Armature Responses

We would like to see what typical bare table responses should look like when operating in the constant current and constant voltage modes. We use the exciter parameters from a Ling Model 201 exciter. The exciter's parameters are:

$L = 0.4 \times 10^{-3}$ henrys

$R = 1.5$ ohms

$K_f = 5.78$ N/ampère

$K_v = 5.78$ volts/meter/second

$m_a = 0.02$ kg

$k_a = 3500$ N/meter

$\eta_a = 0.2$

$\zeta_a = 0.1$ equivalent viscous damping at resonance.

The corresponding calculated parameters for this example are:

$\omega_c = R/L = 3750$ rad/second

$\omega_a = \sqrt{k_a/m_a} = 418$ rad/second

$\beta_1 = \omega_a/\omega_1 = 0.112$

$C_m = (K_f K_v)/R = 22.3$ N-second/meter

$\zeta_e = C_m/2m_a\omega_a = 1.331$

$M_L = m_e/m_a = 0.022$ (a negligible quantity compared to unity)

The dimensionless accelerance is shown as a Bode plot in Fig. 6.5.2 for both current and voltage amplifier modes of operation. In the current mode, the accelerance is similar to that shown in Fig. 6.4.8 for $A_{12}(\omega)$ below the coil resonance. In both Figs. 6.5.2a and 6.4.8, the accelerance increases at a rate of 40 dB/decade below $r = 1$, passes through the armature resonance at $r = 1$, and approaches a constant value of unity for $r \gg 1$. It is seen in Fig. 6.5.2b that the phase angle starts at 180 degrees for $r \ll 1$, passes through 90 degrees at $r = 1$, and goes to zero for $r \gg 1$. This is ideal behavior.

Figure 6.5.3 shows that the input voltage must change by a factor of approximately 13.3 (a value controlled by $\beta_1 + \zeta_e/\zeta_a$ in Eq. (6.5.6)) for this example. This voltage is seen to start at unity for $r \ll 1$, peak at nearly 13.3 at $r = 1$, and return to unity for $r \gg 1$. Such sharp increases in excitation voltage can lead to amplifier clipping when the exciter is driven at high input levels when away from resonance. Hence, we need to constantly monitor input voltages for clipping when operating the power amplifier in the current mode. Otherwise, the current mode of operation may not be taking place.

The dimensionless accelerance for the voltage mode amplifier is also shown in Fig. 6.5.2 for comparison purposes. In this mode, we see that the accelerance is considerably different, particularly due to the high inherent electromagnetic damping C_m. This damping gives a damping ratio $\zeta_e = 1.331$ that is greater than critical damping regardless of armature damping. Figure 6.5.2a shows that the accelerance curve increases more slowly due to the high electromagnetic damping, peaks at unity around $r \cong 4$ to 6, and then begins to drop off at 20 dB/decade for $r > 10$. The 20 dB/decade drop-off is controlled by the $(\beta_1 r)$ term in the denominator of Eq. (6.5.11) since the r^2 term cancels in both numerator and denominator when $r \gg 1$. The inductive inertia has no effect on the results shown.

The voltage phase angle starts at 180 degrees for $r \ll 1$, passes through 90 degrees at $r = 1$, passes through 0 degrees for $r \cong 5$, and continues to

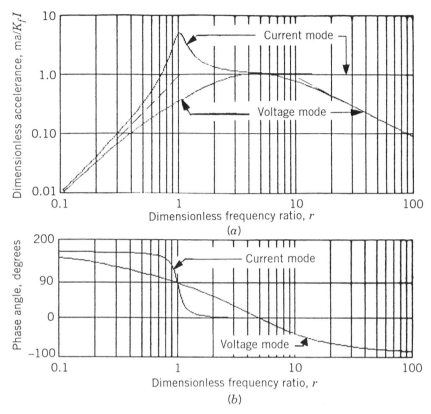

Fig. 6.5.2. Dimensionless bare table accelerance Bode plots. Accelerance and phase are plotted as functions of dimensionless frequency ratio ($r = \omega/\omega_a$) for current and voltage amplitude modes of operation. (*a*) Accelerance. (*b*) Phase.

change until an angle of approximately -90 degrees is reached for large values of r.

This comparison of dimensionless accelerance clearly shows that bare table exciter response is highly dependent on which power amplifier mode of operation is being employed. This major difference between these two modes of behavior is due to the large electrodynamic damping term. The large electrodynamic damping is also nonlinear for large amplitudes of motion that occur at armature resonance, a resonance that usually occurs at quite low frequencies in the order of 20 to 50 Hz. For example, 50 g's at 50 Hz has an amplitude of motion of approximately 0.2 in., which is about 40 percent of typical armature's maximum travel.

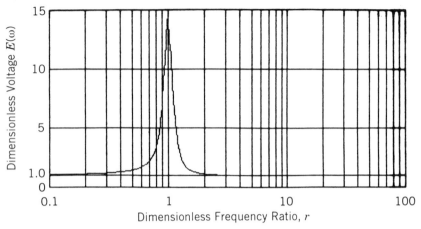

Fig. 6.5.3. Dimensionless voltage required to drive a bare table exciter through armature resonance when driven by a current mode power amplifier.

6.6 INTERACTION OF AN EXCITER AND A GROUNDED SINGLE DOF STRUCTURE

Up to this point we have considered only the vibration exciter's bare table response in order to grasp the significance of various exciter system parameters on its performance specifications. In this section, we attempt to bring in the effects of the SUT itself on test system behavior. It is found that test system performance can be altered significantly by interacting with the SUT. In addition, we find that the force transmitted to the SUT experiences a *force dropout* or *glitch* near the SUT's resonance. Tomlinson,[7] Olsen,[8] and Rao[9] have studied various aspects of these exciter structure interactions.

---

[7] G. R. Tomlinson, "Force Distortion in Resonance Testing of Structures with Electro-Dynamic Vibration Exciters," *Journal of Sound and Vibration*, Vol. 63, No. 3, 1979, pp. 337–350.

[8] N. L. Olsen, "Using and Understanding Electrodynamic Shakers in Modal Applications," *Proceedings of the 4th International Modal Analysis Conference*, Vol. 2, 1986, pp. 1160–1167.

[9] D. K. Rao, "Electrodynamic Interaction between a Resonating Structure and an Exciter," *Proceedings of the 5th International Modal Analysis Conference*, Vol. 2, 1987, pp. 1142–1150.

6.6.1 The Single DOF SUT and Electrodynamic Exciter Model

We are considering the test system shown in Fig. 6.6.1, where the SUT is represented by mass m_s, spring k_s, and damping c_s, while the exciter's armature is represented by inertia m_a, spring k_a, and damping c_a. Usually the armature's mass m_a is greater than its bare table value since additional mass due to connection hardware must be included in m_a. For example, half of the near rigid connection link's mass must be included in m_a.

The two masses are connected by a near rigid link with stiffness k_c. When the link is stiff enough, the structure's motion $x_s(t)$ and the armature's motion $x_a(t)$ are equal. Thus we assume that

$$x(t) = x_s(t) = x_a(t) \tag{6.6.1}$$

If the link stiffness $k_c >$ five times $(k_a + k_s)$, then we can state that Eq. (6.6.1) is reasonably valid for frequencies less than $0.5\omega_c$ where ω_c is given by

$$\omega_c = \sqrt{\frac{k_c(m_a + m_s)}{m_a m_s}} \tag{6.6.2}$$

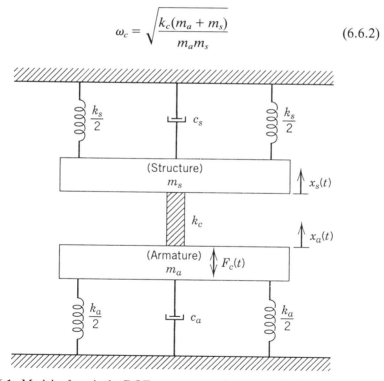

Fig. 6.6.1. Model of a single DOF structure under test and vibration exciter armature connected by a near rigid link.

Frequency ω_c corresponds to inertias m_a and m_s vibrating 180 degrees out of phase with one another as though they were in free space so that springs k_s and k_a do not exist. Thus our simple model is restricted to frequencies below $0.5\omega_c$.

The differential equation of motion describing the behavior of the combined system shown in Fig. 6.6.1 is given by

$$m\ddot{x} + c\dot{x} + kx = F_c(t) = K_f I(t) \qquad (6.6.3)$$

where the mass, damping, and stiffness are given by

$$\begin{aligned} m &= m_a + m_s = m_s(1 + M) \\ c &= c_a + c_s = c_s(1 + C) \\ k &= k_a + k_s = k_s(1 + K) \end{aligned} \qquad (6.6.4)$$

where M, C, and K are dimensionless mass, damping, and stiffness ratios. In working with these equations, we are primarily interested in how the SUT's natural frequency and damping are measured. Consequently, we need to express the exciter's stiffness, damping, and mass in terms of the structure's stiffness, damping, and mass, as shown in Eq. (6.6.4). One convenient parameter is the ratio of the armature's and structure's natural frequencies so that

$$\beta_s = \frac{\omega_a}{\omega_s} = \frac{\sqrt{k_a/m_a}}{\sqrt{k_s/m_s}} \qquad (6.6.5)$$

where ω_a is the armature's natural frequency and ω_s is the structure's natural frequency. Then the stiffness ratio K in Eq. (6.6.4) becomes

$$K = \frac{k_a}{k_s} = \beta_s^2 M \qquad (6.6.6)$$

and the damping ratio C in Eq. (6.6.4) becomes

$$C = \frac{c_a}{c_s} = \frac{2\zeta_a \beta_s M}{2\zeta_s} \qquad (6.6.7)$$

where the mass ratio M is given by

$$M = \frac{m_a}{m_s} \qquad (6.6.8)$$

These dimensionless parameters are useful throughout the rest of this section.

The electrical circuit is the same as that given in Eq. (6.5.2), namely

$$RI + L\dot{I} + K_v \dot{x} = E(t) \tag{6.6.9}$$

We consider both the current and voltage modes of power amplifier operation.

6.6.2 The SUT's Accelerance Response

The structure's displacement, velocity, and acceleration response are dependent on the power amplifier's mode of operation. For convenience, we present the response in terms of a dimensionless accelerance. The accelerance is developed first for the current mode of operation, and then for the voltage mode of operation.

Current Mode of Operation For the current mode of operation, we assume that the reference phasor is current with magnitude I_0. The other phasors have complex magnitudes of X for displacement and E_0 for voltage. Then, the time domain representations of Eqs. (6.6.3) and (6.6.9) become frequency domain equations of

$$X = \frac{K_f I_0}{k - m\omega^2 + jc\omega} = \frac{K_f I_0}{k_s[(1 + \beta_s^2 M) - (1 + M)r^2 + j(2\zeta_s + 2\zeta_a \beta_s M)r]} \tag{6.6.10}$$

for displacement when Eqs. (6.6.4) and (6.6.6) through (6.6.8) are used, and

$$E_0 = R\left[1 + j\frac{L}{R}\omega + j\left\{\frac{K_v K_f}{R}\right\}\left\{\frac{\omega}{k - m\omega^2 + jc\omega}\right\}\right] I_0 \tag{6.6.11}$$

for the required voltage relative to the current phasor I_0 where $r = \omega/\omega_s$. The dimensionless accelerance is obtained from Eq. (6.6.10) as

$$A(\omega) = \frac{m_s(-\omega^2 X)}{K_f I_0} = \frac{-r^2}{(1 + \beta_s^2 M) - (1 + M)r^2 + j(2\zeta_s + 2\zeta_a \beta_s M)r} \tag{6.6.12}$$

while Eq. (6.6.11) gives a dimensionless voltage ratio of

$$E(\omega) = \frac{E_0}{RI_0} = 1 + j\left[\frac{\beta_e}{\beta_s} + \frac{2\zeta_e \beta_s M}{(1 + \beta_s^2 M) - (1 + M)r^2 + j(2\zeta_s + 2\zeta_a \beta_s M)r}\right]r \quad (6.6.13)$$

where β_e is defined in Eq. (6.5.7).
ζ_e is defined in Eq. (6.5.9).
It is clearly evident from Eq. (6.6.12) that the accelerance peaks at

$$r_p = \sqrt{\frac{1 + \beta_s^2 M}{1 + M}} \quad (6.6.14)$$

instead of $r \cong 1$ and that the measured damping is given by

$$2\zeta_m = 2\zeta_s + (\beta_s M)(2\zeta_a) \quad (6.6.15)$$

Equation (6.6.15) shows that the measured damping includes the exciter armature's damping, that is, ζ_a. Equations (6.6.14) and (6.6.15) are clear evidence why it is inappropriate to do current mode testing in order to determine the SUT's natural frequency and damping for these are measured incorrectly, since the exciter system is part of the measured response. *Thus we need to directly measure the force acting on the structure by using a force transducer if we want to measure the structure's true characteristics.*

The dimensionless accelerance from Eq. (6.6.12) is shown in Fig. 6.6.2 for a mass ratio $M = 0.2$ and β_s values of 0.1, 1, and 10. The exciter's parameters are the same as used in Section 6.5. In the $\beta_s = 0.1$ case, the SUT's natural frequency is 10 times the armature's natural frequency and the curve's peak value occurs around $r = 0.91$, as shown. The structure's damping is 1 percent, so that the peak response should be around 50. However, the armature's structural damping η_a is 10 percent, so that using structural damping in place of viscous damping in Eq. (6.6.12) gives a peak value of 38, which is about 32 percent lower than it should be.

Similarly, when $\beta_s = 1$, the correct natural frequency results, since according to Eq. (6.6.14), the value of r_p is unity. However, the damping appears to be 6 percent with a peak value of 16.7, instead of 50 as shown. Hence, the peak is nearly 67 percent lower than it should be. Finally, when the value of β is 10, the structure's natural frequency is increased to a value of 4.18, as given by Eq. (6.6.14), and the effective damping increases to approximately 0.48. However, the peak value is approximately 36, instead of 2, due to the r^2 multiplier. What is happening in

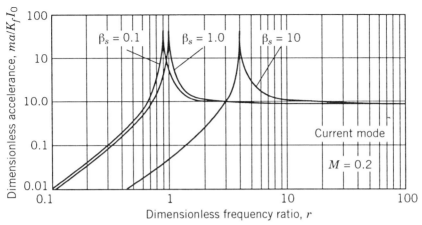

Fig. 6.6.2. A single DOF structure's accelerance when driven by a current mode power amplifier showing effect of rigidly attached exciter armature.

this case is that the armature's spring is controlling the test natural frequency and the armature's damping is dominating the measured peak.

These results clearly indicate the devastating effect that the exciter's armature can have on the measured structural response if we attempt to measure a structure's dynamic properties by using the current as the force transmitted to the structure. This does not mean that we cannot run resonance fatigue tests where we monitor strain in the structure and conduct the resonance dwell test at the test system's natural frequency, that is, the natural frequency of the SUT and armature system combined. Then, a 90 degree phase shift between the current and strain or acceleration is an acceptable control scheme for locking onto test resonance. It is not an acceptable test arrangement when we are looking for the test structure's dynamic properties. These two tests have completely different test objectives, and hence require different test methods that are appropriate to those objectives.

The dimensionless voltages required to drive the exciter under current mode conditions for $M = 0.2$ and $\beta_s = 0.1, 1,$ and 10 are shown Fig. 6.6.3. In this case, the exciter's frequency ratio $\beta_1 = 0.112$ and the electromagnetic damping $2\zeta_e = 2.662$. Here, we find that the voltage increases dramatically for the case when $\beta_s = 0.1$. This large increase is due to driving the exciter's inductance, since it is controlled by the $[(\beta_1/\beta_s)r]$ term in Eq. (6.6.13). The peak corresponding to resonance is not important in this case. In the case when $\beta_s = 1$, there is some peak due to resonance response, but again the inductance requirement is clearly evident at higher frequencies. Finally, in the case when $\beta_s = 10$, the dominant peak is the structural resonance. These curves indicate that inductance and magnetic

420 VIBRATION EXCITERS

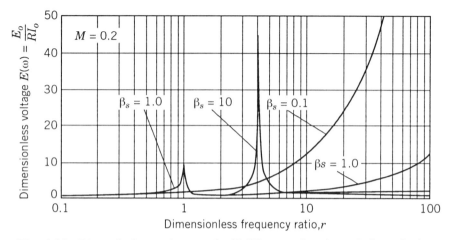

Fig. 6.6.3. Dimensionless voltage ratio E_0/RI_0 as a function of dimensionless frequency ratio for a single DOF structure rigidly attached to the exciter's armature for $\beta_s = 0.1$, 1 and 10 and a mass ratio of $M = 0.2$.

feedback interact in a highly variable manner, dependent on the relative position of the structure's natural frequency and the inductive break frequency ω_ℓ (see Eq. (6.5.8)).

Voltage Mode of Operation Proceeding as before, we can obtain the frequency domain expression for the displacement response as

$$X = \frac{K_f \{E_0/R\}}{k - (m + m_\ell)\omega^2 + j[c + \{L(k - m\omega^2)/R\} + (K_f K_v/R)]\omega} \quad (6.6.16)$$

where there is inductive mass m_ℓ given by

$$m_\ell = \frac{cL}{R} = \frac{(c_a + c_s)L}{R} = \left\{\frac{c_a L}{R}\right\}\left\{1 + \frac{2\zeta_s}{2\zeta_a \beta_s M}\right\} \quad (6.6.17)$$

where the bare table inductive mass is given by $(c_a L/R)$. It is clear from Eq. (6.6.17) that the structure's damping can contribute to the inductive inertia when operating in the voltage mode. Equation (6.6.16) also shows that additional apparent damping has been picked up. The *electrodynamic damping* is given by

$$c_e = \frac{L(k - m\omega^2)}{R} \quad (6.6.18)$$

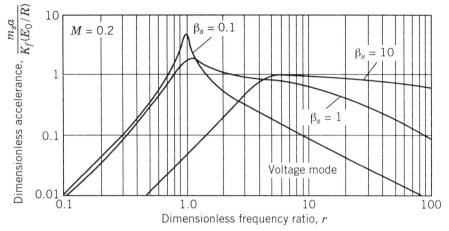

Fig. 6.6.4. A single DOF structure accelerance when driven by a voltage mode power amplifier showing effects of rigidly connected exciter armature on system response.

Equation (6.6.18) shows how inductance, mechanical dynamic response, and electrical resistance combine to give an apparent effect of damping. It should be pointed out that this damping can and will change sign with frequency. Finally, the last damping term involving $K_f K_v/R$ is the electromagnetic damping described previously by Eq. (6.5.10). The dimensionless accelerance is obtained from Eq. (6.6.16) as

$$A(\omega) = \frac{m_s(-\omega^2 X)}{K_f(E_0/R)} = \frac{-r^2}{\Delta(r)} \quad (6.6.19)$$

where

$$\Delta(r) = (1 + \beta_s^2 M) - (1 + M + M_\ell)r^2 \quad (6.6.20)$$
$$+ j\left\{2\zeta_s + \beta_s M(2\zeta_a + 2\zeta_e) + \frac{\beta_\ell}{\beta_s}[1 + \beta_s^2 M - (1 + M)r^2]\right\} r$$

Equation (6.6.20) shows that the exciter's characteristics are even more involved with the measured response than in the current mode case. The additional dominant exciter parameters are the electromagnetic and electrodynamic damping terms.

The dimensionless accelerance is plotted using Eqs. (6.6.19) and (6.6.20) in Fig. 6.6.4. It is seen that the structure's response is totally different from that of Fig. 6.6.2. This dramatic difference is primarily due to the electromagnetic and electrodynamic damping terms that the current mode amplifier removes from test considerations. A controller device must be employed to overcome these combined system characteristics, and the force transmitted to the structure must be measured. These curves are different from those presented by Rao[10] since different exciter parameters were used in the two calculations.

6.6.3 The Force Transmitted to the SUT and Force Dropout

The force transmitted to the structure under test is called $F_s(t)$ and is obtained from the expression

$$m_s \ddot{x} + c_s \dot{x} + k_s x = F_s(t) \tag{6.6.21}$$

The force can be obtained from Eq. (6.6.21) when either Eq. (6.6.10) for the current mode of operation or Eq. (6.6.16) for the voltage mode of operation is inserted for $x(t)$. We consider each case in turn.

Current Mode of Operation Insertion of Eq. (6.6.10) into Eq. (6.6.21) for the current mode excitation case gives a frequency domain expression for the force transmissibility ratio as

$$TR = \frac{F_s}{K_f I_0} = \left[\frac{k_s - m_s \omega^2 + j c_s \omega}{k - m \omega^2 + j c \omega}\right] = \left[\frac{1}{1+M}\right]\left[\frac{1 - r^2 + j 2 \zeta_s r}{r_p^2 - r^2 + j 2 \zeta_t r}\right] \tag{6.6.22}$$

where $r = \omega/\omega_s$, the frequency ratio r_p is given by Eq. (6.6.14), and the test system's damping $2\zeta_t$ is given by

$$2\zeta_t = \frac{2\zeta_s + (\beta_s M)(2\zeta_a)}{1 + M} \tag{6.6.23}$$

It is evident looking at Eq. (6.6.22) that a notch occurs when $r = 1$ and a peak will occur either before or after $r = 1$, dependent on the value of r_p. When r_p is less than one, the peak occurs before the notch at $r = 0.935$, as shown in Fig. 6.6.5 for $\beta_s = 0.5$, a case where the structure's natural frequency is greater than the armature's natural frequency. For this case, the force transmissibility starts at a value of $1/(1 + \beta_s^2 M)$ for small values of r. This force transmissibility is controlled by the spring ratio K as

---

[10] D. K. Rao, "Electrodynamic Interaction Between a Resonating Structure and an Exciter," *Proceedings of the 5th International Modal Analysis Conference*, Vol. 2, 1987, pp. 1142–1150.

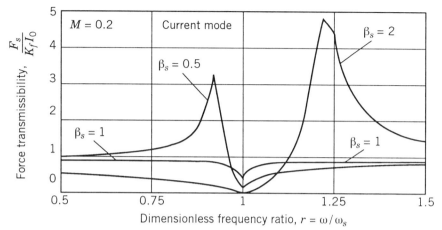

Fig. 6.6.5. Dimensionless force transmissibility versus dimensionless frequency ratio $r = \omega/\omega_s$ for mass ratio $M = 0.2$ and various frequency ratios of $\beta_s = 0.5$, 1, and 2 when using current mode power amplifier.

defined by Eq. (6.6.6). The force transmissibility has a minimum value given by

$$\frac{F_s}{K_f I_0} = \left[\frac{1}{1+M}\right]\left[\frac{j2\zeta_s}{r_p^2 - 1 + j2\zeta_t}\right] \quad (6.6.24)$$

when r is unity. The mass ratio M is an important consideration in minimizing the size of this *notch* or *glitch* in the force transmitted, a consideration that is under the experimenter's control. This ratio should be as small as possible in order to minimize the notch's depth for a given structure's damping. The experimenter has little control over the structure's damping and little direct control over the value of ζ_t, which depends on both the structure's damping and the armature's damping; see Eq. (6.6.23). It is seen that this force transmissibility curve takes on value of $1/(1+M)$ for large values of r, as shown in Fig. 6.6.5.

When $\beta_s = 1$ and r_p is unity, the transmitted force transmissibility in Eq. (6.6.22) starts with the value of $1/(1+M)$, and then has a notch that is controlled exclusively by the system damping, since Eq. (6.6.22) becomes

$$\frac{F_s}{K_f I_0} = \left[\frac{2\zeta_s}{2\zeta_s + (\beta_s M)(2\zeta_a)}\right] \quad (6.6.25)$$

for this case, as shown in Fig. 6.6.5. The force transmissibility is seen to quickly return to the value of $1/(1 + M)$ when r is slightly greater than unity. The reader can easily see that both armature damping and mass ratio M are important to minimize the size of this notch. Of these two parameters, the mass ratio is easiest to control.

In the case when $\beta_s = 2$, $r_p = 1.23$ and the force transmissibility curve is seen to start at the same value of $1/(1 + M)$, as before, and then to decrease slowly to a minimum that is nearly zero, since the $(r_p^2 - r^2)$ term has a value of 0.5 when $r = 1$. This value, along with the exciter's damping, contributes significantly to the size of the notch. This notch is followed quickly by a peak value of nearly 5 at test system resonance, which then returns slowly to a value of $1/(1 + M)$ for large values of r.

The upshot of all of this analysis is that it is desirable to have M as small as possible and to make β_s as small as possible. One way to make β_s zero is to use an exciter that has no armature support springs. In such a case, $r_p^2 = 1/(1 + M)$ and the glitch is nearly nonexistent. The next best condition is to use an armature that has a very low natural frequency and little damping. This works well as long as the structure's natural frequency is not too small.

Voltage Mode of Operation When the voltage mode amplifier response given by Eq. (6.6.16) is inserted into Eq. (6.6.21) and liberal use is made of dimensionless ratios, we obtain

$$TR = \frac{F_s}{K_f(E_0/R)} = \left[\frac{1}{1 + M + M_e}\right]\left[\frac{1 - r^2 + j2\zeta_s r}{r_t^2 - r^2 + j2\zeta_t r}\right] \quad (6.6.26)$$

where M_e is the inductive mass ratio of m_e/m_s with m_e as defined by Eq. (6.6.17).

r_t is the test system's natural frequency, which is given by

$$r_t^2 = \frac{1 + \beta_s^2 M}{1 + M + M_e} \quad (6.6.27)$$

ζ_t, the test system damping ratio, is given by

$$2\zeta_t = \frac{2\zeta_s + \beta_s M\{2\zeta_a + 2\zeta_e\} + \beta_e\{1 + \beta_s^2 M - (1 + M)r^2\}/\beta_s}{1 + M + M_e} \quad (6.6.28)$$

Equation (6.6.28) indicates that the exciter's damping term, $2\zeta_e$, can easily swamp the other terms unless β_s is small, in which case the electrodynamic damping term can cause excessive damping to occur at frequencies away from resonance. Equations (6.6.27) and (6.6.28) show that the inductive mass can be important in some conditions. Again it is important to keep

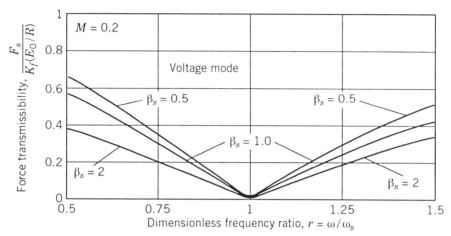

Fig. 6.6.6. Dimensionless force transmissibility versus frequency ratio $r = \omega/\omega_s$ for mass ratio of $M = 0.2$ and frequency ratios of $\beta_s = 0.5$, 1, and 2 when using a voltage mode power amplifier.

the mass ratio M as small as possible in order to reduce the sharpness of the notch that is represented by the numerator of Eq. (6.6.26). Equation (6.6.26) is plotted in Fig. 6.6.6 for $M = 0.2$ and $\beta_s = 0.5$, 1, and 2 for the same exciter used in Section 6.5. The results speak for themselves, for a broad based notch occurs around resonance for each case and the electromagnetic damping is so large that there are no peaks in this range of frequencies, as there were for the current mode of operation case.

We have seen that the exciter's parameters can significantly alter the structure's response. Clearly, the current mode of power amplifier operation is preferred over the voltage mode of operation in order to minimize this interaction. However, there is no basis of hoping to measure structural dynamic properties by using the input coil current as a direct measure of the force into the structure. At really low frequencies, well below structural resonance, the force transmitted to the structure is attenuated by the spring ratio and shows a significant notch at resonance and a peak either before or above resonance. It is clear that the force acting on the structure under test must be measured if accurate structural dynamic characteristics are to be measured.

6.7 INTERACTION OF AN EXCITER AND AN UNGROUNDED STRUCTURE UNDER TEST

We explore the interaction between an electrodynamic vibration exciter and an ungrounded structure under test that is attached directly to the

426 VIBRATION EXCITERS

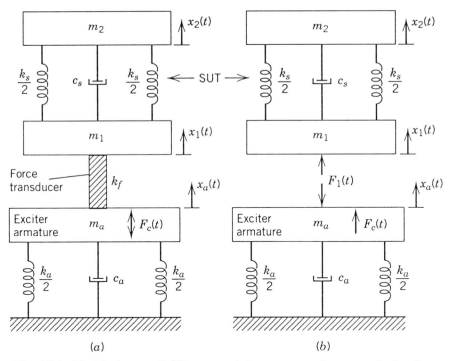

Fig. 6.7.1. Model of a two DOF ungrounded structure under test and vibration exciter armature connected by a near rigid force transducer link. (*a*) Parameter definitions. (*b*) Interface forces.

exciter's armature. For this study, we use a two DOF SUT that is attached to the electro-dynamic vibration exciter, as shown in Fig. 6.7.1. The structure under test is considered to have two masses m_1 and m_2, a stiffness k_s, and damping c_s. A force transducer with stiffness k_f is assumed to be between the SUT's base mass m_1 and the exciter's armature. The armature has mass m_a, stiffness k_a, and damping c_a, as before. The armature has an electro-dynamic force of $F_c(t)$ acting on it, due to the passage of current through its coil. The motion of each item is defined in terms of displacements $x_1(t)$, $x_2(t)$, and $x_a(t)$. The force transducer is assumed to be of sufficient stiffness that we can consider the connection to be rigid between masses m_1 and m_a, so that

$$x_1(t) \cong x_a(t) \tag{6.7.1}$$

The force transducer measures the interface force $F_1(t)$ between mass m_1 and m_a.

We develop a general set of equations in the frequency domain using the accelerance FRF concept. These equations are developed for a general case and then are made specific to the system shown. We look at both the current and voltage modes of amplifier operation.

6.7.1 A General Dynamic Model Using Driving Point and Transfer Accelerance

The structure under test has two acceleration frequency spectra, $A_1(\omega)$ and $A_2(\omega)$. $A_1(\omega)$ is the interface frequency spectrum, while $A_2(\omega)$ is the output frequency spectrum. These acceleration frequency spectra are related to the interface force frequency spectrum $F_1(\omega)$ by

$$A_1(\omega) = A_{11}(\omega) F_1(\omega) \qquad (6.7.2)$$

and

$$A_2(\omega) = A_{21}(\omega) F_1(\omega) \qquad (6.7.3)$$

where $A_{11}(\omega)$ is the SUT's driving point accelerance and $A_{21}(\omega)$ is the SUT's transfer accelerance.

The armature's motion is due both to the coil force $F_c(t)$ and to the interface force $F_1(t)$, which acts in the negative direction, as shown in Fig. 6.7.1b. Thus the armature's acceleration frequency spectrum is given by

$$A_a(\omega) = A_1(\omega) = A_{aa}(\omega) F_c(\omega) - A_{aa}(\omega) F_1(\omega) \qquad (6.7.4)$$

Substitution of Eq. (6.7.2) into Eq. (6.7.4) allows us to solve for the interface force frequency spectrum $F_1(\omega)$ in terms of the system driving point accelerances and the frequency spectrum of the input force $F_c(\omega)$ to obtain

$$F_1(\omega) = \left[\frac{A_{aa}(\omega)}{A_{11}(\omega) + A_{22}(\omega)}\right] F_c(\omega) = \left[\frac{A_{aa}(\omega)}{A_{11}(\omega) + A_{aa}(\omega)}\right] K_f I(\omega) \qquad (6.7.5)$$

since $F_c(\omega) = K_f I(\omega)$. Equation (6.7.5) clearly shows how the force applied to the armature's coil is modified by both the armature's driving point accelerance and the SUT's driving point accelerance. Hence, it should be clear that the test structure can significantly interact with the exciter and alter the input force to the SUT if we only measure the input current.

The SUT's acceleration spectral densities can be obtained by substituting Eq. (6.7.5) into Eqs. (6.7.2) and (6.7.3) to obtain

$$A_1(\omega) = \left[\frac{A_{aa}(\omega)A_{11}(\omega)}{A_{11}(\omega) + A_{aa}(\omega)}\right] K_f I(\omega) \qquad (6.7.6)$$

and

$$A_2(\omega) = \left[\frac{A_{aa}(\omega)A_{21}(\omega)}{A_{11}(\omega) + A_{aa}(\omega)}\right] K_f I(\omega) \qquad (6.7.7)$$

where the coil force is written in terms of the current frequency spectrum. It should be clear from these two equations that the measured output acceleration spectral densities (or auto-spectral densities) show effects of the driving point accelerances of the test item and exciter armature as well as the electromagnetic interactions. The resonances observed in these spectral densities are controlled by the denominator, which includes both driving point accelerances. *Hence, these resonances are test system resonances and not test structure resonances.*

The transmissibility ratio of input acceleration and test structure output acceleration is often used to measure the SUT's dynamic behavior. The transmissibility ratio is given by

$$TR(\omega) = \frac{A_2(\omega)}{A_1(\omega)} = \frac{A_{21}(\omega)}{A_{11}(\omega)} \qquad (6.7.8)$$

which is controlled by the transfer and driving point accelerances. Both of these accelerances share a common denominator, which cancels in Eq. (6.7.8) so that this ratio is controlled by the numerators of these two functions. *Hence the transmissibility ratio is independent of the vibration exciter's dynamic characteristics.*

The voltage current relationship for the exciter is that given by Eq. (6.5.2), which is rewritten here in terms of spectral densities and the SUT's driving point accelerance so that

$$E(\omega) = \left[R + j\left\{L\omega - \frac{K_v K_f A_{11}(\omega)}{\omega}\right\}\right] I(\omega) \qquad (6.7.9)$$

where $-A_{11}(\omega)/\omega$ is the SUT's driving point mobility. It is evident from Eq. (6.7.9) that the electrical parameters of the exciter and mechanical parameters of the SUT can interact in a significant manner. Now we explore a specific two DOF test structure.

6.7.2 Ungrounded Test Structure and Exciter Accelerance Characteristics

The driving point and transfer accelerances for the two DOF system shown in Fig. 6.7.1b are given by

$$A_{11}(\omega) = \frac{k_s - m_2\omega^2 + jc_s\omega}{k_s m - m_1 m_2 \omega^2 + jmc_s\omega} \qquad (6.7.10)$$

and

$$A_{21}(\omega) = \frac{k_s + jc_s\omega}{k_s m - m_1 m_2 \omega^2 + jmc_s\omega} \qquad (6.7.11)$$

where the SUT's total mass m is given by

$$m = m_1 + m_2 \qquad (6.7.12)$$

It is obvious that both accelerances have the same denominator terms but different numerator terms. The driving point accelerance has a deep notch when mass m_2 is acting like a dynamic absorber to base mass m_1 at the absorber frequency

$$\Omega_a = \sqrt{k_s/m_2} \qquad (6.7.13)$$

The test structure has an ungrounded natural frequency when the real part of the denominator is zero, that is,

$$\Omega_n = \sqrt{\frac{k_s m}{m_1 m_2}} \qquad (6.7.14)$$

The transmissibility ratio from Eq. (6.7.8) becomes

$$TR(\omega) = \frac{A_{21}(\omega)}{A_{11}(\omega)} = \frac{k_s + jc_s\omega}{k_s - m_2\omega^2 + jc_s\omega} \qquad (6.7.15)$$

which is the standard form found in all vibration textbooks. Note that the natural frequency contained in the transmissibility concept given by Eq. (6.7.15) is the absorber frequency given by Eq. (6.7.13). This is the same natural frequency that we obtain from Eq. (6.7.14) when the base mass m_1 becomes infinitely large so that there is no base motion. The results in Eqs. (6.7.6), (6.7.7), (6.7.10), (6.7.11), and (6.7.15) show that simply looking at the largest acceleration frequency spectrum, accelerance, or

430 VIBRATION EXCITERS

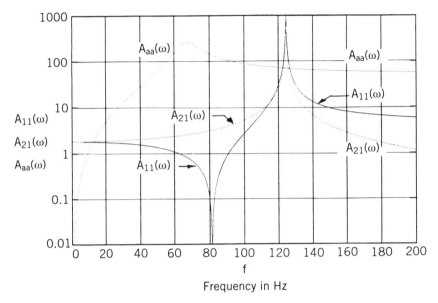

Fig. 6.7.2. Driving point $A_{11}(\omega)$ and transfer $A_{21}(\omega)$ accelerances for a two DOF structure under test and driving point accelerance $A_{aa}(\omega)$ for a one DOF model of the exciter armature.

transmissibility ratio can give drastically different answers to the question: "What defines a structure's natural frequencies?"

The exciter's armature has a driving point acceleration that is given by

$$A_{aa}(\omega) = \frac{-\omega^2}{k_a - m_a\omega^2 + jc_a\omega} \quad (6.7.16)$$

Now we need to calculate these accelerances for a specific system. The test structure is one that we have used in our vibrations laboratory in order to study the dynamics of an ungrounded system as shown in Fig. 6.7.1. The physical parameters for the structure are: $m_1 = 0.242$ kg, $m_2 = 0.352$ kg so that $m = 0.567$ kg, $c_s = 0.2$ N-s/m, and $k_s = 84{,}500$ N/m. The exciter characteristics are as given in Section 6.5.4 of this book.

The magnitudes of the resulting accelerances are shown in Fig. 6.7.2 over a frequency range of 0 to 200 Hz with units of m/s² per newton. The driving point accelerance $A_{11}(\omega)$ is seen to start at a value of $1/m$ near 0 Hz, pass through a valley at 81.25 Hz, pass through a resonance at 124.25 Hz, and then approach a value of $1/m_1$ near 200 Hz, as required by Eq. (6.7.10). This accelerance is seen to have a dynamic range in excess of 100,000 from the valley to the peak. The transfer accelerance $A_{21}(\omega)$

AN EXCITER AND AN UNGROUNDED STRUCTURE UNDER TEST   431

is seen to start at $1/m$ near 0 Hz, pass through a resonance at 124.25 Hz, and then drop off at a rate between $1/\omega$ and $1/\omega^2$, as required by Eq. (6.7.11). The vibration exciter's accelerance $A_{aa}(\omega)$ is seen to start at zero for 0 Hz, to have a resonant peak around 66.6 Hz, and then asymptotically to approach $1/m_a$ for higher frequencies. The armature is seen to have a high amount of damping.

6.7.3 Test Responses for Current and Voltage Mode Power Amplifier Inputs

Now we demonstrate the responses when the power amplifier is used in either the current or the voltage mode of operation.

Current Mode Power Amplifier   For the current mode power amplifier type of operation, the current frequency spectrum $I(\omega)$ is a constant so that we can easily compute the response input-output on a per ampere basis. All variables associated with the current mode of amplifier operation carry an additional subscript I. Hence the interface force frequency spectrum from Eq. (6.7.5) becomes

$$F_{1I}(\omega) = \left[\frac{A_{aa}(\omega)}{A_{11}(\omega) + A_{aa}(\omega)}\right] K_f I(\omega) \qquad (6.7.17)$$

For this case, the interface acceleration frequency spectrum is given by Eq. (6.7.6) and the output acceleration frequency spectrum is given by Eq. (6.7.7). The voltage required to drive the exciter with constant magnitude current is given by Eq. (6.7.9). When the physical variables are substituted into Eqs. (6.7.10) and (6.7.11), the resulting output, in m/s² per ampere, is shown in Fig. 6.7.3 over the 0 to 200 Hz frequency range. The first striking feature of these two curves is that both $A_{1I}(\omega)$ and $A_{2I}(\omega)$ have two apparent resonances. The first resonance is at 12.25 Hz. This is the fundamental natural frequency of a system consisting of the combined SUT and exciter armature mass connected to the armature's support spring k_a. The second common resonance occurs at 122.25 Hz. It is clear that the armature mass has caused a shift in the SUT's apparent natural frequency. This is a 2 Hz reduction in the resonant frequency. The deep valley in the $A_{1I}(\omega)$ curve occurs at 81.25 Hz, where test structure mass m_2 is acting as a dynamic absorber for both mass SUT m_1 and armature mass m_a.

The required voltage $E_I(\omega)$ to produce a constant magnitude input current is calculated from Eq. (6.7.9) and is shown in Fig. 6.7.4, along with the resulting interface force $F_{1I}(\omega)$, which is calculated from Eq. (6.7.17) on a unit current basis so that the units are volts/ampere and

432 VIBRATION EXCITERS

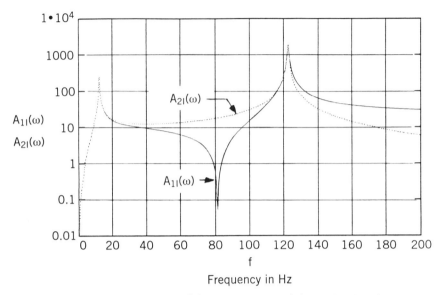

Fig. 6.7.3. A plot of interface $A_{1I}(\omega)$ and output $A_{2I}(\omega)$ acceleration frequency spectra with units of (m/s² per ampere) as a function of frequency for current mode of power amplifier operation.

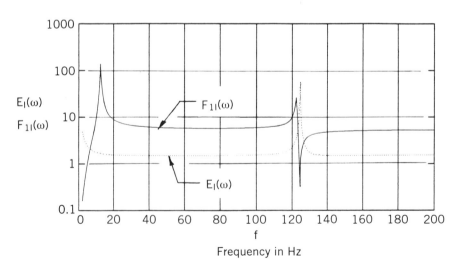

Fig. 6.7.4. Interface force $F_{1I}(\omega)$ (newtons/ampere) and required exciter voltage $E_I(\omega)$ (volts/ampere) frequency spectra for current mode of power amplifier operation.

newtons/ampere, respectively. It is clear that the required voltage is nearly flat until close to test structure ungrounded resonance of 124.25 Hz, where it peaks and then returns to the flat value. The interface force shows a rather interesting set of variations. The first peak value occurs at 12.25 Hz and corresponds to the first test system resonance. This is the large force required to control the SUT's mass m and make it move with the armature's motion. The presence of this large force indicates that the resonance is not due to the test structure itself but rather the entire test system. The second peak force occurs at 122.25 Hz or the combined test system second resonance as seen in Fig. 6.7.3. This peak is due to forcing the test structure to appear to have a resonance at this frequency. However, we see that the interface force drops to a rather deep valley at 124.25 Hz, which is the SUT's ungrounded natural frequency since little force is required to produce the SUT's motion.

Voltage Mode Power Amplifier All variables associated with the voltage mode of power amplifier operation carry an additional V subscript. In order to calculate the behavior of the test system when the power amplifier is operated in the voltage mode, we return to Eq. (6.7.9) and solve for the current to obtain

$$I(\omega) = \frac{E(\omega)}{R + j\{L\omega - (K_v K_f A_{11}(\omega))/\omega\}} \quad (6.7.18)$$

Substitution of this current expression into Eqs. (6.7.6) and (6.7.7) allows us to calculate the acceleration frequency spectra $A_{1V}(\omega)$ and $A_{2V}(\omega)$, while substitution into Eq. (6.7.17) allows us to calculate the interface force $F_{1V}(\omega)$. These two acceleration frequency spectra are plotted in Fig. 6.7.5 with units of m/s^2 per volt, while the actual current $I_V(\omega)$ with units of amperes/volt and interface force $F_{1V}(\omega)$ with units of newtons/volt are plotted in Fig. 6.7.6. It is evident in Fig. 6.7.6 that a significant force and current dropout occur at the SUT's ungrounded natural frequency resonance of 124.25 Hz. This frequency corresponds to the valley in Fig. 6.7.5 as well. The reason there is such a dip in the input current and interface force is due to two simultaneous events. First, the armature motion at resonance causes large back emf voltages to reduce the current passing through the exciter. Second, at structural resonance, little input force is required to cause large motions at 124.25 Hz. Consequently, there is a significant force drop-off at this frequency, which can lead to significant measurement problems due to noise in the signal.

Comparison of Current and Voltage Mode Responses The interface acceleration frequency spectra $A_{1V}(\omega)$ (m/s^2 per volt) for the voltage mode and

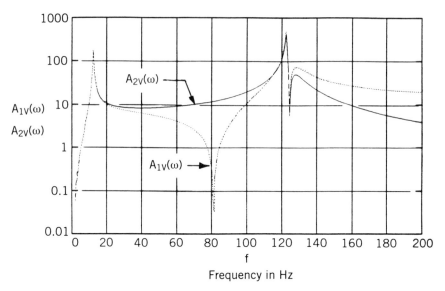

Fig. 6.7.5. A plot of interface $A_{1V}(\omega)$ and output $A_{2V}(\omega)$ acceleration frequency spectra with units of (m/s² per volt) for voltage mode of power amplifier operation.

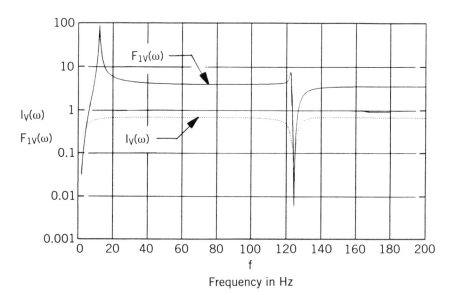

Fig. 6.7.6. Interface force $F_{1V}(\omega)$ (newtons/volt) and resulting exciter current $I_V(\omega)$ (amperes/volt) frequency spectra for voltage mode of power amplifier operation.

$A_{1I}(\omega)$ (m/s² per ampere) for the current mode of power amplifier operation are shown in Fig. 6.7.7a, while the output acceleration frequency spectra $A_{2V}(\omega)$ (m/s² per volt) and $A_{1I}(\omega)$ (m/s² per ampere) are shown in Fig. 6.7.7b. The corresponding interface force frequency spectra $F_{1V}(\omega)$ (N/volt) and $F_{2I}(\omega)$ (N/ampere) are shown in Fig. 6.7.8 for the voltage and current modes, respectively.

From these figures it is clearly evident that the current mode of operation is superior to the voltage mode of operation, due to force dropout that occurs at the SUT's ungrounded natural frequency of 124.25 Hz. In the current mode, the force per ampere goes from about 5.95 N/ampere at 100 Hz to a peak value of 27.0 N/ampere at 122.25 Hz to a valley value of 0.316 N/ampere at 124.25 Hz. The corresponding dynamic range is about 85.4 in this case. Now, in the voltage mode, the force per volt goes from about 3.84 N/volt at 100 Hz to a peak value of 7.14 N/volt at 122.25 Hz to a valley value of 0.0058 N/volt at 124.25 Hz. The corresponding dynamic range in this case is about 1230.

It is clear that force dropout occurs at the SUT's ungrounded natural frequency (124.25 Hz in this case) in both modes of amplifier operation. It is only that the voltage mode dropout is about 15 times worse than the current mode dropout. Consequently, this extra large dropout causes the voltage mode acceleration spectral densities to have an extra valley occur at 124.25 Hz, as is seen in Fig. 6.7.7. These valleys are the source of potential measurement errors due to instrumentation noise. Hence, the coherence function tends to show low values at the true test structure resonance of 124.25 Hz if both the interface force and accelerations frequency spectra are measured.

The acceleration transmissibility is the same regardless of the mode of power amplifier operation, as long as the signal to noise ratio is not a major problem. However, the apparent resonant frequency is really the SUT's ungrounded dynamic absorber frequency and not its natural frequency unless the SUT is attached to a rigid foundation so that mass m_1 cannot move.

6.8 SUMMARY

Many different types of vibration exciter mechanisms are used in vibration testing, including suddenly released static loads, impact loading, mechanical rotating exciters, electrohydraulic exciters, electrodynamic exciters, and the machine's own excitation sources. In all cases, there is an interaction between the exciter and the structure under test (SUT).

A fundamental rule of modal response is always present during vibration excitation in that some modes of vibration are not excited when the generalized excitation force is zero for that mode of vibration. This

436 VIBRATION EXCITERS

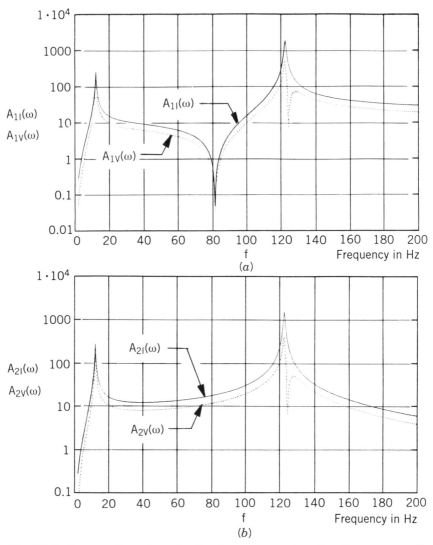

Fig. 6.7.7. A comparison of acceleration frequency spectra when the power amplifier is in either the current or the voltage mode of operation. Note the units are either (m/s² per ampere) or (m/s² per volt). (*a*) Interface acceleration frequency spectra. (*b*) Output acceleration frequency spectra.

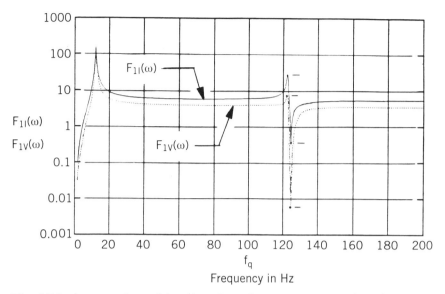

Fig. 6.7.8. A comparison of interface force frequency spectra when the power amplifier is in either the current $[F_{1I}(\omega)]$ or the voltage $[F_{1V}(\omega)]$ mode of operation. Note that the units are either (N/ampere) or (N/volt).

generalized force is zero either when the product of the distributed excitation force times the mode shape adds to zero when summed over the region of excitation or when a concentrated excitation force is acting at a node point for that mode.

Our analysis of the impulse loading method has shown that the force is basically that of a half sine with a peak impact force that is proportional to the product of the momentum of the impacting mass (mv) and the impact frequency, that is, π/T where T is the half sine period. It has also been shown that double impacts easily occur when the mass ratio of the impacting mass to the structure's effective mass is near that of the coefficient of restitution. Note that double impacts are not necessarily bad in impact testing so long as all impacts are properly measured and the data is correctly processed.

The direct drive and rotating unbalance types of mechanical vibration exciters has been analyzed. The effects of table mass and foundation mass on the exciter's behavior have been explored. It was shown that the drive torque varies at twice the rotating frequency and that it becomes large near the SUT's natural frequency. This large variation in torque explains why it is difficult to obtain good vibration data near the SUT's resonances, since the mechanical exciter tends to pop through to a higher frequency due to trying to drive this highly variable torque. The twice rotation frequency explains why twice rpm excitation is observed in the output motion.

Electrohydraulic exciters are generally used for low frequency applications in the range of near 0 to 40 to 100 Hz frequency range, where large amounts of force are required. The hydraulic system employed is quite nonlinear due to servovalve and pipe flow characteristics. Consequently, this system requires considerable care in order to have a stable control system when different test structures are connected to the exciter.

The electrodynamic vibration exciter has been analyzed in great detail since it is used more than any other exciter type. Generally, the exciter has a massive base supported on soft springs. In this configuration, minimum motion is transmitted to any supporting structure, so that a minimum of vibration excitation energy is transmitted to the SUT through a common foundation. It has also been found that the large exciter base inertia gives a foundation that is nearly rigidly mounted for frequencies above the low frequency support natural frequency.

The electrodynamic armature has been shown to have two natural frequencies, one a low frequency that corresponds to total armature mass on its support flexures, and the other one a high frequency out of phase resonance where the current coil and the armature's table top oscillate against one another. The armature's effectiveness in transmitting the coil force to the armature's table drops off rapidly above this second resonance. It has also been seen that this coil resonance decreases with increasing mass being attached to the armature. Hence the usable frequency range is less than that quoted by the exciter manufacturer for a bare table exciter.

The coil force has a nonlinear output relative to current for different coil positions in the gap's magnetic field. The same nonlinear behavior occurs for the back emf that is generated due to table motion. Usually the table motion is small enough so that this nonlinearity is not a serious problem. However, at low frequencies, the table motion is significant and these nonlinear characteristics may manifest themselves in the test results.

The power amplifier that drives the armature coil has two modes of operation: voltage mode and current mode. In the voltage mode, the output voltage applied to the coil circuit is directly proportional to the power amplifier's input voltage. In the current mode, the current through the armature coil is directly proportional to the power amplifier's input voltage. The advantage of the current mode is that the voltage applied to the armature coil is automatically adjusted to counter any back emf that is developed due to coil motion in the magnetic field. In this way we have found that less force dropout occurs at the SUT's natural frequencies when the power amplifier is operating in the current mode than when it is operating in the voltage mode.

The bare table armature shows completely different dynamic behavior depending on whether the power amplifier is in the voltage or current mode of operation. In the voltage mode of power amplifier operation, there is no low frequency armature resonance, due to the high electrodynamic damping caused by energy dissipation that results from the coil circuit's back emf and resistance. However, when operating in the current mode, the low frequency armature resonance is clearly evident, since the input voltage to the coil circuit causes a constant magnitude sinusoidal current (and force) to be applied to the armature. This type of behavior occurs so long as the voltage required to drive the coil at a constant magnitude sinusoidal force does not exceed the power amplifier's output voltage capabilities.

The interactions of the exciter armature with both a simple single DOF grounded structure and a two DOF ungrounded structure have been analyzed for both the current and voltage modes of operation. It is clear from this analysis that a significant interaction takes place between the SUT and the exciter system. The use of either voltage or current modes of power amplifier operation is seen to have a significant impact on the measured results. It is clear from this study that one needs to understand the requirements for a particular test and be clear about what type of information is required. It is one thing to use current input to control a fatigue test, and something else entirely to use current as a SUT's input force and try to measure the SUT's natural frequencies and damping. In this latter case, we are measuring test system behavior instead of the SUT's behavior. It is clearly evident from the ungrounded test simulation that measuring either the interface force or the interface acceleration is a good idea, for the vibration exciter effects are eliminated. In Chapter 7 we shall look at specific test environments and examine how the vibration exciter and other test hardware may interact with the test structure and affect the test results.

REFERENCES

The following references are for general information.
1. Anderson, I. A., "Avoiding Stinger Rod Resonance Effects on Small Structures," *Proceedings of the 8th International Modal Analysis Conference*, Vol. 1, Kissimmee, FL, 1990, pp. 673–678.
2. Clark, A. J., "Sinusoidal and Random Motion Analysis of Mass Loaded Actuators and Valves," *Proceedings of the National Conference on Fluid Power*, Vol. XXXVII, 39th annual meeting, Los Angeles, CA, 1983.
3. Clark, A. J., "Dynamic Characteristics of Large Multiple Degree of Freedom Shaking Tables," *Earthquake Engineering, 10th World Conference*, Balkema, Rotterdam, Holland, 1992, pp. 2823–2828.
4. Foss, G., "Enhancement of Modal Swept Sine Data by Control of Exciting

Forces," *Proceedings of the 8th International Modal Analysis Conference*, Vol. 1, Kissimmee, FL, 1990, pp. 102–108.

5. Harris, C. M. and C. S. Crede (Editors), *Shock and Vibration Handbook*, Vol. 2, McGraw-Hill, New York, Chapter 25 by K. Unholtz, "Vibration Testing Machines," pp. 25-7–25-9.

6. Hunt, F. V., *Electroacoustics: The Analysis of Transduction, and its Historical Background*, American Institute of Physics, for the Acoustical Society of America, Woodbury, NY, 1954.

7. Olsen, N. L., "Using and Understanding Electrodynamic Shakers in Modal Applications," *Proceedings of the 4th International Modal Analysis Conference*, Vol. 2, 1986, pp. 1160–1167.

8. Otts, J. V., "Force Controlled Vibration Tests: A Step Toward Practical Application of Mechanical Impedance," *The Shock and Vibration Bulletin*, No. 34. Part 5, Feb. 1965, pp. 45–53.

9. Rao, D. K., "Electrodynamic Interaction Between a Resonating Structure and an Exciter," *Proceedings of the 5th International Modal Analysis Conference*, Vol. 2, 1987, pp. 1142–1150.

10. Rogers, J. D., "An Introduction to Shaker Shock Simulations," *Proceedings of the 8th International Modal Analysis Conference*, Vol. 2, Kissimmee, FL, 1990, pp. 905–911.

11. Sharton, T. D., "Analysis of Dual Control Vibration Testing," *Proceedings, Institute of Environmental Sciences*, 1990, pp. 140–146.

12. Sharton, T. D., "Dual Control Vibration Tests of Flight Hardware," *Proceedings, Institute of Environmental Sciences*, 1991, pp. 68–77.

13. Smith, R. J., *Electronics: Circuits and Devices*, 3rd ed., John Wiley & Sons, New York, 1987.

14. Smallwood, D. O., "An Analytical Study of a Vibration Test Method Using Extremal Control of Acceleration and Force," *Proceedings, Institute of Environmental Sciences*, 1989, pp. 263–271.

15. Szymkowiak, E. A. and W. Silver, "A Captive Store Flight Vibration Simulation Project," *Proceedings, Institute of Environmental Sciences*, 1990, pp. 531–537.

16. Tomlinson, G. R., "Determination of Modal Properties of Complex Structures Including Non-Linear Effects," Ph.D. Thesis, University of Salford, Salford, UK, May 1979.

17. Tomlinson, G. R., "Force Distortion in Resonance Testing of Structures with Electro-Dynamic Vibration Exciters," *Journal of Sound and Vibration*, Vol. 63, No. 3, 1979, pp. 337–350.

7 The Application of Basic Concepts to Vibration Testing

The 365-ft tall ECOLE vertical axis wind turbine in Canada is rated at 4 megawatts and was tested by Sandia Laboratories by using the step relaxation method to verify the theoretical modal analysis used in its design. (Photo Courtesy of *Sound and Vibration*).

442 THE APPLICATION OF BASIC CONCEPTS TO VIBRATION TESTING

A typical test system consisting of a vibration exciter, a stinger, a force transducer, an accelerometer, a frequency analyzer, and a test item is shown. Note the large mass of the accelerometer on the lightweight disk of the test item, which can cause serious measurement problems. (Photo Courtesy of *Sound and Vibration*).

7.1 INTRODUCTION

In this chapter we examine the behavior of typical vibration test environments where the various elements of the previous concepts are combined in order to see how they affect one another. The goal is to understand how different test elements such as transducers, data processing, test equipment, and test structure can alter the test results. A number of simple but common test environments are used in order to obtain a feel for the type of effects we can experience. Each test setup is described in detail, since the reader is encouraged to set up similar experiments in order to learn by doing.

The test setups that are considered include:

1. Sudden release of a static load, which is often referred to as the step relaxation method (SRM). This method has unique signal conditioning and data processing requirements in order to obtain satisfactory frequency response function (FRF) results.
2. Forced vibrations of a simply supported beam that is directly attached to an exciter. In this case the effects of transducer mass are illustrated. This is a common test environment in the laboratory where we test a structure to meet specific test standards, natural frequencies, and so on.
3. Impulsive testing, where we see that data windowing is extremely important in order to achieve satisfactory results.
4. The use of a vibration exciter to drive a structure with a point load. This experiment uses a free-free beam with a force transducer attached to the beam and a stinger between the exciter and the structure. Different test signals and the effects of data windows are discussed for this application.
5. The measurement of material damping at low frequency reveals subtle data processing problems that result from using a digital approximation of a continuous mathematical process.

A very powerful technique that is used to reconstruct the force acting on a free elastic structure by measuring the acceleration at many locations is called sum of the weighted accelerations Technique, or SWAT. This technique was initially developed at Sandia National Laboratories in the mid 1980s[1] and has been extended by others since. The main idea behind this technique is the principle of motion of the mass center, which relates the resultant force to the motion of the mass center, that is,

$$\bar{R} = m\bar{a}_g = \sum_{i=1}^{N} m_i \bar{a}_i \qquad (7.1.1)$$

where $\bar{R}$ is the resultant force acting on the structure.

---

[1]D. L. Gregory, T. G. Priddy, and D. O. Smallwood, "Experimental Determination of the Dynamic Forces Acting on Non-Rigid Bodies," SAE Technical Paper Series, Paper No. 861791, Aerospace Technology Conference and Exposition, Long Beach, CA, October 1986.

444 THE APPLICATION OF BASIC CONCEPTS TO VIBRATION TESTING

m is the total mass.
m_i is the ith mass.
$\bar{a}_i$ is the ith acceleration.

The main trick is to locate the accelerometers so that signal addition with proper weighting will give the resultant force. While we do not explore this method here, it has proven to be useful in a wide range of studies of elastic bodies subjected to intense dynamic loads, such as projectile penetration studies and drop tests of a nuclear shipping container.

7.2 SUDDEN RELEASE OR STEP RELAXATION METHOD

The step relaxation method (SRM) is often used to test large structures such as space lattice structures,[2] vertical axis wind turbines,[3] and offshore oil platforms.[4] The basic approach is quite simple, as illustrated in Fig. 7.2.1a. We load the structure with a slowly varying loading condition during the T_1 to T_2 time period until the maximum load of F_0 is obtained, the static load is held for some unspecified time period from T_2 to T_3 while test preparations are finished, and then the load is suddenly released in a near step-like manner at time T_3. The sudden release mechanism can be achieved by using either explosive bolts or a replaceable fuse type link, such as a bolt in either tension or shear.

We explore the problems associated with such vibration tests in this

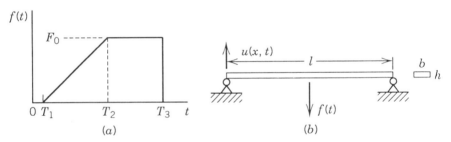

Fig. 7.2.1. Step relaxation method applied to simply supported beam. (*a*) Slowly applied load that is suddenly released. (*b*) Simply supported structure.

---

[2]C. Mutch et al., "The Dynamic Analysis of a Space Lattice Structure via the Use of Step Relaxation Testing," *Proceedings of the 2nd International Modal Analysis Conference*, Orlando, FL., Feb. 1984, pp. 368–377.
[3]J. Lauffer et al., "Mini-Modal Testing of Wind Turbines Using Novel Excitation," *Proceedings of the 3rd International Modal Analysis Conference*, Feb. 1985, pp. 451–458.
[4]M. Martinez and P. Quijada, "Experimental Modal Analysis in Offshore Platforms," *Proceedings of the 9th International Modal Analysis Conference*, Florence, Italy, April 1991, pp. 213–218.

section. We use the simply supported beam shown in Fig. 7.2.1b as our test structure, since the mode shapes and natural frequencies are easily calculated for this case. The static load is applied at the midpoint and the midpoint acceleration is calculated. We then need to determine how we can measure such a transient response and obtain the correct driving point accelerance for this structure from the measured data. Then we apply the static load at the quarter point ($l/4$) and measure the acceleration at the third point ($l/3$). We examine how the measured transfer accelerance compares to the theoretical values. The entire test scheme is simulated using the principles of modal analysis outlined in Sections 3.7 and 3.8.

7.2.1 Theoretical Modal Model

The beam in Fig. 7.2.1b is assumed to be made from steel ($E = 29 \times 10^6$ psi, $\gamma = 0.283$ lb/in.^{3}) with a rectangular cross-section width b ($= 1$ in.) and depth h ($= 0.375$ in.). The length l ($= 41.12$ in.) is chosen so that the first five natural frequencies are 20, 80, 180, 320, and 500 Hz. The corresponding mode shapes are given by sine waves so that the pth mode shape is

$$U_p(x) = \sin\left(\frac{p\pi x}{l}\right) \qquad (7.2.1)$$

The modal loading force Q_p is given by Eq. (3.7.20) as

$$Q_p = \int_0^l P(x) U_p(x)\, dx \qquad (7.2.2)$$

where $P(x)$ is the distributed excitation load per unit length. Since the static force is applied at only one point, the load distribution is described by the Dirac delta function, that is,

$$P(x) = F_i\, \delta(x - x_i) \qquad (7.2.3)$$

where x_i is the location of the input static load application. The implications of Eqs. (7.2.2) and (7.2.3) are that the modal loading force becomes

$$Q_p = F_i \sin\left(\frac{p\pi x_i}{l}\right) \qquad (7.2.4)$$

Then the output response at location x_0 is given by Eq. (3.7.21) as

$$u(x_0, t) = \sum_{p=1}^{N} \frac{Q_p U_p F_i}{k_p - m_p \omega^2 + jc_p \omega} e^{j\omega t} = H_{oi}(\omega) F_i e^{j\omega t} \quad (7.2.5)$$

where $H_{oi}(\omega)$ is the output FRF at location o per unit of force applied at location i.
k_p is the modal stiffness.
m_p is the modal mass.
c_p is the modal damping.
N is the number of modes used in the simulation, five in this case.

7.2.2 Midpoint Excitation and Response

When the excitation and response are measured at the midpoint, the driving point receptance with units of in./lb is calculated from Eq. (7.2.5) and is shown in Fig. 7.2.2a. Here we see that only the odd natural frequencies are present; that is, 20, 180, and 500 Hz, since all even mode values of Q_p are zero, as seen from Eq. (7.2.4). The displacement impulse response function, $h(t)$, for this FRF is shown in Fig. 7.2.2b, from which we see that the motion is completely decayed within 2 seconds of the 3 second time window and that the fundamental frequency dominates this response. Thus this simulation, which uses 4096 data points in the time domain to cover the 3 second data window, satisfies the fast Fourier transform (FFT) requirement for transient analysis of zero initial and final magnitudes and slopes.

Displacement and Acceleration Response to SRM Loading The input force time history is shown in Fig. 7.2.3a, where the maximum force F_i is 25 lbs, the ramp start time T_1 is 36.6 ms, the hold time starts at $T_2 = 437$ ms so that the ramp duration is 400 ms, and the sudden release occurs at 1 second. The ramp time should be a multiple of 50 ms for no oscillations to occur in this system, since its fundamental natural frequency has a period of 50 ms. The midpoint displacement response is shown in Fig. 7.2.3b, where we see that the system tracks the input in a near perfect manner with only a small overshoot oscillation occurring during the hold period. This oscillation quickly decays to zero. In a real test, it is unlikely that we would use such a short rise time and there would be a longer time between T_2 and T_3. All we need in the calculations is to simulate certain characteristics that can be accomplished with the input chosen. The aver-

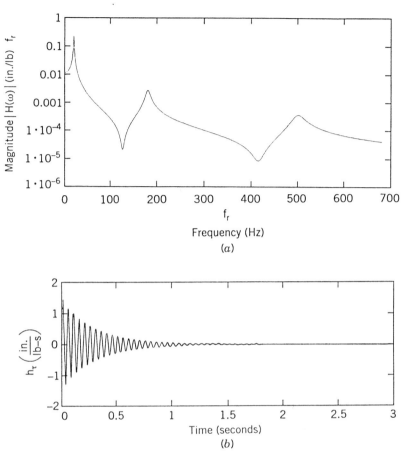

Fig. 7.2.2. Displacement functions for a midpoint excitation of a simply supported beam. (*a*) FRF. (*b*) Input response function.

age deflection during the hold portion is that of static deflection under a 25 lb load ($l^3/48EI \cong 0.284$ in.), as it should be. We see that the sudden release gives rise to a damped displacement oscillation that is dominated by the fundamental frequency.

The acceleration time history (units of g's of acceleration) is shown in Fig. 7.2.3*c*. Here, we see that the ramp causes a small acceleration during the T_1 to T_2 time frame, which is followed by a quiet period during the T_2 to T_3 holding time period, followed in turn by rather severe large amplitude and high frequency accelerations that die out quickly due to the $\zeta_p \omega_p$ term being larger for the higher frequencies than for the fundamental frequency.

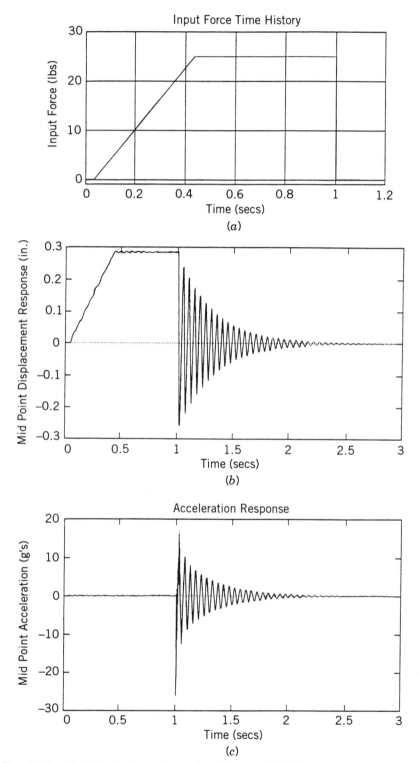

Fig. 7.2.3. (a) Midpoint input force time history. (b) Midpoint displacement response time history. (c) Midpoint acceleration response time history.

7.2.3 Resolving the Measurement Dilemma

If we have the entire time history of the input force and the response (either displacement or acceleration), as shown in Fig. 7.2.3, we are able to calculate either the driving point receptance or accelerance without problems, since we have the entire input and response time histories. In the real world of testing, these time records are often too long to capture. The loading process can extend from minutes to hours as we slowly apply the static load, while we need only a few seconds of record to obtain the response (see Fig. 7.2.3b and 7.2.3c). Thus we are forced to truncate each time history as shown in Figs. 7.2.4a and 7.2.4b, where the initial ramp hold input is removed by the truncation. These truncated records contain 1024 data points in this example, so that the data window is 749 ms long. The corresponding measured accelerance is shown in Fig. 7.2.4c, along with the actual system accelerance. The cause for the many peaks in the measured accelerance is the input force's frequency spectrum, shown in Fig. 7.2.4d, where the many low values cause peaks to occur in the accelerance curve. In fact, this input frequency spectrum is the sinc function discussed in Chapters 2 and 5, since the real input force has been converted into a little rectangular pulse that is about 52 ms long. The corresponding notch frequencies occur about every 29 Hz, as seen in Fig. 7.2.4d.

Lauffer et al.[5] proposed a practical solution to this dilemma where both signals are AC coupled so that the DC offset in the measured load is converted into an exponential impulse. The entire input time histories for both the AC and DC coupled cases are shown in Figs. 7.2.5a and 7.2.5b. It is clear that once sufficient hold time has occurred, the input force starts at a zero value, as shown in region A, so that the beginning time is not so important when we take a subsample of the entire time history. The changes in the beginning acceleration are more subtle to see, but McConnell and Sherman[6] have shown that the same AC coupling must be applied to the acceleration signal as well, so that the AC coupling FRF effects cancel out in the FRF function, since these effects are present in both signals. The original and the AC coupled force frequency spectra are compared in Fig. 7.2.5c for the lower frequencies where it is obvious that the two input frequency spectra are essentially the same. At higher frequencies, these two curves are the same. Hence, the AC coupling trick

---

[5]See footnote 3.
[6]K. G. McConnell and P. J. Sherman, "FRF Estimation under Non-Zero Initial Conditions," *Proceedings of the 11th International Modal Analysis Conference*, Kissimmee, FL, Feb. 1993, pp. 1021–1025.

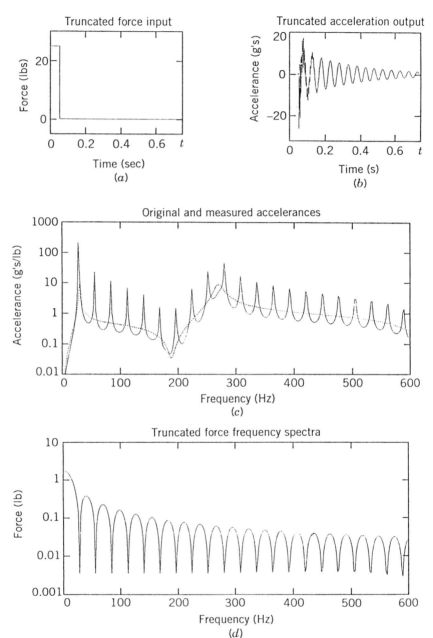

Fig. 7.2.4. (*a*) Truncated input force time history. (*b*) Midpoint acceleration time history. (*c*) Estimated versus actual driving point acceleration. (*d*) Frequency spectrum of truncated input time history.

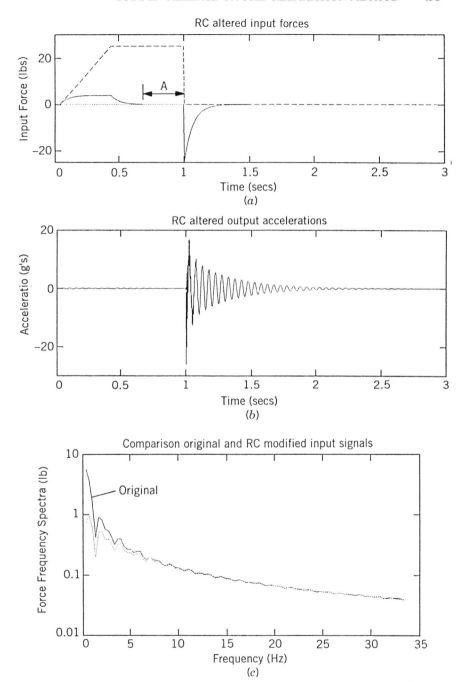

Fig. 7.2.5. (*a*) Comparison of original and AC-coupled input force signal. (*b*) AC-coupled output acceleration signal. (*c*) Comparison of original and modified AC-coupled input frequency spectrum at low frequencies.

only affects the lowest frequencies and the remaining input frequency spectra should be intact.

The typical modified and truncated input force and output acceleration time histories are shown in Fig. 7.2.6a and 7.2.6b, respectively. The output time history shows the potential of filter leakage due to the end point being nonzero with a nonzero slope. If we use the original acceleration time history in Fig. 7.2.6b, we obtain the measured driving point accelerance shown in Fig. 7.2.6c, where the theoretical accelerance is shown for comparison purposes. We see that the measured value matches the theoretical values quite closely near the resonant peaks. However, we see that filter leakage causes ripples in the measured FRF near the valleys.

One way to cure the FRF oscillations is to use an exponential window in order to reduce the amount of leakage. The window applied to the output signal is shown where $e^{-T/\tau_e} = 0.129$ at the end of the 750 ms long window. This means that only 13 percent of the end signal is used and that $\tau_e = 0.362$ second. This values of τ_e translates into 110 percent more damping in the first mode, 27.5 percent more damping for the third mode, and 4.4 percent more damping in the fifth mode. The corresponding driving point accelerance is shown in Fig. 7.2.6d, where the valleys are cleaned up (the oscillations in Fig. 7.2.6c are gone). The price for removing the oscillations is a significant reduction in the peak FRF amplitudes where the lower frequencies are affected more than the higher frequencies, since the effect of the exponential window translates into modal damping given by

$$\zeta_p = \frac{1}{2\pi\tau_e f_p} \quad (7.2.6)$$

where f_p is the pth natural frequency. We clearly see that the use of matched AC coupling for each channel significantly improves the measured FRF, as is shown by comparing the results in Figs. 7.2.4c with 7.2.6c and 7.2.6d.

Accelerance Transfer FRF Response When Excited at l/4 and Output at l/3 The same beam is loaded with the same load time history as shown in Fig. 7.2.2a. This load is applied at the quarter point ($l/4$) and the output motion is measured at the third point ($l/3$). The corresponding AC coupled input force and output acceleration time histories are shown in Figs. 7.2.7a and 7.2.7b. The output shows a second signal is modified by using the same exponential window that was used in the previous example. The measured and theoretical transfer FRF curves that result when the raw

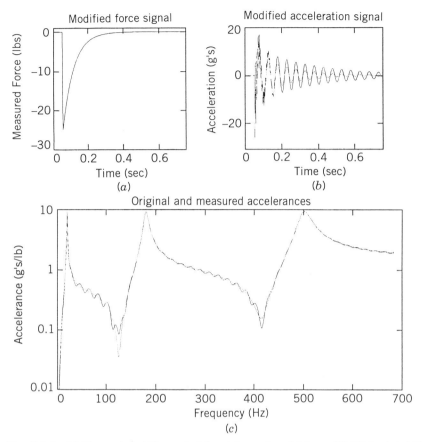

Fig. 7.2.6. (*a*) Truncated AC-coupled input force time history. (*b*) Midpoint AC-coupled output acceleration signal with and without exponential window to reduce filter leakage. (*c*) Comparison of theoretical and measured accelerance with filter leakage. (*d*) Comparison of theoretical and measured accelerance without filter leakage.

AC coupled data is used are shown in Fig. 7.2.7*c*, and those that result when the exponential window is used are shown in Fig. 7.2.7*d*. The characteristic filter ripple is evident in Fig. 7.2.7*c* due to filter leakage, while the lowered peak magnitudes due to the exponential window are found in Fig. 7.2.7*d*. In both figures, the measured and theoretical FRF's are quite good.

A striking feature of Figs. 7.2.7*c* and 7.2.7*d* compared to Figs. 7.2.6*c* and 7.2.6*d* is the different resonant frequencies that are present. In Figs. 7.2.6*c* and 7.2.6*d*, the natural frequencies are 20, 180, and 500 Hz, which

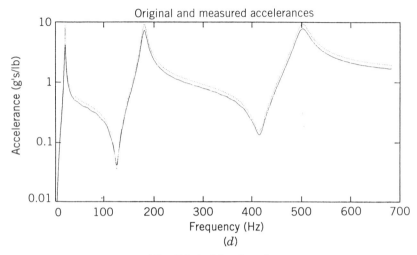

Fig. 7.2.6. (*Continued*).

correspond to the 1st, 3rd, and 5th modes of vibration while the even modes are absent. In Figs. 7.2.7c and 7.2.7d, the natural frequencies are 20, 80, and 500 Hz, which correspond to the 1st, 2nd, and 5th natural frequencies while the 3rd and 4th natural frequencies are absent. These results are in accordance with Eqs. (7.2.1) and (7.2.4). From Eq. (7.2.1), we see that the 3rd natural frequency is absent, due to the accelerometer being placed at the 3rd mode's node point. The 4th natural frequency is absent due to the excitation being applied at a node point given by Eq. (7.2.4). Thus we clearly see that absence of a resonance can be caused by two events in this type of testing: the input force and/or the output accelerometer being placed at a node point. Consequently, one needs to apply the excitation at more than one point in order to obtain the desired natural frequencies in most structures.

7.2.4 Results from an Actual Test

The results presented here are adapted from a student project,[7] where the SRM loading was applied to a portal frame as shown in Fig. 7.2.8. The steel girder was 1.5 in. wide, 0.75 in. deep, and 36 in. long between the column support points. The steel columns were 0.75 in. in diameter and 48 in. long from the floor to the center of the girder. The column

---

[7] J. Gruening, C. Clover, and S. Rittmueller, "Exciting a Portal Frame by Applying and Releasing Static Loads," Term Project Report, Engineering Mechanics 545x, Iowa State University, Ames, IA, Spring 1994.

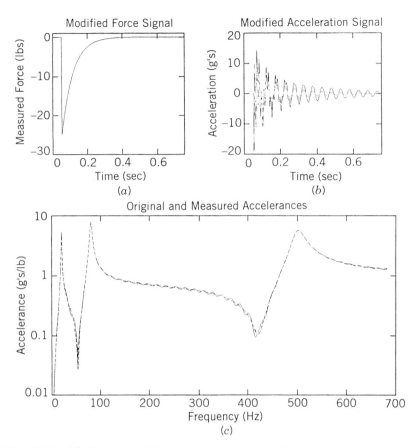

Fig. 7.2.7. (*a*) Truncated AC-coupled input force time history at the quarter point. (*b*) Third point AC-coupled output acceleration signal with and without exponential window to reduce filter leakage. (*c*) Comparison of theoretical and measured accelerance with filter leakage. (*d*) Comparison of theoretical and measured accelerance without filter leakage.

base was attached to a large cast iron tie down plate. The static excitation force was generated by using a 50 lb steel weight, which was attached to a force transducer through a hook and heavy duty string that was cut by a sharp knife to release the weight. The digital oscilloscope used had no anti-aliasing filter, so that a 5 kHz sample rate was used to gather all data reported in order to minimize the amount of anti-aliasing effects encountered. Each time sample used contained 16,384 data points, which were transferred to a personal computer for processing.

Frequency Domain Effects of AC Coupling on the Force Signal   The effects

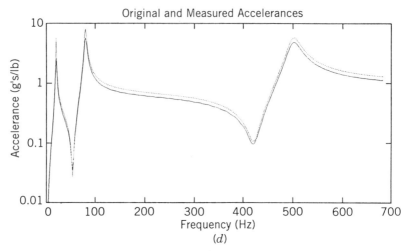

Fig. 7.2.7. (*Continued*).

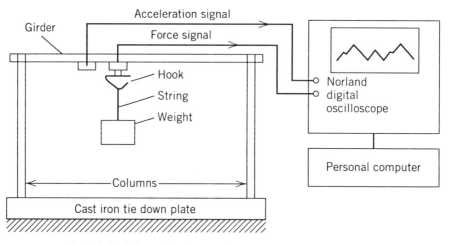

Fig. 7.2.8. Schematic of portal frame SRM test arrangement.

of AC signal coupling on frequency domain of an actual force signal are examined here. The force transducer used was of the dual time constant type that had an effective time constant of 21.7 seconds when DC coupled to the digital oscilloscope. When the transducer is connected to the AC coupled input of the digital oscilloscope, the effective time constant is 0.1 second (this corresponds to a low end 3 dB cutoff frequency of 1.6 Hz). Both signals were recorded simultaneously, as shown in Figs. 7.2.9a and 7.2.9b. We see the static load ramp is applied in about 0.5 second, this

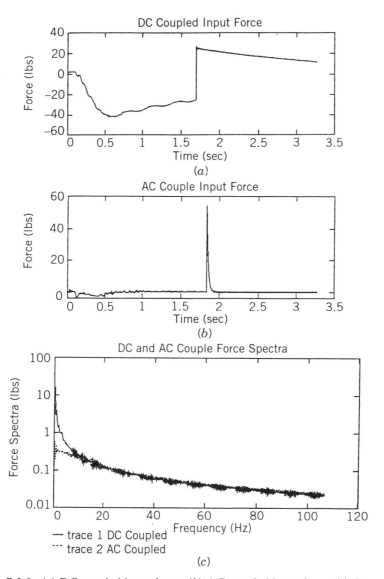

Fig. 7.2.9. (*a*) DC-coupled input force. (*b*) AC-coupled input force. (*c*) Comparison of DC- and AC-coupled input force frequency spectra at low frequencies. (Adapted from J. Gruening, C. Clover, and S. Rittmueller, "Exciting a Portal Frame by Applying and Releasing Static Loads," Term Project Report, Engineering Mechanics 545x, Iowa State University, Ames, IA, Spring, 1994.)

458 THE APPLICATION OF BASIC CONCEPTS TO VIBRATION TESTING

ramp is followed by the static load condition for about 1.2 seconds, during which the force transducer's dual time constant is causing a slowly decaying signal. Then the cord is cut and the load is suddenly released so that a near step occurs, giving an overshoot that slowly decays exponentially toward zero with the transducer's dual time constant. The AC coupled signal is basically zero up to the instant the cord is cut, at which time the exponential pulse occurs, which quickly dies out at the AC coupled time constant. The frequency spectra of both signals are compared in Fig. 7.2.9c over the 0 to 110 Hz frequency range. The DC coupled signal has higher low frequency components and a significant filter leakage ringing that is superimposed on the general curve. The AC coupled signal is much smoother and matches the DC signal very closely for frequencies higher than 6 Hz. In fact, a comparison to 2 kHz confirmed this close match throughout. Thus AC coupling can produce an acceptable estimate of the actual force frequency spectrum, as shown previously.

Test Results Typical test results are shown in Fig. 7.2.10, where Fig. 7.2.10a shows the input force frequency spectrum, Fig. 7.2.10b shows the output acceleration frequency spectrum, and Fig. 7.2.10c shows the acceleration transfer FRF between input at the third point and output at the quarter point of the girder. The input force frequency spectrum in Fig. 7.2.10a shows a number of little glitches that correspond to the larger resonant peaks in Fig. 7.2.10b. These glitches are due to the mass of the hook used to support the weight, which the force transducer senses due to the large girder accelerations. We also note that the force spectrum becomes noisy above 600 Hz (or below a magnitude of 0.0001) and has a low spot around 800 Hz. This noisiness comes from the fact that we are using a 12 bit A/D converter that has a dynamic range limit around 5×10^{-5}.

The first five natural frequencies of the girder when treated as a simply supported beam are shown in Table 7.2.1. The first five flexural natural frequencies for the columns are also shown, where the lower values correspond to clamped-pinned boundary conditions and the higher values correspond to clamped-clamped boundary conditions. These two boundary conditions should give a rough range for column natural frequencies.

Fig. 7.2.10. (a) Third point input force frequency spectrum. (b) Quarter point output acceleration frequency spectrum. (c) Transfer FRF accelerance frequency spectrum for portal frame. (Adapted from J. Gruening, C. Clover, and S. Rittmueller, "Exciting a Portal Frame by Applying and Releasing Static Loads," Term Project Report, Engineering Mechanics 545x, Iowa State University, Ames, IA, Spring, 1994.)

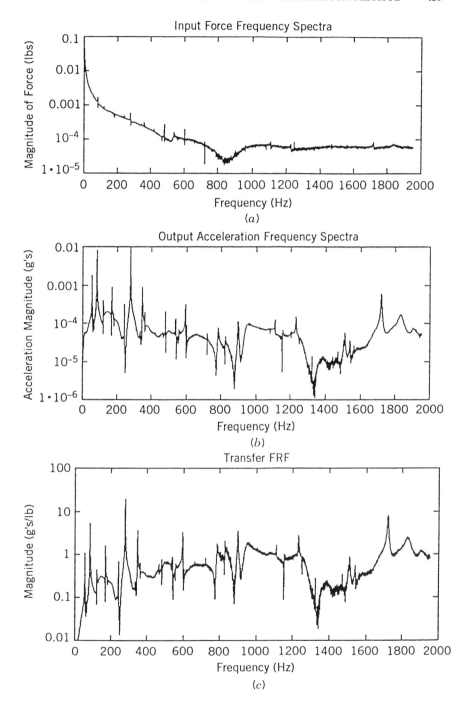

TABLE 7.2.1. Theoretical and Measured Natural Frequencies for the Portal Frame Under Test

Mode Number	Girder Motion Vertical		Column Horizontal Motion	
	Theoretical (Hz)	Measured (Hz)	Theoretical Range (Hz)	Measured (Hz)
1	70	84.5[a]	40–57.7	56.9
2	281	280[a]	129–159	123
3	632	600	268–312	243 and 347
4	1124	1119[a]	459–515	484
5	1756	1728	701–771	—

[a]Largest responses.

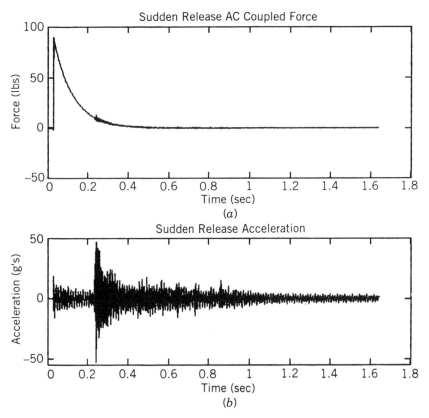

Fig. 7.2.11. (*a*) Measured input force time history when SRM weight bounces on tie down foundation. (*b*) Output acceleration time history for the same event. (Adapted from J. Gruening, C. Clover, and S. Rittmueller, "Exciting a Portal Frame by Applying and Releasing Static Loads," Term Project Report, Engineering Mechanics 545x, Iowa State University, Ames, IA, Spring, 1994.)

In addition, the column has a fundamental axial (fixed-free boundary conditions) natural frequency around 1035 Hz. The corresponding measured resonant peaks are shown as well. We see from Table 7.2.1 that the 1st, 2nd, and 5th girder modes are most dominant while the 3rd and 4th modes are heavily reduced, as would be expected from the force and accelerometer transducer positions and our theory. Generally, the column vibration modes are quite a bit smaller than the girder vibration modes and generally fit into the predicted frequency range.

The Bounce Results The 50 lb weight was supposed to land on a soft pad to cushion its shock loading on the tie down plate, which it did most times. Once, the cushion was accidentally moved so that the steel weight landed directly on the cast iron tie down plate. The AC coupled force and acceleration signals were captured and are shown in Figs. 7.2.11a and 7.2.11b, respectively. It is clear from both signals that significant response begins to occur about 0.25 second into each signal. This response is due to energy being transmitted from the foundation up the columns to the girder. The bouncing weight certainly excited the girder more than when it was suddenly released. This example clearly shows how forces into the foundation can be the cause of a vibration response so that proper exciter isolation from the foundation must always be a concern.

The students were delighted with these measurement results, for the results gave them an opportunity to show how coherence can indicate major testing problems since they had a significant response due to an unmeasured input. They were shocked when the calculated coherence function showed a unity value at all frequencies. How could this result be correct? After much review of calculation procedures and thinking, one of the students showed that the coherence result should be unity in this case, since only one data set was used in its calculation. Recall that coherence is a statistical concept. When we have only a single set of data, there are no other data sets to challenge the results obtained, so all terms cancel out and unity results. Hence, we relearned the important lesson that coherence requires several data set averages before it can have any meaning at all.

7.3 FORCED RESPONSE OF A SIMPLY SUPPORTED BEAM MOUNTED ON AN EXCITER

We often run vibration tests by attaching the test item to the armature of a vibration exciter, and then driving the armature to meet some specified vibration input that may require either a fixed sine, a slowly swept sine, a random, or a shock input. The physical system shown in Fig. 7.3.1 represents the basic features of such a test where a simply supported beam

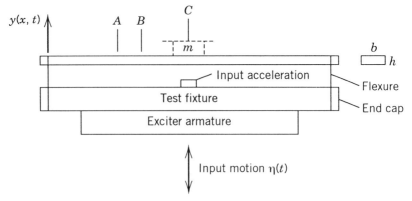

Fig. 7.3.1. Schematic of test setup employed for this example.

is attached to the exciter's armature through a test fixture. In this case the test fixture consists of a 1 in. × 1 in. × $l = 11.75$ in. long steel bar that is bolted to the exciter head in four places and steel flexures at each end that are $h = 0.05$ in. thick × $b = 0.5$ in. wide × 1.26 in. long. The simply supported aluminum beam is also $l = 11.75$ in. long × $h = 0.25$ in. deep × $b = 0.75$ in. wide. The flexures are attached to the beam and the steel bar by bolting through end caps as shown.

The input acceleration of the exciter armature is measured with an accelerometer attached to the test fixture as shown. The output acceleration is measured either at the midpoint ($l/2$), called C, or the third point ($l/3$), called B. The accelerometer is an Endevco Model 2222, which has a mass of about 1 gm. Different amounts of mass (shown as a dashed block) are attached at either the midpoint C or the quarter point ($l/4$)A. We look at a number of theoretical and experimental results to illustrate the basic behavior of such systems in order to establish the kinds of events that can occur while conducting such tests.

7.3.1 Modal Model of the Test Environment

This test environment is the practical implementation of the vibration concepts outlined in Sections 3.7 and 3.8 and applied to a beam.

Dynamic Excitation Load We need to establish what the driving mechanism is and what effects this mechanism can have on our test results. First, the beam's motion needs to be described. This is done by using the kinematic diagram of Fig. 7.3.2, where absolute beam motion $y(x, t)$ is related to the test fixture motion $n(x, t)$ and the relative beam motion $u(x, t)$ by

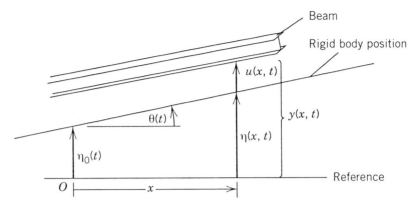

Fig. 7.3.2. Kinematic diagram showing relationships among input motions $\eta(x,t)$ (or $\eta_0(t)$ and $\theta(t)$) and beam relative motion $u(x,t)$.

$$y(x,t) = \eta(x,t) + u(x,t) \qquad (7.3.1)$$

The test fixture motion can also be written in terms of the center point motion $\eta_0(t)$ and the armature's rotation angle $\theta(t)$ to give

$$\eta(x,t) = \eta_0(t) + x\theta(t) \qquad (7.3.2)$$

when x is measured from the exciter's center. Second, beam Equation (3.8.1) is rewritten here in terms of the y-coordinate so that

$$\rho\frac{\partial^2 y}{\partial t^2} + C\frac{\partial y}{\partial t} + \frac{\partial^2}{\partial x^2}\left[EI\frac{\partial^2 y}{\partial x^2}\right] = f(x,t) \qquad (7.3.3)$$

where $\rho\,(=\rho(x))$ is the mass per unit length.
  C is the damping per unit length.
  EI is the beam bending stiffness.
  $f(x,t)$ is the external excitation per unit length, which is zero for this case.

Then, substitution of Eq. (7.3.1) into Eq. (7.3.3) gives

$$\rho\frac{\partial^2 u}{\partial t^2} + C\frac{\partial u}{\partial t} + \frac{\partial^2}{\partial x^2}\left[EI\frac{\partial^2 u}{\partial x^2}\right] = -\rho\frac{\partial^2 \eta}{\partial t^2} \qquad (7.3.4)$$

where we have assumed that the major damping contribution comes from internal energy dissipation mechanisms so that $C\dot{y}(x,t)$ is secondary to $C\dot{u}(x,t)$. Note that $\partial^2\eta/\partial x^2 = 0$ since $\eta(x,t)$ is the rigid body motion so it is a linear function of x. Equation (7.3.4) shows us that the driving excitation per unit length for the relative motion of the beam $u(x,t)$ is the inertia force due to base acceleration; that is, $\rho\ddot{\eta}(x,t)$. If we substitute Eq. (7.3.2) into Eq. (7.3.4) we obtain

$$\rho\frac{\partial^2 u}{\partial t^2} + C\frac{\partial u}{\partial t} + \frac{\partial^2}{\partial x^2}\left[EI\frac{\partial^2 u}{\partial x^2}\right] = -\rho(\ddot{\eta}_0(t) + x\ddot{\theta}(t)) \qquad (7.3.5)$$

which clearly shows the effect of exciter armature rotation on the input. We see that, when mounting a test item on a vibration exciter, we can have two inputs superimposed where one is constant over position along the beam and is due to $\ddot{\eta}_0(t)$, while the other input is a linear function of position x due to exciter armature rotation $\ddot{\theta}(t)$. Since it is impossible to design an armature support system that has no angular motion, we need to expect the possibility of dual input that is frequency dependent.

Ideal Response The ideal response is obtained by using the modal analysis approach of Section 3.7. We start by assuming that the excitation can be separated into a spatial and temporal product of functions (see Eq. (3.7.16)) so that

$$P(x)f(t) = -\rho(a_0 + x\alpha_0)f(t) \qquad (7.3.6)$$

where a_0 is the magnitude of the input linear acceleration.
α_0 is the magnitude of the input angular acceleration.
$f(t)$ is the common time variation input function.

Then we assume that the beam's response can be written in terms of mode shapes $U_p(x)$ and modal generalized coordinates $q_p(t)$ (see Eq. (3.7.17)) so that

$$u(x,t) = \sum_{p=1}^{N} U_p(x)q_p(t) \qquad (7.3.7)$$

When Eq. (7.3.7) is substituted into Eq. (7.3.5) and the resulting equation is integrated over the beam's length l and we take orthogonality into consideration, we obtain the differential equation for the modal generalized coordinate to be (see Eq. (3.7.19))

$$m_p\ddot{q}_p + C_p\dot{q}_p + k_p q_p = Q_p f(t) \qquad (7.3.8)$$

FORCED RESPONSE OF A SIMPLY SUPPORTED BEAM   465

where Q_p is the pth modal excitation force that is given by (see Eq.(3.7.20))

$$Q_p = \int_0^l P(x)U_p(x)\,dx \qquad (7.3.9)$$

m_p is the pth modal mass and is given by (see Eq. (3.8.8))

$$m_p = \int_0^l \rho(x)U_p^2(x)\,dx \qquad (7.3.10)$$

ω_p is the pth natural frequency for a simply supported beam and is given by (see Eq. (3.8.12))

$$\omega_p = (p\pi)^2 \sqrt{\frac{EI}{\rho l^4}} \qquad (7.3.11)$$

k_p is the pth modal stiffness, which is given by

$$k_p = \omega_p^2 m_p \qquad (7.3.12)$$

Then, the beam's relative motion for harmonic excitation at frequency ω becomes (see Eq. (3.7.21))

$$u(x,t) = \sum_{p=1}^N U_p(x)H_p(\omega)Q_p\,e^{j\omega t} \qquad (7.3.13)$$

where $H_p(\omega)$ is the pth generalized coordinate FRF given by

$$H_p(\omega) = \frac{1}{k_p - m_p\omega^2 + jc_p\omega} \qquad (7.3.14)$$

The mode shapes for a simply supported beam are given by Eq. (3.8.11) as

$$U_p(x) = \sin\left(\frac{p\pi x}{l}\right) \qquad (7.3.15)$$

and are shown in Fig. 3.7.2 along with the constant and linear inertial excitation distributions. For our purposes, we assume that α_0 is zero. Then the excitation force for the constant magnitude loading per unit length is obtained from Eq. (7.3.9) (see Eq. (3.7.23) as well) as

$$Q_p = -\frac{pla_0}{p\pi}(1 - \cos(p\pi)) = \begin{cases} 0 & \text{for } p \text{ even} \\ -2pla_0/p\pi & \text{for } p \text{ odd} \end{cases} \quad (7.3.16)$$

Equation (7.3.16) clearly shows that all even modes are not excited by this technique so that only half of the natural frequencies are excited. If the test item is a simply supported uniform plate, this technique will excite only one-fourth of the natural frequencies! Thus this excitation method has the potential for providing us with an unknown (and limited) vision of the natural frequencies that could be of concern.

The natural frequencies for this simply supported beam are calculated from Eq. (7.3.11) to obtain the first three values as 161.4, 646, and 1453 Hz. Then, using the beam's physical parameters in the above equations, the acceleration transmissibility ratio FRF is calculated from

$$H_b(\omega) = 1 - \omega^2 \sum_{p=1}^{N} \frac{U_p(x)H_p(\omega)Q_p}{a_0} \quad (7.3.17)$$

for any location x along the beam. This theoretical transmissibility ratio is compared to the measured transmissibility ratio[8] in Fig. 7.3.3 for the case when $x = l/2$ (midpoint) where the fundamental resonant frequency is predicted very well, 158 Hz versus 161.4 Hz. The second natural frequency around 650 Hz is not present since the second mode is not excited by the inertia force ($Q_2 = 0$ in Eq. (7.3.16)) and the accelerometer is mounted at a second mode node point. Thus the second mode's participation is suppressed by two dynamic response characteristics, the excitation and the transducer location. The third resonant frequency is measured at 1372 Hz, compared to the predicted value of 1453 Hz. This 5.6 percent reduction could be due to various causes such as axial spring behavior of the end flexures.

The small glitch around 1060 Hz is caused by the accelerometer's position in the b direction (see Fig. 7.3.1). If the accelerometer is mounted so its mass center is over the beam center, this glitch goes away. If the accelerometer is moved too far, the glitch returns but with the peak and valley positions reversed. Consequently, we see that the position of the 1 gm accelerometer can cause a small peak-valley pair to occur that may not be repeatable if we remove the accelerometer and then remount it.

---

[8]All experimental data presented here was obtained by Professor Maximo Alphonso and Mr. P. Varoto during a 1994 visit by Professor Alphonso to Iowa State University.

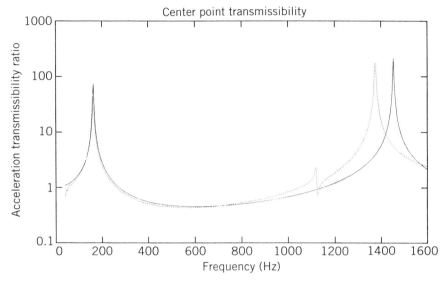

Fig. 7.3.3. A comparison of the measured (dashed line) and theoretical (solid line) acceleration transmissibility ratio between the midpoint of a simply supported beam and the exciter armature.

7.3.2 The Effect of Accelerometer Mass on Measurements

The accelerometer's 1 gm mass causes the measured natural frequency to be less than it actually is. In order to show this effect, a second 1 gm accelerometer was attached to the beam at the location of the original accelerometer. When this was done, the fundamental natural frequency dropped from 157.25 to 156 Hz ($\cong 0.8$ percent) when the analyzer had a frequency resolution of 0.25 Hz, and the third natural frequency dropped from 1372 Hz to 1360 Hz ($\cong 0.9$ percent) when the analyzer had a 2 Hz frequency resolution.

The mass loading effect was addressed by Dossing.[9] The basic assumption is that the mass addition is so small that the original mode shapes are not significantly altered. Consequently, the modal stiffness k_p remains unaltered while the modal mass is altered by adding the mass. For this case, the beam mass distribution is given by

$$\rho(x) = \rho + m\delta(x - x_0) \tag{7.3.18}$$

where ρ is the beam's mass per unit length.

---

[9] Ole Dossing, "The Enigma of Dynamic Mass," *Sound and Vibration*, Nov. 1990, pp. 16–21.

468 THE APPLICATION OF BASIC CONCEPTS TO VIBRATION TESTING

$\delta(x - x_0)$ is the Dirac delta function.
x_0 is the location of added mass m.

Then, substituting Eqs. (7.3.18) and (7.3.15) into Eq. (7.3.10), we obtain the modal mass as

$$m_p = \frac{\rho l}{2} + m \sin^2\left(\frac{p \pi x_0}{l}\right) \qquad (7.3.19)$$

where ρ is a constant over the beam's length. For our physical system, $x_0 = l/2$, $\rho l = 100$ gm, and $m = 1.0$ gm so that the original mass is $50 + 1 = 51$ gm when we measure the natural frequencies as 157.25 and 1372 Hz. When the second accelerometer is added, the mass becomes $51 + 1 = 52$ gm and the measured natural frequencies are 156 and 1360 Hz. From Eq. (7.3.12), we predict that

$$f_1 = \sqrt{51/52}\, 157.3 = 155.7 \text{ Hz} \quad \text{and} \quad f_3 = \sqrt{51/52}\, 1372 = 1359 \text{ Hz}$$

These frequencies are very close to the values measured. Note that we can use mass ratio in the correction process so it is permissible to use gram units. Similarly, we would predict that the actual natural frequencies are

$$f_1 = \sqrt{51/50}\, 157.3 = 158.8 \text{ Hz} \quad \text{and} \quad f_3 = \sqrt{51/50}\, 1372 = 1386 \text{ Hz}$$

Thus we see that we can estimate the effect of the accelerometer mass on the measured frequencies by adding a second transducer as near as possible to the actual accelerometer and noting the changes in natural frequency. Next we examine the limitations of this simple formula for predicting changes in natural frequency.

7.3.3 The Limits for Modal Mass Correction to Natural Frequencies

We need to investigate the implications of adding significant amounts of mass to the beam at the midpoint. First, we observe that the excitation comes about from inertial loading, so we need to recalculate the modal excitation force Q_p. We assume that there is no angular acceleration in the input motion so that α_0 is zero as before. Then, using $\rho(x)$ from Eq. (7.3.18) and the mode shape function from Eq. (7.3.15) in Eq. (7.3.9), we obtain

$$Q_p = -\left\{\frac{\rho l}{p \pi}\{1 - \cos(p\pi)\} + m \sin\left(\frac{p \pi x_0}{l}\right)\right\} a_0 \qquad (7.3.20)$$

where a_0 is the magnitude of the input acceleration at frequency ω. When $x_0 = l/2$, the first three odd modal excitation forces become

$$Q_1 = -\left\{\frac{2\rho l}{\pi} + m\right\} a_0 \qquad Q_3 = -\left\{\frac{2\rho l}{3\pi} - m\right\} a_0$$

$$Q_5 = -\left\{\frac{2\rho l}{5\pi} + m\right\} a_0 \tag{7.3.21}$$

and the first three even excitation forces become

$$Q_2 = 0 \qquad Q_4 = 0 \qquad Q_6 = 0 \tag{7.3.22}$$

The added mass m does not cause the even modes to be excited, since this mass is attached at a node point for each even mode. We also see that the odd modes show that the added mass increases the first and fifth modal excitation forces and reduces the third excitation force.

The modal mass from Eq. (7.3.19) becomes

$$m_1 = m_3 = m_5 = \frac{\rho l}{2} + m \tag{7.3.23}$$

for the first three odd modes and

$$m_2 = m_4 = m_6 = \frac{\rho l}{2} \tag{7.3.24}$$

for the first three even modes. Hence, the addition of mass at the midpoint does not alter the even mode's modal mass, since this added mass is attached at one of the even mode's nodes.

The experimental (diamonds) and theoretical (solid line) first and third natural frequency variations as functions of added mass at the midpoint are shown in Fig. 7.3.4. The experimental and theoretical variations (as calculated from Eq. (7.3.12) when the above modal mass values are used) show close agreement, in Fig. 7.3.4a, for masses that are 1.6 times the beam's actual mass or 3.2 times the beam's modal mass. However, the experimental natural frequencies are much greater than those predicted

470 THE APPLICATION OF BASIC CONCEPTS TO VIBRATION TESTING

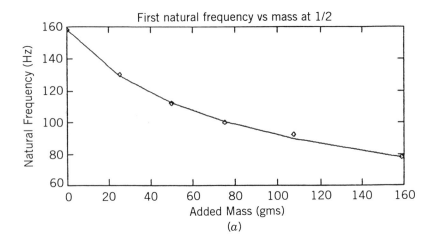

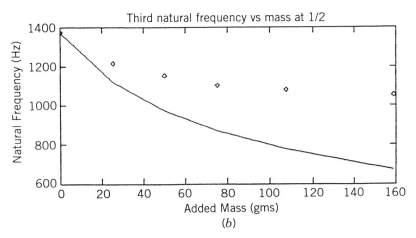

Fig. 7.3.4. A comparison of experimental (diamonds) and theoretical (solid line) natural frequency variations with mass added at the simply supported beam's midpoint. (*a*) First natural frequency. (*b*) Third natural frequency. (From a report by Maximo Alphonso and P. Varoto, 1994, during a visit to Iowa State University, Ames, IA.)

by theory for the third natural frequency, as shown in Fig. 7.3.4*b*. Why such a difference?

The first natural frequency mode shape is that of a half sine wave. If we apply a static load to the center of a simply supported beam, the static deflection curve is parabolic in shape, which is very close to that of half sine function. Consequently, the first mode shape changes little and the modal stiffness k_1 remains essentially constant regardless of the mass added. This is not the case for the third mode, for the mode shape is significantly altered by the addition of the added mass. Consequently, both the modal stiffness k_3 and the modal mass m_3 are altered in this case. We can show that this is indeed the case by solving the theoretical problem with mass m attached at the center. We find that the first mode shape remains essentially intact with no significant changes, the even modes are not affected at all, and the higher odd mode shapes have a significant changes in mode shape, modal mass, and modal stiffness. Hence, the formulas suggested by Dossing (and presented here wherein the modal stiffness is assumed to be unaffected) have significant limitations where the added mass should be less than a few percent of the structure's mass, probably on the order of 10 percent or less.

7.3.4 Added Mass at the Quarter Point

In order to illustrate the effects of changing the added mass location, we attach mass m at $x_0 = l/4$. Then the modal excitation force from Eq. (7.3.20) becomes

$$Q_p = -\left\{\frac{\rho l}{p\pi}\{1 - \cos(p\pi)\} + m \sin\left(\frac{p\pi}{4}\right)\right\} a_0 \qquad (7.3.25)$$

which gives the odd mode modal excitation forces of

$$Q_1 = -\left\{\frac{2\rho l}{\pi} + 0.707m\right\} a_0 \qquad Q_3 = -\left\{\frac{2\rho l}{3\pi} + 0.707m\right\} a_0$$

$$Q_5 = -\left\{\frac{2\rho l}{5\pi} - 0.707m\right\} a_0 \qquad (7.3.26)$$

and even mode modal excitation forces of

$$Q_2 = -ma_0 \qquad Q_4 = 0 \qquad Q_6 = ma_0 \qquad (7.3.27)$$

A comparison of Eqs. (7.3.26) and (7.3.27) with Eqs. (7.3.21) and (7.3.22) shows that the second and sixth modes are excited in this case while the fourth mode is not, since a node point occurs where the added mass is attached.

The modal mass is obtained from Eq. (7.3.19) when $x_0 = l/4$ is substituted to give

$$m_p = \frac{\rho l}{2} + m \sin^2\left(\frac{p\pi}{4}\right) \qquad (7.3.28)$$

This gives modal masses of

$$m_1 = m_3 = m_5 = \frac{\rho l}{2} + \frac{m}{2} \qquad m_2 = m_6 = \frac{\rho l}{2} + m$$

$$m_4 = \frac{\rho l}{2} \qquad (7.3.29)$$

where we see that only the fourth modal mass is unaltered by the addition of mass at one of its node points. Thus we anticipate that the 4th, 8th, 12th, and so on, modes also have constant modal masses of $\rho l/2$. The odd modes all have the same modal mass, while the other even modes have the largest modal mass.

The accelerometer is mounted at the third point (point B in Fig. 7.3.1). The corresponding measured and calculated acceleration transmissibility is shown in Fig. 7.3.5a, where the first resonance is clearly present around 158 Hz, a small glitch appears in the experimental results where the second mode is expected just above 600 Hz, and a rather significant third mode resonance occurs just below 1400 Hz. The ideal theoretical response shows only the fundamental response in this frequency range, since the accelerometer is mounted at a node point for the third mode. Why should there be such a large difference between the theory and experiment?

The theoretical model used 120 points to describe the beam's length so that the distance between theoretical points is 0.0979 in. The acceleration transmissibility is calculated for one point to the left and one point to the right of the node point and the results are compared to the ideal third point transmissibility in Fig. 7.3.5b. Here, we see that the point to the left of the third point has a valley occur before the resonant peak, while the point to the right of the third point has a valley occur after the resonant peak. Comparing these results with the experimental results in Fig. 7.3.5a shows that the accelerometer is located to the right of the third point node, since a valley occurs after the peak. Hence, we find that the results appear to be quite sensitive to the accelerometer's location as far as the third mode is concerned, since the transducer is located near the third mode's node point. As we go to higher natural frequencies, the distance between nodes becomes smaller so that precise transducer location becomes a more important consideration.

The effect of mass loading at the quarter point on measured natural

FORCED RESPONSE OF A SIMPLY SUPPORTED BEAM 473

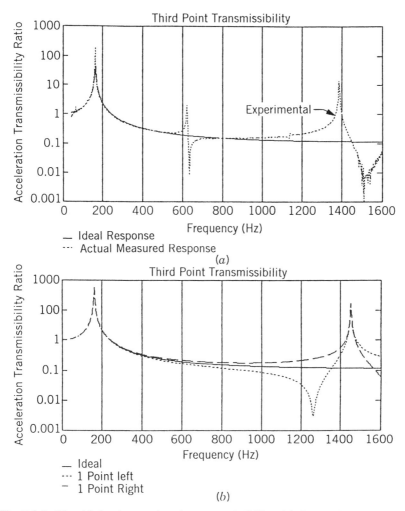

Fig. 7.3.5. The third point acceleration transmissibility. (*a*) Comparison of experimental and theoretical values. (*b*) Comparison of three theoretical transmissibilities when a transducer is slightly mispositioned. (From a report by Maximo Alphonso and P. Varoto, 1994, during a visit to Iowa State University, Ames, IA.)

frequency is examined next. The modal mass is estimated from Eq. (7.3.29) as $m_1 = m_3 = 50.0 + 50.1/2 = 75.1$ gm when 50.1 gm is attached at the quarter point. The fundamental natural frequency is estimated from Eq. (7.3.12) as

$$f_1 = \sqrt{50/75.1}\ 158 = 129.1 \text{ Hz} \qquad (7.3.30)$$

474   THE APPLICATION OF BASIC CONCEPTS TO VIBRATION TESTING

which compares favorably to the measured value of 128 Hz and

$$f_3 = \sqrt{50/75.1}\, 1374 = 1124\,\text{Hz} \qquad (7.3.31)$$

which compares poorly with the measured frequency of 1302 Hz. Again, the first mode frequency change due to added mass is quite accurately predicted (within 0.9 percent), while the third mode natural frequency change has a significant error (13.6 percent).

An intriguing acceleration transmissibility was obtained when the accelerometer was at the midpoint and a 50.1 gm added mass was attached at the quarter point, as shown in Fig. 7.3.6a. Here we find two widely spaced resonant peaks (one near 1000 Hz and the other near 1350 Hz) where we anticipated only the third natural frequency. The 50.1 gm mass was composed of two 25 gm masses connected with a short 10–32 brass screw that had a 0.007 in. shoulder that caused masses m_1 and m_2 to be separated as shown in Fig. 7.3.6b. When the masses were reassembled so that the gap between masses was gone, the dashed line labeled "rigid" in Fig. 7.3.6a was obtained, which shows a single resonance around 1300 Hz. It is evident that the beam's angular motion at the quarter point caused mass m_1 to resonate in bending the brass screw, which acted like a flexural spring at a lower frequency. Once the masses were reassembled so that the gap was eliminated, masses m_1 and m_2 formed a rigid body.

It should be apparent from these simple tests that little details can cause significant changes. First, for transducer mass on the order of 1 percent of the structure, it is advisable to check on the significance of this transducer mass by adding a second equal mass at the transducer location. If the change in resonant frequencies is acceptable, then we can proceed with the test. If the change is too large, then we need to select a smaller transducer. We see in this example that the higher natural frequencies appear to be less sensitive to transducer mass than the fundamental natural frequency. The correction formula given by Dossing can be inadequate for the higher frequencies and larger masses since it is assumed that the mode shapes and modal spring constant are unaffected by the presence of the mass.

Second, transducer location can be a significant factor that needs to be carefully monitored. If we move an accelerometer around the structure for different tests, we can change not only the natural frequencies and mode shapes but also the measured vibration response. These mass loading and transducer location considerations are usually more important when dealing with relatively small and lightweight structures.

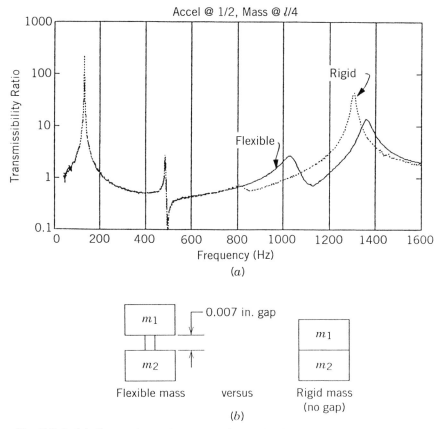

Fig. 7.3.6. (*a*) Comparison of two experimental midpoint acceleration transmissibility ratios when mass at quarter point is flexible and rigidly connected. (*b*) Physical arrangement of 50.1 gm mass with flexible and rigid conditions. (From a report by Maximo Alphonso and P. Varoto, 1994, during a visit to Iowa State University, Ames, IA.)

7.4 IMPULSE TESTING

Impulse vibration testing is a technique where the structure under test is excited by striking it with an instrumented impacting mass or hammer, as described in Section 6.1, in order to generate short time duration force. This test method is often used in experimental modal analysis of known linear structural systems since it is easy to move the input excitation source from location to location while the output accelerometers remain at fixed locations. Consequently we can generate the large number of required input-output FRFs in a minimum amount of time. We explore the requirements to conduct impulse tests here and in Section 7.5.

7.4.1 Impulse Requirements

When an impact hammer strikes a linear structure, the resulting force time history is basically a half sine, as shown in Fig. 7.4.1a. This force is characterized by its peak value F_0 and time duration τ. This pulse generates the input frequency spectrum in Fig. 7.4.1b, so it is important to understand its limitations relative to the time window used in data collection.

For our example in Fig. 7.4.1, we have assumed that $F_0 = 100$ lb (or newtons), that $\tau = 0.01$ second, and that the sample rate is 1023 times per second so that the time between samples is 0.9775 ms. Then a 1024 sample data window has a window period T of 1 second, so that the frequency analysis has a frequency resolution $\Delta f = 1$ Hz. It is clear that the corresponding frequency spectrum in Fig. 7.4.1b has a number of zero (or near zero) values. The first zero occurs at approximately $1.5/\tau$ (or 150 Hz in

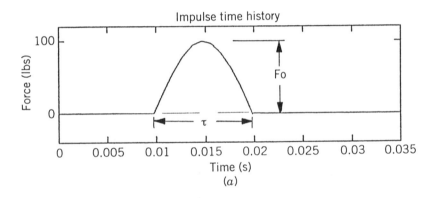

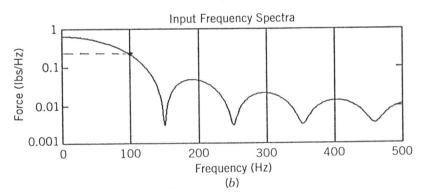

Fig. 7.4.1. Ideal half-sine impulse function. (a) Time history showing peak value F_0 and duration τ. (b) Corresponding frequency spectrum.

this case), with the other zero values equally spaced after the first one at increments of approximately $1/\tau$, that is, at 250, 350 Hz, and so on. These zero (or near zero) values cause measurement and/or interpretation problems in our results.

A zero frequency component indicates that no energy is available to excite the structure at that frequency. This means that observation of any measured output acceleration frequency spectra will not clearly indicate a resonance condition when one is actually present in the structure. This zero frequency component also affects the calculated FRF since any signal noise in the output frequency spectra will be divided by a near zero value from the input frequency spectra. This situation causes a "ghost" resonant peak to occur that does not exist in reality. Both of these outcomes due to zero input frequency components are undesirable.

These two near zero frequency spectrum problems can be avoided by making the impact duration τ smaller so that the first zero frequency is pushed outside the frequency analyzer's operating range. Two parameters are available to reduce the impact time for a given structure: tip stiffness and hammer mass, as discussed in Section 6.1. A practical choice is to select these parameters so $1/\tau > f_{max}$ where f_{max} is the frequency analyzer's range. This choice causes the input to be significant, as shown in Fig. 7.4.1b for the 0 to 100 Hz frequency range, where the input ranges from 0.634 to 0.231 lb/Hz when $\tau = 0.01$ second. Consequently, for $f_{max} = 500$ Hz, shown in Fig. 7.4.1b, we see that $\tau \leq 0.002$ second. Thus we can make the input frequency spectrum have significant values for our range of analysis, and one source of error and confusion is removed from our results.

7.4.2 The Input Noise Problem

A significant noise problem can occur during impact testing, for over 99 percent of the input signal should be zero! This comes about since impulse duration τ is on the order of $0.002T$. This means that the signal can be highly contaminated by the instrumentation noise floor during the time that there is no impulse signal. Figure 7.4.2a shows an ideal half sine impulse with a time duration of $\tau = 0.01$ second and a peak value of 100 lb (or newtons). A uniformly distributed random signal of ± 10 lb (or newtons) is generated as a background noise that is due to the instrumentation noise floor. This noise level is assumed to be rather high at 10 percent of the peak force so that the significance of high noise data is evident. This noise is shown in Fig. 7.4.2a and is added to the ideal signal to obtain a realistic input signal of noise plus ideal signal.

A frequency analysis of the ideal and ideal plus noise signals is shown in Fig. 7.4.2b. We see that the ideal frequency has the zero at 150 Hz, so we consider that only the 0 to 100 Hz frequency range is of practical

478 THE APPLICATION OF BASIC CONCEPTS TO VIBRATION TESTING

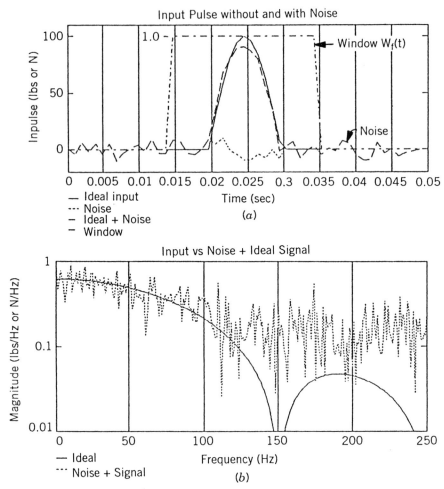

Fig. 7.4.2. Input impulse and instrument noise analysis. (*a*) Ideal impulse and noise with window function. (*b*) Comparison of frequency spectra for ideal and noise and ideal signal. (*c*) Comparison of ideal and windowed noise plus ideal signal.

interest for us. In this frequency range, we see that the noise causes significant variation to occur on the input frequency spectra so that we would have to use a large number of averages to reduce the effect of this random noise on our measured results. Is there a simple way to remove this noise?

In order to overcome this noise problem, a rectangular transient window function $W_f(t)$ in Fig. 7.4.2a is applied to the noise plus ideal signal.

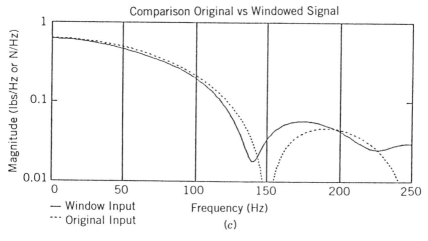

Fig. 7.4.2. (*Continued*).

This window function is simply called the *transient window* in order to distinguish it from the rectangular window of period T, and is defined by

$$W_f(t) = \begin{cases} 0 & \text{for} \quad 0 \text{ ms} < t < 13.7 \text{ ms} \\ 1 & \text{for } 14.7 \text{ ms} < t < 34.2 \text{ ms} \\ 0 & \text{for } 35.2 \text{ ms} < t < 1 \text{ second} \end{cases} \quad (7.4.1)$$

In this way, only the noise next to the pulse and during the pulse is included in the measured signal.

The original and transient windowed ideal plus noise signal frequency spectra are compared in Fig. 7.4.2c, where we see that they are very close to one another over the 0 to 100 Hz frequency range. In fact, repeated application of the random noise generator gave results where, in some instances, the windowed spectrum was lower than the ideal, as shown, while in other instances it was a little higher than the ideal. It is clear that the situation of Fig. 7.4.2c is superior to that in Fig. 7.4.2b and that a few averages should be adequate. Generally, if the window width is closer to the pulse duration, these two curves are even-closer. Hence, we see that it is advantageous to use the transient window to reduce uncertainty in the input due to instrument noise. The correct procedure for setting up this rectangular window relative to trigger levels, slope, and so on, is discussed in detail in Section 7.5.

7.4.3 The Output Leakage Problem

When a lightly damped structure is impacted, the resulting vibration tends to persist for a period of time that is much longer than the data window, as shown in Fig. 7.4.3a. This signal truncation leads to filter leakage that causes the resulting frequency spectrum to contain frequency components that are required to handle the sudden termination of the signal. One

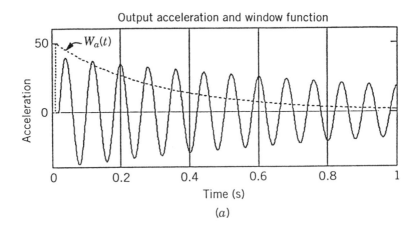

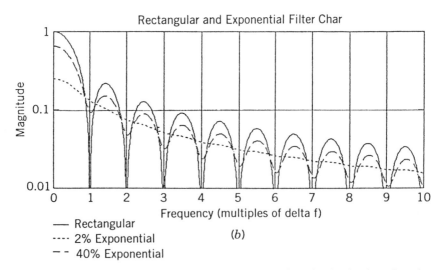

Fig. 7.4.3. (a) Output signal and exponential window function in the time domain. (b) Comparison of digital filter characteristics for rectangular window and exponential window functions.

method to overcome this filter leakage is to use the exponential window function $W_a(t)$ shown in Fig. 7.4.3a and described by

$$W_a(t) = \begin{cases} 0 & \text{for } 0 < t < t_0 \\ e^{-(t-t_0)/\tau_e} & \text{for } t_0 < t < T \end{cases} \qquad (7.4.2)$$

where t_0 is the time when the window begins.
τ_e is the window decay time constant.
The exponential window is rated by the percentage P of the signal that is left at its end; that is, a 2 percent window is one where $e^{-(T-t_0)/\tau_e} = 0.02$. Hence, the decay time constant is related to percentage P by

$$\tau_e = -\frac{(T-t_0)}{\ln(P)} \qquad (7.4.3)$$

where t_0 is usually small compared to T so that we can drop it from Eq. (7.4.3).

Comparison of Exponential and Rectangular Window Function Digital Filters Previously, we have seen that it is required to treat both the input and output signals with the same window functions so that their frequency spectra are convolved with identical functions. Here, we see that we are trying to mix two different window functions.

If we perform a Fourier transform of Eq. (7.4.2), we obtain

$$W_a(\omega) = \frac{\tau_e}{1 + j\omega\tau_e} [e^{-j\omega t_0} - e^{(t_0-T)/\tau_e} e^{-\omega T}] \qquad (7.4.4)$$

This filter function is compared to the rectangular window's sinc function in Fig. 7.4.3b, where their magnitudes are plotted against the relative frequency given in multiples of Δf. In this case all three window functions have the same window duration T (=1 second), while the exponential windows are assumed to start at $t_0 = 5$ ms. One exponential window has 2 percent leakage ($P = 0.02$), while the other has 40 percent leakage ($P = 0.4$). The 2 percent window starts at a value of about 0.249 at zero relative frequency, and then dies off just below the sinc function peaks with increasing relative frequency, while the 40 percent window starts at a value of 0.652 at zero relative frequency, and then oscillates about the 2 percent line in a manner similar to the sinc function. We should also suspect that the output peak values will be attenuated by the exponential window since the zero relative frequency component is less than unity. It is clear that the rectangular window and the exponential window filter characteristics drop off at 6 dB/octave. However, the exponential window has more filter leakage, due to the lack of zeros that occur at every Δf. Consequently, the output valleys should be increased due to this extra amount of filter leakage. Now we explore these exponential window effects on measured damping.

Effects of Exponential Window on Modal Damping The measured damping for the *p*th natural mode is too large, due to the exponential window function since each mode's exponential damping term is multiplied by the window's exponential function such that

$$e^{-t/\tau_e} e^{-\zeta_p \omega_p t} = e^{-(1/\tau_e + \zeta_p \omega_p)t} = e^{-(\zeta_w + \zeta_p)\omega_p t} \qquad (7.4.5)$$

Hence the measured damping ratio ζ_m is related the *p*th mode's damping ratio ζ_p and the window damping ratio ζ_w by

$$\zeta_m = \zeta_p + \zeta_w \qquad (7.4.6)$$

where ζ_w is related to window decay time τ_e by

$$\zeta_w = \frac{1}{\tau_e \omega_p} \qquad (7.4.7)$$

for *p*th natural frequency ω_p. Note again that these correction equations apply to each mode separately and not globally to the entire FRF.

7.4.4 Application of Impulse Testing to a Free-Free Beam

A cold rolled steel free-free beam that is 93 in. long is shown in Fig. 7.4.4*a* where it is suspended by two lightweight cords attached about $0.23l$ from each end. The beam's cross section is $b = 1.25$ in. $\times$ $h = 1$ in. An accelerometer is attached at each location labeled 1 (for the midpoint), 2 (for the left end), and 3 (for the right-hand end) in order to measure the output acceleration when the beam is impacted at either location 1 or location 2.

Theoretical Considerations The theoretical natural frequencies are calculated from

$$\omega_p = (\beta_p l)^2 \sqrt{EI/\rho l^4} \qquad (7.4.8)$$

where ρ is the beam's mass per unit length.
        EI is its bending stiffness.
        $\beta_p l$ are the eigenvalues given in Chapter 3.

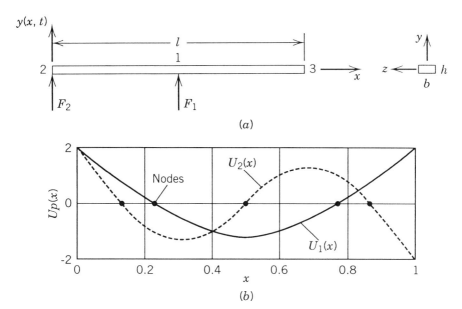

Fig. 7.4.4. Free-free beam used in test example. (*a*) Schematic of test setup (top view). (*b*) First two nonrigid body mode shapes.

For this beam, the natural frequencies for vibration in the horizontal y-direction in the 0 to 800 Hz frequency range are 0, 24.3, 67.6, 132.4, 218, 327, 457, 608, and 781 Hz. We see that the first natural frequency is zero. This means that there can be a rigid body motion before the flexural vibrations occur. These rigid body motions are those of a rigid slender bar on two long support strings, so that there are two low frequency rigid body modes, one for rigid body translation in the y-direction and one for rigid body rotation about the vertical z-axis. We are not concerned about these modes of vibration since they are below 2 Hz in this case.

The first two flexible mode shapes are shown in Fig. 7.4.4*b*. The first mode shape $U_1(x)$ is seen to be a symmetrical shape with respect to the midpoint where the end points move in phase and the midpoint has zero slope. The second mode shape $U_2(x)$ is seen to be a nonsymmetrical shape with respect to the midpoint where the ends move out of phase and the midpoint has zero motion (node point). These characteristics are also present in the higher modes where the odd numbered modes have the ends moving in phase with zero midpoint slope, while the even numbered modes have the ends moving out of phase with zero motion at the midpoint. It is obvious that exciting the beam at the midpoint removes all even modes, since all even modes have a node point at location 1, and thus are not excited. Similarly, we see that the end points are ideal since no node points occur there.

The transfer acceleration for this beam can be calculated from the modal acceleration model

$$H_{ab}(\omega) = -\omega^2 \sum_{p=1}^{N} \frac{U_p(a) U_p(b)}{k_p - m_p \omega^2 + jc_p \omega} \qquad (7.4.9)$$

where $x = a$ is the output location.
$x = b$ is the input location.
k_p is the modal stiffness.
c_p is the modal damping.
m_p is the modal mass.
$U_p(x)$ is the pth mode shape.
N is the number of modes used in the model.

Equation (7.4.9) is used to calculate the theoretical midpoint driving point acceleration $H_{11}(\omega)$, as shown in Fig. 7.4.5a, the transfer acceleration $H_{12}(\omega)$ between points 1 and 2, as shown in Fig. 7.4.5b, and the driving point acceleration at the end $H_{22}(\omega)$, as shown in Fig. 7.4.5c. In both Figs. 7.4.5a and 7.4.5b only the odd natural frequencies are seen to exist, since point 1 is node point for all even natural mode shapes. Only in Fig. 7.4.5c are all natural frequencies evident since there are no node points at location 2. These results reinforce the fact that a limited set of tests can leave some resonances not being measured, even when they are excited, since the accelerometer is mounted at a node point as was the case for $H_{12}(\omega)$. Now we look at our experimental results.

Comparison of Experimental Driving Point Accelerances The beam was excited at midpoint 1 and end 2 with both an impact hammer and a vibration exciter that is driven with a pseudorandom signal. The impact data used a value of τ_e such that about 2 percent leakage could occur; that is, $e^{-T/\tau_e} \cong 0.02$. The resulting driving point accelerances are shown in Figs. 7.4.6a ($H_{11}(\omega)$) and 7.4.6b ($H_{22}(\omega)$).

The general character of each response agrees with the theoretical curves shown in Fig. 7.4.5. It is apparent in Fig. 7.4.6 that the impulse data has lower peak values (marked by the horizontal dashes) and higher valley values (marked by the horizontal dashes) than the pseudorandom experimental results. Any attempt to correct the FRF's peak values directly for the exponential window damping gives less than satisfactory results in general. There are several reasons for this. One is that the force transducer used in the random tests is attached directly to the structure and is sensitive to the beam's bending moments so that the reference values from the pseudorandom test may be in error by being too large or

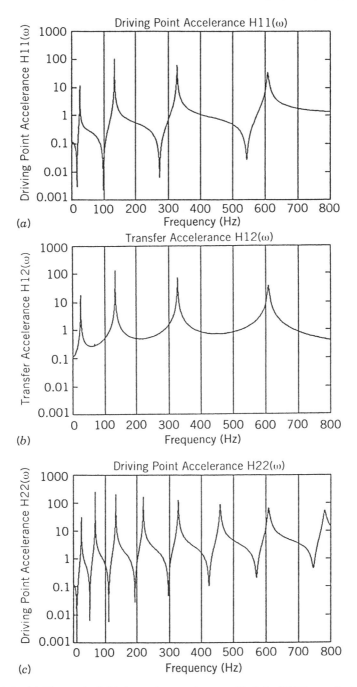

Fig. 7.4.5. (*a*) Theoretical driving point accelerance $H_{11}(\omega)$. (*b*) Theoretical transfer accelerance $H_{12}(\omega)$. (*c*) Theoretical driving point accelerance $H_{22}(\omega)$.

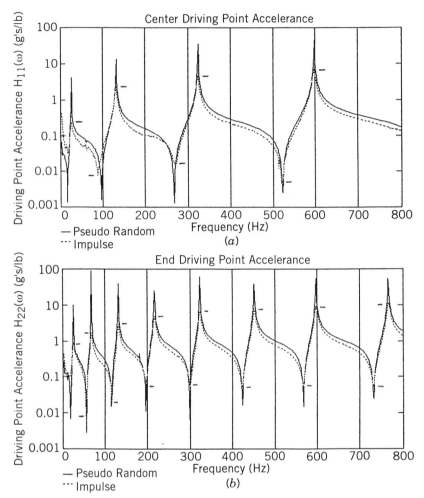

Fig. 7.4.6. Comparison of driving point accelerances obtained from impulse and pseudorandom excitation methods. (a) $H_{11}(\omega)$. (b) $H_{22}(\omega)$.

too small depending on the force transducer's orientation on the beam.[10] A second reason is that the correction scheme of Eq. (7.4.6) applies to the modal damping and not to the entire FRF. Section 7.5 will address the errors that can result from using a mixture of rectangular and exponential windows, improper trigger levels, and trigger slope selections, as well as recommend a proper procedure for setting up the windows.

---

[10]P. Cappa and K. G. McConnell, "Base Strain Effects on Force Measurements," *Proceedings*, *Spring Society for Experimental Mechanics*, Baltimore, MD, June 1994, pp. 520–530.

7.5 SELECTING PROPER WINDOWS FOR IMPULSE TESTING

Impulse testing of moderately sized structures has a number of advantages, particularly in experimental modal analysis and general vibration surveys. One of the earlier papers by Halvorsen and Brown[11] describes a number of issues about impulse testing, including windowing and nonlinear effects. However, this technique requires that test personnel carefully select the data window variables according to certain rules, since we are mixing two types of windows, rectangular for the input pulse and exponential for the output vibration. How these two windows are set up can significantly affect the measured results. Clark et al.,[12] Cafeo and Trethewey,[13] Trethewey and Cafeo,[14] and McConnell and Varoto[15] have published recent papers addressing these issues. This section attempts to consider all parameters involved in properly setting up the data windows so that no significant information is lost and/or no distorted results obtained. It is assumed that matched anti-aliasing filters are used in both data channels before the signal reaches the analog to digital (A/D) converter. The effects demonstrated here are due to processing ideal noise free data so that no measurement errors are involved, and the information is taken from the paper by McConnell and Varoto.

7.5.1 Window Parameters

The transient window parameters that we can usually select are shown in Fig. 7.5.1a for the input pulse with a main lobe of duration τ. Time duration τ controls the first zero in the input's frequency spectrum, as was discussed in Section 7.4. The anti-aliasing filters modify the actual pulse by rounding out the initial slope and generating a filter ringing after the pulse so that total pulse time $T_p > \tau$. The filter ringing has been attributed to two sources, the residual vibration of the impact hammer and anti-

---

[11] W. G. Halvorsen and D. L. Brown, "Impulse Technique for Structural Frequency Response Testing," *Sound and Vibration*, Nov. 1977, pp. 8–21.

[12] R. L. Clark, A. L. Wicks, and W. J. Becker, "Effects of an Exponential Window on the Damping Coefficient," *Proceedings of the 7th International Modal Analysis Conference*, Vol. 1, Las Vagas, NV, 1989, pp. 83–86.

[13] J. A. Cafeo and M. W. Trethewey, "Impulse Test Truncation and Exponential Window Effects on Spectral and Modal Parameters," *Proceedings 8th International Modal Analysis Conference*, Vol. 1, Orlando, FL, 1990, pp. 234–240.

[14] M. W. Trethewey and J. A. Cafeo, "Tutorial: Signal Processing Aspects of Structural Impact Testing," *The International Journal of Analytical and Experimental Modal Analysis*, Vol. 7, No. 2, 1992, pp. 129–149.

[15] K. G. McConnell and P. S. Varoto, "The Effects of Window Functions and Trigger Levels on FRF Estimations from Impact Tests," *Proceedings of the 13th International Modal Analysis Conference*, Nashville, TN, Feb. 1995.

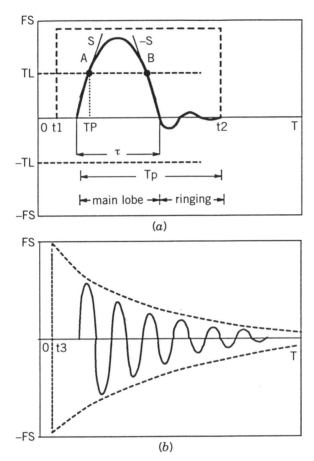

Fig. 7.5.1. The transient data window parameters. (*a*) The input transient window. (*b*) The output exponential window. (From K. G. McConnell and P. S. Varoto, "The Effects of Window Functions and Trigger Levels on FRF Estimations from Impact Tests," *Proceedings of the 13th International Modal Analysis Conference*, Society for Experimental Mechanics.)

aliasing filter ringing. Modern hammers are designed to minimize the residual vibration, but anti-aliasing filter ringing will always occur since we must use the filters for the digitizing processes. Hence, we always need to include filter ringing in our considerations.

The selectable parameters for a transient impulse measurement in Fig. 7.5.1a are window duration T, trigger level TL, slope S (either positive at point A or negative at point B), trigger point TP, transient window beginning time t_1 and ending time t_2, and A/D converter full scale setting FS. The window duration T controls the frequency component spacing since $\Delta f = 1/T$, while the number of data points controls the frequency analysis range. Trigger level TL and slope S determine which input signal point will become the trigger point that is stored at time location TP in the data window. Trigger point TP is selected so that the leading edge of the input is sufficiently long (0 to TP) to ensure that the input data is valid by starting at or near zero for a time period before the impulse and that the output is at or near zero as well during this initial time period of 0 to TP. Otherwise, we could have output acceleration contamination due to either a previous input that has not died away or an unknown input source that is not measured.

In order for a trigger to occur, two conditions must be met. The input voltage must be greater than trigger level TL and slope S must be positive or negative according to which condition is selected by the user. It is obvious that the transient window duration $(t_2 - t_1)$ must be greater than T_p in order to capture the significant parts of the input signal. However, if we set the slope to be negative, the trigger point is location B and time t_1 must be before TP by an amount greater than $\tau/2$. Otherwise the transient window would remove a portion of the input data.

The output signal is shown in Fig. 7.5.1b, along with the exponential window. The exponential window has parameters of window starting time t_3 and exponential window decay time constant τ_e that controls the percent of window leakage P. The window's starting time t_3 should be equal to t_1 so that the entire output signal is captured since the exact location of the signal's beginning is not precisely controlled. The role of the exponential window is to reduce filter leakage by forcing the response signal to near zero values at the end of the data window. We explore the effects of improper use of these window parameters during the capture of transient data.

7.5.2 Modeling the Data Process

We use a three DOF modal model to explore the effects of each transient window parameter. The modal system parameters were selected to require only the 0 to 100 Hz frequency range and are given in Table 7.5.1. Each modal damping is selected to have a product of $\zeta_p \omega_p = 0.18$ so that each

TABLE 7.5.1. Model Parameters for Simulating FRF

Mode	A_p	f_n (Hz)	ζ_n	$\zeta_p f_p$
1	1	20	0.0090	0.18
2	1	40	0.0045	0.18
3	1	72	0.0025	0.18

mode decays at the same rate over the time window. The system's ideal FRF is calculated from the modal model formula

$$H(\omega) = \sum_{p=1}^{N} \frac{A_p}{\omega_p^2 - \omega^2 + j2\zeta_p\omega_p\omega} \tag{7.5.1}$$

while the input pulse is calculated from

$$f(t) = \begin{cases} 0 & 0 \le t \le t_i \\ F_0 \sin((\pi/\tau)t) & t_i \le t \le t_i + \tau \\ 0 & t_i + \tau \le t \end{cases} \tag{7.5.2}$$

For our calculations, we assume F_0 is unity, duration time $\tau = 10$ ms, and pulse starting time $t_i = 0.2$ second. There are 4096 data points in the 5 second data window so that the time between data points is 1.221 ms and the frequency spectral resolution is 0.2 Hz. The original input pulse is shown in Fig. 7.5.2a as the dashed curve. The output motion frequency spectrum $X(\omega)$ is obtained by taking the Fourier transform of the input signal $F(\omega)$ and multiplying it by the system FRF from Eq. (7.5.1) so that

$$X(\omega) = H(\omega)F(\omega) \tag{7.5.3}$$

The output time history $x(t)$ is obtained from the inverse Fourier transform of $X(\omega)$. This time data is passed through an anti-aliasing filter by a process described in the following paragraph and shown in Fig. 7.5.2b. It is clear that this vibration has ceased by the end of the 5 second window.

In practice, the input and output time data must pass through matched anti-aliasing filters. The anti-aliasing filters were generated by using a second order system with a 200 Hz break frequency and 68 percent critical damping. This second order filter characteristic was then raised to the 10th power to generate a filter that drops off at approximately 120 dB/octave; that is,

$$H_{aa}(\omega) = \left[\frac{1}{1 - (\omega/\omega_0)^2 + j2\zeta(\omega/\omega_0)}\right]^{10} \tag{7.5.4}$$

where $\omega_0 = (200) \times (2\pi)$. Then the original time history frequency spectrum is multiplied by $H_{aa}(\omega)$ in order to generate the actual time history signals that are available in the frequency analyzer. Thus the measured force frequency spectrum becomes

SELECTING PROPER WINDOWS FOR IMPULSE TESTING 491

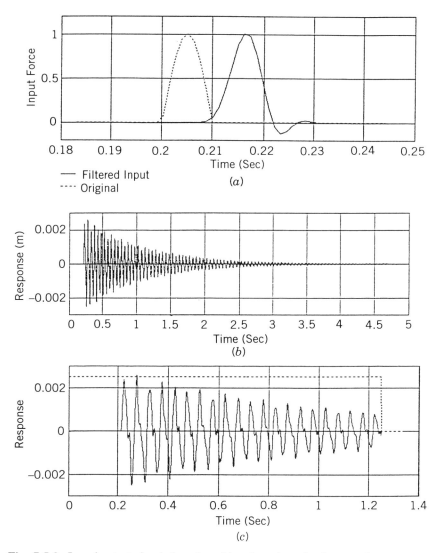

Fig. 7.5.2. Impulse test simulation time histories when the input pulse starts at 0.2 second. (*a*) Input impulse (original—dashed curve—and filtered by anti-aliasing filter—solid curve). (*b*) Filtered output time history (4096 data points). (*c*) Truncated output signal of 1024 data points. (From K. G. McConnell and P. S. Varoto, "The Effects of Window Functions and Trigger Levels on FRF Estimations from Impact Tests," *Proceedings of the 13th International Modal Analysis Conference*, Society for Experimental Mechanics.)

492 THE APPLICATION OF BASIC CONCEPTS TO VIBRATION TESTING

$$F_m(\omega) = H_{aa}(\omega)F(\omega) \tag{7.5.5}$$

and the measured output motion frequency spectrum becomes

$$X_m(\omega) = H_{aa}(\omega)X(\omega) \tag{7.5.6}$$

where $F_m(\omega)$ is the filtered frequency spectrum of the original input signal and $X_m(\omega)$ is the filtered frequency spectrum of the original output signal. The inverse Fourier transform of these input and output frequency spectra is used to generate the 4096 point time data set that would appear as the sampled data in an ideal 4096 data point frequency analyzer. These ideal noise free time histories are shown in Figs. 7.5.2a and 7.5.2b and are called $f_m(t)$ and $x_m(t)$, respectively.

It is clear in Fig. 7.5.2a that the ideal measured pulse (solid curve) is different from to the original pulse (dashed curve), which is altered by the anti-aliasing filter so that it is time shifted about 9 ms, its leading edge is rounded out, its peak value is approximately 1 N, and its filter ringing is present after the main pulse. The anti-aliasing filter effects are nearly impossible to detect in the output signal since any filter ringing is mixed in with the many oscillations of several frequencies.

Now, we assume that our analyzer will capture only 1024 of these 4096 time data points that are separated by 1.221 ms so that the data window's length is reduced to 1.25 seconds and the resulting frequency analysis will have a frequency resolution of 0.8 Hz. There are no problems capturing all significant input data points, since most of the input signal is zero other than the impulse shown in Fig. 7.5.2a. However, Fig. 7.5.2c shows that there is a potential for significant filter leakage when the data sample is truncated at 1.25 seconds.

The measured input force time history contains 1024 data points that are operated on by the window function $W_f(t)$ so that we have

$$p(t) = W_f(t)x_m(t) \tag{7.5.7}$$

as our effective input time history, as shown by the solid curve in Fig. 7.5.2a. This time history has a frequency spectrum of

$$P(\omega) = \mathcal{F}[p(t)] \tag{7.5.8}$$

The corresponding measured output time history contains the 1024 data points that are shown in Fig. 7.5.2c. This time history is operated on by the exponential window function $W_e(t)$ so that

SELECTING PROPER WINDOWS FOR IMPULSE TESTING   493

$$y(t) = W_e(t)x_m(t) \qquad (7.5.9)$$

This time history has an output frequency spectrum that is given by

$$Y(\omega) = \mathscr{F}[y(t)] \qquad (7.5.10)$$

It should be clear that $P(\omega)$ and $Y(\omega)$ include the effects of the matched anti-aliasing filters and the two different window functions $W_f(t)$ and $W_e(t)$. However, we found in Section 7.4 (Fig. 7.4.3b) that the digital filter characteristics of the rectangular window (sinc function) and the exponential window are similar, so that convolution in the frequency domain should have similar but not precisely equal effects.

The measured input-output FRF is then calculated from the $H_1(\omega)$ definition

$$H_m(\omega) = \frac{G_{PY}(\omega)}{G_{PP}(\omega)} \qquad (7.5.11)$$

where $G_{PP}(\omega)$ and $G_{PY}(\omega)$ are the input auto-spectral density and the cross-spectral density, respectively. The auto- and cross-spectral densities are calculated from

$$G_{PP}(\omega) = 2P_m^*(\omega)P_m(\omega) \qquad (7.5.12)$$

$$G_{PY}(\omega) = 2P_m^*(\omega)Y_m(\omega) \qquad (7.5.13)$$

We should note that in the absence of noise, both $H_1(\omega)$ and $H_2(\omega)$ give the same results so we need to calculate only one.

Now we explore the effects of using various window variations, starting with the exponential window. Then the input window parameters will be changed one at a time. In this way, we can show the effects of each window parameter on our experimental result, and we can establish the required rules in order to achieve good test results.

7.5.3 Truncation and Exponential Window Effects

We need to see the effects of when we start the impulse in the window as well as how truncation without and with the exponential window affects our results. Thus we need to identify when the pulse starts and when the window is applied for each test simulation. A 2 percent exponential window is used in all cases involving such windows.

Truncation of Signal (TP ≅ 0.2 Second) **Without** *Exponential Window* In this case, the recorded 1024 point signals start to have significant values about 0.21 second, as shown in Figs. 7.5.2a and 7.5.2c, where we note that the signal is zero for about 16 percent of the 1.25 second window. The calculated FRF magnitude and phase for these two signals are compared to the original FRF (calculated from Eq. (7.5.1)) in Figs. 7.5.3a and 7.5.3b, respectively, where $\Delta f = 0.8$ Hz. It is clear that truncation leads to significant filter leakage that causes oscillation of the FRF magnitudes and phase information so that the results are difficult to interpret and are unacceptable. The three resonant peak values are compared in Table 7.5.2, where we see that the measured magnitude peaks are about 70 percent of the original peaks.

Truncation of Signal ($TP \cong 0.2$ *Second*) **With** *Exponential Window* The original truncated signal in Fig. 7.5.2c is multiplied by the 2 percent ($\tau_e = 0.2684$ second) exponential window that begins at $t_3 = 0.2$ second. The resulting FRF magnitude and phase are compared to the original FRF in Figs. 7.5.3c and 7.5.3d, respectively. It is clear from these plots that the filter leakage oscillations are removed, the peak values are low (marked by a dash), and the valleys are higher (also marked by a dash). The values in Table 7.5.2 show that each measured peak is about 22 percent of the original peak values. However, these errors can be corrected for by using Eqs. (7.4.5) through (7.4.7). We note that each peak has essentially the same percent of the original peak, since the product of $\zeta_p f_p$ is the same for each vibration mode. This equal percentage is not true in general.

Truncation of Signal ($TP \cong 0.02$ *Second*) **Without** *Exponential Window* In this case, the recorded 1024 point signals start to have significant values about 0.029 second instead of 0.21 second, so that the input and output signals are zero for about 1.6 percent of the 1.25 second window. The calculated FRF magnitude and phase for these two signals are compared to the original FRF (calculated from Eq. (7.5.1)) in Figs. 7.5.4a and 7.5.4b, respectively, where $\Delta f = 0.8$ Hz. It is clear that truncation leads to significant filter leakage, which causes more of a distortion of the FRF magnitudes and phase information rather than the oscillations obtained in Figs. 7.5.3a and 7.5.3b. The results in this case are less difficult to interpret

---

Fig. 7.5.3. FRF calculated from truncated 1024 data points showing filter leakage oscillations when pulse start point is 0.2 second. (a) Magnitude for no exponential window. (b) Phase for no exponential window. (c) Magnitude when 2 percent exponential window is used. (d) Phase when 2 percent exponential window is used. (From K. G. McConnell and P. S. Varoto, "The Effects of Window Functions and Trigger Levels on FRF Estimations from Impact Tests," *Proceedings of the 13th International Modal Analysis Conference*, Society for Experimental Mechanics.)

SELECTING PROPER WINDOWS FOR IMPULSE TESTING   495

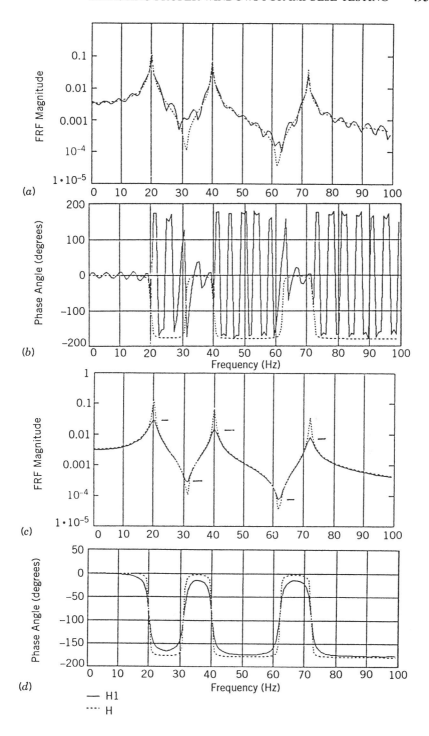

TABLE 7.5.2. Peak Values for Truncated and Exponential FRF Analysis Showing Percent of Ideal Peak

Frequency (Hz)	Original $H(\omega)$	Truncated[a] $H_t(\omega)$	%	Exponential[b] $H_e(\omega)$	%	Truncated[c] $H_t(\omega)$	%	Exponential[c] $H_e(\omega)$	%
20	1.389	0.959	69.0	0.304	21.9	1.043	75.1	0.346	24.9
40	0.695	0.478	68.6	0.152	21.9	0.525	75.5	0.173	24.9
72	0.371	0.260	70.2	0.083	22.4	0.283	76.3	0.095	25.5

[a] All table FRF values $H(\omega)$, $H_t(\omega)$, and so on, are multiplied by 10.
[b] Case where $TP = 0.20$ second.
[c] Case where $TP = 0.02$ second.

but are still undesirable. The three resonant peak values are compared in Table 7.5.2, where we see that the measured magnitude peaks are about 75 percent of the original peaks.

Truncation of Signal ($TP \cong 0.02$ *Second*) *With Exponential Window* The original truncated signal is multiplied by the 2 percent ($\tau_e = 0.3144$ second) exponential window that begins at $t_3 = 0.02$ second. The resulting FRF magnitude and phase are compared to the original FRF in Figs. 7.5.4c and 7.5.4d, respectively. It is clear from these plots that the filter leakage distortions of Figs. 7.5.4a and 7.5.4b are effectively removed, the peak values are low (marked by a dash), and the valleys are higher (also marked by a dash). The values in Table 7.5.2 show that each measured peak is about 25 percent of the original peak values. However, these errors can be corrected for by using Eqs. (7.4.5) through (7.4.7).

These results show that the exponential window does indeed clean up the filter leakage problems that occur from simply using the truncated data. We see that the filter leakage oscillation problems are less when the zero portion of the signal occupies less than 2 percent of the data window. Hence, one of the rules for setting up the window is to select t_3 and trigger point (TP) so that we have a minimum of zero data at the beginning.

Fig. 7.5.4. FRF calculated from truncated 1024 data points showing filter leakage oscillations when pulse start point is 0.02 second. (*a*) Magnitude for no exponential window. (*b*) Phase for no exponential window. (*c*) Magnitude when 2 percent exponential window is used. (*d*) Phase when 2 percent exponential window is used. (From K. G. McConnell and P. S. Varoto, "The Effects of Window Functions and Trigger Levels on FRF Estimations from Impact Tests," *Proceedings of the 13th International Modal Analysis Conference*, Society for Experimental Mechanics.)

SELECTING PROPER WINDOWS FOR IMPULSE TESTING 497

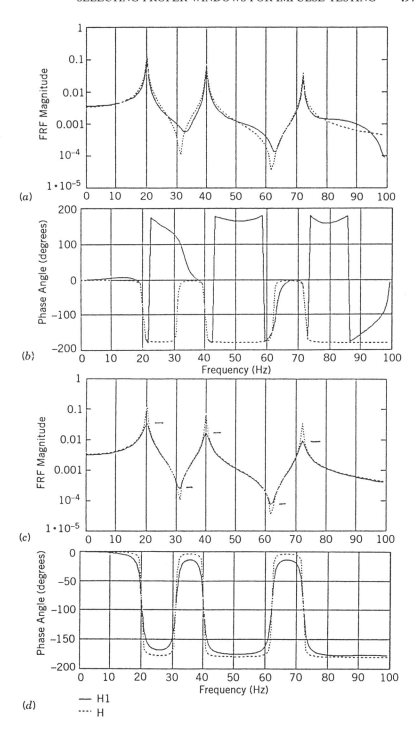

7.5.4 The Effects of the Input Transient Window

In this subsection we investigate the type of effects that an improperly set transient window can have on the measured FRF. In this calculation, we use all 4096 input and output data points so that there is no truncation window leakage to be concerned with and the only effect on the calculated FRF is due to how we are using the transient window function. Consider the input signal situation shown in Fig. 7.5.5a, where the original signal that has passed through the anti-aliasing filter and been recorded in the analyzer is shown as a solid line: this is the same signal as the solid line in Fig. 7.5.2a. The transient window is shown by a dotted line, while the product of the recorded signal with the transient window is the data that is analyzed and is shown by the squares. This situation is equivalent to using negative slope S in Fig. 7.5.1a (point B) where the positive peak force just exceeds the trigger level TL and the transient window's start time t_1 is equal to the trigger point time TP. Consequently, all input to the left of the trigger point is zero and only part of the impulse and the filter ringing are analyzed as inputs.

The magnitude and phase of the FRF are shown in Figs. 7.5.5b and 7.5.5c, respectively. The measured magnitude (solid line) is too large compared to the theoretical FRF (dotted line). It is clear that the measured FRF becomes closer to the theoretical FRF at higher frequencies, and that they actually cross near 90 Hz. This excessive amplitude is due to the output frequency spectrum being too small, due to elimination of nearly the front half of the input signal. The phase angle starts out being very close at low frequencies and begins to shift toward more positive angles as the frequency increases. In fact, we see that this shift is a straight line increasing with frequency.

The implication of this phase shift is seen more clearly in the Nyquist plot of Fig. 7.5.5d. In this case, the first resonance circle with a peak at A is rotated counterclockwise about 15 degrees (285 degrees) while the magnitude is severely distorted. The second resonance circle with a peak at B is rotated counterclockwise by about 28 degrees (298 degrees), and the magnitude appears to be distorted. The third resonance circle with a peak at C is rotated about 50 degrees counterclockwise (320 degrees), and the magnitude is also distorted. This rotation of the resonance circle in

---

Fig. 7.5.5. (a) Comparison of original filtered input pulse to one cut off by transient window. (b) Resulting FRF magnitude. (c) Resulting phase. (d) Nyquist plot. (From K. G. McConnell and P. S. Varoto, "The Effects of Window Functions and Trigger Levels on FRF Estimations from Impact Tests," *Proceedings of the 13th International Modal Analysis Conference*, Society for Experimental Mechanics.)

SELECTING PROPER WINDOWS FOR IMPULSE TESTING 499

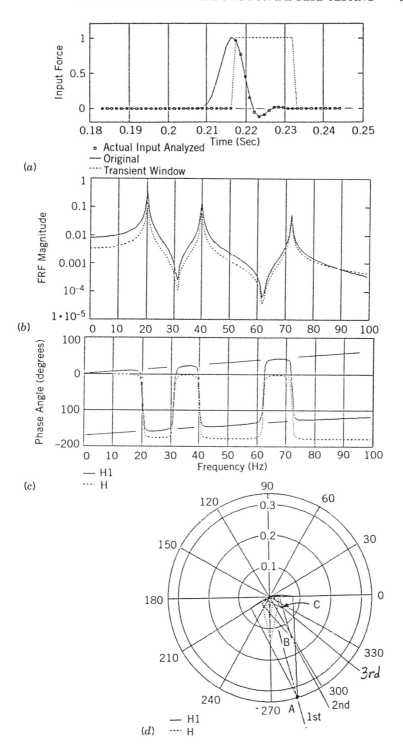

the Nyquist plot is typical of experiences that have occurred during field measurements.[16] The actual resonance circle should be centered on the imaginary axis, either on the positive side or the negative side as shown by the dashed lines. Thus the rotation of the resonance circle in the Nyquist plot (in either direction) is a clear indication that the transient window may be chopping off either the leading or trailing portion of the input pulse.

7.5.5 Recommended Procedure for Setting Window Parameters

In order to obtain the best performance from a given frequency analyzer when conducting impulse tests, the following nine steps and considerations must be taken into account.

1. Decide the frequency range that you need to work with and adjust the hammer mass and impact tip so that there are no near zeros in the input frequency spectra. Then measure the impact duration τ.

2. Set time t_1 to be in the first 2 to 5 percent of the data window. Then set $t_1 = t_3$ so that the transient window and the exponential window start at the same time. When everything else is properly set up, there should be no significant data before these times in any measured signal.

3. Set the exponential time constant τ_e to achieve the desired percentage of signal leakage, usually in the 2 to 10 percent range.

4. Next, set the trigger point $TP \geq t_1 + \tau/2$ so that no portion of the leading edge of the input can be cut off by the transient window function.

5. If possible, try and determine the amount and duration of filter ringing that occurs when the input pulse is measured. Then set the upper transient window function time $t_2 \cong TP +$ duration of significant filter ringing (often on the order of 3 to 10τ).

6. Set the trigger level TL so that $0.4FS < TL < 0.6FS$. This high range ensures consistent impact inputs and good signal to noise ratios if the FS is properly set as noted below.

7. We need to set the slope. To do this, we need to know if the input pulse is positive as shown in Fig. 7.5.2a, or negative. We want to trigger on the pulse's leading edge, that is, point A rather than point B in Fig. 7.5.1a. Hence, we use positive slope S with a positive impulse, as shown at point A, and a negative slope S with a negative impulse. If we break this rule, then we need to change TP in step 4 to $TP \geq t_1 + \tau$, so that no part of the impulse is cut off by the transient window.

8. We need to establish the noise floor for each data channel in deciding

---

[16]Private conversation, Prof. David Ewins, Imperial College, London, U.K., Summer 1991.

what the *FS* setting should be. This consideration is more important for the output signal compared to the input signal, for we have the transient window to reduce input instrument noise floor effects. This choice is not available for the output signal. We have limited dynamic range in our A/D converters, that is 12 bit range of ±2028 counts and 16 bit range of ±32,768 counts. Hence, we should have useful signal of about 2000 counts for 12 bit converters and around 30,000 counts for the 16 bit converter instruments. To determine whether to use the full 32,000 count range of a 16 bit converter, we can run a preliminary set of data and establish the dynamic range of the signal. However we come to a decision, the full scale voltage selected must take the instrument noise floor into account, for in most tests we need all or nearly all of the dynamic range that is available to us. If the noise is too high, we need to reexamine our instrumentation choice for this particular test.

9. Take a preliminary set of data and look for any of the telltale signs of truncation in the output, transient window chopping off part of the input pulse (look at the Nyquist diagram for higher frequency resonance circle rotating away from the imaginary axis), or near zeros occurring in the input frequency spectra (or auto-spectral density). I believe there are many test results that have rotation in the Nyquist diagram where this rotation is blamed on either nonlinear behavior or hysteric damping rather than on improperly set transient window or improperly used instruments.

7.6 VIBRATION EXCITER DRIVING A FREE-FREE BEAM WITH POINT LOADS

A common test method is to drive a structure with one or more electrodynamic vibration exciters using either swept sine, chirp, pseudorandom, random, or sine on random signals. Each type of excitation has its own advantages and disadvantages dependent on test objectives.

7.6.1 Selecting the Excitation Signal

Sinusoidal Test Signal The sine dwell or slowly swept sine has distinct advantages in identifying closely spaced natural frequencies and mode shapes, the presence of nonlinear behavior, and so on, that may not be resolved using any other methods. In the opinion of Professor Lallement,[17] the sinusoid is the only signal to use when exploring the precise characteristics of an unknown structure that may be nonlinear, have multiple closely

---

[17]Private conversation during a visit to Université De Besançon, Besançon, France, Spring 1991.

THE APPLICATION OF BASIC CONCEPTS TO VIBRATION TESTING

spaced resonances, and so on. Its largest disadvantage is the long test times that are required.

Chirp Test Signal The chirp[18] is an impulsive type of input that can have excitation over a wide range of frequencies and avoids impulse loading problems where we need a zero net impulse in order to keep the exciter's armature from bottoming out. The chirp is generated by selecting the lower f_1 and upper f_u frequencies that act over the chirp's time period T_c. The chirp is generated by the function

$$x(t) = A \sin[2\pi(f_1 + \mu t/2)t + \alpha] \qquad (7.6.1)$$

where A is the amplitude.
α is an arbitrary phase angle.
μ is defined by

$$\mu = \frac{f_u - f_1}{T_c} \qquad (7.6.2)$$

We want the signal to have minimum filter leakage. This requires that the signal start and end with the same value and have the similar slope that is either positive or negative. These requirements will be met if phase angle $\alpha = 0$ and we have an integer of m complete center frequency (f_c) cycles in time period T_c so that

$$f_c T_c = m \qquad (7.6.3)$$

where the center frequency is defined by

$$f_c = \frac{f_u + f_1}{2} \qquad (7.6.4)$$

A typical chirp signal is shown in Fig. 7.6.1a for the conditions $f_1 = 100$ Hz, $f = 400$ Hz, $A = 100$, and $T_c = 0.1$ second. The corresponding impulse frequency spectrum is shown in Fig. 7.6.1b, where we see that the significant input occurs in the 100 to 400 Hz frequency range, the center frequency is 250 Hz, and integer $m = 25$. The frequency spectrum of Fig. 7.6.1b is not achieved in practice since the power amplifier, exciter,

---

[18]R. C. Wei, "Structural Wave Propagation and Sound Radiation Study Through Time and Spatial Processing," Ph.D. Thesis, Iowa State University, Ames, IA, 1994.

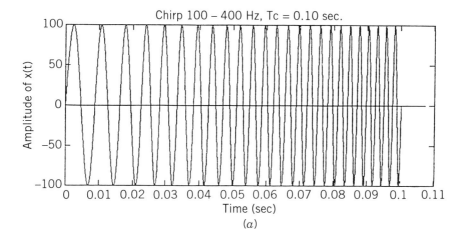

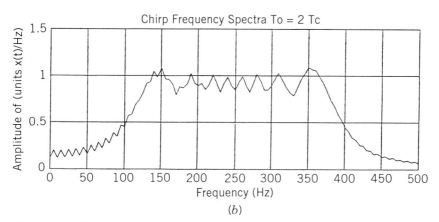

Fig. 7.6.1. (*a*) Chirp excitation signal $f_1 = 100$ Hz, $f_u = 400$ Hz, and $T_c = 0.1$ seconds in time domain. (*b*) Impulse frequency spectrum.

and structure FRFs modify this spectrum; hence we must measure this input force.

Since the chirp input lasts for a significant portion of the data window period T, the exciter adds significant damping to the test structure's response after the chirp is applied. This additional damping can be helpful in that the structure's response will die out more quickly and there is less leakage in the output data window for a given measurement period T.

Pseudorandom Test Signal The pseudorandom signal has many advantages[19] for obtaining both a quick look and an accurate estimate of a structure's response over a broad range of frequencies. The pseudorandom signal easily generates a specified input auto-spectral density (which includes a sine or random type of excitation), uses rectangular windows on the input and output signals without filter leakage since the signals are periodic in window period T, requires only one average if we wait for start up transients to disappear, has good signal to noise ratios, is helpful in quickly identifying that we are dealing with a nonlinear system, and gives the same information that we obtain by using a random signal and waiting for many averages in order to remove uncertainty in the measured input-output FRF estimate. This need for an average of one or two time records saves a lot of test time; and hence, is extremely cost effective.

The pseudorandom signal has two known disadvantages. One disadvantage is that fatigue test results can be different in that the loading sequence is repeatable and there can be peak resonances that are not excited because a resonance does not correspond to one of the pseudorandom frequencies. Experience[20] with fatigue tests of lightly damped test specimens excited with either broad-band random or sinusoidal signals with a Rayleigh peak statistical distribution suggests that equivalent fatigue lives may be obtained when using pseudorandom excitation signals. However, only direct experimentation can resolve this question. A second disadvantage is the *French onion soup* phenomenon (i.e., your mom's soup doesn't taste like my mom's soup so it must not be French onion soup), since a pseudorandom signal does not sound like a random signal when both have the same auto-spectral density. This is the experimenter's psychological response coming into play, which may or may not be valid. Only successful experience will remove resistance to using pseudorandom signals in vibration testing when it is an appropriate testing signal.

We should carefully select our input signals to be most efficient in meeting our test objectives.

7.6.2 Test Setup

The 93 in. long free-free beam described in Section 7.4 and Fig. 7.4.4a is used in this experiment[21] and the data reported here comes from this

---

[19] K. G. McConnell and P. S. Varoto, "Pseudo-Random Excitation Is More Cost Effective than Random Excitation," *Proceedings of the 12th International Modal Analysis Conference*, Vol. 1, Honolulu, HA, 1994, pp. 1–11.

[20] R. B. Thakkar, "Exact Sinusoidal Simulation of Fatigue Under Guassian Narrow Band Random Loading," Ph.D. Thesis, Iowa State University, Ames, IA, 1972.

[21] A. Cooper, A. Assadi, and P. S. Varoto, "Driving a Free Free Structure with an Exciter," Student Project Report, Engineering Mechanics 545x, Department of Aerospace Engineering and Engineering Mechanics, Iowa State University, Ames, IA, Spring 1994.

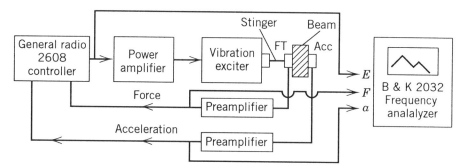

Fig. 7.6.2. Test setup used to measure free-free beam response.

student experimental project. The force transducer (Fig. 7.6.2) is attached to the beam in Fig. 7.4.4a at either its midpoint (location 1) or the left end (location 2). A stinger is used to connect the force transducer (FT) to the exciter as shown in Fig. 7.6.2. Accelerometers are mounted at each location (1, 2, and 3). The General Radio Model 2608 vibration controller is used to generate a random voltage signal E that goes to the power amplifier to drive the exciter. The force signal F is used as feedback so that the controller's output signal varies with frequency to cause the force signal frequency spectrum to be nearly constant over the frequency range of near 0 to 800 Hz. The Bruel & Kjaer Model 2032 frequency analyzer is used to monitor various input and output signals for generating the required FRFs.

7.6.3 Theoretical Exciter Structure Interaction

The interaction of the exciter with the test structure follows the lines developed in Chapter 6. The free body diagram (FBD) of the beam and exciter armature are shown in Fig. 7.6.3. The frequency domain driving point response $Y_s(\omega)$ is related to the structure's driving point receptance $H_s(\omega)$ and the driving point force $F_s(\omega)$ by

$$Y_s(\omega) = H_s(\omega) F_s(\omega) \qquad (7.6.5)$$

The corresponding frequency domain voltage current relationship from Sections 6.4, 6.5, and 6.7 is given by

$$(R + jL\omega)I(\omega) + jK_v\omega Y_a(\omega) = E(\omega) \qquad (7.6.6)$$

where R is the armature circuit's resistance.
L is the armature circuit's inductance.
$I(\omega)$ is the circuit current.

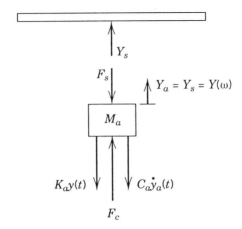

Fig. 7.6.3. FBD of beam and exciter armature.

K_v is the velocity sensitive back emf constant.
$E(\omega)$ is the voltage applied to the circuit.
$Y_a(\omega)$ is the armature motion.

The armature's dynamics are described by

$$[K_a - M_a\omega^2 + jC_a\omega]Y_a(\omega) = K_f I(\omega) - F_s(\omega) \qquad (7.6.7)$$

where K_a is the armature support stiffness.
C_a is the armature damping.
M_a is the armature mass.
K_f is the force current constant.

Now we assume that the beam and armature motions are the same so that

$$Y_s(\omega) = Y_a(\omega) = Y(\omega) \qquad (7.6.8)$$

and we define the armature dynamic stiffness as

$$G(\omega) = K_a - M_a\omega^2 + jC_a\omega \qquad (7.6.9)$$

so that Eq. (7.6.7) becomes

$$G(\omega)Y(\omega) = K_f I(\omega) - F_s(\omega) \qquad (7.6.10)$$

We are interested in the force that acts on the beam and is measured by the force transducer as well as the current and voltage relationships under both voltage and current modes of amplifier behavior. We find that

$$D(\omega) = 1 + G(\omega)H_s(\omega) \qquad (7.6.11)$$

terms occur often when we formulate these equations from Eqs. (7.6.5), (7.6.6), and (7.6.10).

Voltage Mode of Operation When we solve for the current voltage relationship from Eqs. (7.6.5), (7.6.6), and (7.6.10), we obtain

$$\frac{I(\omega)}{E(\omega)} = \frac{D(\omega)}{RD(\omega) + j\omega[LD(\omega) + K_vK_fH_s(\omega)]} \qquad (7.6.12)$$

The force on the structure and the input voltage are related by

$$\frac{F_s(\omega)}{E(\omega)} = \frac{K_f}{RD(\omega) + j\omega[LD(\omega) + K_vK_fH_s(\omega)]} \qquad (7.6.13)$$

These two equations show clearly how the structure's driving point FRF $H_s(\omega)$ and the armature's dynamic stiffness $G(\omega)$ influence the current and the force acting on the structure.

Current Mode of Operation The voltage current relationship is the reciprocal of Eq. (7.6.12), so that

$$\frac{E(\omega)}{I(\omega)} = \frac{RD(\omega) + j\omega[LD(\omega) + K_vK_fH_s(\omega)]}{D(\omega)} \qquad (7.6.14)$$

The force acting on the structure per unit of current is given by

$$\frac{F_s(\omega)}{I(\omega)} = \frac{K_f}{D(\omega)} \qquad (7.6.15)$$

Equations (7.6.12) through (7.6.15) are used to predict the system's theoretical behavior.

7.6.4 Comparison of Experimental and Theoretical Results

The theoretical and experimental driving point accelerances are compared in Fig. 7.6.4 where the solid curve is the theoretical and the dashed curve is the experimental result. The theoretical driving point accelerances were calculated using Eq. (7.4.9), while the experimental driving point accelerances were obtained using pseudorandom excitation and the Bruel & Kjaer Model 2032 frequency analyzer. The end driving point accelerance

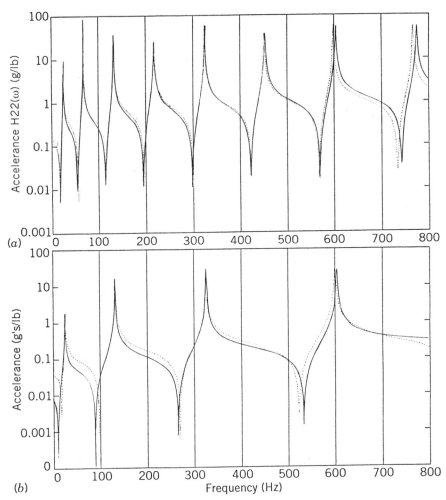

Fig. 7.6.4. Comparison of experimental (dashed) and theoretical (solid) driving point accelerances. (a) Endpoint $H_{22}(\omega)$ of a free-frequency beam. (b) Midpoint $H_{11}(\omega)$ of a free-free beam. (From A. Cooper, A. Assadi, and P. S. Varoto, "Driving a Free Free Structure with an Exciter," Student Project Report, Engineering Mechanics 545x, Iowa State University, Ames, IA, Spring 1994.)

$H_{22}(\omega)$ is shown in Fig. 7.6.4a, where the theoretical damping is adjusted to match the experimental results for each resonant condition. Overall, a good fit between the two curves is achieved.

The midpoint driving point accelerance $H_{11}(\omega)$ is shown in Fig. 7.6.4b. The two curves do not fit nearly as well as they do for $H_{22}(\omega)$ in Fig. 7.6.4a. One reason for this discrepancy is the force transducer environment. At location 2 the force transducer is exposed to more bending moments, due to the angular motion of the beam's end, and these bending moments are known to cause some measurement errors.[22] At the midpoint, the force transducer is exposed to large bending strains that are known to cause additional sensitivity that is in phase (either positively or negatively) with the resonant condition.[23] Consequently, we see that even under the best of conditions, the effects of instruments on the data cannot be underestimated.

The controller in Fig. 7.6.2 uses force as the feedback and adjusts the output voltage E in order to drive the structure at its midpoint with a measured force spectrum that is constant over the 0 to 800 Hz frequency range. The corresponding input voltage E was measured, along with the force $F_s(\omega)$, in order to determine the force dropout that can occur when operating in the voltage mode. This data was used to measure the output force per unit of input voltage (to the amplifier) type of FRF, which can be compared to the theoretical results from Eq. (7.6.13). There is an unknown scale factor between these two sources since Eq. (7.6.13) refers to the input voltage supplied to the exciter's armature circuit, not to the power amplifier's input voltage. Thus the theoretical results from Eq. (7.6.13) must be adjusted so that the two curves will have the same values in the 60 to 80 Hz frequency range as shown in Fig. 7.6.5.

It is clear from Fig. 7.6.5 that significant force dropout occurs at each resonance and theory predicts the occurrence of this dropout. We also see that the shape of the experimental dropout is different at each resonance since the amount the curve moves up before the notch increases at higher frequencies. The amount of dropout from peak to valley also increases with the higher frequencies. This potential for rapid changes in the force with frequency around the resonant condition means that the apparent damping may be incorrectly measured, since the structure's response is less than it should be due to the reduction in the input force, while damping is often amplitude dependent as well. Then if we add to the force dropout effects the effects of force transducer sensitivity to bending moments and base strain, it is little wonder that it is difficult to achieve repeatability of damping in our test results if we change force transducers or disassemble and reassemble the test setup. For in disassembly, the orientation of the force transducer with respect to the beam's bending strains and bending moments on the force transducer will alter the output force signal.

---

[22] See Section 4.9.5.
[23] P. Cappa and K. G. McConnell, "Base Strain Effects on Force Measurements," *Proceedings, Spring Society for Experimental Mechanics*, Baltimore, MD, June 1994, pp. 520–530.

510 THE APPLICATION OF BASIC CONCEPTS TO VIBRATION TESTING

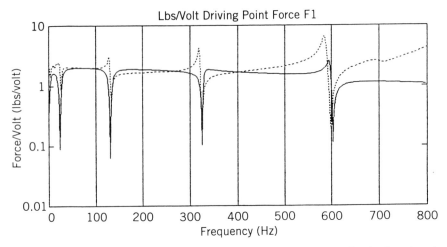

Fig. 7.6.5. Comparison of theoretical (solid) and experimental (dashed) ratio of force on structure per input volt showing interaction of structure with exciter to cause force dropout in the vicinity of structure resonances. (From A. Cooper, A. Assadi, and P. S. Varoto, "Driving a Free Free Structure with an Exciter," Student Project Report, Engineering Mechanics 545x, Iowa State University, Ames, IA, Spring 1994.)

7.7 WINDOWING EFFECTS ON RANDOM TEST RESULTS

While conducting random vibration tests of a free-free beam, students obtained a rather surprising result.[24] They had been running a driving point accelerance test at the midpoint of the 93.0-in.-long free-free beam described in Section 7.6.2 while using pseudorandom excitation with rectangular window functions on both input and output signals. The rectangular windows were being used, since a pseudorandom signal is periodic in

---

[24]A. Cooper, A. Assadi, and P. S. Varoto, "Driving a Free Structure with an Exciter," Student Project Report, Engineering Mechanics 545x, Department of Aerospace Engineering and Engineering Mechanics, Iowa State University, Ames, IA, Spring 1994.

Fig. 7.7.1. Measured driving point accelerance $H_{11}(\omega)$ with pseudorandom excitation when (*a*) Hanning window is used on both signals (H_1 solid and H_2 dots). (*b*) $H_1(\omega)$. (*c*) $H_2(\omega)$ when rectangular window is used on the output signal and Hanning window on the input signal (comparison curve from (*a*) above). (From A. Cooper, A. Assadi, and P. S. Varoto, "Driving a Free Free Structure with an Exciter," Student Project Report, Engineering Mechanics 545x, Iowa State University, Ames, IA, Spring 1994.)

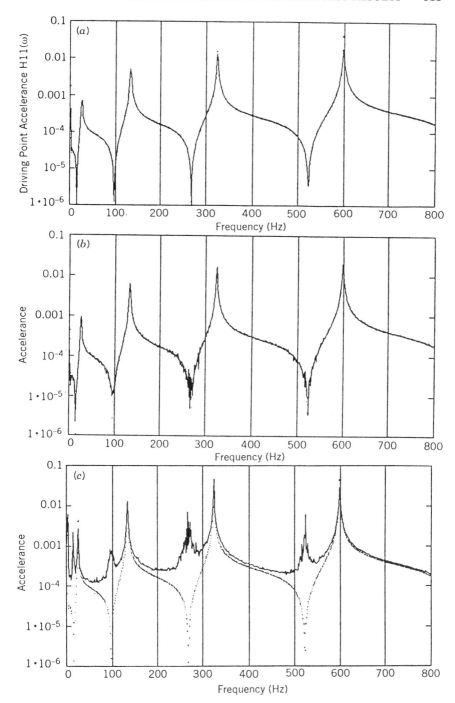

the data window. They obtained the results shown in Fig. 7.7.1a, where $H_1(\omega)$ is the solid line, $H_2(\omega)$ is the dotted line, and $H_2(\omega) \geq H_1(\omega)$ as usual.

Then they switched to broad band-random excitation and began to repeat the test. The $H_1(\omega)$ acceleration FRFs shown in Fig. 7.7.1b were obtained when both the random input and output were processed using a Hanning window (dotted line), the input was processed using the Hanning window, and the output was processed using the rectangular window (solid line). It is clear that the rectangular window results are more uncertain below values of 10^{-4} while they match closely at the peaks. Then the students displayed the $H_2(\omega)$ acceleration FRF that is shown in Fig. 7.7.1c as the solid line. These results are compared to the results shown in Fig. 7.7.1a (dotted line). It is evident that the valleys have become peaks with a lot of uncertainty, while the peaks are in reasonable agreement with the Hanning windowed results. An investigation of the frequency analyzer showed that the output data was being processed using a rectangular window while the input data was being processed with a Hanning window; in other words, we have a case of mixed data windows. The reason for this discrepancy was discussed in a recent paper[25] and the results are presented here as a reminder of what to check for when this happens to experimental data.

7.7.1 A Model of Window Function Filter Leakage Characteristics

The digital filter characteristics for the Hanning and rectangular window functions are shown in Fig. 7.7.2a and 7.7.2b, respectively. Their filter characteristic magnitudes are plotted on a log scale versus the log scale of the relative frequency $(f - f_c)$ that is given in multiples of Δf where f_c is the filter's center frequency. It is clear that the Hanning window is fatter around f_c than the rectangular window, but it falls off at a very rapid rate. It is the shape of these digital filter skirts that cause the dramatic change in $H_2(\omega)$ shown in Fig. 7.7.2a, since the FRF response peaks adjacent to the valleys are leaking into the valley regions. The side lobes also alternate in sign so that any magnitude envelope function will distort any estimates made using these functions alone.

The filter skirts are modeled as straight lines in the log-log plots and are unity at the center frequency. Convenient functions that fit these general filter characteristics were found to be of the form

---

[25]K. G. McConnell and P. S. Varoto, "The Effects of Windowing on FRF Estimates for Closely Spaced Peaks and Valleys", *Proceedings of the 13th International Modal Analysis Conference*, Nashville, TN, Feb. 1995.

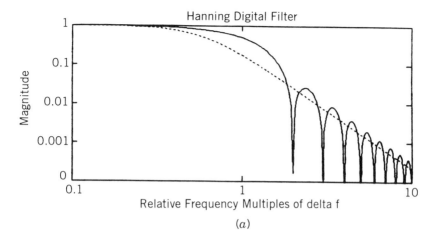

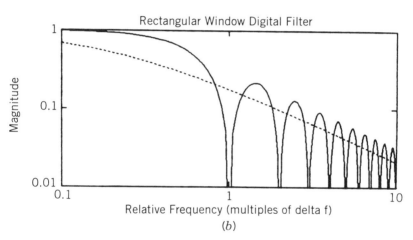

Fig. 7.7.2. Comparison of digital filter characteristics (solid line) with their envelope function (dashed line). (*a*) Hanning. (*b*) Rectangular windows. (Courtesy of the *Proceedings of the 13th International Modal Analysis Conference*, Society for Experimental Mechanics.)

$$E_h(\omega - \omega_c) = E_h(f - f_c) = \frac{\beta}{\beta + |f - f_c|^3} \qquad (7.7.1)$$

for the Hanning window and

514   THE APPLICATION OF BASIC CONCEPTS TO VIBRATION TESTING

$$E_r(\omega - \omega_c) = E_r(f - f_c) = \frac{\alpha}{\alpha + |f - f_c|} \qquad (7.7.2)$$

for the rectangular window. The values of α and β were determined so that the the area under the magnitude of the Hanning characteristic from the first notch and Eq. (7.7.1) are the same while the area under the rectangular filter characteristic above the first notch is the same for the magnitude of the sinc function and Eq. (7.7.2). The required values for α and β are $\alpha = 0.215$ and $\beta = 0.216$. The corresponding filter skirt functions are shown as a dashed curve in Fig. 7.7.2.

7.7.2 Estimating FRF Errors Due to Leakage

The FRF Error Equations   The FRF Errors are estimated using the single input-output model shown in Fig. 5.9.4, where $f(t)$ is the actual input force, $n(t)$ is the input signal's noise, $x(t)$ is the measured input, $o(t)$ is the actual output motion due to $f(t)$, $m(t)$ is the output signal's noise, and $y(t)$ is the measured signal. We are going to assume here that the output noise $m(t)$ is dominated by digital filter leakage in the region of the valleys. This noise is related to the output signal $o(t)$ that occurs at the resonant peaks. Hence, we can use the $\hat{H}_2(\omega)$ error estimator given by Eq. (5.9.26), namely,

$$\hat{H}_2(\omega) = H(\omega)\left(1 + \frac{\hat{G}_{mm}(\omega)}{|H(\omega)|^2 \hat{G}_{ff}(\omega)}\right) \qquad (7.7.3)$$

where we assume that output signal noise $\hat{G}_{mm}(\omega)$ is dominated by the filter leakage from the adjacent resonant peaks. Thus, we estimate that

$$\hat{G}_{mm}(\omega) = \sum_{p=1}^{N_p} |H(\omega_p)|^2 G_{ff}(\omega_p) E^2(\omega - \omega_p) \qquad (7.7.4)$$

where $H(\omega_p)$ is the pth peak value and $E(\omega - \omega_p)$ is the leakage error function from either Eqs. (7.7.1) or (7.7.2). Substitution of Eq. (7.7.4) into Eq. (7.7.3) gives

$$\hat{H}_2(\omega) = H(\omega)\left(1 + \sum_{p=1}^{N_p} \frac{|H(\omega_p)|^2 E^2(\omega - \omega_p) \hat{G}_{ff}(\omega_p)}{|H(\omega)|^2 \hat{G}_{ff}(\omega)}\right) \qquad (7.7.5)$$

where we have assumed that $\hat{G}_{ff}(\omega)$ is not equal $\hat{G}_{ff}(\omega_p)$ due to force drop out at resonance, as explained in Section 7.6. Now Eq. (7.7.5) should underpredict the errors due to filter leakage, since only the peak values are being used instead of all values adjacent to the valleys. It is also apparent from Eq. (7.7.5) that the valleys could easily become peaks as $H_2(\omega)$ becomes small. We will use Eq. (7.7.5) to predict the values of $\hat{H}_2(\omega)$ for comparison to the experimental results when both the rectangular and Hanning window functions were employed.

If we return to Section 5.9 and analyze the output for $\hat{H}_1(\omega)$ for this case where we assume that filter leakage is the source of error, we find that $\hat{H}_1(\omega)$ is approximated by

$$\hat{H}_1(\omega) \cong H(\omega)\left(1 + \sum_{p=1}^{N_p} \frac{H(\omega_p)E_\eta(\omega - \omega_p)F(\omega_p)}{H(\omega)F(\omega)}\right) \quad (7.7.6)$$

in the absence of any input noise $n(t)$. Again, we see that the ratio of force frequency spectra $F(\omega)$ provides a multiplying factor. Equation (7.7.6) is used to estimate the leakage effects on the measured value of $\hat{H}_1(\omega)$ for both the rectangular and Hanning window functions. It is noted that Eqs. (7.7.5) and (7.7.6) are gross approximations of a complex calculation process that takes place during each window of data. We want to demonstrate that filter leakage is a significant error source for the conversion of valleys into peaks.

Comparison of FRF Window Errors with Experimental Results Now we will compare the experimental results from using $\hat{H}_2(\omega)$ with rectangular and Hanning windows on the output signal to the values predicted by Eq. (7.7.5) in Figs. 7.7.3a and 7.7.3b, respectively. The values of $\hat{H}_2(\omega)$ estimated from Eq. (7.7.5) are shown as the dotted line, whereas the actual measured experimental values are shown as a solid line. The ratio of $G_{ff}(\omega_p)/G_{ff}(\omega)$ is assumed to be 0.1 in order to account for force dropout at the resonant peaks. In Fig 7.7.3a, it is seen that the rectangular window function causes the experimental valleys to appear as peaks, while Eq. (7.7.5) shows that the valleys are also converted to peaks when the rectangular error function from Eq. (7.7.2) is used. There is a significant difference between the measured and calculated results, as indicated by the cross-hatched area between the two curves. Remember that Eq. (7.7.5) is based on only the leakage from the adjacent peaks and not on any values closer to the valleys. Clearly, filter leakage from only the peak values is significant, due to the fact that the rectangular window falls off too slowly. The results in Fig. 7.7.3b are due to using the Hanning error function from Eq. (7.7.1) in Eq. (7.7.5). It is clear that the values of $H(\omega)$ are predicted quite well, with only the peak values being the source of

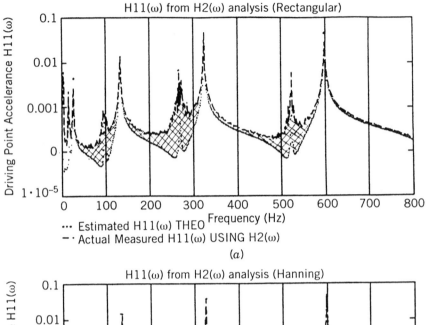

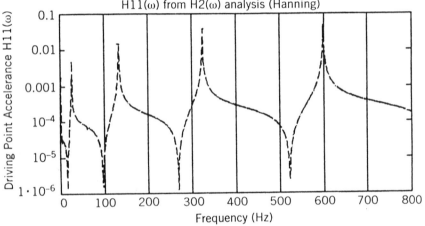

Fig. 7.7.3. Comparison of measured $H_2(\omega)$ to error estimation from Eq. (7.7.5) when using (*a*) rectangular window and (*b*) Hanning window functions. (Courtesy of the *Proceedings of the 13th International Modal Analysis Conference*, Society for Experimental Mechanics.)

error. Hence, we see that filter leakage in this case is not sufficient to cause serious errors on the order to convert valleys into peaks.

When we compare the $H_1(\omega)$ experimental data and the predictions from Eq. (7.7.6), we obtain the plots shown in Figs. 7.7.4a and 7.7.4b. When the output has a rectangular window, the use of the envelope function from Eq. (7.7.2) in Eq. (7.7.6) causes the values of $\hat{H}_1(\omega)$ to be too high at the valleys. The force frequency spectra ratio of $F(\omega_p)/F(\omega) \cong 0.3$ is due to force dropout at each resonant peak. However, it is clear that this filter leakage is the primary source for the noise and raising of the valleys. Again, when the Hanning window is used experimentally and, the Hanning envelope function from Eq. (7.7.1) is used it is obvious in Fig. 7.7.4b that the two curves are very close. Hence, we have demonstrated that the ability of the Hanning digital filter to fall off at such a high rate is important in obtaining reasonable results near the valleys.

7.7.3 Theoretical Simulation of Leakage and Its Effects

In order to establish the experimental behavior in the absence of noise and to study other combinations of mixing the window functions, a simple 2-degree-of-freedom system is used with the modal properties given in Table 7.7.1. The corresponding driving point FRF $H(\omega)$ is then obtained from

$$H(\omega) = \sum_{n=1}^{N} \frac{A_n}{\omega_m^2 - \omega^2 + j2\zeta_m \omega_m \omega} \qquad (7.7.7)$$

Equation (7.7.7) is used to calculate the output response spectra for N different random input frequency spectra $F(\omega)$. The corresponding N numbers of measured time histories are multiplied by either the rectangular or the Hanning window functions. Then, the N corresponding frequency spectra are averaged and used to estimate $\hat{H}_1(\omega)$ and $\hat{H}_2(\omega)$ according to

$$\hat{H}_1(\omega) = \frac{\sum_{n=1}^{N} (\hat{G}_{fo}(\omega))_n}{\sum_{n=1}^{N} (\hat{G}_{ff}(\omega))_n} \qquad (7.7.8)$$

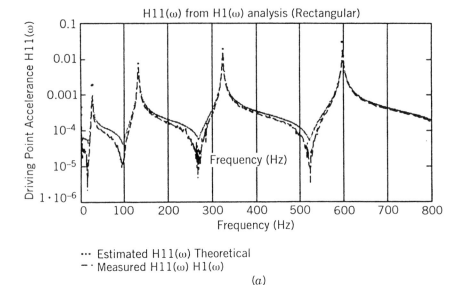
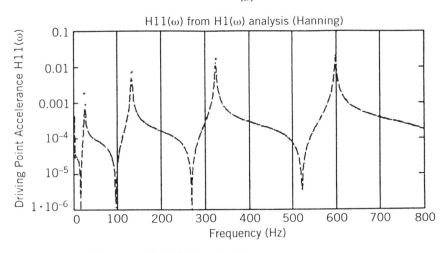

Fig. 7.7.4. Comparison of measured $H_1(\omega)$ to error estimation from Eq. (7.7.6) when using (*a*) rectangular window and (*b*) Hanning window functions. (Courtesy of the *Proceedings of the 13th International Modal Analysis Conference*, Society for Experimental Mechanics.)

TABLE 7.7.1. Theoretical Model Modal Parameters

Mode	A_n	f_n (Hz)	ζ_n
1	1.0	25.0	0.005
2	1.0	50.0	0.007

and

$$\hat{H}_2(\omega) = \frac{\sum_{n=1}^{N} (\hat{G}_{oo}(\omega))_n}{\sum_{n=1}^{N} (\hat{G}_{fo}(\omega))_n} \qquad (7.7.9)$$

Four different cases are studied. First, both input and output signals are processed using Hanning windows. The resulting FRFs as estimated from $\hat{H}_1(\omega)$ and $\hat{H}_2(\omega)$ using 100 data samples are shown in Fig. 7.7.5a. The results here are standard fare, since $\hat{H}_2(\omega) \geq \hat{H}_1(\omega)$ for all data points displayed.

Second, both the input and output signals are processed using the rectangular window function, with the resulting FRFs shown in Fig. 7.7.5b. In this case, we see that the $\hat{H}_1(\omega)$ curve shows the existence of a valley, even though it has a high amount of uncertainty, while $\hat{H}_2(\omega)$ shows the false peak in the region of the valley. It is clear that the rectangular window on the output signal is more harmful to $\hat{H}_2(\omega)$ than to $\hat{H}_1(\omega)$ in this case.

Third, the input signal is processed using a rectangular window, whereas the output is processed using a Hanning window, with the resulting FRFs shown in Fig. 7.7.5c. In this case, the magnitude of $\hat{H}_2(\omega)$ is always greater than the magnitude of $\hat{H}_1(\omega)$. The general characteristics of the two curves are very close, with both peaks and valleys being clearly defined. Unless $\hat{H}_2(\omega)$ and $\hat{H}_1(\omega)$ are plotted together, we would not easily notice this slight difference in results.

Fourth, the input signal is processed using a Hanning window while the output signal is processed using a rectangular window, with the results shown in Fig. 7.7.5d. In this case, $\hat{H}_1(\omega)$ shows more uncertainty for values less than 10^{-2}, with the valley showing the greatest uncertainty. The values of $\hat{H}_2(\omega)$ agree reasonably well for the two major peak values, but $\hat{H}_2(\omega)$ inverts the valley, as shown experimentally. It is clear that the values of $\hat{H}_2(\omega)$ are greater than those of $\hat{H}_1(\omega)$ and that this excess is due to rectangular window leakage.

The leakage process is not as simple as we have shown using Eq. (7.7.7). The envelope filter leakage envelopes are not the same as the digital filter characteristics of the sinc function and the Hanning window

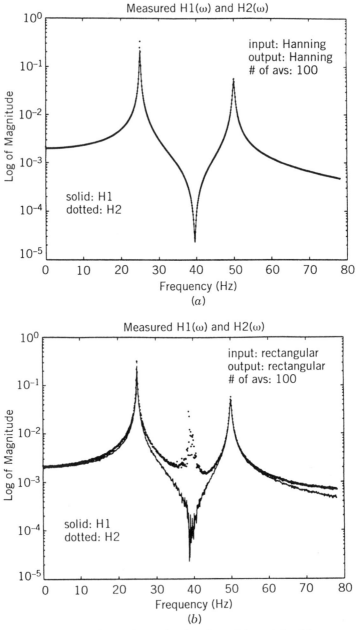

Fig. 7.7.5. Theoretical FRFs from simulated modal model without any signal noise. (*a*) Hanning window both input and output. (*b*) Rectangular window on both input and output. (*c*) Input rectangular and output Hanning. (*d*) Input Hanning and output rectangular. (Courtesy of the *Proceedings of the 13th International Modal Analysis Conference*, Society for Experimental Mechanics.)

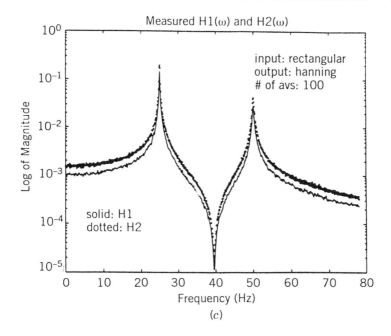

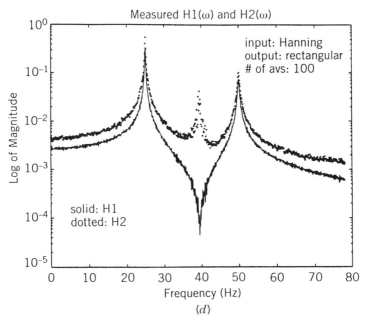

Fig. 7.7.5. (*Continued*).

function, since they have positive and negative values that affect the results given by Eq. (7.7.5). However, we have seen that the general trends are due to the filter leakage term.

7.7.4 Recommendations to Check for This Filter Error

When we are working with a large number of data channels, it is possible for human error to cause the windows on the input and output data channels to be mixed instead of being set the same with Hanning windows. Thus, we need to have a check list that allows us to evaluate quickly whether or not we have mixed the rectangular and the Hanning windows. The results in Fig. 7.7.5 give us the best indicators when we are working with random signals.

1. Look at the uncertainty in the valley region. This uncertainty can be due to two causes. First, we need to check to see that we are not having signal to noise problems with either the input or output signals. Second, we need to check to see that both channels have the same Hanning window.
2. Compare the FRF results obtained from $H_1(\omega)$ and $H_2(\omega)$. First, if they match well as in Fig. 7.7.5a, then any signal to noise problem is under control and both window functions are the same. Second, if the results show the valleys in $H_1(\omega)$ are peaks in $H_2(\omega)$, then we have a window mismatch as shown in Figs. 7.7.5b and 7.7.5d and we need to check the type of windows being used. Third, if the results are similar to those of Fig. 7.7.5c, we need to check that the window functions match and that there are no signal noise problems in the output signal since both mixed windows and output noise problems can cause $H_2(\omega)$ to be significantly larger than $H_1(\omega)$.

7.8 LOW FREQUENCY DAMPING MEASUREMENTS REVEAL SUBTLE DATA PROCESSING PROBLEMS

Not all data collection problems occur at high frequency. Digital processing of experimental data from 1 Hz signals that are used to calculate a structure's energy loss per cycle revealed two subtle errors that invalidated the test results.[26] One error was found to be a hardware error that occurred under a particular set of operating conditions of a digital processing oscillo-

---

[26]P. D. Holst, "A Method for Predicting and Verifying Damping in Structures Using Energy Loss Factor," M.S. Thesis, Iowa State University, Ames, IA, 1994.

DAMPING MEASUREMENTS REVEAL DATA PROCESSING PROBLEMS   523

scope. The second error involved a software error that resulted from a standard built-in program routine that was used to calculate the experimental results. We examine both errors and see how they affected the measured results. We need to keep in mind that many of our data processing concepts come from the calculus of continuous functions, but we end up processing the data with digital techniques that are discrete processes that may not necessarily be exactly the same as the continuous functions.

7.8.1 The Test Setup

It is desired to measure the energy dissipated per cycle in a simple structure at low frequencies. The structure that was used to evaluate the technique for measuring the energy dissipation per cycle consisted of a simply supported beam of length l, as shown in Fig. 7.8.1a, where load F is applied at the midspan and the resulting midspan deflection y is measured as shown. The simple supports consisted of knife edges, a mating rotation plate, and rollers in order to minimize support energy dissipation. It was later found that this support system had too much damping.

The basic idea is to determine the enclosed area in a force deflection diagram when the structure is subjected to a load of one or more cycles, as shown in Fig. 7.8.1b. The procedure for calculating the energy loss per cycle is rather straightforward, as outlined in a number of vibration books such as Thomson.[27] In this case, the energy dissipated per cycle is given by the integral over a complete cycle that is defined by

$$\Delta E = \int F \, dy = \int_0^T F\dot{y} \, dt \qquad (7.8.1)$$

where T is the cycle's period. It is evident from Eq. (7.8.1) that *all* we need to do is to measure force F and defection y, differentiate y with respect to time, multiply force F times velocity $\dot{y}$, and then integrate over one cycle.

A simulated plot of the measured force F (dotted line) and the force velocity product of $F\dot{y}$ (solid line) for typical data ranges is shown in Fig. 7.8.2 as a function of time. The continuous integration procedure over one period of T seconds in Eq. (7.8.1) can be extended to N cycles where the integration period becomes NT. The integration can be estimated by using Q data points separated by Δt seconds, so that digital summation over $Q - 1$ data points replaces the continuous integral. Thus Eq. (7.8.1)

---

[27]William T. Thomson, *Theory of Vibrations with Applications*, 4th ed., Prentice Hall, Englewood Cliffs, NJ, 1993.

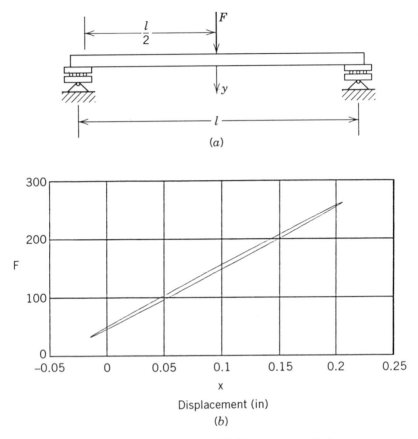

Fig. 7.8.1. (*a*) Schematic of test setup. (*b*) Force versus displacement to show hysteresis loop due to damping for phase shift $\phi = 2$ degrees.

becomes

$$\Delta E = \frac{1}{N} \int_0^{NT} F\dot{y}\, dt = \frac{\Delta t}{N} \sum_{i=1}^{Q-1} (F\dot{y})_i \qquad (7.8.2)$$

It should be clear from Eq. (7.8.2) and Fig. 7.8.2, that one should pick points A and B as the integration region, as opposed to points C and D since A and B have values near zero while the values at C and D are large. Thus an error of picking Q one point too large or too small at points A and B is insignificant when compared to the same error of plus or minus one point in selecting Q when using points C and D.

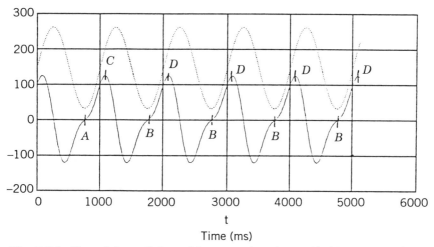

Fig. 7.8.2. Plot of force F (pounds) versus time (seconds) (dashed curve) and instantaneous power ($f\dot{y}$ in.lb/s) (solid curve) for phase shift $\phi = 2$ degrees.

7.8.2 The Hardware Error

Tests were run on an aluminum beam, a steel beam, and a fiberglass beam. Each beam had nearly the same cross section with the same nominal outside dimensions of 1×1 in. (25.4×25.4 mm). The tests were all run at the same frequency of 1 Hz. The support span length l was 28.74 in. (730 mm). The initial test results showed that the damping was nearly the same for each beam so it became apparent that something was wrong. One reason could be that the supports provided so much damping that support damping overshadowed the beam's damping. Another reason might be that something was wrong with the measurement technique and/or the data processing procedure.

In order to evaluate the measurement technique/data processing procedure, the sample rate was changed. When this was done, a plot of the force versus displacement showed larger hysteresis loops with increasing time between data samples, a result that makes absolutely no sense. What could be wrong?

The measurement system consisted of a Norland 3001 digital oscilloscope that measured the force voltage and the linear variable differential transformer (LVDT) voltage using the simultaneously sampling 12 bit A/D converters associated with channels 3 and 4. The data was scaled into force and displacement units, and then processed using Eq. (7.8.2). Note that channels 1 and 2 were not used since they had 10 bit A/D converters.

In order to investigate what was happening, a 1 Hz, 0.8 volt sine wave

526 THE APPLICATION OF BASIC CONCEPTS TO VIBRATION TESTING

was the input for both channels 3 and 4, which were DC coupled. When channel 3 data was plotted against channel 4 data, the hysteresis loop became larger with larger sample times and smaller with smaller sample times, a totally unexpected result. A closer examination of the data showed that channel 4 was shifted one sample time to the left compared to channel 3. When channel 4 was shifted one sample time to the right, there was no hysteresis loop, as one would expect for correlated in phase signals.

This particular instrument had been used for over 12 years and no one had observed such behavior. Further experimentation revealed that this behavior would occur only under a special set of circumstances. This instrument has an *active-inactive* switch for each data channel, in order to inform the instrument which channels have active data on them. For this case, both channels 1 and 2 were in the *inactive* position while channels 3 and 4 switches were in the *active* position, since only two channels of data were being utilized. When these switches were in these positions, channel 4 data was time shifted one Δt position relative to channel 3 data. However, if either of channel 1 or 2 had its switch set on *active*, the time shift did not occur. Thus the *hardware* problem was identified and eliminated by selecting a proper set of *active-inactive* switch settings.

7.8.3 A Software Problem

When the experiments were repeated using the aluminum beam with different sample times, it was found that the results were still sample time dependent, even though less dependent than previously before the hardware problem was solved.

In order to investigate the reason for the continued sensitivity to sample time, the data processing procedure was simulated digitally in order to understand what could be happening.

The measured force is represented in the digital domain by

$$F_p = F_0 + F_1 \sin(\omega t_p) \qquad (7.8.3)$$

where F_0 is the offset force.
    F_1 is the magnitude of the force variation (range/2).
    ω is the frequency of 2π rad/s.
    t_p ($=\Delta t\, p$) is the *p*th time.
    Δt is the sample time between samples.
    p is the index variable that ranges from 0 to 1023, just as it does in the digital oscilloscope.

Similarly, the beam deflection is represented digitally by

$$y_p = y_0 + y_1 \sin(\omega t_p - \phi) \qquad (7.8.4)$$

where y_0 is the offset displacement.

y_1 is the magnitude of motion (range/2).

ϕ is the phase shift of the response relative to the input force that is a measure of the system damping.

The time derivative of Eq. (7.8.4) is given theoretically by

$$v_p = (\omega y_1) \cos(\omega t_p - \phi) \qquad (7.8.5)$$

where (ωy_1) is the magnitude of the velocity.

The Norland computes the time derivative using a standard digital formula for velocity that is given by

$$u_p = \frac{y_{p+1} - y_p}{\Delta t} \qquad (7.8.6)$$

It is clear from Eq. (7.8.6) that it predicts the velocity at time t_p based on what is going to happen and is thus biased in a consistent manner toward a slope that occurs in the future. Similarly, if we use $(y_p - y_{p-1})$ in place of $(y_{p+1} - y_p)$ in Eq. (7.8.6), we would have a slope that is biased to the past. In either case, the velocity is either biased toward the past or the future but not the present. We can also compute the slope, and hence the velocity, by using a more balanced center point slope where

$$w_p = \frac{y_{p+1} - y_{p-1}}{2 \Delta t} \qquad (7.8.7)$$

Equation (7.8.7) is more balanced and should not have a consistent bias error since it is based on the average slope at time t_p.

To check out this error proposition, the error between the actual velocity from Eq. (7.8.5) and the Norland's estimate from Eq. (7.8.6) is computed using

$$EN_p = v_p - u_p \qquad (7.8.8)$$

while the error between the actual valocity from Eq. (7.8.5) and center point estimate of velocity as given by Eq. (7.8.7) is given by

$$Ec_p = v_p - w_p \qquad (7.8.9)$$

Equations (7.8.8) and (7.8.9) are plotted in Fig. 7.8.3a, from which it is seen that Eq. (7.8.7) is by far the better velocity estimate, compared to the standard formula of Eq. (7.8.6). The force F and the Norland formula error E_N (actually 2000 times E_N) from Eq. (7.8.8) are plotted versus

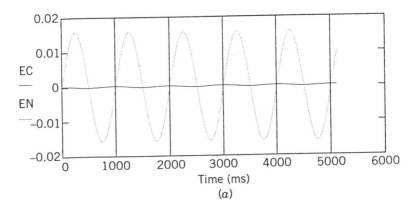

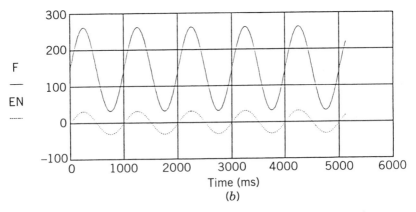

Fig. 7.8.3. (a) Comparison of velocity errors E_c (solid) and E_N (dashed) compared to 0.691 ips. (b) Force F (solid) and 2000 times E_N (dashed) versus time to show correlation and biased error with force.

time in Fig. 7.8.3b. It is clear that the force and derivative error are in phase so that the error in the product of $F\dot{y}$ must always have the same sign. This means that an integration in time over one cycle will not average out the error term, so its effects will be quite evident in the integrated results.

Then the energy dissipated per cycle was computed using the theoretical velocity from Eq. (7.8.5), the center point velocity from Eq. (7.8.7), and the Norland's velocity estimate from Eq. (7.8.6). The results are shown in Table 7.8.1 for sample interval times ranging from 1 to 20 ms. It is clearly seen that the center point and theoretical velocity based results in columns 2 and 3 agree with one another, while the Norland's formula causes significant variation and errors to occur. The approximate maxi-

TABLE 7.8.1. Energy Dissipated per Cycle and Maximum Velocity Error

Sample time Δt (ms)	Energy Dissipated per Cycle			Max. Vel. Error
	$E_{D7}{}^a$ (in.-lb)	$E_{D5}{}^b$ (in.-lb)	$E_{D6}{}^c$ (in.-lb)	$E_N{}^d$ (in./second)
1	1.285	1.285	1.161	0.003
5	1.285	1.285	0.661	0.015
10	1.285	1.285	0.0371	0.030
20	1.282	1.285	−1.210	0.060

[a]Based on average velocity from Eq. (7.8.7).
[b]Based on theoretical velocity from Eq. (7.8.5).
[c]Based on biased velocity from Eq. (7.8.6).
[d]Maximum velocity in biased velocity.

mum value of E_N for each sample interval time is shown in last column. It is clear that the error starts at 0.003 when $\Delta t = 1$ ms per sample interval and that there are about 1000 data points in a cycle. This error increases 20 times when $\Delta t = 20$ ms per sample interval so that only 50 data points describe a single cycle. It is clear that this systematic error causes the integrated results to reduce with increasing values of Δt and actually change sign when $\Delta t = 20$ ms. Thus the method employed in computing derivatives is extremely important in this case.

When the Norland was reprogrammed to use Eq. (7.8.7) in order to correct for this software error and the channel *activity-inactivity* switches were properly set, the test results showed remarkable consistency of energy lost per cycle for a given beam under a given load cycle regardless of the sample time employed. Thus the *software* error is clearly identified and corrected for. It was only under these new calculation conditions that is was possible to properly evaluate the support problem.

7.8.4 Another Common Measurement Error Source

Often we purchase multichannel data gathering systems to plug into our computer motherboards that sample from one channel to the next in a sequential manner while employing a multiplexer. This means that we do not have data that is time correlated between channels since each channel is sampled at a slightly different time. How can this time shift problem be avoided? Note that we cannot shift all data one data point in this case, as we could with the Norland, since the data is not sampled at the same time as in the case of the Norland instrument, but rather, at slightly different times.

When we have only a few data channels, as in this example of measuring damping, we can employ double sampling of one or more of the data channels. As an example of double sampling, consider that we first sample the displacement, then the force, and then the displacement again. Now, if we average the two displacement readings, namely, one on either side of the force reading, we have the best estimate of the position when the force is read. Then we should be able to proceed as outlined above, and will have this time shift problem well in hand.

Whether sequential sampling of channels is a problem or not depends on the channel sample rate, as opposed to how often the channels are sampled, as well as what we plan to do with the data. For example, 100,000 samples/second is a rather common sample rate. This means that there is a 10 μs time between data samples in a given data set. The data set of, say, 64 channels may be gathered at 100 times/second; that is, the data set is taken every 10 ms. However, with 64 data channels, there is a 0.64 ms time difference between the first and last channel. This may be sufficient time difference to cause serious data analysis problems and cannot be ignored. On the other hand, if we were using six data channels, the time difference of 0.06 ms between the first and last channels may not be significant compared to the 10 ms between data sets. The significance of this data delay depends on how the data is used in any data processing.

This testing experience again reveals the problems that can arise as we go from the continuous conceptual models learned in calculus to digital data processing. In principle, it appears as though we are doing the calculations correctly. However, a systematic bias error, as in the velocity estimate, was found to be particularly detrimental in this case where we were trying to find a small quantity in the midst of large numbers. The systematic bias error was in phase with the signal and caused large errors to occur.

This example indicates that we need to be vigilant in using our equipment since even well functioning hardware can contain the seeds of improper functioning under certain conditions that were not foreseen by its designer. Thus users must be alert to what they are doing.

7.9 SUMMARY

The goal of this chapter has been to put all of the elements discussed in the previous chapters to work in typical test environments. In that context, we have considered the sudden release or step relaxation method of excitation, which required careful signal conditioning in order to overcome the limitation of the length of time that history can be recorded in many instances, as well as the requirements for the FFT calculation process. It has also been demonstrated that unwanted inputs can cause interesting results of perfect coherence to occur when we take only one set of data.

This experience shows that coherence is a statistical concept that requires several sets of data before it starts to have meaning.

The forced response of a simple support beam mounted on a vibration exciter has shown us that the type of excitation is proportional to the test item's mass distribution. The effects of transducer mass on the measurements has been explored, and a published correction method has been shown to have significant limitations. It has been shown that the best way to determine the effect of transducer mass on the measured natural frequencies is to add a second mass that is equivalent to that of the accelerometer. In this way, the effect of both mass and mass moment of inertia are accounted for. If the frequency shifts are too large, then smaller transducers must be employed.

When doing impulse testing, it was found that impulse duration must be short enough to generate significant input over the range of measured frequencies. We have found that input and output signal windowing is an important consideration in reducing input noise and output leakage. The exponential window causes each mode to have too much damping, which must be accounted for in interpreting the end results. This section was followed by one that spent considerable effort exploring the effects of different windows and the selection of trigger levels and slope on the measured results. It has been demonstrated that improper use of windows can cause measured data to be rotated in the Nyquist plots. This section ended with a set of recommended procedures to follow when conducting such tests.

The implementation of a vibration exciter driving a structure has been demonstrated using a free-free beam with point loads. One of the first choices is the type of excitation signal to be used in order to achieve the desired test results. The sinusoidal, chirp, and pseudorandom signals have been considered. Consideration of theoretical and experimental models has shown that force dropout occurs at each resonance when the power amplifier is operated in the voltage mode. It has been shown that the force transducer's base bending strains, when mounted at the midpoint, and bending moment sensitivity, when mounted at the end of the beam, are the most likely sources of discrepancy between the two test results. In addition, it has been found that the improper selection of windows causes the FRF valleys to become peaks. Recommendations for detecting this misapplication of data windows have been made.

Finally, a subtle data processing problem that is associated with using digital techniques to simulate the derivative of a signal has been explored. We need to keep in mind that many of our data processing concepts come from the calculus of continuous functions but we calculate the results using discrete digital processes.

In nearly every one of these examples, the proper use of the exciter and the frequency analyzer has proven to be crucial in obtaining reliable results—if it is assumed that the instrumentation is properly selected and implemented. However, it should be clear that all elements must be taken into account in trying to validate the quality of the experimental data that is obtained.

REFERENCES

1. Cafeo, J. A. and M. W. Trethewey, "Impulse Test Truncation and Exponential Window Effects on Spectral and Modal Parameters," *Proceedings of the 8th International Modal Analysis Conference*, Vol. 1, Orlando, FL, 1990, pp. 234–240.
2. Cappa, P. and K. G. McConnell, "Base Strain Effects on Force Measurements," *Proceedings of the Spring Society for Experimental Mechanics*, Baltimore, MD, June 1994, pp. 520–530.
3. Clark, R. L., A. L. Wicks, and W. J. Becker, "Effects of an Exponential Window on the Damping Coefficient," *Proceedings of the 7th International Modal Analysis Conference*, Vol. 1, Las Vegas, NV, 1989, pp. 83–86.
4. Dossing, Ole, "The Enigma of Dynamic Mass," *Sound and Vibration*, Nov. 1990, pp. 16–21.
5. Halvorsen, W. G. and D. L. Brown, "Impulse Technique for Structural Frequency Response Testing," *Sound and Vibration*, Nov. 1977, pp. 8–21.
6. Holst, P. D., "A Method for Predicting and Verifying Damping in Structures Using Energy Loss Factor," M.S. Thesis, Iowa State University, Ames, IA, 1994.
7. Lauffer, J., et al., "Mini-Modal Testing of Wind Turbines Using Novel Excitation," *Proceedings of the 3rd International Modal Analysis Conference*, Feb. 1985, pp. 451–458.
8. Martinez, M. and P. Quijada. "Experimental Modal Analysis in Offshore Platforms,", *Proceedings of the 9th International Modal Analysis Conference*, Florence, Italy, Apr. 1991, pp. 213–218.
9. McConnell, K. G. and P. J. Sherman, "FRF Estimation Under Non-Zero Initial Conditions," *Proceedings of the 11th International Modal Analysis Conference*, Kissimmee, FL, Feb. 1993, pp. 1021–1025.
10. McConnell, K. G. and P. S. Varoto, "Pseudo-Random Excitation Is More Cost Effective than Random Excitation," *Proceedings of the 12th International Modal Analysis Conference*, Vol. 1, Honolulu, HA, 1994, pp. 1–11.
11. McConnell, K. G. and P. S. Varoto, "The Effects of Window Functions and Trigger Levels on FRF Estimations from Impact Tests," *Proceedings of the 13th International Modal Analysis Conference*, Nashville, TN, Feb. 1995.
12. McConnell, K. G. and P. S. Varoto, "The Effects of Windowing on FRF Estimates for Closely Spaced Peaks and Valleys," *Proceedings of the 13th International Modal Analysis Conference*, Nashville, TN, Feb. 1995.
13. Mutch, C., et al., "The Dynamic Analysis of a Space Lattice Structure via

the Use of Step Relaxation Testing," *Proceedings of the 2nd International Modal Analysis Conference*, Orlando, FL, Feb. 1984, pp. 368–377.
14. Thakkar, R. B., "Exact Sinusoidal Simulation of Fatigue under Gaussian Narrow Band Random Loading," Ph.D. Thesis, Iowa State University, Ames, IA, 1972.
15. Trethewey, M. W. and J. A. Cafeo, "Tutorial: Signal Processing Aspects of Structural Impact Testing," *The International Journal of Analytical and Experimental Modal Analysis*, Vol. 7, No. 2, 1992, pp. 129–149.
16. Wei, R. C., "Structural Wave Propagation and Sound Radiation Study Through Time and Spatial Processing," Ph.D. Thesis, Iowa State University, Ames, IA, 1994.

Some Swat Method References

17. Bateman, V. I., T. G. Carne, and D. M. McCall, "Force Reconstruction for Impact Tests of an Energy-Absorbing Nose," *International Journal of Analytical and Experimental Modal Analysis*, Vol. 7, No. 1, Jan. 1992.
18. Carne, T. G., V. I. Bateman, and C. R. Dohrmann, "Force Reconstruction Using the Inverse of the Mode-Shape Matrix," *Proceedings of the 13th Biennial Conference on Mechanical Vibration and Noise*, DE-Vol. 38, ASME, Sept. 1991, pp. 9–16.
19. Carne, T. G., R. L. Mayes, and V. I. Bateman, "Force Reconstruction Using the Sum of Weighted Accelerations Technique,", *Proceedings of the 12th International Modal Analysis Conference*, Honolulu, HA, Feb. 1994, pp. 1054–1062.
20. Gregory, D. L., T. G. Priddy, and D. O. Smallwood, "Experimental Determination of the Dynamic Forces Acting on Non-Rigid Bodies," SAE Technical Paper Series, Paper No. 861791, Aerospace Technology Conference and Exposition, Long Beach, CA, October 1986.
21. Mayes, R. L., "Measurement of Lateral Launch Loads on Re-Entry Vehicles Using SWAT," *Proceedings of the 12th International Modal Analysis Conference*, Honolulu, HA, Feb. 1994, pp. 1063–1068.
22. Priddy, T. G., D. L. Gregory, and R. G. Coleman, "Strategic Placement of Accelerometers to Measure Forces by the Sum of the Weighted Accelerations," SAND87-2567.UC.38, National Technical Information Service, 5285 Port Royal Road, Springfield, VA 22162, Feb. 1988.
23. Priddy, T. G., D. L. Gregory, and R. G. Coleman, "Measurement of Time-Dependent External Moments by the Sum of the Weighted Accelerations," SAND88-3081.UC.38, National Technical Information Service, 5285 Port Royal Road, Springfield, VA 22162, Jan. 1989.
24. Steven, K. K., "Force Identification Problems—An Overview," *Proceedings of the 1987 SEM Spring Conference on Experimental Mechanics*, Houston, TX, June 1987, pp. 838–844.

8 General Vibration Testing Model: From the Field to the Laboratory

8.1 INTRODUCTION

The objective of a test program is to simulate a specified environment at some known overtest. While this statement sounds simple, the process of measuring field data, generating a satisfactory test specification, and conducting a suitable laboratory test is not simple. There are many reasons

A U.S. Navy SH-2G helicopter is undergoing field vibration tests in order to establish operating vibration levels that are needed for laboratory testing of electronic and mechanical subsystems. (Photo Courtesy of *Sound and Vibration*).

536 GENERAL VIBRATION TESTING

The dynamic response of the wheel suspension system is being determined in the laboratory with frequency response functions and multiple coherences being displayed on the computer-aided testing system. The required laboratory input must come from appropriate field tests for certain laboratory tests. (Photo Courtesy of *Sound and Vibration*).

for running vibration tests. Some reasons are: (1) to qualify a test item to meet a set of specifications, (2) to verify a theoretical model of a test item, (3) to determine the test item's modes of failure, (4) to develop adequate production quality assurance test methods, and (5) to develop dynamic inputs from field data for use with finite-element analysis or other dynamic design tools. It is the objective of this chapter to identify some of the issues that we are involved with in this process of simulation of a field vibration environment in the laboratory.

These issues were dramatically brought home to me while reviewing the shock and vibration testing procedures at Sandia National Laboratories during the summers of 1988 and 1989. In addition, I had the honor of presenting the *Murray Lecture* to the 1994 spring Society for Experimental Mechanics meeting in Baltimore, MD, where I presented a subset of this

INTRODUCTION 537

topic. This chapter is composed directly from notes that I developed while at Sandia Laboratories[1] and from the materials developed for the Murray Lecture.[2]

I first became aware of some issues that are involved in going from field test to laboratory simulation at a Shock and Vibration Symposium over 20 years ago. The essence of the story is that the U.S. Navy wanted a successful installation of an electronic device mounted on top of a fairly large gun turret. Apparently, Navy personnel measured the vertical acceleration on top of the turret where the electronic device was to be mounted while several shells were fired from the gun. These acceleration records were used to develop the *dynamic environment*. The electronic device was designed, tested, and passed this dynamic environment. Yet when it was installed, the device was quickly destroyed by shock loading when the gun was fired. Obviously, the actual dynamic environment was not translated into an adequate test specification. Two possible reasons why there were problems are: (1) the vertical acceleration on top of the turret was probably significantly less than the horizontal acceleration resulting from the reaction to firing the gun at a low elevation, and (2) the electronic device was mounted so that it presented a significant frontal area to the gun's air blast. Clearly, the procedures employed were inadequate and caused unnecessary grief to all parties concerned.

Rogers et al.[3] describe a situation where the test specifications outlined in MIL-STD-810D[4] were inappropriate for a situation where the structure to be tested (called test item) was either an approximately 2000 or 4000 lb shipping container that was to be carried on board CH-47D helicopters. Field measurements with the shipping containers on board showed a reduction in the random auto-spectral density (ASD) in the 10 to 500 Hz frequency range from $2.88 g_{RMS}$ to $0.277 g_{RMS}$, a reduction of over 10 to 1. Similarly, the test specification has four discrete sinusoidal frequency inputs that were reduced by factors of 2.8, 16, 25, and 42, respectively. Obviously, the test item would have been needlessly overdesigned to meet the original dynamic environment that consisted of the floor vibration of an unloaded helicopter.

While visiting an automotive parts supplier, I noticed an interesting

---

[1] K. G. McConnell, "A General Vibration Testing Model: From the Field to the Laboratory," unpublished personal notes developed during summers of 1988 and 1989.

[2] K. G. McConnell, "From Field Vibration Data to Laboratory Simulation," *Experimental Mechanics*. Vol. 34, No. 3, Sept. 1994, *pp*. 1–13.

[3] J. D. Rogers, D. B. Beightol, and J. W. Doggett, "Helicopter Flight Vibration of Large Transportation Containers—A Case for Test Tailoring," *Proceedings, Institute of Environmental Sciences*, 1990, pp. 515–521.

[4] MIL-STD-810D, "Environmental Test Methods and Engineering Guidelines," July 19, 1983.

diagram mounted on the wall of the vibration exciter control room. This diagram showed the process for obtaining field acceleration data as an input source to the vibration exciter in order to test the company's product. Basically, the process was to mount an accelerometer on the surface of a vehicle where the company's product was to be mounted, and then, make a long time record. This time record was analyzed to obtain an ASD that was played back as an input to the test item. The supplier's engineering staff assumed that installing their product had no effect when the test item is attached to the vehicle. We find that there are situations where this process is valid and others where it is invalid. This type of results, with success in one case and failure in another, has led to considerable confusion. As we see, the assumption that this is a valid procedure is often incorrect.

These stories illustrate that significant questions need to be answered in planning vibration tests. These questions include:

1. What field test conditions should be used?
2. Where should transducers be placed on the test item to achieve the maximum useful information over a given frequency range?
3. How should field data be stored for future use?
4. What is the effect of changing boundary conditions between field and laboratory test environments?
5. Which testing procedures are best suited in simulating a given field environment?

This chapter explores these questions in an indirect manner by developing a simple linear model, then applying this model to a simple two DOF structural system in both the field environment and the laboratory testing environment, where several possible testing procedures are used, and finally developing a more general model to describe the more general case. The goal is to illuminate the kinds of testing problems that may occur as well as to understand the role that the dynamic characteristics of the test item, vehicle, and laboratory test equipment have on the field and laboratory vibration responses. The general input-output relationships that are employed in this chapter are now outlined.

8.1.1 General Linear System Relationships

The purpose of this section is to present the general linear input-output relationships for linear structures as shown in Fig. 8.1.1. The notation used here corresponds to the frequency-domain representation commonly employed in modal analysis, where we describe the input-output relationships as a function of frequency ω that is called the frequency response

INTRODUCTION 539

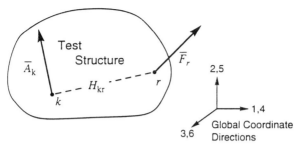

Fig. 8.1.1. Locations of input-output positions for measuring experimental accelerance FRF $[H(\omega)]$ of a structure in terms of global coordinates. (Courtesy of *Experimental Mechanics*, Society for Experimental Mechanics.)

function, or simply FRF. Hence, $F(\omega)$ is written as F, $X(\omega)$ as X, while $x(t)$ is written simply as x. These frequency components contain both real and imaginary parts so that magnitude and phase information is preserved.

The input at location r in Fig. 8.1.1 consists of two vectors—a force vector and a moment vector. This input is represented by the symbol $\bar{\mathbf{F}}_r$ $(=\bar{\mathbf{F}}_r(\omega))$. There are six potential components for each input location. These components are written in terms of the global directions of 1–3 and 4–6 as shown, so that F_1, F_2, F_3 are the force components and F_4, F_5, F_6 are the moment components.

Similarly, the structure's output motion at location k can be described in terms of two acceleration vectors, one for linear acceleration and the other for angular acceleration. This output acceleration is represented by the symbol $\bar{\mathbf{A}}_k$ $(=\bar{\mathbf{A}}_k(\omega))$. There are six potential components at each location k. These output components are written in terms of the global directions so that A_1, A_2, A_3 are the linear acceleration components and A_4, A_5, A_6 are the angular acceleration components. Each force component and each acceleration component are complex (magnitude and phase) functions that are frequency dependent. The relationship between these six input-output components can be written in matrix form as

$$\{A\}_k = [H_{pq}]_{kr}\{F\}_r \qquad (8.1.1)$$

for each value of k and r where p and q range from 1 to 6. Thus H_{kr} $(=H_{kr}(\omega))$ in Eq. (8.1.1) has the potential for 36 input-output FRF relationships for *each input* and *output location pair* used in describing a structure.

Additional implications of Eq. (8.1.1) can be seen if we assume that we have one accelerometer attached at location k with its primary sensing axis in the global 2 direction and a force transducer attached at location

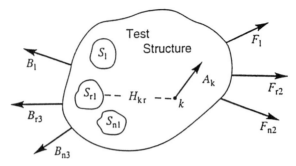

Fig. 8.1.2. Schematic of general test structure showing various sources of vibration excitation. (Courtesy of *Experimental Mechanics*, Society for Experimental Mechanics.)

r that measures the force in the global 1 direction. Then, from Eq. (8.1.1), the sensed acceleration becomes

$$A_{2_k} = H_{21_{kr}}F_{1_r} + \overbrace{H_{22_{kr}}F_{2_r} + H_{23_{kr}}F_{3_r} + H_{24_{kr}}F_{4_r} + H_{25_{kr}}F_{5_r} + H_{26_{kr}}F_{6_r}}^{\text{unaccounted for terms}}$$
(8.1.2)

where the $H_{xy_{kr}}$ ($= H_{xy}(\omega)$) is the FRF between global directions x and y for the points k and r. However, we usually interpret Eq. (8.1.2) according to a single relationship that is given by

$$A_k \cong H_{kr}F_r \qquad (8.1.3)$$

Clearly, the measured input is F_{1_k}, the output is A_{2_k}, and the assigned FRF function H_{kr} is contaminated by unknown and unaccounted for terms *since the measured response is due to all inputs*. Although we use the simpler form given by Eq. (8.1.3), we need to keep in mind that H_{kr} can be any one of 36 input-output relationships, as indicated by Eq. (8.1.1). Thus the situation is potentially far more complicated than we are treating it here. In addition, we should note that we use selected discrete points such as k and r to represent a continuous structure, so that our knowledge is further restricted by the potential sparsity of spatial data points.

Further insight into the type of problems we are dealing with can be obtained from Fig. 8.1.2, where a general structure is shown that has internal (S_r), external (F_r), and connector (interface) (B_r) excitation forces. In each case, these excitation sources can be either forces or moments. Each of these inputs contributes to the acceleration measured at location k so that applying Eq. (8.1.3) to each excitation source gives

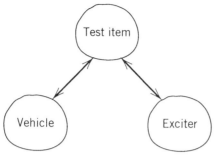

Fig. 8.1.3. Three structures involved in the process of going from field environment to laboratory environment. (Courtesy of *Experimental Mechanics*, Society for Experimental Mechanics.)

$$A_k = \sum_{r1}^{n1} H_{kr}S_r + \sum_{r2}^{n2} H_{kr}F_r + \sum_{r3}^{n3} H_{kr}B_r \qquad (8.1.4)$$
$$\text{internal} + \text{external} + \text{connector}$$

where index r varies to include all excitation sources of each type. Equation (8.1.4) shows us that the measured response in the field can be due to many types of sources. When the boundary conditions are changed, it is clear that the connector forces change and so must the output. When the structure has internal vibration sources, shutting them off changes the output acceleration. Hence, all three excitation sources may contribute to differences that occur between the field and laboratory environments. For convenience, we combine the internal and external excitation sources as nonconnector (or external) forces compared to connector (interface) forces.

8.1.2 Three Structures Involved in the Process

Three major structural systems are involved in most test situations: (1) test item, (2) vehicle, and (3) vibration exciter as shown in Fig. 8.1.3. The test item is attached to the vehicle in the field and to the exciter in the laboratory. The interaction of the test item with its field (vehicle) and laboratory (exciter) environments is the concern of this chapter. We define the dynamic characteristics of each structure in the frequency domain.

Finally, the reader is warned that there are a large number of equations in this chapter. At first glance, the chapter may appear to be simply a mathematical exercise. However, the intent is quite the contrary, for many equations are needed to describe the various options that are available in conducting vibration tests in the field and the laboratory environments. It is only through a comparison of the various idealized responses that the

542 GENERAL VIBRATION TESTING

implications of doing a test one way versus another become clear. In addition the reciting of results would leave the reader completely devoid of the rationale, as well as a feel for why certain variables play an important role.

8.2 A TWO POINT INPUT-OUTPUT MODEL OF FIELD AND LABORATORY SIMULATION ENVIRONMENTS

In this section, a simple two input-output linear model of each structure is used to describe field conditions in Fig. 8.2.1 and model laboratory conditions in Fig. 8.2.2. In the Fig. 8.2.1 model, the vehicle has two motions (Y_1 and Y_2) and two excitation forces (P_1 and P_2), while the test item has two motions (X_1 and X_2) and two excitation forces (F_1 and F_2). Similarly, in the Fig. 8.2.2 model, the exciter has two motions (Z_1 and Z_2) and two forces (Q_1 and Q_2), while the test item has two motions (U_1 and U_2) and two forces (R_1 and R_2). Subscript 1 is used to denote the interface (connection) point, while subscript 2 is used to identify the noninterface point. The interface connections share a common force and motion. These common forces and motions create the required boundary conditions for each environment. However, these boundary motions and forces can be quite different in the field and laboratory environments.

8.2.1 Notation Scheme

The complexity of dealing with three structures that can interact in two combinations can make it difficult to keep track of which situation is being talked about. The following scheme is adopted for this chapter, as illustrated in Figs. 8.2.1 and 8.2.2.

Fig. 8.2.1. Schematic depicting field environment.

Fig. 8.2.2. Schematic depicting laboratory environment.

FIELD AND LABORATORY SIMULATION ENVIRONMENTS   543

Vehicle   The vehicle is described by

$$\{Y\} = [V]\{P\} \qquad (8.2.1)$$

where $\{Y\} = \{Y(\omega)\}$ is the motion frequency spectrum in terms of either displacement, velocity, or acceleration.
$[V] = [V(\omega)]$ is the vehicle's input-output FRF matrix in terms of either receptance, mobility, or accelerance depending on $\{Y\}$ being displacement, velocity, or acceleration.
$\{P\} = \{P(\omega)\}$ is the input force frequency spectrum.
The vehicle's input-output FRF matrix corresponds to the bare vehicle condition, that is the test item is not attached.

Test Item   The test item needs to be described by a field model and a laboratory model since these structures can have slightly different characteristics due to manufacturing tolerances and so on. Consequently, we use

$$\{X\} = [T]\{F\} \qquad (8.2.2)$$

for the field environment
where $\{X\} = \{X(\omega)\}$ is the test item motion (displacement, velocity, or acceleration).
$[T] = [T(\omega)]$ is the freely supported test item FRF (receptance, mobility, or accelerance).
$\{F\} = \{F(\omega)\}$ is the input force frequency spectrum.
Similarly, for the laboratory environment we have

$$\{U\} = [T^*]\{R\} \qquad (8.2.3)$$

where $\{U\} = \{U(\omega)\}$ is the test item motion (displacement, velocity, or acceleration).
$[T^*] = [T^*(\omega)]$ is the freely supported test item FRF (receptance, mobility, or accelerance).
$\{R\} = \{R(\omega)\}$ is the input force frequency spectrum.
It is assumed that the number of input-output points is the same for both test item environments.

Exciter   The exciter is described by

$$\{Z\} = [E]\{Q\} \qquad (8.2.4)$$

where $\{Z\} = \{Z(\omega)\}$ is the motion frequency spectrum (displacement, velocity, or acceleration).

$[E] = [E(\omega)]$ is the exciter's input-output FRF matrix (receptance, mobility, or acceleration.
$\{Q\} = \{Q(\omega)\}$ is the input force frequency spectrum.

8.2.2 The Field Environment

The vehicle's interface motion at location 1 is obtained from Eq. (8.2.1) as

$$Y_1 = V_{11}P_1 + V_{12}P_2 = V_{11}P_1 + Ye_1 \qquad (8.2.5)$$

where Ye_1 is the interface motion due to the nonconnector forces such that

$$Ye_1 = V_{12}P_2 \qquad (8.2.6)$$

in this case. It is clear from Eqs. (8.2.5) and (8.2.6) that the vehicle's interface motion is due to interface (connector) force P_1 and non-connector force P_2. Similarly, the test item's interface motion is obtained from Eq. (8.2.2) as

$$X_1 = T_{11}F_1 + T_{12}F_2 = T_{11}F_1 + Xe_1 \qquad (8.2.7)$$

where Xe_1 is the interface motion due to nonconnector forces so that

$$Xe_1 = T_{12}F_2 \qquad (8.2.8)$$

It is clear from Eqs. (8.2.7) and (8.2.8) that the interface motion is due to connector force F_1 and nonconnector force F_2.

The interface boundary conditions between the vehicle and test item can be written as

$$X_1 = Y_1 \qquad (8.2.9)$$

for motions and

$$P_1 = -F_1 \qquad (8.2.10)$$

since the forces are positive in the direction of positive motion as shown in Fig. (8.2.1). Now, substituting Eq. (8.2.10) into Eq. (8.2.5) and equating Eqs. (8.2.5) and (8.2.7) according to Eq. (8.2.9) gives the corresponding single interface force as

$$F_1 = \frac{V_{12}P_2 - T_{12}F_2}{T_{11} + V_{11}} = \frac{Ye_1 - Xe_1}{FI} \qquad (8.2.11)$$

which is also written in terms of the original interface motions from Eqs. (8.2.6) and (8.2.8). We define the *field interface dynamic characteristic FI* as

$$FI = T_{11} + V_{11} \qquad (8.2.12)$$

This dynamic characteristic is the sum of the two driving point characteristics, one from each object before they are connected. Note, all FRF functions used in these equations must be based on the same dynamic characteristic such as receptance, mobility, or acceleration.

The interface motion of the combined (connected) system is then given by

$$X_1 = T_{11}F_1 + T_{12}F_2 = \frac{T_{11}}{FI}(Ye_1 - Xe_1) + Xe_1 \qquad (8.2.13)$$

which can also be written as

$$X_1 = \frac{T_{11}}{FI} Ye_1 + \frac{V_{11}}{FI} Xe_1 \qquad (8.2.14)$$

These two simple equations clearly illustrate the subtle interaction of the vehicle and test item dynamic characteristics in terms of either forces or motions. For example, in Eq. (8.2.13), the relative interface motion $(Ye_1 - Xe_1)$ is due to the nonconnector forces, a motion that exists when the two structures are not connected together. This motion is multiplied by the test item's driving point characteristic T_{11} and is divided by the *field interface* dynamic characteristic *FI*. We also see in Eq. (8.2.14) that the test item characteristic T_{11} is multiplied by the motion Ye_1 that is induced by the external force P_2 acting on the vehicle. Similarly, the vehicle characteristic V_{11} is multiplied by the test item motion Xe_1 that is induced by the external force F_2 acting on the test item. When $F_2 = 0$, then the interface motion is due only to the vehicle excitation sources.

The motion at point 2 in the test item is found by substituting Eq. (8.2.11) into Eq. (8.2.2) to obtain

$$X_2 = T_{21}F_1 + T_{22}F_2 = \frac{T_{21}}{FI}(Ye_1 - Xe_1) + Xe_2 \qquad (8.2.15)$$

where Xe_2 is the motion at location 2 due to the external force F_2, namely, $Xe_2 = T_{22}F_2$. It is obvious that the external force F_2 is the cause of two motions in Eq. (8.2.15); Xe_1 and Xe_2.

If the ratio of X_2 to X_1 is solved for from Eqs. (8.2.13) and (8.2.15), there results

$$\frac{X_2}{X_1} = \frac{(T_{21}/FI)(Ye_1 - Xe_1) + Xe_2}{(T_{11}/FI)(Ye_1 - Xe_1) + Xe_1} \quad (8.2.16)$$

Equation (8.2.16) clearly shows that the results are dependent only on the test item characteristics T_{11} and T_{21} if $F_2 = 0$ so that Xe_1 and Xe_2 are both zero. In this special case, the field measured FRF shows the field dynamic characteristics of the test item alone since FI cancels when Xe_1 and Xe_2 are zero. In this case, Eq. (8.2.16) gives the traditional transmissibility ratio that is given by

$$TR_{21} = \frac{T_{21}}{T_{11}} \quad (8.2.17)$$

This result shows that it is possible to determine the test item's transmissibility ratio characteristics only in the absence of external forces. If F_2 is not equal to zero, then Eq. (8.2.16) clearly shows that the FRF is dependent on the external load as well as the vehicle and test item dynamic characteristics of V_{11} and T_{11}, and no general conclusions can be drawn.

8.2.3 Laboratory Environment

Following the steps of Section 8.2.2 above, the interface motions resulting from external forces on the exciter and the test item are given by

$$Ze_1 = E_{12}Q_2 \quad (8.2.18)$$

for the exciter and

$$Ue_1 = T_{12}^* R_2 \quad (8.2.19)$$

for the test item. The resulting interface force when these two are connected together becomes

$$R_1 = \frac{E_{12}Q_2 - T_{12}^* R_2}{T_{11}^* + E_{11}} = \frac{Ze_1 - Ue_1}{LI} \quad (8.2.20)$$

where the denominator term is the *laboratory dynamic characteristic* of the connection interface given by

$$LI = T_{11}^* + E_{11} \qquad (8.2.21)$$

The laboratory interface dynamic characteristic depends on the driving point dynamic characteristics of both the test item used in the laboratory and the exciter employed.

The interface motion U_1 in the laboratory is given by

$$U_1 = T_{11}^* R_1 + T_{12}^* R_2 = \frac{T_{11}^*}{LI}(Ze_1 - Ue_1) + Ue_1 \qquad (8.2.22)$$

which can also be written in a form similar to Eq. (8.2.14):

$$U_1 = \frac{T_{11}^*}{LI} Ze_1 + \frac{E_{11}}{LI} Ue_1 \qquad (8.2.23)$$

Similarly, the motion at point 2 in the test item becomes

$$U_2 = T_{21}^* R_1 + T_{22}^* R_2 = \frac{T_{21}^*}{LI}(Ze_1 - Ue_1) + Ue_2 \qquad (8.2.24)$$

where Ue_1 and Ue_2 are due to the external force R_2 acting on the test item. The ratio of motion at location 2 relative to the interface motion at location 1 becomes

$$\frac{U_2}{U_1} = \frac{(T_{21}^*/LI)(Ze_1 - Ue_1) + Ue_2}{(T_{11}^*/LI)(Ze_1 - Ue_1) + Ue_1} \qquad (8.2.25)$$

The external force R_2 is the excitation source for motions Ue_1 and Ue_2 in Eq. (8.2.25). Consequently, when R_2 is zero, the motion ratio reduces to the test item FRFs T_{11}^* and T_{21}^*, which are the same transmissibility relationships described by Eq. (8.2.17) for the field environment. Note that LI cancels under these conditions. Hence, it is possible to compare the field transmissibility ratio to the laboratory transmissibility ratio if the external force F_2 in the field and R_2 in the laboratory are both zero. Otherwise, it is clear from Eqs. (8.2.16) and (8.2.25) that there is a mixture of information that may not be the same in the laboratory and field environments.

8.2.4 Discussion of Elementary Results

We have begun to see the complexity of the interaction that takes place between the test item and its two mating structures, that is, vehicle and exciter. It is clear from these equations that driving point and transfer FRFs combine in interesting ways to predict the combined system behavior where the connection forces and motions are dramatically different from the unconnected motions. Hence, taking bare vehicle data may not be a good idea unless the vehicle's driving point FRF is obtained as well.

Recently, Sweitzer[5] completed a study of the basic issues developed here from a slightly different point of view. He thoroughly explored some additional ideas but his basic results and those contained here are in agreement once the fact that he used Thevenin's theorem is taken into account.

8.3 LABORATORY SIMULATION SCHEMES BASED ON THE ELEMENTARY MODEL

Six different testing and control scenarios are considered in this section. The simple two variable linear input-output dynamic model used in Section 8.2 and shown in Figs. 8.2.1 and 8.2.2 is used in this section. In the first set of three scenarios, the external forces are zero in both the field and laboratory environments. This set of conditions corresponds to the situation where the interface forces and motions are dominant over any external loads. In the second set of three scenarios, an external force exists in the field environment while no external force is acting in the laboratory environment. This set of conditions corresponds to trying to match field environments with a single laboratory input while ignoring the external loads, a situation that occurs far too often.

A typical simple vibration test system with a feedback control scheme is shown in Fig. 8.3.1. This system consists of a controller (either analog or digital), a power amplifier, an exciter, a test item, transducers to measure interface force FI, interface acceleration A_1, and output acceleration A_2. In the test scenarios that follow, we refer to this as our test system where the controller drives the power amplifier and exciter so that the specified interface force F_1, interface acceleration A_1, or output

---

[5]K. A. Sweitzer, "Vibration Models Developed for Subsystem Test," M.S. Thesis, Syracuse University, Utica, NY, May 1994.

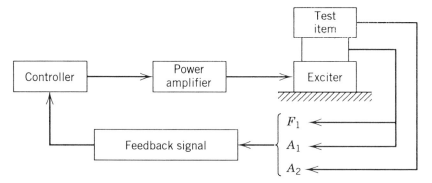

Fig. 8.3.1. Schematic of simple test system showing main components and various feedback choices for six different test scenarios.

acceleration A_2 is achieved dependent on which variable is selected as the feedback variable.

Each scenario corresponds to one of the common control schemes that are employed in vibration testing. The first scenario is to control the test system so that the interface motion is the same in the field and laboratory environments. The second scenario is to control the test system so that the interface force is the same in the laboratory and field environments. The third scenario is to control the test system so that the output response at point 2 in the test item structure is the same in the field and laboratory environments. In addition, the field external force F_2 is to be zero for the first three test scenarios and to be nonzero for the second set of three test scenarios. The laboratory external force R_2 is zero for all scenarios considered here. In this way, the effect of the lack of external force in the laboratory testing becomes evident.

8.3.1 Test Scenario Number 1—Matched Interface Motions and No External Forces

In this scenario, the external forces F_2 (field) and R_2 (laboratory) are zero. The interface motion is controlled to be the same in the field and laboratory environments so that the imposed testing condition is

$$U_1 = X_1 \tag{8.3.1}$$

Then, using Eqs. (8.2.14), (8.2.18), and (8.2.22) along with Eq. (8.3.1), we obtain the required bare table exciter motion as

$$Ze_1 = E_{12}Q_2 = \begin{bmatrix} LI \\ T_{11}^* \end{bmatrix} X_1 = \begin{bmatrix} LI \\ FI \end{bmatrix}\begin{bmatrix} T_{11} \\ T_{11}^* \end{bmatrix} Ye_1 \qquad (8.3.2)$$

where it is noted that $Xe_1 = 0$ because $F_2 = 0$ and $Ue_1 = 0$ because $R_2 = 0$. The bare table motion is the motion that exists at the exciter connector interface when the test item is not attached to the exciter system. Equation (8.3.2) shows the relationship between the various field and laboratory test parameters. X_1 is the actual measured field motion. The exciter bare table motion Ze_1 is related to the unloaded vehicle motion Ye_1 (the vehicle motion that occurs when the test item is not present) by the two interface dynamic properties FI and LI as well as by the driving point dynamic characteristic of the test item when in the field and the laboratory. These dynamic characteristics can vary widely between their peaks and valleys over the frequency range of interest.

The required exciter driving force Q_2 is obtained from Eq. (8.3.2) as

$$Q_2 = \begin{bmatrix} LI \\ T_{11}^* \end{bmatrix}\frac{X_1}{E_{12}} = \begin{bmatrix} T_{11}^* + E_{11} \\ T_{11}^* E_{12} \end{bmatrix} X_1 \qquad (8.3.3)$$

The power amplifier, the vibration exciter, and the control system must be capable of producing this amount of force over the range of test frequencies in order to simulate the field interface motion X_1. Recall that there are electrical limitations where force dropout occurs at test system resonance due to the back emf (back electromotive force) generated by the exciter table motion, so the required force may not be available.

When Eqs. (8.2.20) and (8.3.2) are combined, the required laboratory interface force becomes

$$R_1 = \frac{Ze_1}{LI} = \frac{X_1}{T_{11}^*} = \begin{bmatrix} T_{11} \\ T_{11}^* \end{bmatrix}\begin{bmatrix} Ye_{11} \\ FI \end{bmatrix} = \begin{bmatrix} T_{11} \\ T_{11}^* \end{bmatrix} F_1 \qquad (8.3.4)$$

From Eq. (8.3.4) it is clearly seen that laboratory interface force R_1 will be equal to the field interface force F_1 when the dynamic characteristics T_{11} and T_{11}^* are equal. Again, there can be large mismatches when the peaks and valleys of T_{11} do not match those of T_{11}^*. The importance of the mismatch in the T_{ij} FRFs, as well as how enveloping can be effectively employed in vibration testing, needs to be investigated.

The test item motion at location 2 is obtained by substitution of Eq. (8.3.2) into Eq. (8.2.24). This gives

$$U_2 = T_{21}^* R_1 = \left[\frac{T_{21}^*}{T_{11}^*}\right] X_1 = \left[\frac{T_{11}}{T_{11}^*}\right]\left[\frac{T_{21}^*}{FI}\right] Ye_1 \qquad (8.3.5)$$

for the laboratory motion. This laboratory motion should match the field motion described by Eq. (8.2.15), which can be modified for the field test conditions. Combining Eqs.(8.2.15) and (8.3.5) gives

$$X_2 = T_{21} F_1 = \left[\frac{T_{21}}{FI}\right] Ye_1 = \left[\frac{T_{11}^*}{T_{11}}\right]\left[\frac{T_{21}}{T_{21}^*}\right] U_2 \qquad (8.3.6)$$

It is apparent that these two equations will be identical if T_{11} and T_{21} are the same at all frequencies in both the field and test environment.

It should be evident that three major assumptions are required for the above results to occur. These assumptions are: (1) the exciter system must be capable of producing the required exciter force Q_2 at all input frequencies, (2) the control system must be capable of controlling the exciter system over the dynamic range of the test, and (3) the test item dynamic characteristics as described by T_{11} and T_{21} must be the same in the field and in the laboratory.

8.3.2 Test Scenario Number 2—Matched Interface Forces and No External Forces

In this scenario, the external forces acting on the test item, F_2 (field) and R_2 (laboratory), are zero. The interface force is controlled to be the same in both the field and laboratory environments so that the imposed testing condition is

$$R_1 = F_1 \qquad (8.3.7)$$

Then, substitution of Eqs. (8.2.11) and (8.2.20) into Eq. (8.3.7) gives the required exciter interface motion

$$Ze_1 = E_{12} Q_2 = \left[\frac{LI}{FI}\right] Ye_1 = LI\, F_1 \qquad (8.3.8)$$

The exciter driving force Q_2 required to produce the interface force described by Eq. (8.3.8) is obtained from Eq. (8.3.8) to be

$$Q_2 = \left[\frac{LI}{E_{12} FI}\right] Ye_1 = \frac{LI}{E_{12}} F_1 \qquad (8.3.9)$$

Note that Eq. (8.3.9) is automatically satisfied if the control system and exciter can satisfy the requirements of Eq. (8.3.7).

Substitution of the results from Eqs. (8.3.7) and (8.3.8) into Eq. (8.2.22) gives the interface motion to be

$$U_1 = T_{11}^* R_1 = \left[\frac{T_{11}^*}{LI}\right] Ze_1 = \left[\frac{T_{11}^*}{FI}\right] Ye_1 \qquad (8.3.10)$$

which, when compared to Eq. (8.2.13), shows an identical result with the field if the T_{11} FRF is the same in the laboratory and the field environments. Similarly, the test item motion at point 2 in the laboratory environment can be obtained from Eqs. (8.2.20), (8.2.24), and (8.3.8). This gives

$$U_2 = T_{21}^* R_1 = T_{21}^* F_1 = \left[\frac{T_{21}^*}{FI}\right] Ye_1 \qquad (8.3.11)$$

which, when compared to Eq. (8.2.15), shows that true simulation will occur provided that $T_{21}^* = T_{21}$.

The second scenario shows that a single force controlled excitation will perform in a satisfactory manner, provided that the same three assumptions of test system performance from the previous scenario are satisfied. A major problem with force controlled scenarios is that often the force transducers are sensitive to shear forces, to base strains, to misalignment of the transducer, and to bending moments, as well as to the desired axial forces they are designed to measure.

8.3.3 Test Scenario Number 3—Matched Test Item Motion and No External Forces

In this scenario, the external forces, F_2 (field) and R_2 (laboratory), are zero. The motion at point 2 in the test item is controlled to be the same in both the field and laboratory environments so that the imposed testing condition becomes

$$U_2 = X_2 \qquad (8.3.12)$$

Equation (8.3.12) states that the exciter system is adjusted to produce a desired motion at a given point in the test item, which should match that found in the field. In order to obtain the desired laboratory operating conditions Eqs. (8.2.15), (8.2.18), and (8.2.24) are combined with Eq. (8.3.12) to obtain the bare table exciter motion, which is given by

SIMULATION SCHEMES BASED ON THE ELEMENTARY MODEL   553

$$Ze_1 = \begin{bmatrix} LI \\ T_{21}^* \end{bmatrix} X_2 = E_{12} Q_2 = \begin{bmatrix} LI \\ FI \end{bmatrix} \begin{bmatrix} T_{21} \\ T_{21}^* \end{bmatrix} Ye_1 \qquad (8.3.13)$$

from which the required exciter driving force becomes

$$Q_2 = \begin{bmatrix} LI \\ E_{12} T_{21}^* \end{bmatrix} X_2 \qquad (8.3.14)$$

The corresponding interface motion can be obtained by combining Eqs. (8.2.13), (8.2.22), and (8.3.13), which results in

$$U_1 = T_{11}^* R_1 = \begin{bmatrix} T_{11}^* \\ T_{21}^* \end{bmatrix} X_2 = \begin{bmatrix} T_{21} \\ T_{21}^* \end{bmatrix} \begin{bmatrix} T_{11}^* \\ FI \end{bmatrix} Ye_1 \qquad (8.3.15)$$

A comparison of Eq. (8.3.15) with Eq. (8.2.13) shows that $U_1 = X_1$, provided that $T_{21} = T_{21}^*$ and $T_{11} = T_{11}^*$. The corresponding interface force (note that $F_1 = R_1$) is given by Eqs. (8.2.15) and (8.2.24) as

$$R_1 = \frac{U_2}{T_{21}^*} = \frac{X_2}{T_{21}} \qquad (8.3.16)$$

It should be apparent that the third scenario is dependent on the same three test system capability assumptions required by the first two scenarios.

These single point input results show that potentially it is possible to obtain good simulation under these restricted conditions. It is doubtful that such good results will be obtained when there is more than one input force to the test item in the field and the laboratory environments. The next three scenarios will show that external forces cause testing problems that are not easily reconciled without significant expenditures of time and resources.

8.3.4 Test Scenario Number 4—Matched Interface Motion with Field External Force but No Laboratory External Force

In this scenario, the field external force F_2 is nonzero while the laboratory external force R_2 is zero. The interface motion is controlled to be the same in the laboratory and field environments. This test condition is described by Eq. (8.3.1), that is,

$$U_1 = X_1 \qquad (8.3.1)$$

Recall that $Ue_1 = Ue_2 = 0$ since $R_2 = 0$ (see Eqs. (8.2.19) and (8.2.24)).

554   GENERAL VIBRATION TESTING

When Eqs. (8.2.18) and (8.2.22) are combined with Eq. (8.3.1), the required bare table exciter interface motion becomes

$$Ze_1 = E_{12}Q_2 = \left[\frac{LI}{T_{11}^*}\right]X_1 \qquad (8.3.17)$$

which, when combined with Eq. (8.2.13), gives

$$Ze_1 = \left[\frac{LI}{FI}\right]\left[\frac{T_{11}}{T_{11}^*}\right](Ye_1 - Xe_1) + \left[\frac{LI}{T_{11}^*}\right]Xe_1 \qquad (8.3.18)$$

Once the bare table exciter interface motion is known, the other quantities of interest can be obtained.

The interface force required to move the test item during laboratory testing is obtained by combining Eqs. (8.2.20) and (8.3.18). This gives

$$\begin{aligned}R_1 &= \left[\frac{T_{11}}{T_{11}^*}\right]\left[\frac{Ye_1 - Xe_1}{FI}\right] + \frac{Xe_1}{T_{11}^*} \\ &= \left[\frac{T_{11}}{T_{11}^*}\right]F_1 + \frac{Xe_1}{T_{11}^*}\end{aligned} \qquad (8.3.19)$$

Equation (8.3.19) clearly shows that the field and the laboratory interface forces are not the same, even when T_{11} and T_{11}^* are the same. This result is due to the fact that there is no way to include the effect of Xe_1 in the laboratory when $R_2 = 0$. Hence, an error term must exist. The amount of error is often frequency dependent, depending on the relative size of each term in Eq. (8.3.19) that is also frequency dependent.

The laboratory simulated motion at location 2 in the test item is obtained from Eq. (8.2.24) when Eq. (8.3.18) is substituted. The result is

$$U_2 = \left[\frac{T_{11}}{T_{11}^*}\right]\left[\frac{T_{21}^*}{FI}\right](Ye_1 - Xe_1) + \left[\frac{T_{21}^*}{T_{11}^*}\right]Xe_1 \qquad (8.3.20)$$

from which, in comparison with Eq. (8.2.15) for the field conditions,

$$X_2 = \left[\frac{T_{21}}{FI}\right](Ye_1 - Xe_1) + Xe_1 \qquad (8.3.15)$$

it is obvious that the motions are not the same. The first terms are equal provided $T_{11} = T_{11}^*$ and $T_{21} = T_{21}^*$. The second term will not be equal unless by coincidence, that is, T_{11}^* is equal to T_{21}^* or Xe_1 is quite small

SIMULATION SCHEMES BASED ON THE ELEMENTARY MODEL 555

compared to the other terms in the experiment. The same three fundamental assumptions of test system performance as required in the first scenario are required in this case. Thus controlling the test so that the interface connector motions are equal does not guarantee that the other parameters, such as interface forces and motion at another point in the test structure, match as well. The laboratory results can be severely distorted depending on the relative magnitude of the various forces and motions!

8.3.5 Scenario Number 5—Matched Interface Forces with Field External Force but No Laboratory External Force

In this scenario, the field external force F_2 is nonzero while the laboratory external force R_2 is zero. The interface connection forces are controlled to be the same in the field and the laboratory so that Eq. (8.3.7) is satisfied, that is, so that

$$R_1 = F_1 \tag{8.3.7}$$

Recall from Eq. (8.2.19), that $Ue_1 = Ue_2 = 0$. Then, combining Eqs. (8.2.11), (8.2.18), and (8.2.20) with Eq. (8.3.7) gives the bare table exciter motion as

$$Ze_1 = \left[\frac{LI}{FI}\right](Ye_1 - Xe_1) = E_{12}Q_2 \tag{8.3.21}$$

The resulting laboratory interface motion can be estimated by substituting Eq. (8.3.21) into Eq. (8.2.23) to obtain

$$U_1 = \left[\frac{T^*_{11}}{LI}\right]Ze_1 = \left[\frac{T^*_{11}}{FI}\right](Ye_1 - Xe_1) + 0 \tag{8.3.22}$$

where the zero motion that results from R_2 being zero is emphasized by the plus zero term on the right-hand side of Eq. (8.3.22). The test item laboratory interface motion given by Eq. (8.3.21) can be compared to the field motion, as expressed by Eq. (8.2.13), that is,

$$X_1 = \left[\frac{T_{11}}{FI}\right](Ye_1 - Xe_1) + Xe_2 \tag{8.3.13}$$

The first term in each equation should be nearly equal. However, the second term is completely missing in Eq. (8.3.22). This suggests that the interface motion due to an external force cannot be obtained by matching the interface forces alone as is done in this case.

The motion of any point 2 in the test item can be obtained by combining Eqs. (8.2.24) and (8.3.21) to obtain

$$U_2 = \left[\frac{T_{21}^*}{FI}\right](Ye_1 - Xe_1) + 0 \qquad (8.3.23)$$

where the zero is used to emphasize the missing motion due to R_2 being zero in the laboratory. This laboratory simulated result must compare to the motion that occurred in the field as given by Eq. (8.2.15), that is,

$$X_2 = \left[\frac{T_{21}}{FI}\right](Ye_1 - Xe_1) + Xe_2 \qquad (8.2.15)$$

Again, a comparison shows that the first term of each equation may match up quite well when $T_{21} = T_{21}^*$. The second term is completely missing in the laboratory simulation. In fact, both simulated motions (X_1 and X_2) suffer the same defect under the force control scenario, since there is no motion that is equivalent to Xe_2 since the nonconnector force $R_2 = 0$ in the laboratory simulation. These simulation results are dependent on the same three assumptions about test system behavior described with scenario number 1.

8.3.6 Scenario Number 6—Matched Test Item Motion with Field External Force but No Laboratory External Force

In this scenario, the external force F_2 is nonzero, while the laboratory external force R_2 is zero. The motion in the test item at location 2 is controlled to be the same in both the laboratory and field environments. This test condition is described by Eq. (8.3.12), that is,

$$U_2 = X_2 \qquad (8.3.12)$$

Again, recall that $Ue_1 = Ue_2 = 0$. When Eqs. (8.2.15), (8.2.18), and (8.2.24) are combined with Eq. (8.3.12), the required exciter bare table interface motion becomes

$$Ze_1 = E_{12}Q_2 = \left[\frac{LI}{FI}\right]\left[\frac{T_{21}}{T_{21}^*}\right](Ye_1 - Xe_1) + \left[\frac{LI}{T_{21}^*}\right]Xe_2 \qquad (8.3.24)$$

The corresponding laboratory interface force is obtained by substituting

Eq. (8.3.24) into Eq. (8.2.20) and using the definition of F_1 from Eq. (8.2.11) to obtain

$$R_1 = \left[\frac{T_{21}}{T_{21}^*}\right]\left[\frac{Ye_1 - Xe_1}{LI}\right] + \frac{Xe_2}{T_{21}^*} = \left[\frac{T_{21}}{T_{21}^*}\right]F_1 + \frac{Xe_2}{T_{21}^*} \qquad (8.3.25)$$

It is obvious from Eq. (8.3.25) that the laboratory and field interface forces are different even when $T_{21} = T_{21}^*$, since Xe_2 is nonzero in the field. In the laboratory, Xe_2 is accounted for by the simulation control scenario described by Eq. (8.3.12).

When Eq. (8.3.24) is substituted into Eq. (8.8.22), the corresponding laboratory interface motion becomes

$$U_1 = \left[\frac{T_{21}}{T_{21}^*}\right]\left[\frac{T_{11}^*}{FI}\right](Ye_1 - Xe_1) + \frac{T_{11}^*}{T_{21}^*}Xe_2 \qquad (8.3.26)$$

This motion can be compared to that given by Eq. (8.2.13):

$$X_1 = \left[\frac{T_{11}}{FI}\right](Ye_1 - Xe_1) + Xe_1 \qquad (8.2.13)$$

where it is obvious that the first terms are matched when $T_{21} = T_{21}^*$ and $T_{11}^* = T_{11}$. However, the second terms can never match (unless by accident), since they involve distinctly different terms, namely Xe_2 and Xe_1.

8.3.7 Summary of Six Test Scenarios

Six different testing and control scenarios are considered in this section. In scenarios 1 through 3, there are no external forces acting on the test item in the field and laboratory environments. Each of these control scenarios is found to have the potential to be satisfactory within the ability of the control and exciter systems to implement the control condition and the matching of the test item FRF in the field and the laboratory. This result explains why certain laboratory simulations are quite successful since the dominant inputs come from the connection points while the external load effects are minimal. Three assumptions need to be satisfied if these three control scenarios are to work properly. These assumptions are: (1) the exciter system must be capable of producing the required exciter force Q_2 at all input frequencies, (2) the control system must be capable of controlling the exciter system over the dynamic range of the test, and (3) the test item dynamic characteristics as described by T_{11} and T_{21} must be the same in the field and in the laboratory.

In scenarios 4 through 6, the field external force is not simulated in the laboratory. Each control scenario is found to be inadequate under these conditions, since it is impossible to simulate the external force with some kind of equivalent interface motion or force as well as the motion at some internal point of the test item. This result begins to explain why test simulation of field tests where there are significant external force excitations are poorly simulated when the external force is not simulated in the laboratory. This type of behavior is described in a paper by Szymkowiak and Silver.[6] The conditions required for proper implementation of the external forces have not been explored at this point, but one can easily guess that the cross-correlation between signals in the laboratory and field must be the same. Otherwise, the output would not be the same as indicated by Eqs. (8.2.14) and (8.2.23).

8.4 AN EXAMPLE USING A TWO DOF TEST ITEM AND A TWO DOF VEHICLE

In order to demonstrate the practical aspects of Sections 8.2 and 8.3, a relatively simple experiment was conducted using a two degrees of freedom (DOF) test item and a two DOF vehicle as shown in Fig. 8.4.1. The vehicle is simulated by using a portal frame attached to an electrodynamic vibration exciter. The portal frame has a central mass mv_1 and a base mass mv_2 that is attached to the exciter's armature. Note that the symbol V is used with mass, stiffness, damping and acceleration to designate that these variables belong to the vehicle while the symbol T is used to designate the test item's variables. The significant vehicle accelerations for this example are measured by using accelerometers labeled Av_1 and Av_2. The vehicle has a spring constant kV and a viscous damping constant Cv. The interface surface between the two systems is between the force transducer and vehicle mass mv_1 as shown in Fig. 8.4.1a. The force transducer's mass is considered to be part of the test item.

The test item consists of a symmetric double cantilever beam with masses mT'_2 attached at each end that also include accelerometer mass of AT_2. The central clamp, interface force transducer, and accelerometer AT_1 have a total of mass mT_1. The two masses mT'_2 are modeled as a combined single mass mT_2, as shown in Fig. 8.4.1b. The test item has a spring constant kT (twice the single cantilever spring constant) and viscous damping CT.

---

[6]E. A. Szymkowiak and W. Silver, II, "A Captive Store Flight Vibration Simulation Project," *Proceedings of the 36th Annual Institute of Environmental Sciences*, New Orleans, LA, Apr. 1990, pp. 531–538.

AN EXAMPLE USING A TWO DOF TEST ITEM AND VEHICLE

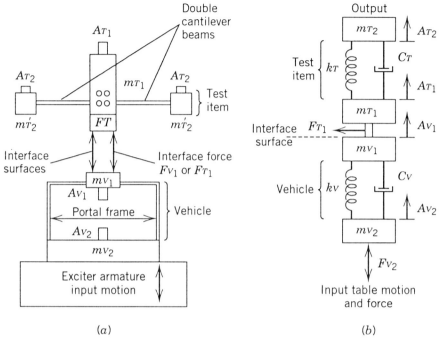

Fig. 8.4.1. (*a*) Physical model. (*b*) Equivalent spring mass damper model. (Courtesy of *Experimental Mechanics*, Society for Experimental Mechanics.)

8.4.1 Test Item and Vehicle Dynamic Characteristics

Theoretical Considerations The driving point accelerance FRF can be obtained from any vibrations text[7] by proper interpretation of a standard two DOF forced vibration response. This interpretation gives the freely suspended driving point accelerance for the test item at point 1 as

$$T_{11} = \frac{AT_1}{FT_1} = \frac{kT - m_{T2}\omega^2 + jCT\omega}{kTmT - (m_{T1}m_{T2})\omega^2 + jCTmT\omega} \quad (8.4.1)$$

where mT is the total test item mass given by

$$mT = (m_{T1} + m_{T2}) \quad (8.4.2)$$

The transfer-accelerance response spectrum at point 2 due to input force

---

[7]Rao, S. S., *Mechanical Vibrations*, 2nd ed., Addison-Wesley, Reading, MA, 1990, pp. 269–420.

frequency spectrum at point 1 or the response spectrum at point 1 due to input force frequency spectrum at point 2 is given by

$$T_{12} = T_{21} = \frac{AT_2}{FT_1} = \frac{kT + jCT\omega}{kTmT - (mT_1 mT_2)\omega^2 + jCTmT\omega} \tag{8.4.3}$$

The driving point acceleration for the test item at point 2 is given by

$$T_{22} = \frac{AT_2}{FT_2} = \frac{kT - mT_1\omega^2 + jCT\omega}{kTmT - (mT_1 mT_2)\omega^2 + jCTmT\omega} \tag{8.4.4}$$

These three accelerance FRFs completely define this simple test item's dynamic possibilities. Note that these FRFs correspond to those of a semidefinite system that has free boundaries and can move as a rigid body.

In like manner we can write down similar FRFs for the vehicle. However, as will be seen shortly, only the vehicle's driving point accelerance FRF is required for this experiment. The vehicle has a driving point accelerance characteristic that is given by

$$V_{11} = \frac{AV_1}{FV_1} = \frac{-\omega^2}{kV - mV_1\omega^2 + jCV\omega} \tag{8.4.5}$$

when the base is fixed. A second important FRF function for this test is the transmissibility of acceleration from the vehicle's base at location 2 to the interface at location 1. This transmissibility FRF is designated by TRV_{12} and is given by

$$TRV_{12} = \frac{AV_1}{AV_2} = \frac{kV + jCV\omega}{kV - mV_1\omega^2 + jCV\omega} \tag{8.4.6}$$

where the symbol v is used to indicate the vehicle's transmissibility.

Measured Characteristics A structure is defined by its driving point and transfer accelerance FRFs. Equation (8.4.1) gives the driving point accelerance of the test structure. In this case, mass mT_2 acts as a dynamic absorber at frequency f_a when the real part of the numerator goes to zero. Similarly, Eq. (8.4.1) shows that T_{11} is a maximum at a resonance condition f_n when the real part of the denominator is zero. The measured

AN EXAMPLE USING A TWO DOF TEST ITEM AND VEHICLE

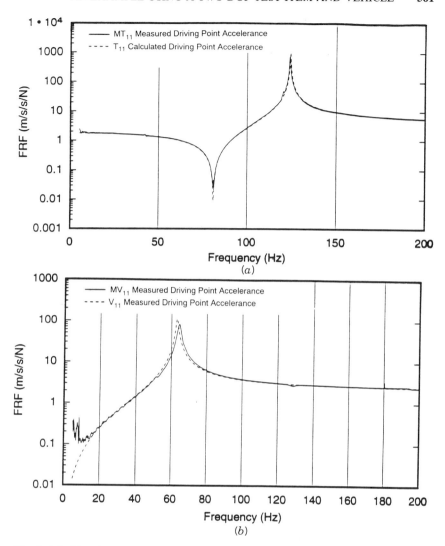

Fig. 8.4.2. Test item and vehicle driving point accelerances that define each structure's parameter. (*a*) Test item. (*b*) Vehicle. (Courtesy of *Experimental Mechanics*, Society for Experimental Mechanics.)

driving point accelerance MT_{11} of the test item is shown in Fig. 8.4.2*a* where the M symbol preceding the T_{11} indicates that it comes from *measured* vibration data. A detailed analysis of this experimental data while using Eq. (8.4.1) gives the physical constants shown in Table 8.4.1 for the test item. The valley occurs at 81.25 Hz while the peak occurs at 123.75 Hz. Using the parameters of Table 8.4.1 in Eq. (8.4.1) allows us to calculate

562 GENERAL VIBRATION TESTING

TABLE 8.4.1. Test System Parameters

Test Item Parameters			Vehicle Parameters		
Variable	Value	Units	Variable	Value	Units
m_{T1}	0.2481	kg	m_{V1}	0.447	kg
m_{T2}	0.3251	kg	k_V	7.04×10^4	N/m
k_T	8.45×10^4	N/m	C_V	3.50	N s/m
C_T	0.20	N-s/m	f_N	63.25	Hz
f_a	81.25	Hz	f_a — Absorber frequency		
f_n	123.75	Hz	f_n — Natural frequency		

the theoretical accelerance T_{11}, which is also shown in Fig. 8.4.2a. Clearly, a reasonably good fit occurs between the experiment and the model when the damping value is selected to provide a reasonable fit to both the peaks and the valleys.

The measured driving point accelerance MV_{11} of the vehicle is shown in Fig. 8.4.2b. The theoretical expression for V_{11} is given by Eq. (8.4.5). The effective mass is estimated at 200 Hz, while the resonance is seen to occur at 63.25 Hz. Using these values in Eq. (8.4.5) allows us to estimate the vehicle's physical parameters as given in Table 8.4.1. The theoretical curve is calculated from Eq. (8.4.5) by using these values for k_V, C_V, and m_{V1}. Again, a reasonably good fit occurred. However, the values selected must also satisfy the acceleration transmissibility function described by Eq. (8.4.6). Thus, although the natural frequency does not fit the curve shown exactly, it does fit the more accurately determined transmissibility resonance frequency. The slight frequency shift between the driving point accelerance peak frequency and the acceleration transmissibility peak frequency is due to the finite size of the armature mass.

8.4.2 Laboratory Test Setup Employed During Tests

The main laboratory equipment employed in these experiments is shown in Fig. 8.4.3. A General Radio Model 2608 digital vibration controller is used to control the Unholtz–Dickie Model T 206 vibration exciter. The feedback can be either an exciter table (armature) acceleration or a force acting as input to the test item. The vibration controller's output signal drives the Unholtz–Dickie Model TA 250 power amplifier. This power amplifier and exciter combination can produce a maximum bare armature random root mean square (RMS) acceleration in excess of 40 g_{RMS} in the 5 to 3 kHz frequency range.

Pcb Model 302A02 accelerometers measured the vehicle base acceler-

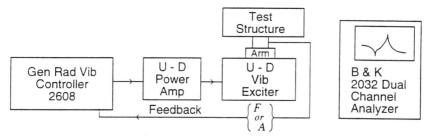

Fig. 8.4.3. Schematic of vibration test setup showing main instruments. (Courtesy of *Experimental Mechanics*, Society for Experimental Mechanics.)

ation AV_2 and test item interface acceleration AV_1, while a Pcb Model 208A03 force transducer measured the interface force F_{T_1}. The test item's output acceleration AT_2 was measured using an Endevco Model 2222C miniature accelerometer and a Kistler Model 504A charge amplifier. The Bruel & Kjaer Model 2032 dual channel frequency analyzer was used to measure various input-output FRFs. The frequency analyzer results were transferred to a personal computer (PC) where this data was plotted and theoretical curves were calculated using MathCAD© 4.0 software.

8.4.3 Field Simulation Results

There are two field simulation cases. First, the vehicle is simulated as the portal frame attached to the exciter armature as shown in Fig. 8.4.1a, and the test item is not attached. We call this situation the bare *vehicle interface* case. Second, the test item is attached to the vehicle at the interface. We call this situation the *connected field* test case. In the description that follows, the ASD is used. The general relationship between input-output ASDs and an input-output FRF $H(\omega)$ is given by

$$G_{\text{output}} = |H(\omega)|^2 G_{\text{input}} \qquad (8.4.7)$$

where it is clear that all phase information is lost in the squaring process. We shall use Eq. (8.4.7) as the basis for calculating the experimental and theoretical results.

Vehicle Base Input Acceleration ASD, G_{A_0} In both of the field cases, the digital controller generated a random signal in the 0 to 200 Hz bandwidth

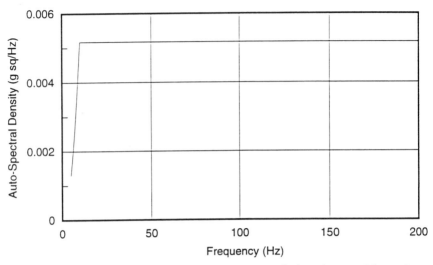

Fig. 8.4.4. Input acceleration ASD, GA_0, that is applied to the portal frame base to generate field vehicle motions. (Courtesy of *Experimental Mechanics*, Society for Experimental Mechanics.)

by using the ASD function GA_0 as shown in Fig. 8.4.4. This spectrum increases with a slope of 40 dB/decade from 0 to 10 Hz, at which point the curve becomes flat at 0.00517 g²/Hz. This spectrum gives an overall 1 g_{RMS} input vibration level, a level that is easily within the exciter's 40 g_{RMS} capabilities.

Bare Vehicle Interface Acceleration ASD, GA_{V_1} The bare vehicle interface acceleration ASD, GA_{V_1}, with units of g²/Hz, is shown in Fig. 8.4.5. The experimental interface acceleration ASD is obtained from the measured acceleration transmissibility FRF, $MTR_{V_{12}}$. Here, $MTR_{V_{12}}$ is obtained from the Bruel & Kjaer analyzer where the vehicle interface acceleration A_{V_1} and the vehicle input motion A_{V_2} are the inputs. Equation (8.4.7) requires that the magnitude of this measured FRF be squared and multiplied by the input ASD, GA_0 in this case, as defined in Fig. 8.4.4. This gives

$$GA_{V_1} = |MTR_{V_{12}}|^2 GA_0 \qquad (8.4.8)$$

The theoretical interface acceleration ASD, CGA_{V_1}, is obtained from Eq. (8.4.6) and the vehicle parameters in Table 8.4.1 that are used to calculate $TR_{V_{12}}$, which is then squared and multiplied by GA_0 as required by Eq. (8.4.7), so that

AN EXAMPLE USING A TWO DOF TEST ITEM AND VEHICLE 565

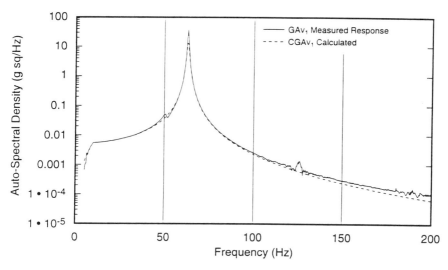

Fig. 8.4.5. Measured and calculated bare vehicle interface acceleration ASD, GA_{V_1}. (Courtesy of *Experimental Mechanics*, Society for Experimental Mechanics.)

$$CGA_{V_1} = |TR_{V_{12}}|^2 GA_0 \qquad (8.4.9)$$

Note that the symbol C precedes GA_{V_1} to indicate that this is a calculated or theoretical result. The plots in Fig. 8.4.5 show a very good fit between the experimental vehicle interface acceleration ASD and that calculated from theory.

Combined Vehicle and Test Item Response ASDs, GF_{T_1}, GA_{T_1}, and GA_{T_2}
The interface force, the interface acceleration, and the test item output acceleration ASDs are shown in Fig. 8.4.6 when the vehicle is excited with the input ASD shown in Fig. 8.4.4. These curves are described as follows.

The experimental interface force ASD, GF_{T_1}, is shown in Fig. 8.4.6a. The measured interface force FRF, MF_{T_1}, is obtained from inputs of force F_{T_1} and vehicle base acceleration A_{V_2} into the frequency analyzer. This FRF is squared and multiplied by GA_0 as required by Eq. (8.4.7) so that

$$GF_{T_1} = |MF_{T_1}|^2 GA_0 \qquad (8.4.10)$$

The theoretical interface force ASD, CGF_{T_1}, is calculated by combining

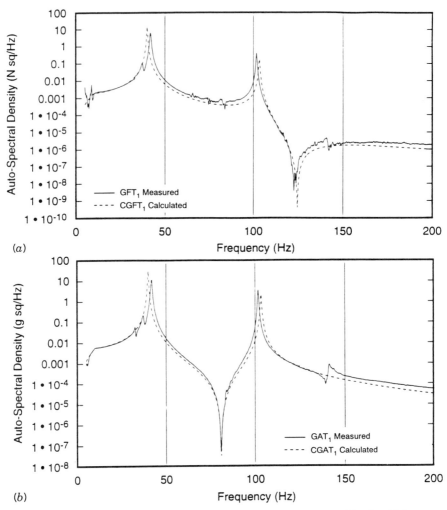

Fig. 8.4.6. A comparison of the measured and theoretical combined vehicle-test item field ASD responses. (*a*) Interface force. (*b*) Interface acceleration. (*c*) Test item output acceleration. (Courtesy of *Experimental Mechanics*, Society for Experimental Mechanics.)

Eqs. (8.4.6), (8.4.7), and (8.2.11) to obtain

$$CGF_{T_1} = \left| \frac{TRV_{12}}{T_{11} + V_{11}} \right|^2 GA_0 \qquad (8.4.11)$$

Note that these two curves follow one another quite closely. This indicates

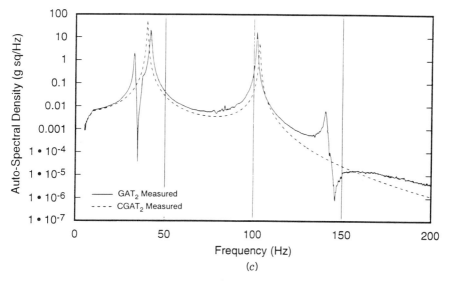

Fig. 8.4.6. (*Continued*).

that the theoretical model based on vehicle V_{11} and test item T_{11} driving point accelerances along with the transmissibility ratio FRF TRV_{12} provides a good prediction equation for this situation.

We need to consider the data's large dynamic range. The Bruel & Kjaer analyzer employs 16 bit analog to digital (A/D) converters that give a dynamic range of 1×10^9 in the ASD plot. Clearly, the experimental data shows significant noise at the 1.0×10^{-8} level, which is nine log cycles below the peak values. Hence, we are seeing the analyzer's dynamic range limits in the deep valley near 125 Hz.

The measured interface acceleration ASD, GAT_1, with units of g^2/Hz, is shown in Fig. 8.4.6b for the connected case. The interface acceleration transmissibility FRF, MHV_0, is obtained from the Bruel & Kjaer analyzer by using acceleration inputs of AV_2 and AT_1 so that

$$GAT_1 = |MHV_0|^2 GA_0 \qquad (8.4.12)$$

The theoretical interface acceleration ASD, $CGAT_1$, is calculated from Eqs. (8.2.13) and (8.4.7) giving

$$CGAT_1 = \left| \frac{T_{11} TR_{12}}{T_{11} + V_{11}} \right|^2 GA_0 \qquad (8.4.13)$$

These two curves follow one another over a dynamic range in excess of eight log cycles. Clearly the driving point accelerances play a significant role in predicting this interface motion.

The experimental output acceleration ASD, GA_{T2}, with units of g^2/Hz, is shown in Fig. 8.4.6c. The measured output acceleration transmissibility FRF, MH_T, is obtained by using acceleration inputs of A_{T2} and A_{V2}. This FRF is squared and multiplied by GA_0 as required by Eq. (8.4.7), giving

$$GA_{T2} = |MH_T|^2 GA_0 \qquad (8.4.14)$$

The theoretical output ASD, CGA_{T2}, is calculated from Eqs. (8.4.3) and (8.4.11), giving

$$CGA_{T2} = \left|\frac{T_{21}TR_{12}}{T_{11} + V_{11}}\right|^2 GA_0 \qquad (8.4.15)$$

Although there is general agreement between the measured and the theoretical values, additional events occur in the field-measured results. The first event is a double peak below 50 Hz where only one peak occurs theoretically. This double peak is caused by the low frequency rocking motion due to the roller support of the double cantilever beam (see Fig. 8.4.1a). The second event occurs just below 150 Hz and may be a torsional resonance due to inadequate flexural stiffness of the portal frame support or bending moment sensitivity of the force transducer. Generally, there is good agreement between the theoretical and the experimental results, indicating that Eq. (8.4.15) is a reasonable method for predicting this output.

8.4.4 Laboratory Simulation

The test item is attached directly to the exciter's armature at the force transducer. In this case, we can use either of two possible feedback signals with the vibration controller. These signals are either the test item input acceleration A_{T1} (this corresponds to test scenario number 1 in Section 8.3) or test item input force F_{T1} (this corresponds to test scenario number 2 in Section 8.3). We did not use the test item's output motion as a feedback control in this example.

Three different inputs can be applied to the test item in the laboratory simulation when using the digital vibration controller. These inputs are (1) the bare vehicle interface acceleration ASD, GA_{V1}, (2) the connected interface acceleration ASD, GA_{T1}, and (3) the connected interface force ASD, GF_{T1}. In each case, we can measure the appropriate laboratory responses. For example, if the controlled test item input is the bare vehicle interface acceleration, then the appropriate measured responses are the

TABLE 8.4.2 Laboratory Simulation Choices of Test Structure

Input to Test Structure	Measured Lab Response	Comparison	
		Theoretical	Measured[a]
GA_{V_1}	VGF_{T_1}	CGF_{T_1}	GF_{T_1}
GA_{V_1}	VGA_{T_2}	CGA_{T_2}	GA_{T_2}
GA_{T_1}	AGF_{T_1}	CGF_{T_1}	GF_{T_1}
GA_{T_1}	AGA_{T_2}	CGA_{T_2}	GA_{T_2}
GF_{T_1}	FGA_{T_1}	CGA_{T_1}	GA_{T_1}
GF_{T_1}	FGA_{T_2}	CGA_{T_2}	GA_{T_2}

[a] Measured when test item and vehicle are connected in the field.

test item input force and output acceleration. These measured responses can be compared with appropriate theoretical and measured field responses for the combined system. The test combinations used are shown in Table 8.4.2.

Test Item Excited with Bare Vehicle Interface Acceleration ASD GA_{V_1} The bare vehicle's output ASD, GA_{V_1}, with units of g²/Hz as shown in Fig. 8.4.5, is applied to the test item input by the vibration exciter. The resulting output ASDs are shown in Figs. 8.4.7a and 8.4.7b. The prefix V symbol is used to indicate that the spectral densities are due to the bare vehicle interface acceleration input GA_{V_1}.

The interface force ASD, VGF_{T_1}, with units of N²/Hz, is shown in Fig. 8.4.7a along with the theoretical interface force CGF_{T_1} and field interface force GF_{T_1}, which is obtained when vehicle and test item are connected in the field. The theoretical force is calculated for the actual laboratory test conditions by using Eqs. (8.4.1), (8.4.6), and (8.4.7) to give

$$CGF_{T_1} = \left|\frac{1}{T_{11}}\right|^2 GA_{V_1} \qquad (8.4.16)$$

Clearly, the measured interface force VGF_{T_1} is dramatically different from that obtained under field conditions GF_{T_1} in both magnitude and natural frequencies. We also see that the modified theory of Eq. (8.4.16) can predict the measured interface force for this case by using only the test item's driving point accelerance, T_{11}. Hence, the interface force is over 1000 times too small at points A and D and is over 1000 times too large at points B and C, and the test item is severely undertested and overtested at different frequencies.

The output acceleration ASDs are shown in Fig. 8.4.7b. Again we see that these output accelerations differ significantly from the field acceler-

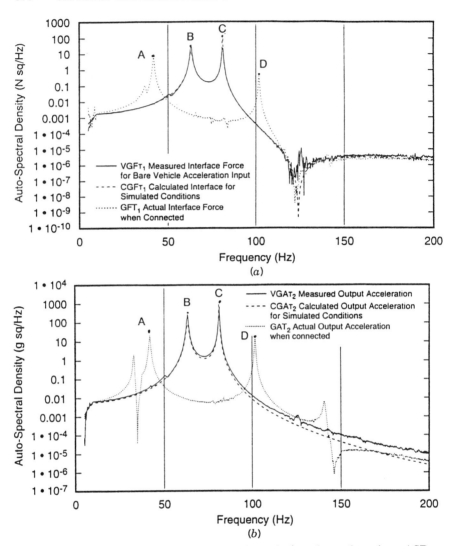

Fig. 8.4.7. A comparison of the measured, theoretical, and actual test item ASDs. (*a*) Interface force. (*b*) Output acceleration when the test item is excited by the bare vehicle interface acceleration ASD, GAv_1. (Courtesy of *Experimental Mechanics*, Society for Experimental Mechanics.)

ations in magnitude and frequency. The theoretical ASD, $CGAT_2$, is calculated by using Eqs. (8.4.1), (8.4.3), (8.4.6), and (8.4.7) to give

$$CGAT_2 = \left|\frac{T_{21}}{T_{11}}\right|^2 GAv_1 \qquad (8.4.17)$$

Again the modified theoretical equation can predict the test results. Also note that only T_{11} is involved in the denominator in Eqs. (8.4.16) and (8.4.17) while both T_{11} and V_{11} are involved in Eqs. (8.4.11), (8.4.13), and (8.4.15). Also note that GAV_1 in Eqs. (8.4.16) and (8.4.17) is calculated by using Eq. (8.4.9). In this case, points A and D are over 1000 times too small, while points B and C are 10,000 times too large. Thus, we see that points A and D are undertested while points B and C are severely overtested.

The results in Fig. 8.4.7 show that it is inappropriate to excite the test item in the laboratory with the bare vehicle's acceleration ASD since we obtain incorrect natural frequencies and vibration levels. For example, the first resonant peak occurs at 63.25 Hz and corresponds to the vehicle's natural frequency in Fig. 8.4.5, while the second resonant peak occurs at 81.25 Hz and corresponds to the dynamic absorber frequency in Fig. 8.4.2a. Neither of these frequencies corresponds to the actual field resonances. Clearly, there is severe overtesting at 63.25 and 81.25 Hz, and severe undertest occurring around 43 and 103 Hz. However, the use of bare vehicle field data for test item inputs is a common method for obtaining field data for laboratory simulation and is the same method used by the automotive parts supplier mentioned in the introduction, who obtained similarly inaccurate results.

Test Item Excited with Connected Interface Acceleration ASD, GAT_1 The results obtained when the field interface acceleration ASD, $GATV_1$, is used to control the vibration exciter are shown in Fig. 8.4.8. These test results carry the leading symbol of A to indicate the input is the connected interface acceleration ASD, $GATV_1$.

The interface force ASD, $AGFT_1$, with units of N^2/Hz, is shown in Fig. 8.4.8a. In this case, the laboratory simulation and the field interface forces are nearly identical at points A and B, while the theoretical values of $CGFT_1$ were calculated from Eq. (8.4.11) and also agree well with both the field and laboratory results. Again note the force's large dynamic range so that the valley at C is contaminated by the A/D converter's lack of dynamic range.

The output acceleration ASD, $AGAT_2$, with units of g^2/Hz, is shown in Fig. 8.4.8b. Again the laboratory simulation and the field output accelerations match up well over the entire dynamic range. The theoretical curve, given by Eq. (8.4.15), predicts the two major resonances in magnitude and frequency, but the same two discrepancies, one below 50 Hz and the other below 150 Hz, occur for the same reasons noted before in regard to Fig. 8.4.6c.

The results in Fig. 8.4.8 indicate that realistic test item simulation occurs when the field-measured interface acceleration ASD (obtained when both vehicle and test item are connected) is used as the controller's input. This

572 GENERAL VIBRATION TESTING

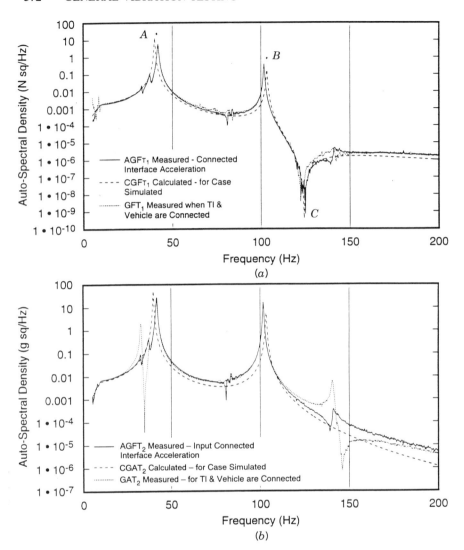

Fig. 8.4.8. A comparison of the measured, theoretical, and actual test item ASDs. (*a*) Interface force. (*b*) Output acceleration when the test item is excited by the field interface acceleration ASD, GA_{T_1}. (Courtesy of *Experimental Mechanics*, Society for Experimental Mechanics.)

behavior corresponds to that predicted by test scenario number 1 in Section 8.3.

Test Item Excited with Combined Interface Force ASD, GF_{T_1} The measured results when the interface force ASD, GF_{T_1}, is used to control the

AN EXAMPLE USING A TWO DOF TEST ITEM AND VEHICLE

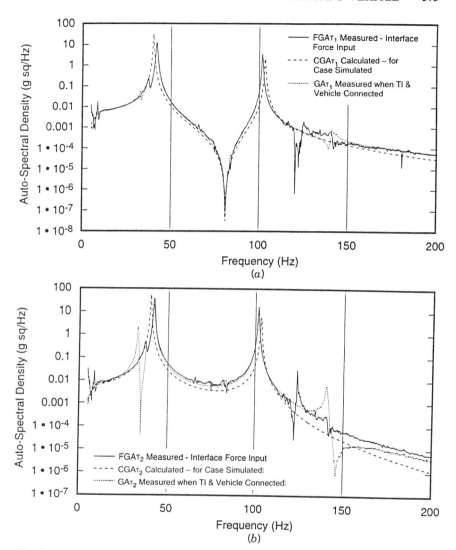

Fig. 8.4.9. A comparison of the measured, theoretical, and actual test item ASDs. (a) Interface acceleration. (b) Output acceleration when the test item is excited by the field interface force ASD, GF_{T_1}. (Courtesy of *Experimental Mechanics*, Society for Experimental Mechanics.)

test item input force are shown in Fig. 8.4.9. These measured results carry a prefix symbol of F to indicate the input excitation is the connected interface force ASD. Both cases show that the simulated and field-measured results are reasonably close in agreement. Likewise, the theoretical predictions, $CGAT_1$ from Eq. (8.4.13) and $CGAT_2$ from Eq. (8.4.15), are in close agreement. This agreement is closer for the interface acceleration than for the output acceleration. This behavior corresponds to that predicted for test scenario number 2 in Section 8.3.

The results from Figs. 8.4.8 and 8.4.9 indicate that we can produce reasonable simulations of the field test conditions in the test item when either the actual field interface acceleration or force is properly applied to the test item. In addition, we see that the simple theory is effective in predicting the response for every test instance, namely,

1. The bare vehicle responses, see Fig. 8.4.5
2. The connected vehicle and test item, see Fig. 8.4.6
3. The incorrectly simulated laboratory environment where the input was the bare vehicle interface motion, see Fig. 8.4.7
4. The correctly simulated laboratory environment where either the interface acceleration or interface force inputs were used, see Figs. 8.4.8 and 8.4.9

We have been quite successful in simulating the field dynamic environment in this case when using either the interface acceleration or force ASDs. The wide dynamic range found in the measured interface ASDs of GFT_1 and GAT_1 in Figs. 8.4.6a and 8.4.6b are known to cause controller problems near the deep valleys, a phenomenon that caused us problems as well for a while. A comparison of Figs. 8.4.9a and 8.4.9b show that the acceleration valley and the force valley occur at different frequencies so that one signal is strong when the other essentially disappears and is easily dominated by the instrumentation noise floor. Also, we have seen here that both signals can produce good test results in this type of situation away from the valleys. Thus it is not surprising that Smallwood[8] and Sharton[9] (14) have achieved good success when using a combination of interface force and acceleration signals as their control signals.

Now the question is, Can this simple theory predict what the interface force and acceleration ASDs should be for use in the design stage of the test item when all we have available is the bare vehicle interface acceleration ASD? Let us take a moment to explore this question.

---

[8]D. O. Smallwood, "An Analytical Study of a Vibration Test Method Using Extremal Control of Acceleration and Force," *Proceedings, Institute of Environmental Sciences*, 35th Annual Meeting, "Building Tomorrow's Environment," Anaheim, CA, May 1–5, 1989, pp. 263–271.

[9]T. D. Sharton, "Force Limited Vibration Testing at JPL," *Proceedings, 14th Aerospace Testing Seminar*, Manhattan Beach, CA Mar. 1993, pp. 241–251.

8.4.5 Predicting Interface Forces and Accelerations from Bare Vehicle Interface Acceleration ASD Data

Some reflection on these results suggests that we can obtain the proper test item inputs even though all we have is the bare vehicle interface acceleration ASD. The required interface force ASD is given by Eq. (8.4.11), which can be rewritten in terms of G_{AV_1} so that

$$CGF_{T_1} = \left|\frac{1}{T_{11} + V_{11}}\right|^2 G_{AV_1} \qquad (8.4.18)$$

Similarly, the required interface acceleration ASD is given by Eq. (8.4.13), which can be written in terms of G_{AV_1} so that

$$CGA_{T_1} = \left|\frac{T_{11}}{T_{11} + V_{11}}\right|^2 G_{AV_1} \qquad (8.4.19)$$

Equations (8.4.18) and (8.4.19) show that in addition to the bare vehicle acceleration ASD, G_{AV_1}, we need the driving point accelerance of the vehicle and of the test item. The vehicle's driving point accelerance can be measured at the time G_{AV_1} is obtained with additional effort. The test item's accelerance can be measured in the laboratory if the test item exists. However, if the test item only exists as a designer's finite-element model, then T_{11} can be estimated from that finite-element model. An additional feature of Eq. (8.4.18) is that it tells the designer the range of forces that the interface attachment bolt must be able to carry. The estimates given by Eqs. (8.4.18) and (8.4.19) can be used as a basis for drawing more realistic test input ASD envelopes.

8.4.6 Summary and Conclusions for This Simple Example

We have shown that three structures, each with unique dynamic characteristics, are involved in vibration testing: the vehicle, the test item (or finite-element model), and the vibration exciter. However, when these structures are connected in different combinations, we find that they create new structures with significantly different dynamic characteristics.

Experience has shown that the process of interpreting field data in order to generate adequate input for either laboratory tests or finite-element estimates has often produced undesirable results. Many factors contribute to these undesirable results. First, we have shown that there are 36 potential input-output FRFs that may be involved between any two points in a structure so that use of single variable transducers can lead to seriously contaminated data. Second, the motion at any point in the structure is due to boundary forces, internal excitation forces, and external excitation forces. Thus proper boundary conditions and external forces must be provided in the laboratory setting, a requirement that is often difficult to achieve. Third, we use incorrect methods to convert bare vehicle data into a suitable test input.

Our experience in performing this set of elementary experiments shows that the two DOF models adequately predicted this system's behavior over the limited frequency range of 0 to 200 Hz because a *single boundary interface force (or acceleration) ASD* is involved. Clearly, we see that the driving point accelerances of *both* the vehicle and the test item play an enormously important role in describing what happens when the test item is attached to the vehicle in the field at a single point. These two driving point accelerances also play an important role in converting bare vehicle acceleration frequency spectra into excitation spectra for either laboratory or finite-element simulation when a single interface boundary point is involved.

However, if we look at Fig. 8.1.2 for the more general case, we see that we are dealing with the potential of multiple boundary interface forces, internal excitation forces, and external excitation forces. It should be clear that the driving point accelerances of the vehicle and test item at the boundary interface connections control the combined system's dynamic behavior in the field. In turn, this dynamic behavior controls how the test item responds to both internal and external excitation, as predicted by Eq. (8.1.4), since the transfer FRFs are determined by these boundary forces.

Similarly, when we mount the test item on to a test fixture in the laboratory, the interface driving point accelerances of the test item and the test fixture (as installed on the exciter) control the test item's dynamic characteristics in the laboratory environment. Often the test fixture is constructed to be "as rigid as possible" over the range of test frequencies regardless of the vehicle's driving point accelerance characteristics. This design concept is defended by the fact that we want to control the exciter's armature acceleration to match either the vehicle's interface acceleration at one point or average acceleration from several interface points. Unfortunately, with multiple boundary forces as well as one or more internal and external excitation forces, it is little wonder that we have serious simulation problems when we violate the interface boundary conditions since the test item does not have the correct natural frequencies and mode shapes while in the laboratory environment. Thus the lament, "I seem to overtest at some frequencies and undertest at other frequencies!" Improper test fixture design contributes significantly to this problem. Consequently, in many cases, the input ASD is enveloped and increased to cover up the real problems of testing, which in turn, produces excessively rugged test items in some frequency ranges and weak test items in other frequency ranges.

8.5 THE GENERAL FIELD ENVIRONMENT MODEL

In this section, the general dynamic description of a connected test item and vehicle is examined in a more general sense where there are multiple interface connections and multiple external (nonconnecting) excitation forces that include internal forces as well. It is assumed that there are Nc connection points between the two structures. The connection points are selected in such a way that these points are labeled 1 through Nc in both structures so that a connection point carries the same number in each structure. A schematic diagram of the test item attached to a vehicle is shown in Fig. 8.5.1, where three interface connections (labeled A, B, and C) are indicated. The test item's response motion spectrum at location q is called X_q $(= X_q(\omega))$ and can be either displacement, velocity, or acceleration. X_q can be a vector quantity that can have three scalar components, but for our purposes, we shall treat it symbolically as a single scalar variable. The test item's input force spectrum at location p is called F_p $(= F_p(\omega))$ and includes all connector, external, and internal forces. The vehicle's response motion spectrum at location q is called Y_q $(= Y_q(\omega))$ and can be either displacement, velocity, or acceleration. The vehicle's input force spectrum at location p is called P_p $(= P_p(\omega))$ and includes all connector, external, and internal forces.

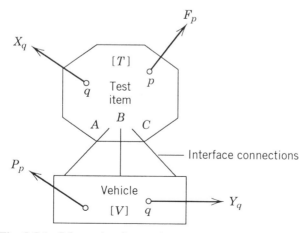

Fig. 8.5.1. Schematic of a test item mounted on a vehicle.

578   GENERAL VIBRATION TESTING

The vehicle and test item are described by their general input-output FRF relationships, which were given by Eqs. (8.2.1) and (8.2.2), that is,

$$\{Y\} = [V]\{P\} \tag{8.5.1}$$

for the vehicle and

$$\{X\} = [T]\{F\} \tag{8.5.2}$$

for the test item. The vehicle has Nv number of input-output points, while the test item has NT number of input-output points. We use this model in developing our general field environment equations.

8.5.1 Basic Motions Due to Interface and Noninterface Forces

In this section Eqs. (8.5.1) and (8.5.2) are expanded in detail to describe the motion of the vehicle and test item in general. Then, in the following section, the two structures are connected together to form the field test environment.

Test Item For the scheme shown in Fig. 8.5.1, the motion response in the test item at location q is given by

$$X_q = \underbrace{\sum_{p=1}^{N_c} T_{qp} F_p + \sum_{p=N_c+1}^{N_T} T_{qp} F_p}_{\text{Connector + External}} \qquad q = 1 \text{ to } NT \tag{8.5.3}$$

where the motion is conveniently broken into two parts. One part is due to the connector forces and the other part is due to external (nonconnector) forces acting on the test item. Recall that *connector forces* are those that act at the Nc interface connection points. The nonconnector forces are those that act at points $Nc + 1$ through NT in the test item. Consequently, Eq. (8.5.3) can be written as

$$X_q = Xc_q + Xe_q \qquad q = 1 \text{ to } NT \tag{8.5.4}$$

where *c* stands for motion related to the connector forces.
        e stands for motion due to the external forces.
Thus, the motion at location q due to the *connector forces* is given by

THE GENERAL FIELD ENVIRONMENT MODEL 579

$$Xc_q = \sum_{p=1}^{Nc} T_{qp} F_p \quad \text{for } q = 1 \text{ to } NT \tag{8.5.5}$$

and the motion at q due to the *external forces* is given by

$$Xe_q = \sum_{p=Nc+1}^{NT} T_{qp} F_p \quad \text{for } q = 1 \text{ to } NT \tag{8.5.6}$$

In view of Eqs. (8.5.5) and (8.5.6), Eq. (8.5.3) can be written as

$$\{X\} = \{Xc\} + \{Xe\} \tag{8.5.7}$$

where the identity of the force group causing test item motion is clearly evident.

Vehicle In a similar manner, the motion response at location q in the vehicle can be written in forms similar to those of Eqs. (8.5.3) thorough (8.5.7). These equations become

$$Y_q = \underbrace{\sum_{p=1}^{Nc} V_{qp} F_p}_{\text{Connector}} + \underbrace{\sum_{p=Nc+1}^{NT} V_{qp} P_p}_{\text{External}} \quad \text{for } q = 1 \text{ to } Nv \tag{8.5.8}$$

where the motion is conveniently broken into two parts as before. One part is due to the connector forces acting on the vehicle and the other part is due to the external forces acting on the vehicle. Thus, Eq. (8.5.8) can be written as

$$Y_q = Yc_q + Ye_q \quad \text{for } q = 1 \text{ to } Nv \tag{8.5.9}$$

The motion at q due to the *connector forces* is given by

$$Yc_q = \sum_{p=1}^{Nc} V_{qp} P_p \quad \text{for } q = 1 \text{ to } Nv \tag{8.5.10}$$

and the motion at q due to the *external forces* is given by

$$Ye_q = \sum_{p=Nc+1}^{Nv} V_{qp} P_p \quad \text{for } q = 1 \text{ to } Nv \tag{8.5.11}$$

Equation (8.5.8) can also be written as

$$\{Y\} = \{Yc\} + \{Ye\} \qquad (8.5.12)$$

where the identity of the force group causing the motion is clearly evident.

The importance of Ye_q is that this is the reference motion that occurs at the qth location when only external loads are acting on the structure without any interconnection forces. This is the motion that a transducer would measure on the vehicle at location q *if the test item were not present.* This motion is carefully separated from the motion that results due to the connector (interface) forces; that is, Yc_q.

8.5.2 Interface Boundary Conditions and Resulting Forces and Motions

The next step is to attach the two structures to one another as is done in the field and indicated schematically in Fig. 8.5.1. In order to connect them together, the interface boundary conditions need to be stated and applied to the equations. A number of different boundary conditions can be used. One possibility is to use an input-output frequency response relationship for each connector structure. Another possibility is to use a simple interface boundary condition where any connector structure is assumed to be part of the vehicle. We use the simpler connector boundary condition in this section. The interface boundary conditions are described by motions and forces. Thus we have

$$X_q = Y_q \qquad q = 1 \text{ to } Nc \qquad (8.5.13)$$

for the interface motions and

$$F_p = -P_p \qquad \text{for } p = 1 \text{ to } Nc \qquad (8.5.14)$$

for the interface connection forces. When Eqs. (8.5.3) and (8.5.8) are substituted into Eq. (8.5.13) and the definitions of Eqs. (8.5.6) and (8.5.11) are used, the result becomes

$$\sum_{p=1}^{Nc} T_{qp} F_p + Xe_q = \sum_{p=1}^{Nc} V_{qp} P_p + Ye_q \qquad \text{for } q = 1 \text{ to } Nc \quad (8.5.15)$$

which, on substitution of Eq. (8.5.14), reduces to

THE GENERAL FIELD ENVIRONMENT MODEL 581

$$\sum_{p=1}^{Nc} (T_{qp} + V_{qp})F_p = (Ye_q - Xe_q) \quad \text{for } q = 1 \text{ to } Nc \quad (8.5.16)$$

Equation (8.5.16) can also be rewritten in matrix form as

$$[T + V]\{F\} = \{Ye - Xe\}_c \quad (8.5.17)$$

where the subscript c is used on the relative motion vector $\{Ye - Xe\}$ to emphasize that these are the relative connector motions that are due to external loads on the vehicle and test item, respectively. The diagonals of the $Nc \times Nc$ matrix $[T + V]$ are composed of the sum of the driving point FRFs at each connector point. The off diagonal terms of $[T + V]$ are the sum of the transfer FRFs between the various connector points. Thus, the $[T + V]$ matrix is square and symmetrical about the diagonal.

The connector forces can be obtained from Eq. (8.5.17) by solving for the interface force vector $\{F\}$ (which has Nc terms). This force vector becomes

$$\{F\} = [T + V]^{-1}\{Y_e - X_e\}_c \quad (8.5.18)$$

which can be written conveniently as

$$\{F\} = [TV]\{Y_e - X_e\}_c \quad (8.5.19)$$

where

$$[TV] = [T + V]^{-1} \quad (8.5.20)$$

is the inverse of the sum of the vehicle and the test item connector point driving point and transfer FRFs. The $[TV]$ matrix is $(Nc \times Nc)$ in size, namely, the number of connecting points between the test item and the vehicle.

Equation (8.5.19) shows that the interface force is dependent on the relative motion between the connecting points; that is, $\{Ye - Xe\}_c$. This relative motion is due to the external forces that act on the test item (see Eq. (8.5.6) for motion Xe_q) and the external forces that act on the vehicle (see Eq. (8.5.10) for motion Ye_q). Recall that the external forces are independent of the connector force group. It is clear from Eq. (8.5.19) that the connector forces depend only on the vehicle connecting point motions (Ye_q for $q = 1$ to Nc) when no external forces are acting directly on the test item.

8.5.3 Test Item Motions

Now that we have the expression for the required interface forces, we can determine the resulting test item motion when the vehicle and the test item are connected. The connector force, as given in Eq. (8.5.19), can be written as

$$F_p = \sum_{k=1}^{Nc} TV_{pk}(Ye_k - Xe_k) \qquad (8.5.21)$$

Then, if we combine Eqs. (8.5.3), (8.5.4), and (8.5.6) with Eq. (8.5.21), the test item motion at location q becomes

$$X_q = \sum_{k=1}^{Nc} FE_{qk}(Ye_k - Xe_k) + Xe_q \quad \text{for } q = 1 \text{ to } N_T \qquad (8.5.22)$$

where

$$FE_{qk} = \sum_{p=1}^{Nc} T_{qp} TV_{pk} \qquad (8.5.23)$$

is the *field environment dynamical response*. Since FE_{qk} is multiplied by the connector point motion due to the external forces on each structure (test item and vehicle), these motions carry the double lower case symbols. The e symbol means these are connector motions at location k due to all of the external loads on that structure as though the structures were not connected together. Thus the relative motion at the connectors is due to the external loads on the two structures. Equation (8.5.22) can be written in matrix form as

$$\{X\} = [FE]\{Ye - Xe\}_c + \{Xe\} \qquad (8.5.24)$$

where $[FE]$ is a matrix of order $(Nc \times N_T)$ and the others are of order $(1 \times N_T)$. Equation (8.5.24) clearly shows the role of the connector motions in producing motion at any point in the structure.

If, for example, there are no external loads on the test item in the field environment then, $\{Xe\} = \{0\}$, and Eq. (8.5.24) reduces to

$$\{X\} = [FE]\{Ye\}_c \qquad (8.5.25)$$

which shows that the vehicle/test item connector motion is the only excitation source in this type of field environment. This happens to be an important situation, since the chances of a successful laboratory simulation are high for this case. Similarly, if the qth connector motion of the vehicle

THE GENERAL FIELD ENVIRONMENT MODEL 583

and the qth connector motion of the test item are the same in magnitude and phase, then no connector forces will be due to that motion, since $Ye_q - Xe_q = 0$.

If, on the other hand, the motions are equal but out of phase with one another, then the force would be twice as great as it would be with only vehicle connector motion since $(Ye_q - Xe_q) = 2Ye_q$. Thus correct correlation relationships of these two motions can play a significant role in the proper simulation of a given environment.

Test Item Transmissibility Ratio One idea that is employed to study test item dynamic characteristics is to measure the frequency response ratio of two different motions within the test item. From Eq. (8.5.22), the FRF representing the ratio of two output motions can be expressed as

$$\frac{X_q}{X_p} = \frac{\sum_{k=1}^{Nc} FE_{qk}(Ye_k - Xe_k) + Xe_q}{\sum_{k=1}^{Nc} FE_{pk}(Ye_k - Xe_k) + Xe_p} = TR_{pq}(\omega) \qquad (8.5.26)$$

where $TR_{qp}(\omega)$ is the transmissibility ratio between points p and q. It is obvious that this ratio contains a great deal of information that involves the dynamic characteristics of the vehicle and test item combined, compared to the vehicle alone or the test item alone. This is the field condition that we need to be able to simulate!

The problem is that Eq. (26) contains too much information in general. It is seen that the ratio is highly dependent on the field environmental matrix, the connector point relative motions, and the external forces (other than the interconnecting forces) acting on the test item. Even in the special case when the external forces are zero ($F_p = 0$ for $p = Nc + 1$ to NT), the ratio becomes

$$\frac{X_q}{X_p} = \frac{\sum_{k=1}^{Nc} FE_{qk}Ye_k}{\sum_{k=1}^{Nc} FE_{pk}Ye_k} = TR_{qp}(\omega) \qquad (8.5.27)$$

This equation reduces to a simple form independent of the input motion only when $Nc = 1$, that is, only when there is only one connector point between the vehicle and the test item. This is a very special case.

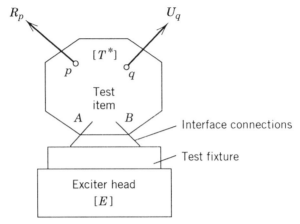

Fig. 8.6.1. Schematic of test item attached to the vibration exciter.

8.6 THE GENERAL LABORATORY ENVIRONMENT

In this section, the general dynamic description of a connected test item and vibration exciter is examined in a more general sense where there are multiple interface connections and multiple nonconnector excitation forces, including external and internal forces. A schematic diagram of the test item mounted on the exciter is shown in Fig. 8.6.1 where it is assumed that there are Ec connection points between the two structures. The connection points are selected in such a way that these points are labeled 1 through Ec for both structures, and that matching connection points have the same identification number. In the laboratory, the test item motion (displacement, velocity, or acceleration) at location q are designated by U_q $(= U_q(\omega))$, while the force at location p is designated by R_p $(= R_p(\omega))$, where R_p represents all force sources (connector, external, and internal). Similarly, the exciter motion at location q is designated by Z_q $(= Z_q(\omega))$, while the force at location p is designated by Q_p $(= Q_p(\omega))$, where Q_p represents all force sources (connector, external, and internal). While the forces and motions are vector quantities, they are treated as single variables in these equations. The two interface connections are labeled A and B in Fig. 8.6.1.

The test item and exciter are described by their general input-output FRF relationships, as given by Eqs. (8.2.3) and (8.2.4), that is,

$$\{U\} = [T^*]\{R\} \qquad (8.6.1)$$

for the test item and

$$\{Z\} = [E]\{Q\} \qquad (8.6.2)$$

for the exciter. The test item has N_L input-output points and the exciter has N_E input-output points. We use these models in developing our general laboratory environment, following the steps of Section 8.5.

8.6.1 Basic Motions Due to Interface and Noninterface Forces

In this subsection Eqs. (8.6.1) and (8.6.2) are examined in detail to describe the motion of the test item and the exciter in general. Then, in Section 8.6.2, the two structures will be connected to form the laboratory test environment.

Test Item For the scheme shown in Fig. 8.6.1, the motion response at location q in the test item due to the laboratory imposed forces R_p is given by

$$U_q = \underbrace{\sum_{p=1}^{Ec} T^*_{qp} R_p}_{\text{connector}} + \underbrace{\sum_{p=Ec+1}^{N_T} T^*_{qp} R_p}_{\text{external}} \qquad \text{for } q = 1 \text{ to } N_T \qquad (8.6.3)$$

Recall that N_T is to total number of motion points considered on the test item. The laboratory test item's dynamic characteristic may differ from the field test item's. Thus T^*_{qp} is used to distinguish the laboratory test item from the field test item.

The motion is conveniently broken into two parts: one part is due to the connector interface forces, while the other part is due to the external forces acting on the test item. Recall that external forces include all nonconnector forces. Thus Eq. (8.6.3) can be written as

$$U_q = Uc_q + Ue_q \qquad \text{for } i = 1 \text{ to } N_T \qquad (8.6.4)$$

where c stands for connector related excitation forces.
   e stands for external related excitation forces.
The motion at location q due to the *connector forces* is given by

$$Uc_q = \sum_{p=1}^{Ec} T^*_{qp} R_p \qquad \text{for } q = 1 \text{ to } N_T \qquad (8.6.5)$$

The motion at location q due to the *external forces* is given by

$$Ue_q = \sum_{p=Ec+1}^{N_T} T^*_{qp} R_p \quad \text{for } q = 1 \text{ to } N_T \qquad (8.6.6)$$

In view of Eqs. (8.6.5) and (8.6.6), Eq. (8.6.3) can be written as

$$\{U\} = \{Uc\} + \{Ue\} \qquad (8.6.7)$$

where the identity of the force group causing each motion group is clearly evident.

Exciter In a similar manner, the motion at location q on the exciter can be written in forms similar to those of Eqs. (8.6.3) thorough (8.6.7), giving

$$Z_q = \underbrace{\sum_{p=1}^{Ec} E_{pq} Q_p}_{\text{connector}} + \underbrace{\sum_{p=Ec+1}^{N_E} E_{qp} Q_p}_{\text{external}} \quad \text{for } q = 1 \text{ to } N_E \qquad (8.6.8)$$

where Q_p are the forces acting on the exciter.
E_{qp} is the exciter FRF.
Recall that N_E is the total number of motion points used with the exciter system. This motion is conveniently broken into two distinguishable parts: one part is due to the connector forces, while the other part is due to the nonconnector (external and/or internal) excitation forces acting on the exciter. Thus Eq. (8.6.8) can be written as

$$Z_q = Zc_q + Ze_q \quad \text{for } q = 1 \text{ to } N_E \qquad (8.6.9)$$

The motion at location q due to the *connector forces* is given by

$$Zc_q = \sum_{p=1}^{Ec} E_{qp} Q_p \quad \text{for } q = 1 \text{ to } N_E \qquad (8.6.10)$$

The motion at location q due to the *external forces* is given by

$$Ze_q = \sum_{p=Ec+1}^{N_E} E_{qp} Q_p \quad \text{for } q = 1 \text{ to } N_E \qquad (8.6.11)$$

Equation (8.6.9) can also be written as

THE GENERAL LABORATORY ENVIRONMENT

$$\{Z\} = \{Zc\} + \{Ze\} \tag{8.6.12}$$

where the identity of the force group causing the motion is clearly evident.

The importance of Ze_q is that this is the reference motion that occurs at the qth location when only external loads are acting on the exciter structure without any interconnection forces. This is the motion that a transducer would measure on the exciter *if the test item were not present*. This motion is carefully separated from the motion that results due to the interface forces.

8.6.2 Interface Boundary Conditions and Resulting Forces and Motions

The next step is to attach the two structures to one another in a manner similar to that in the field and indicated schematically in Fig. 8.6.1. In order to connect them, the interface boundary conditions need to be stated and applied to the equations. A number of different boundary conditions can be used. One possibility is to use an input-output frequency response relationship for each connector structure. Another possibility is to use a simple interface boundary condition where the connector is assumed to be part of the exciter. In other words, any test fixture is assigned to be part of the exciter's dynamic characteristics. We use the simpler connector boundary condition, that is, the interface boundary conditions are

$$U_q = Z_q \quad \text{for } q = 1 \text{ to } Ec \tag{8.6.13}$$

for the motions and

$$R_p = -Q_p \quad \text{for } p = 1 \text{ to } Ec \tag{8.6.14}$$

for the connection forces. When Eqs. (8.6.3) and (8.6.8) are substituted into Eq. (8.6.13) and the definitions of Eqs. (8.6.6), (8.6.11), and (8.6.14) are used, the result is

$$\sum_{p=1}^{Ec} (T_{qp}^* + E_{qp}) R_p = (Ze_q - Ue_q) \quad \text{for } p = 1 \text{ to } Ec \tag{8.6.15}$$

Equation (8.6.15) can be rewritten in matrix form as

$$[T^* + E]\{R\} = \{Ze - Ue\}_c \tag{8.6.16}$$

where the symbol c is used on the relative motion vector to emphasize that these are connector motions that are caused by external (nonconnector) forces acting on the exciter and test item. The connector forces can be obtained from Eq. (8.6.16) by solving for the connector force vector $\{R\}$ (which has Ec terms). This force vector becomes

$$\{R\} = [T^* + E]^{-1}\{Ze - Ue\}_c \qquad (8.6.17)$$

which can be written conveniently as

$$\{R\} = [TE]\{Ze - Ue\}_c \qquad (8.6.18)$$

where

$$[TE] = [T^* + E]^{-1} \qquad (8.6.19)$$

is the inverse of the sum of the frequency response functions of the exciter and the test item at the connector points. The $[TE]$ matrix is ($Ec \times Ec$) in size, namely, the number of connecting points between the test item and the exciter.

Equation (8.6.18) shows that the interface force is dependent on the relative motion between the connecting points, that is, $\{Ze - Ue\}_c$. This relative motion is due to external forces that act on the test item (see Eq. (8.6.6) for motion Ue_q) and those external forces that act on the exciter (see Eq. (8.6.10) for motion Ze_q). Recall that these external forces are those forces that do not belong to the connector force group. It is clear that the connector forces depend only on the exciter connecting point motions (Ze_q for $q = 1$ to NE) when there are no external forces acting directly on the test item.

8.6.3 Test Item Motions

Now that we have the expressions for the interface forces, we can determine the resulting test item motion when it is attached to the vibration exciter. The connector force, as given in Eq. (8.6.18), can be written as

$$R_p = \sum_{k=1}^{Ec} TE_{pk}(Ze_k - Ue_k) \qquad \text{for } p = 1 \text{ to } Ec \qquad (8.6.20)$$

Then, if we combine Eqs. (8.6.3), (8.6.4), and (8.6.6) with Eq. (8.6.20), the test item motion at location q becomes

$$U_q = \sum_{k=1}^{Ec} LE_{qk}(Ze_k - Ue_k) + Ue_q \quad \text{for } q = 1 \text{ to } NT \quad (8.6.21)$$

where

$$LE_{qk} = \sum_{k=1}^{Ec} T_{qp}^* TE_{pk} \quad (8.6.22)$$

is the *laboratory environment dynamical response*. Since LE_{qk} is multiplied by the connector motion due to the external forces on each component (test item and exciter), these motions carry double symbols. The symbol e means these are connector motions at location k due to all of the external loads on that structure. Thus the relative motion at the connectors is due to the external loads on the two structures. Equation (8.6.21) can be written in matrix form as

$$\{U\} = [LE]\{Ze - Ue\}_c + \{Ue\} \quad (8.6.23)$$

where $[LE]$ is a matrix of order $(Ec \times NT)$ and the others are of order $1 \times NT$). Equation (8.6.23) shows the role of the connector motions, as well as external force driven motions, in producing motion at any other point in the structure.

If, for example, there are no external loads on the test item in the laboratory environment, so that $Ue_q = 0$ for $q = 1$ to NE, then Eq. (8.6.23) reduces to

$$\{U\} = [LE]\{Ze\}_c \quad (8.6.24)$$

which shows that the exciter/test item connector motion is the only excitation source in this environment. This is an important result, since we can have a high probability of success in a simulation that requires only connector motions to be controlled.

Similarly, if the qth connector motion of the exciter and the qth connector motion of the test item are equal in magnitude and phase, then there is no connector force due to that motion, since $Ze_q - Ue_q = 0$. If the motions are equal but out of phase with one another, then the force would be twice as great as it would be with only exciter connector motion since $(Ze_q - Ue_q) = 2Ue_q$. Thus a correct correlation relationship between these two motions plays a significant role in the proper simulation of a given environment in the laboratory, since it is seen that the term $(Ze_q - Ue_q)$ can vary from zero to twice the loading when both motions are equal in magnitude.

590 GENERAL VIBRATION TESTING

Test Item Transmissibility Ratio One idea that is employed to study test item dynamic characteristics is to measure the frequency response ratio of two different motions within the test item. From Eq. (8.6.21), the FRF representing the ratio of two motions in the test item becomes

$$\frac{U_q}{Y_p} = \frac{\sum_{k=1}^{Ec} LE_{qk}(Ze_k - Ue_k) + Ue_q}{\sum_{k=1}^{Ec} LE_{pk}(Ze_k - Ue_k) + Ue_p} = TR^*_{qp}(\omega) \qquad (8.6.25)$$

where $TR^*_{qp}(\omega)$ is the transmissibility ratio between points p and q. It is obvious that this ratio contains a great deal of information concerning the dynamic characteristics of the test item and the exciter taken together, not the test item or the exciter alone. The problem is that this equation contains too much information in general. It is seen that this ratio is dependent on the *laboratory environmental matrix* terms (LE_{pk}), the connector point relative motions ($Ze_k - Ue_k$) that are due to the external loads, and the external forces (other than the interconnecting forces) acting on the test item. Even in the special case when the external forces are zero ($R_p = 0$ for $p = Ec + 1$ to NT), the ratio becomes

$$\frac{U_q}{U_p} = \frac{\sum_{k=1}^{Ec} LE_{qk} Ze_k}{\sum_{k=1}^{Ec} LE_{pk} Ze_k} = TR^*_{qp}(\omega) \qquad (8.6.26)$$

This equation reduces to a simple form independent of the input motion only when $Ec = 1$, that is, only when there is only one connector point between the exciter and the test item. This is a limited special case in which the exciter's characteristics drop completely out of the test environment.

8.7 COMPARISON OF FIELD AND LABORATORY ENVIRONMENTS

We need to compare the field environment from Section 8.5 with the laboratory environment from Section 8.6 in order to grasp some of the issues that we are confronted with when attempting to simulate the field environment in the laboratory while using vibration exciters. We look at the general case first, and then try to understand what happens if we have a four point structure where two of the points are connector points, one point is an external load point, and the fourth point is any other point in the test item. While it would be wonderful to give a solid answer in this section, the best we can do is to recognize some of the issues that are involved and wait for active research to reveal better methods and understanding. It will be seen that knowing the forces acting on the test item under field conditions is certainly one promising path to follow in achieving a satisfactory laboratory test specification.

8.7.1 Comparison of Theoretical Results for the General Case

In Section 8.5 we found that the field interface force is given by Eq. (8.5.19) as

$$\{F\} = [TV]\{Y_e - X_e\}_c \qquad (8.7.1)$$

where

$$[TV] = [T + V]^{-1} \qquad (8.7.2)$$

is the interface dynamic characteristic that is dependent on the vehicle and test item interface driving point and transfer FRFs. The connector motions Ye and Xe in Eq. (8.7.1) are due to external forces acting on each structure so that

$$Ye_q = \sum_{p=Nc+1}^{Nv} V_{qp} P_p \qquad \text{for } q = 1 \text{ to } Nc \qquad (8.7.3)$$

and

$$Xe_q = \sum_{p=Nc+1}^{N_T} T_{qp} F_p \qquad \text{for } q = 1 \text{ to } Nc \qquad (8.7.4)$$

In Section 8.6, we found that the laboratory interface force is given by Eq. (8.6.18) as

$$\{R\} = [TE]\{Ze - Ue\}_c \qquad (8.7.5)$$

where

$$[TE] = [T* + E]^{-1} \qquad (8.7.6)$$

is the laboratory interface dynamic characteristic that depends on the test item and vibration exciter driving point and transfer FRFs. The interface motions Ze and Ue are due to external loads acting on each structure

592 GENERAL VIBRATION TESTING

so that

$$Ze_q = \sum_{p=Ec+1}^{N_E} E_{qp}F_p \quad \text{for } q = 1 \text{ to } Nc \tag{8.7.7}$$

and

$$Ue_q = \sum_{p=Ec+1}^{N_T} T^*_{qp}R_p \quad \text{for } q = 1 \text{ to } Nc \tag{8.7.8}$$

The test item motions in the field are given by Eq. (8.5.24) as

$$\{X\} = [FE]\{Ye - Xe\}_c + \{Xe\} \tag{8.7.9}$$

where the *field environment* dynamic characteristic $[FE]$ contains terms described by Eq. (8.5.22) as

$$FE_{qk} = \sum_{p=1}^{Nc} T_{qp}TV_{pk} \tag{8.7.10}$$

The test item motion at location q due to the *external forces* is given by

$$Xe_q = \sum_{p=Nc+1}^{N_T} T_{qp}F_p \quad \text{for } q = 1 \text{ to } N_T \tag{8.7.11}$$

Similarly, in the laboratory, the corresponding test item motion becomes

$$\{U\} = [LE]\{Ze - Ue\}_c + \{Ue\} \tag{8.7.12}$$

where the *laboratory environment* dynamic characteristic is given by

$$LE_{qk} = \sum_{p=1}^{Ec} T^*_{qp}TE_{pk} \tag{8.7.13}$$

The test item motion at location q due to external forces is given by

$$Ue_q = \sum_{p=Ec+1}^{N_V} T^*_{qp}F_p \quad \text{for } q = 1 \text{ to } N_T \tag{8.7.14}$$

A comparison of these equations shows that the same test item motion and interface forces will occur in the field and laboratory environments if the following conditions apply:

1. The test items have identical FRF characteristics $T_{pq} = T^*_{pq}$.
2. The vehicle and exciter have identical FRF characteristics $V_{pq} = E_{pq}$. This is a nearly impossible requirement for test fixture design.
3. The external forces are identical in both environments so that all phasing (or cross correlation characteristics) are the same between external forces so that motions $Ye_q = Ue_q$, and so on.

Since it is impractical to satisfy all of these conditions, the question becomes: How do I use a vibration controller to achieve satisfactory results?

First, from Eqs. (8.7.9) and (8.7.12), we see that we must adjust the external forces so that $\{Xe\} = \{Ue\}$. The problem with this requirement is that we do not know what $\{Xe\}$ is from field measurements, since we only measure the complete response $\{X\}$ at each point of interest. The other possibility is to look at Eqs. (8.7.1) and (8.7.5), which control the interface forces. In this case we can begin by adjusting a controller to make the interface forces the same, so that $\{R\} = \{F\}$. The difficulty with this approach is that the external forces must be adjusted correctly in order to generate the test item interface motions $\{Ue\}_c$ before we can generate the correct interface forces. This leads to a difficult observation that the controller may have to look at all test item motions and forces in order to come up with the correct inputs. In the following subsection, we shall explore a four point test item to discern what might be a successful approach.

8.7.2 The Four Point Test Item Model

In order to determine a way to overcome the complexities presented above, we shall consider a simple interpretation of an armament store type of test item that is attached to an aircraft, as shown in Fig. 8.7.1a. The simple force system is shown in Fig. 8.7.1b. In this case, points 1 and 2 are interface points while point 3 represent the external loading point near the test item's nose and point 4 represents any other point in the test item. Then we write four field input-output equations as

$$\left.\begin{aligned} X_1 &= T_{11}F_1 + T_{12}F_2 + T_{13}F_3 \\ X_2 &= T_{21}F_1 + T_{22}F_2 + T_{23}F_3 \end{aligned}\right\} \text{connector}$$
$$X_3 = T_{31}F_1 + T_{32}F_2 + T_{33}F_3 \quad \text{external loading}$$
$$X_4 = T_{41}F_1 + T_{42}F_2 + T_{43}F_3 \quad \text{any test item point}$$
(8.7.15)

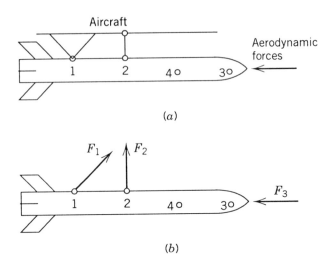

Fig. 8.7.1. Schematic of an armament mounted on an aircraft wing. (a) Actual arrangement. (b) Assumed force loading.

Similarly, we can write four laboratory input-output equations as

$$\left.\begin{array}{l} U_1 = T_{11}^*R_1 + T_{12}^*R_2 + T_{13}^*R_3 \\ U_2 = T_{21}^*R_1 + T_{22}^*R_2 + T_{23}^*R_3 \end{array}\right\} \text{connector}$$
$$U_3 = T_{31}^*R_1 + T_{32}^*R_2 + T_{33}^*R_3 \quad \text{external loading}$$
$$U_4 = T_{41}^*R_1 + T_{42}^*R_2 + T_{43}^*R_3 \quad \text{any test item point}$$
(8.7.16)

When we examine Eqs. (8.7.15) we observe that we can solve for F_q if we know T_{qp} of the test item and at least three motions X_q. It is assumed that the connection points are two of the motions that are measured in this case. Then, we can use either X_3 or X_4 as the third motion. In addition, we can check on the accuracy of our work either by measuring the interface forces F_1 and F_2 or by measuring all four motions, and then, solving for the three forces via different sets of simultaneous equations. Then, we would have forces based on combinations of X_1, X_2, X_3 or X_1, X_2, X_4, or X_2, X_3, X_4. Once we know these forces and motions from the field environment, we can compute the cross spectral densities so that we can establish what the phase relationships should be between each of the forces as well as each of the motions. This theory sounds simple. In practice, the matrix inversion can be easily singular if there are any significant errors in determining T_{pq}. The pseudoinverse method shows some promise in this case.

It is one thing to measure the motions and solve for the forces and it is quite another thing to simulate them in the laboratory. First, we require three exciters and force measuring transducers. Second, we need to establish some type of control program and feedback scheme that takes the exciter characteristics into account and can produce force signals and motions (at locations 1, 2, and 3 in this case) that have the same cross-spectral density as the field forces and motions. These laboratory forces are R_1, R_2, and R_3, as seen in Eqs. (8.7.16).

It is apparent in Eqs. (8.7.16) that the test item may have slightly different FRF characteristics unless $T_{qp} \cong T_{qp}^*$. Consequently, one needs to bring the system up to a low input and response level (say 1/8 full scale) so that engineering judgments and control adjustments can be made in order to obtain a satisfactory output response before the control system is allowed to be brought up to full scale testing.

It is apparent from this subsection that forces (and/or moments) are extremely important in securing a rational scheme for laboratory simulation of complex structures. I believe that research along these lines will provide a reasonable path to achieve satisfactory results for test items where external input points like point 3 can be identified. If there are any significant external forces that act at unspecified locations, then this technique cannot be made to work.

8.8 SUMMARY

This chapter has dealt with a difficult topic of how to go from a field environment to a satisfactory laboratory test. It has clearly been shown that there are potentially 36 input-output FRFs between any two points in the structure. Often we measure only one input and one output, so that our measurement practices contribute to poorly defined FRFs.

We have shown that three types of structures are involved in the process. These structures are the vehicle and test item in the field environment and the vibration exciter and test item in the laboratory environment. A simple model consisting of a two point vehicle, a test item, and an exciter model has been used to illustrate the types of motions and interface forces that occur and the type of FRF interactions and complexities that are involved. It became apparent that driving point FRFs are important in the simplest case. Then, these simple theoretical results have been applied to six possible test scenarios, from which it has become apparent that there is no way to cause the same motion to occur in the test item in the laboratory as occurs in the field when the field external force is not used in the laboratory tests.

An example using a two DOF test item and a two DOF vehicle has shown that the theoretical and experimental results can be easily predicted when there are no external forces acting on the test item. The only force in this case is a single connector force. Under these conditions dynamic ranges on the order of 10^9 have easily been handled. It has been shown that either interface motion or interface force frequency spectra from the combined vehicle and test item can be used in the control scheme to achieve very good simulation. We have also shown that it is possible to convert bare vehicle interface motion into proper interface force and interface motion for use in the laboratory. This process is limited to the situation where there is only a single interface connecting point. However, we have shown that severe overtesting and undertesting can occur if the bare vehicle interface motion is used in the laboratory without modification that takes the test item's driving point acceleration into account.

Both a general field environment model and a laboratory environment model have been developed that show how we can address the motions due to interface forces, compared to those caused by external (noninterface) forces. It is clear that the driving point and transfer FRFs of the vehicle and test item interact to form the *field environment* dynamic characteristics, while the driving point and transfer FRFs of the vibration exciter and test item form the *laboratory environment* dynamic characteristics. It is clear from these equations that the motions and forces interact significantly at the interface between the test item and either the vehicle or the vibration exciter. Thus going from a single interface point to multiple interface points significantly raises the level of complexity that we are dealing with.

Finally, we have attempted to understand the requirements for having a satisfactory laboratory simulation. It soon became apparent that working on the basis of forces acting on the test item holds great promise for helping us develop adequate test specifications that may be more universal. It should be apparent from this chapter that satisfactory laboratory simulation is a difficult task and no simple answers are found in this book for the general case, only for the special case where a single interface force and motion dominate the field environment and the laboratory environment. I am actively involved in resolving these issues.

REFERENCES

1. ANSI Standards S 2.31, S 2.32, S 2.34, AND S 2.45.
2. Bootle, R., "Aircraft Mission Profile—Vibration Levels," *Proceedings, Institute of Environmental Sciences*, 1990, pp. 528–530.
3. Caruso, H., "Correlating Reliability Growth Vibration Test and Aircraft Mission Profile Vibration Loads and Effects," *Proceedings, Institute of Environmental Sciences*, 1990, pp. 702–707.
4. Charles, D., "Procedures to Estimate Vibration Severities of Stores on Helicopters," *Proceedings, Institute of Environmental Sciences*, 1990, pp. 694–701.

5. Fletcher, J. N., "Global Simulation: A New Technique for Multiaxis Test Control," *Proceedings, Institute of Environmental Sciences*, 1990, pp. 147–152.
6. Frey, R., "The Vibration and Shock Environment for Commercial Computer Systems," *Proceedings, Institute of Environmental Sciences*, 1990, pp. 658–662.
7. Hansen, K., "On the Relation Between Vibration Input and Local Strain," *Proceedings, Institute of Environmental Sciences*, 1990, pp. 610–617.
8. McConnell, K. G., "From Field Vibration Data to Laboratory Simulation," *Experimental Mechanics*, Vol. 34, No. 3, Sept. 1994, pp. 1–13.
9. MIL-STD-167-2(Ships), May 1, 1974.
10. MIL-STD-810D, "Environmental Test Methods and Engineering Guidelines," July 19, 1983.
11. Richards, D. P., "The Derivation of Procedures to Estimate Vibration Severities of Airborne Stores," *Proceedings, Institute of Environmental Sciences*, 1990, pp. 679–687.
12. Rogers, J. D., D. B. Beightol, and J. W. Doggett, "Helicopter Flight Vibration of Large Transportation Containers—A Case for Test Tailoring," *Proceedings, Institute of Environmental Sciences*, 1990, pp. 515–521.
13. Sharton, T. D., "Force Limited Vibration Testing at JPL," *Proceedings of the 14th Aerospace Testing Seminar*, Manhattan Beach, CA, Mar. 1993, pp. 241–251.
14. Sharton, T. D., "Analysis of Dual Control Vibration Systems," *Proceedings, Institute of Environmental Sciences*, 1990, pp. 140–146.
15. Singal, R. K. and I. K. Maynard, "Vibration Validation of a Spacecraft Container," *Proceedings, Institute of Environmental Sciences*, 1990, pp. 509–514.
16. Smallwood, D. O., "An Analytical Study of a Vibration Test Method Using Extremal Control of Acceleration and Force," *Proceedings, Institute of Environmental Sciences*, 35th Annual Meeting, "Building Tomorrow's Environment," Anaheim, CA, May 1–5, 1989, pp. 263–271.
17. Sweitzer, K. A., "Vibration Models Developed for Subsystem Test," M.S. Thesis, Syracuse University, Utica, NY, May 1994.
18. Szymkowiak, E. A. and W. Silver, II, "A Captive Store Flight Vibration Simulation Project," *Proceedingsof the 36th Annual Meeting, Institute of Environmental Sciences*, New Orleans, LA, Apr. 1990, pp. 531–538.
19. Torstensson, H. O. and T. Trost, "The Transportation Environment and Its Characterization," *Proceedings, Institute of Environmental Sciences*, 1990, pp. 624–634.
20. Wall, J. S., "Standard Test—Tailoring for Test—or Both?," *Proceedings, Institute of Environmental Sciences*, 1990, pp. 522–527.

INDEX

Absorber, dynamic, 125
AC circuits, 170
Accelerance, 74, 83
Accelerometer:
  angular, 182
  cross-axis sensitivity, 201
    correcting voltage readings, 205
    FRF contamination removal, 206
    resonance, 210
    single axis model, 202
    tri-axial model, 203
  FRF, 181
  gravity effects, 166
  mechanical model, 164
  positioning error, 473
  sensitivity, 173, 175, 179, 181
  transient response, 189
Accelerometer calibration:
  constant input, 235
  sinusoidal input, 235
  transient input, 237
Admittance, 74
Aliasing, 267
  anti-aliasing filter, 269
  spatial, 131
Amplifier, 170
  charge, 173
  operational, 170
  power, 405
  summing, 170
  unity gain, 170
Amplitude rms, 15
Analyzers, Fourier:
  dual channel, 333
  single channel, 263
Angular frequency, 18, 66
Auto-correlation function:
  periodic signals, 42, 46
  random signals, 48
  transient signals, 46
Auto-spectral density:
  periodic signals, 46
  random signals, 49
  transient signals, 47
Averaging:
  impacts, 475
  overlapping windows, 303
  temporal mean square, 15

Back to back calibration, 235
Bandwidth, 264, 312
Base excitation, 164, 459
Base strain, errors, 253
Beam attached to exciter, 461
  accelerometer mass effects, 467
  correction limits, 468
  dynamic load, 462
  ideal response, 464
Beam model, 140, 150
Bias error, 299
Bode plots, FRF:
  accelerance, 83
  mobility, 82
  receptance, 79
Boundary conditions, 130, 134, 144
Break frequency, 176
Broad band random process, 53

Cable dynamics, 155
Cable noise, electrical, 253
Calibration:
  back to back, 235
  force transducers, 240
  gravimetric, 237
Cantilever, 145, 147
Capacitance voltage, 170
Characteristic frequency equation, 65, 105, 117, 130, 141, 143
Charge:
  piezoelectric sensitivity:
    displacement sensitivity, 172
    unit sensitivity, 173
  voltage and capacitance, 170
Charge amplifier, 173
  circuit, 173
  characteristics, 175
  response to shock signal, 198
Charge sensitivity model, 172
Chaotic, 113
Chirp excitation, 328, 501
Circle fit of SDOF plot, 82
Clamped-free, boundary conditions:
  beam, 145
  rod, 134
Closed loop control, 370, 505, 549
Coherence, 337, 341, 342, 461
Common mode rejection, 170

599

600 INDEX

Complex:
  conjugates, 22
  exponential method, phasor, 19
  Fourier coefficients, 18, 32
  stiffness, 78
Compliance, 75
Condition:
  amplifier, signal, 170
  boundary, 130, 134, 144
  initial, 67
Continuous system models:
  beam, 141
  rod, axial and torsion, 127
  string, 127
Convolution, 266
Correlation, 35
  auto, 42, 46, 48
  cross, 38, 44, 54
  multiple random process, 55
  normalized coefficient, 37
  periodic time, 38
  transient, 44
Coulomb damping, 68, 75
Crest factor, 58
Critical damping ratio, 67
Cross-axis:
  accelerometer sensitivity, 201
  force moments, 247
  resonance, 210
Cross-correlation function:
  periodic signals, 38
  random signals, 54
  transient signals, 44
Cross-spectral density function:
  periodic signals, 38
  random signals, 54
  transient signals, 44
Cycle:
  limit, 110
  stable, 105

Damped systems:
  continuous:
    beams, 4th order, 140
    rods, etc., 127
  SDOF, 64
  MDOF, 114
Damping:
  coefficient, 65
  coulomb friction, 75
  critical damping ratio, 67
  equivalent viscous, 75
  estimates, 75
  hysteretic, 75
  line, 80
  loss factor, 75
  modal, 120, 136
  structural, 75

Decrement, log, 68
Degrees-of-freedom, 59, 64, 114, 130, 301
Density:
  auto-spectral, 48
  cross-spectral, 54
  function, probability, 57
  power spectral, 48
Digital filtering characteristics, 271
Digital frequency analyzer, 265
  calculation relationships, 280
  display scaling, 282
  operating principles, 278
Digital data processing errors, 522
  hardware type problem, 525
  software type problem, 526
Dirac delta function, 53, 94, 267
Dirt, humidity–cables, 173
Displacement charge sensitivity, 172
Drive rod, 228. *See also* Stinger
Dual channel analyzer, 333
  coherence, 337
    nonlinear effects on, 341
    resolution bias error, 342
    signal time delays, 342
  FRF estimates, actual:
    auto-spectral averaging, 340
    cross-spectral averaging, 340
  FRF estimates, ideal:
    $H_a(\omega)$, 337
    $H_1(\omega)$, 336
    $H_2(\omega)$, 337
  inputs:
    correlated, 354
    uncorrelated, 353
  input–output relationships, 335
  more than one input source, 352
  signal noise effects:
    input and output noise, 350
    input only noise, 345
    output only noise, 348
Dual time constants, 180
Duffing's equation, 107
Duhamel's integral, 94
Dynamic absorber effect, 125
Dynamic calibration:
  accelerometer, 235
  force transducers, 240
Dynamic stiffness, 116, 121, 136
Dynamic environment, 3

Effective time constants, 198
Eigenvalues, 143
Eigenvectors, 143
Electrohydraulic exciters, 382
  components, 383
  earthquake simulator, 387
Electromagnetic exciters, 391
  armature, 392

INDEX **601**

Electromagnetic exciters (*contd*)
   dynamics, 399
   resonance, 402
  bare table characteristics:
   current mode, 408
   electromagnetic damping, 409
   electromechanical, 407
   inductive mass, 410
   mechanical, 402
   voltage mode, 410
  electromechanical coupling, 403
   nonlinear coupling, 404
  field coil, 392
  grounded structure response, 414
   current mode operation, 417
   force drop out grounded:
    current mode, 422
    voltage mode, 424
   voltage mode operation, 420
  power amplifier, 405
   current mode, 406
   voltage mode, 405
  support dynamics, 393
   isolation from supports, 399
  test system resonance, 428
  ungrounded structure, 425
   current mode, 431
    force dropout, 432, 437
   voltage mode, 433
    force dropout, 434, 437
Environmental error sources:
  base strain, 251
  cable noise, 253
   electromagnetic, 253
   ground loops, 253
   triboelectric effects, 253
  humidity and dirt, 173, 254
  mounting the transducer, 254
  nuclear radiation, 255
  temperature, 255
  transducer mass, 256, 467
  transverse sensitivity, 256
Equation of motion:
  continuous system:
   2nd order, 127
   4th order, 140
  Duffing's equation, 107
  MDOF system, 115
  modal, 120
  SDOF system, 65
Equivalent time constant, 198
Exciter:
  attachment to structure, 460
  controlling, 370
  electrohydraulic, 382
  electromagnetic, 391
  free-free beam experiment, 501
  mechanical, 371

  direct drive, 372, 377
  drive torque considerations, 381
  reaction type, 379
  rotating unbalance, 379
  periodic, 501
  pseudorandom, 501
  random, 501
  sine sweep, 501
  single point, 487
  static, 364, 444
  transient, impulse, 366
   coefficient of restitution, 368
Exponential window:
  impulse testing, 480, 488, 493, 500
  SRM testing, 445, 455

Fast Fourier Transform (FFT):
  Fourier series, 23
   periodic functions, 24
   transient signals, 32
Field to laboratory, 535
  comparison of field and laboratory, 590
  example of two DOF system, 558
   bare vehicle predictions, 575
   conclusions, 575
   field simulation, 561
   laboratory simulation, 568
  general field environment, 577
   [FE] matrix, 582
   [TV] matrix, 581
  general laboratory environment, 584
   [LE] matrix, 589
   [TE] matrix, 588
  linear model, 538
   connector forces, 540
   external forces, 540
   internal forces, 540
   unaccounted for terms, 540
  three structures in process, 541
  two point input–output model, 542
   field environment, 544
   field interface dynamic characteristic, 545
   example structure and test results, 558
   laboratory environment, 546
   laboratory interface dynamic characteristic, 547
   six laboratory excitation scenarios, 548
   summary observation, 557
Filters:
  anti-aliasing, 267
  center frequency, 272
  commonly used digital, 288
   flat top, Hanning, Kaiser-Bessel, rectangular, 288
  leakage, 273
Fixed boundary condition:
  beam, 145

602 INDEX

Fixed boundary condition (*cont'd*)
  rod, 134
Flexuaral vibrations, 140
Force:
  balance, 65, 115, 127
  damping, 65
  drop out, 443, 437, 509
  impulse, 475
  inertia, 65, 115, 127
Force transducer:
  calibration, 240
  correcting for seismic mass:
    electronic correction, 230
    driving point acceleration FRF, 226
    phase shift errors are big, 232
    transfer acceleration, FRF, 228
  base strain effects, 251
  bending moment effects, 247
  environmental effects, 251
  general model, 211
    attached to fixed foundation, 215
    attached to impulse hammer, 216
    attached to SUT and exciter, 217
  impedance head, 222
  sensitivity, 214
Forced response:
  direct solution, 119
  Duffing's equation, 107
  frequency domain approach, 91
  linear system input–output, 96
  single DOF system, 72
Formula:
  continuous, 127, 140
  Euler, 19
  Parseval's, 26
Fourier analyzer, 261
Fourier series:
  periodic signals, 24
  random signals, 48
  transient signals, 29
Fourier transform, transient, 32
Free-free beam experiment, 501
  exciter structure interaction, 505
    current mode amplifier, 507
    force drop out, 509
    voltage mode amplifier, 507
  selecting excitation signal, 501
  windowing effects on random tests, 510
Free vibration response, 66, 67, 444
Frequency:
  damped and undamped natural, 66
  Nyquist, 269
  peak, resonant, 79
  spectral density, 32
  spectrum, 16
Frequency analyzer, 261
  basic process, 263

  dual channel, 333
  single channel, 285
Frequency response functions (FRF), 6, 73
  acceleration, 74, 75
  admittance, 75
  apparent mass, 75
  driving point, 119
  dynamic compliance, 75
  dynamic flexibility, 73
  inertance, 75
  mechanical impedance, 75
  mobility, 74, 75
  overall accelerometer, 181
  receptance, 73
  transfer, 119
Function:
  auto-correlation, 42, 46, 48
  auto-spectral density, 48
  coherence, 337, 341, 461
  cross-correlation, 38, 44, 54
  cross-spectral density, 38, 44, 54
  dirac delta, 53, 94, 267
  impulse response, 94
  probability density, 57
  ramp hold, 100
  sinc, 27, 30
  step, 100, 444
  transfer:
    acceleration, 74
    mobility, 75
    receptance, 73, 126
  window, 265

Gain, amplifier, 170, 405
Generalized coordinates, 135
Gravimetric calibration of accelerometers, 237
Grounded structure, exciter performance, 414

Half-power points, 286
Hammer excitation, 216
Hanning window, 288
Heterodyning, 315
Humidity and dirt, 254. *See* Environmental error sources
Hysteretic damping, 75

Impulse excitation:
  frequency content, 476
  multiple hits, 477
  sensitivity to hammer mass, 216, 366
  waveform, 366
Impedance head, 222
Impedance:
  electrical, 170
  mechanical, 75

INDEX **603**

Impulse response function, 4–96
Impulse testing, 475
  free-free beam, 482
  impulse requirements, 476
    input noise problems, 477
  output leakage problems, 480
    effect on damping, 482
    exponential vs. rectangular window, 481
  selecting proper windows, 487
  window parameters, 487
    duration, 488
    modeling data processes, 489
    pulse duration, 487
    recommended procedure, 500
    slope, 488
    trigger level, 488
    truncation and exponential windows, 493
    wrong trigger point, 498
Inductance, voltage, 170
Inertance, 75
Inertia matrix, 116
Initial conditions, 67
Input–output model, 91
  Duhamel, Faltung, convolution, 94
  frequency domain, 92
  impulse response function, 94
  random, 98
  receptance impulse response relationship, 96
  shock response spectra, 98
  time domain, 94
Inverse fourier transform:
  periodic, 24, 86
  transient, 32, 86
Integral:
  convolution, 94
  Duhamel, 94

Leakage, filter, 273
Lenz's law, 404
Limit cycle, 110
Linear input–output model, 91
  convolution, 94
  Duhamel, 94
  Faltung, 94
  frequency domain, 92
  impulse response, 94
Load cells, *see* Force transducers
Loading mass, 467
Logarithmic decrement, 68
Longitudinal vibration, 127
Loss factor, 75
Lumped parameter systems, *see* Multi-degree-of-freedom

Mass:
  compensation, 223
  line, 79
  loading, 223, 467
  matrix, 116
  modal, 117, 135, 141
  seismic, 164
Mathematical expectation, 36
Matrix:
  damping, 115
  mass, 115
  stiffness, 115
Mean, temporal, 15, 34
Mean square, 15, 34
  RMS (root mean square), 15
Mean square spectral density, *see* Auto-spectral density
Mechanical impedance, 75
Mobility, 75, 82
Modal:
  analysis, 116, 130
  damping, 121
  distributed excitation, 137, 138
  equations, 116, 128, 140
  excitation, 120, 135, 137, 138
  loading coefficient, 120
  mass, 133, 142
  model, 149
  orthogonality, 133
  point loads, 137, 139
  response, 122
  shapes, 120, 131, 141
  stiffness, 135, 142
  summation, 120, 133
Mode shape:
  beam, 142
  continuous model, 131
  discrete model, 118
Model, general, 5
Motion:
  axial, 127
  base, 164
  caused by boundary and external forces, 5
  critically damped, 67
  lateral, transverse, 140
  longitudinal, 127
  torsional, 127
  transverse, lateral, 140
  undamped, 66
Multi-degree-of-freedom (MDOF):
  FRF plots, 126
  modal model, 120, 149
  response model, 119, 120, 122
  spatial model, 127, 140
  system, 115, 127, 140
  undamped analysis, 117, 130, 141

604 INDEX

Multi-degree-of-freedom (MDOF) (cont'd)
   viscously damped, 117, 130, 141
Multi-point excitation, 577, 585

Natural frequency:
   angular, 66
   continuous systems, 130, 141
   damped MDOF, 117
   damped SDOF system, 67
   orientation (gravity) effects, 71
   undamped MDOF system, 117
   undamped SDOF system, 66
Narrow-band random process, 53
Node points, 131
Noise:
   dual channel analyzer, 344
   spectral line uncertainty, 297
Nonlinear systems, 103
   coherence, effect on, 341
   cubic stiffness, 105, 107
   Duffing forced response, 107
   jump phenomena, 108
   limit cycles, 110
   Mathieu equation, 113
   phase plane, 104
   relaxation oscillations, 111
   simple pendulum, 106
   van der Pol equation, 110
Nyquist:
   circle fit, 85
   frequency, 269
   plots, 84
      accelerance, 87
      mobility, 86
      receptance, 84

Operational amplifier, 170
Orthogonality of modes, 133
Overall system calibration, 234

Parseval's formula:
   frequency analyzer, 282
   periodic signals, 26
   transient signals, 34
   random signals, 50
Periodic functions, 12, 16, 17, 18, 23
Periodic fourier series, 24
Phase angle, 18, 73
Phase plane, see also Nonlinear systems
   focus point, 105
   separatrix, 107
   vortex point, 105
Phase shift:
   lag, 20, 73
   lead, 20
   RC high pass filter, 175
   SDOF, 73
Phasor, 19

derivatives, 20
real valued functions, 20
Piezobeam sensor:
   angular acceleration, 186
   linear acceleration, 186
Piezoelectric sensor circuits:
   built-in voltage follower, 177
   charge amplifier, 173
   charge sensitivity model, 172
   equivalent time constant, 197
   transient response, 195
   undershoot, 196
Pinned boundary condition, 145
Plots:
   Bode, 80–85
   Nyquist, 86–91
   real and imaginary, 86–91
Point mobility, see Mobility
Power amplifier:
   current mode, 406
   voltage mode, 405
Power spectral density (PSD), see Auto-
   spectral density
Probability density function, 57
Pseudoinverse, 594

Ramp-hold excitation:
   step relaxation method (SRM), 444
   transducer response, 193
Random signal classification, 12
Random vibration:
   analysis, 48
   excitation 501, 510
   input–output relationships, 98, 335
   pseudorandom, 503
   window leakage problems, 273, 512
Ratio, damping, 67
Real and imaginary plots:
   accelerance, 87
   mobility, 86
   receptance, 84
Receptance FRF:
   continuous, 135, 149
   MDOF, 126
   SDOF, 73
Resistance, voltage, 170
Resonance:
   cross-axis, 210
   frequency, 66, 73, 75, 119, 135, 149
   region (FRF), 73
Response:
   forced, 72, 107, 119, 135
   free, 66, 104, 117, 130
   impulse, 94
   steady state, see Response, forced
   transient 67, 94
Response function:
   frequency, see FRF

INDEX 605

Response function (cont'd)
  impulse, 94
  step, 94
Root mean square (RMS), 15
Rotating unbalance, 371, 379

Sample function, 265
Scan analysis:
  averaging, 323, 326, 327
  shift factor, 324
Seismic transducer model, 163
Seismic transducers:
  acceleration type, 163–200
    angular accelerometers, 182
    linear accelerometers, 163
  force type, 211–233
Semidefinite system, 115
Sensitivity:
  accelerometers, 172, 173, 176, 181
  bending moment, 247, 509
  cross-axis, 201
  force transducers, 211
Servo valves, 383
Shaker, *see* Exciter
Shock response spectra (SRS), 98
  maxi-max SRS, 99
  primary SRS, 99
  residual SRS, 99
Signal analysis, 11, 29
Signal classification, 12
Signal conditioner circuits:
  amplifier circuits, 170
  charge amplifier, 173
  voltage follower, 176
Signal filtering:
  anti-aliasing, 267
  digital characteristics, 271
  digital (zoom), 315
Sinc function, 27, 30
Single channel analyzer:
  bias errors in the estimate, 299
  common window functions, 288
  comparison sinusoidal signals, 292
  flat top, 288
  Hanning, 288
  Kaiser-Bessel, 288
  rectangular, 288
Filter characteristics, 286
    bandwidth, 286
    center frequency, 286
    ripple, 288
    selectivity, 288
  overlapping signal analysis, 303
  uncertainty, 309
  real time analysis, 310
  recommended window usage, 302
  sample function, 265
  spectral line uncertainty, 297

spectral smearing, 314, 323
time sampling process, 264
window function, 265
Single degree of freedom, (SDOF):
  acceleration, 74
  basic model, 64
  coulomb damped response, 68
  free damped response, 67
  forced response, 72
    damping models, 75
  frequency response function, 73
    acceleration, 74
    mobility, 74
    receptance, 73
  impulse response, 70
  initial condition response, 67
  log decrement, 68
  mobility, 75
  modal model, 120, 135, 149
  natural frequency, 66
  ramp hold response, 190
  receptance, 73
  step response, 191
  transient response, 94, 191
  undamped, 66
  viscously damped, 67
Sinusoidal functions, *see* Periodic functions
Square wave, 27
Spectral Density:
  auto (ASD) periodic signals, 42
  auto (ASD) random signals, 50
  auto (ASD) transient signals, 46
  cross (CSD) periodic signals, 38
  cross (CSD) random signals, 54
  cross (CSD) transient signals, 44
Spectrum analyzers, *see* Frequency analyzer
Spectral smearing, 314, 323, 329
Spring:
  line, 79
  linear, 65
  nonlinear, 104
Static excitation schemes, (*see* Step relaxation method), 364
Statistical distribution functions:
  chi-squared, 58
  Gaussian, normal, 57
  Rayleigh, 58
Step relaxation method (SRM), 444
  actual test on portal frame, 454
  bounce test and coherence, 461
  measurement dilemma, 449
    AC coupling, 452
    comparison of inputs, 452, 457
    false resonances, 450
  theoretical model, 445
Stiffness, *see also* Spring
  cubic, 105
  nonlinear, 103

606   INDEX

Stiffness (cont'd)
  matrix, 116
  modal, 120, 135, 142
Stinger, 392, 505
Strain sensitivity, 252
Structural damping, 76. See also Hysteretic damping
Superposition, see Modal; Convolution
SWAT, 443

Test tailoring, 535. See also Field to laboratory
Temperature, 255
Temporal:
  auto-correlation, 42, 46, 48
  cross-correlation, 38, 44, 54
  mean, 15
  mean square, 15
  root mean square, 15
Time constant, 175
  equivalent, 198
  internal, 179
Time sampling process, 264
Two DOF vibration model, 114
  characteristic frequency equation, 117
  dynamic absorber, 125
  dynamic stiffness, 116
  equations of motion, 116
  modal
    damping, 120
    dynamic stiffness, 121
    generalized excitation, 120, 121
    mass, 120
    matrix, 118
    stiffness, 120
  mode shapes, 117
  natural frequencies, 117
  receptance
    driving point, 119
    transfer, 119
  semidefinite, 115
Transducers, 161
  accelerometers, 163
  built-in voltage followers, 177
  force, 211
  internal time constants, 179
  linear and angular accelerometers, 182
  overall FRF, 181
Transducer error sources, see also Environmental error sources
  base strain sensitivity, 252
  location, 473
  mass effects, 256
  moment sensitivity, 247

Transient response, transducers:
  electrical, 194
  mechanical, 189
Transient signal analysis, 29
  difference to periodic, 29
  Fourier transform, 32
  mean, 34
  mean square, 34
  Parseval's formula, 34
Transient vibration:
  analysis, 67, 86, 94
  by impulse, 94, 237, 243, 475
  by SRM, 444
Transverse sensitivity, see Cross-axis
Triboelectric noise, 253

Undamped system:
  analysis linear, 66, 86, 117, 130, 141
  analysis nonlinear, 103
  modes of vibration, 117, 130, 132, 141
  natural frequencies, 66, 117, 130, 141
Undershoot, signal, 196

Variables, separation of, 130, 141
Vibration concepts, 63
Vibration exciters, 363. See also Exciters
Vibration testing:
  basic cases, 441
  definition, 9
  test tailoring, 535
Voltage:
  capacitance, 170
  followers, 177
  inductance, 170
  resistance, 170

Wax attachment of transducers, heavy grease, 254
Wave speed, 130
Window function coefficients, 289
Window functions, 265
  digital filter characteristics, 271
  exponential, 480, 493
  flat top, 288
  Hanning, 288
  Kaiser-Bessel, 288
  leakage, 273
  rectangular, 288

Zoom analysis, 313
  heterodyning method, 315
  long time record method, 317
  sample tracking, 321
  spectral smearing, 314, 329

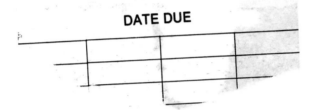